AREA HANDBOOK
for the
PERSIAN GULF STATES

Coauthors

Richard F. Nyrop

Beryl Lieff Benderly
Laraine Newhouse Carter
William W. Cover
Darrel R. Eglin
Robert A. Kirchner
Philip W. Moeller
William A. Mussen, Jr.
Clarence Edward Pike
Rinn-Sup Shinn

Research completed on January 3, 1977

First Edition

Published 1977

DA Pam 550–185

Library of Congress Cataloging in Publication Data

Nyrop, Richard F
Area handbook for the Persian Gulf States.

"DA pam 550–185."
Bibliography: p. 411–427.
Includes index.
1. Persian Gulf States. I. American University, Washington, D.C. Foreign Area Studies. II. Title.
DS247.A13N97 953.6 77–23854

First Edition, First Printing, 1977

For sale by the Superintendent of Documents, U.S. Government Printing Office
Washington, D.C. 20402

Stock No. 008-020-00682-3

FOREWORD

This volume is one of a series of handbooks prepared by Foreign Area Studies (FAS) of The American University, designed to be useful to military and other personnel who need a convenient compilation of basic facts about the social, economic, political, and military institutions and practices of various countries. The emphasis is on objective description of the nation's present society and the kinds of possible or probable changes that might be expected in the future. The handbook seeks to present as full and as balanced an integrated exposition as limitations on space and research time permit. It was compiled from information available in openly published material. An extensive bibliography is provided to permit recourse to other published sources for more detailed information. There has been no attempt to express any specific point of view or to make policy recommendations. The contents of the handbook represent the work of the authors and FAS and do not represent the official view of the United States government.

An effort has been made to make the handbook as comprehensive as possible. It can be expected, however, that the material, interpretations, and conclusions are subject to modification in the light of new information and developments. Such corrections, additions, and suggestions for factual, interpretive, or other change as readers may have will be welcomed for use in future revisions. Comments may be addressed to:

The Director
Foreign Area Studies
The American University
5010 Wisconsin Avenue, N.W.
Washington, D.C. 20016

PREFACE

The *Area Handbook for the Persian Gulf States* is an attempt to provide a comprehensive study of the dominant aspects of the five societies of Bahrain, Kuwait, Qatar, Oman, and the United Arab Emirates and to identify patterns of behavior characteristic of their people. In view of the fact that Iran, Iraq, and Saudi Arabia also border on the Persian Gulf and Oman only partially does so, the title clearly is one of convenience. Moreover the governments of the countries included in this study officially use the term *Arabian Gulf;* the authors decided, however, to use *Persian Gulf,* the term used by most cartographers and Western governments. This study and the *Area Handbook for the Yemens* replace the *Area Handbook for the Peripheral States of the Arabian Peninsula.*

The study results from the combined efforts of a Foreign Area Studies multidisciplinary team of researchers assisted by the organization's support staff. The team was supervised by Richard F. Nyrop, who wrote chapter 1 and coordinated the contributions of the other authors. Beryl Lieff Benderly wrote chapters 3 and 4 and the Geographic and Demographic Setting section for chapter 6; Laraine Newhouse Carter wrote chapter 2 and the Geographic and Demographic Setting and the Government and Politics sections for chapter 10; William W. Cover wrote the National Defense and Internal Security sections for chapters 6, 9, and 10; Darrel R. Eglin wrote chapter 5 and the Economy sections for chapters 6, 7, 8, and 9; Robert A. Kirchner wrote the Government and Politics sections for chapters 6 and 7 and the National Defense and Internal Security section for chapter 7; Philip W. Moeller wrote the Geographic and Demographic Setting sections for chapters 7, 8, and 9; William A. Mussen, Jr., wrote the Government and Politics and the National Defense and Internal Security sections for chapter 8; Clarence Edward Pike wrote the Economy section for chapter 10; and Rinn-Sup Shinn wrote the Government and Politics section for chapter 9. The chapters represent the work of the authors and do not represent the official view of the United States government.

The authors are grateful to individuals in various agencies of the United States government who gave of their time, documents, and special knowledge to provide data and perspective. The authors are also grateful to several Arab scholars and diplomats residing in Washington, D.C., who were most helpful.

Sources of information used included scholarly studies, official reports of governments and international organizations, foreign and domestic newspapers, and numerous periodicals. Brief comments on some of the more valuable sources as possible further reading appear at the end of all chapters except chapter 1. Economic and demographic data ranged from excellent to unavailable. Except for data on oil production and revenues, which are relatively up-to-date and reliable, the data should be used with caution. Unless otherwise cited, weights are presented in metric tons.

The transliteration of Arabic words and phrases posed a particular problem. For many words—such as Muhammad, Muslim, Quran, and shaykh—the authors followed a modified version of the system adopted by the United States Board on Geographic Names and the Permanent Committee on Geographic Names for British Official Use, known as the BGN/PCGN system. In numerous instances, however, the names of people or places were so well known by another spelling that to have used the BGN/PCGN system might have caused confusion. The reader will therefore find Mecca rather than Makkah, Oman rather than Uman, Bahrain rather than Bahrayn, United Arab Emirates rather than Amirates, and Sultan Qaboos rather than Sultan Qabus. A glossary is included for the reader's convenience.

Arab names are sometimes confusing to the Western reader; they usually contain the patrilineal genealogy for three generations. Generally, indigenous Arabs along the Persian Gulf use one of two kinds of name forms: the first contains the father's and grandfather's names; the second substitutes a tribal name for that of the grandfather. For example, the name of a son of a Qatari ruler could be written as either Abd al Rahman ibn Qasim ibn Muhammad or Abd al Rahman ibn Qasim Al Thani. In the first example, Abd al Rahman is the man's identifying personal name, similar to the use of a given name in the West. He would be addressed as Mr. Abd al Rahman or Shaykh Abd al Rahman, as the case might be. *Ibn* and *bin* both mean son of; they are used interchangeably and are not a necessary part of the name, since the second name (when there are three)—in this case, Qasim—always indicates the father's name. Muhammad is the grandfather's name. Thus a Western rendition of the first example would be: Abd al Rahman, son of Qasim, who is son of Muhammad. In the second example Al Thani is used instead of Muhammad, indicating the family or tribal clan to which Abd al Rahman belongs. This usage is common among illustrious families, particularly ruling ones, and among nonurbanized gulf Arabs.

The second example is the most common name form in the Arabian Peninsula. The Western rendition of the second example is: Abd al Rahman, son of Qasim, belonging to or of the family of the Thani. The *al* in the middle of the first name should not be confused with the *Al* that precedes the tribal name. For the convenience of the reader the authors have used the uppercase *Al* with the family name. In other works, however, including government lists, sometimes this *Al* is lowercase. Abd al Rahman and other names like it, for instance Abd al Aziz and Abd al Wahid, are very common among gulf Arabs. *Abd al* means slave of; *Abd al Rahman,* for instance, means Servant of the Merciful (God). Rahman, Aziz, Wahid are designations of the attributes of God. The authors have transliterated these kinds of names as they would be written in Arabic. In speech, however, these names are usually elided; for instance, Abd al Rahman is pronounced Abdur Rahman and Abd al Wahid is pronounced Abdul Wahid. In the Western press the names may appear in the elided, that is, the spoken, form.

A gulf Arab woman does not take her husband's name at marriage. Women's names, like men's, usually indicate the father's and the grandfather's or the tribal name. The Arabic word for daughter is *bint,* but this seldom appears in the name form.

The Islamic hijra calendar, which has 354 days divided into twelve months, continues in wide popular use in the Persian Gulf states, although official business is conducted according to the Gregorian calendar. Conversions from the hijra year to a Gregorian date appear simple but actually are quite complicated. To assist the reader a conversion table is provided (see table A). Farmers, fishermen, and others working outdoors, however, apparently use the stars to calculate planting times and fishing seasons.

Table A. Conversion of Hijra Years to Gregorian Dates

Gregorian Date	Hijra Year (Muharam through Dhu al Hijja)	Gregorian Date	Hijra Year (Muharam through Dhu al Hijja)
June 25, 1960* ..	1380	February 26, 1971 .	1391
June 14, 1961 ...	1381	February 15, 1972*	1392
June 4, 1962	1382	February 4, 1973 ..	1393
May 25, 1963 ...	1383	January 23, 1974 ..	1394
May 13, 1964* ..	1384	January 13, 1975 ..	1395
May 1, 1965	1385	January 2, 1976* ..	1396
April 21, 1966 ...	1386	December 22, 1977 .	1397
April 11, 1967 ...	1387	December 11, 1978 .	1398
March 30, 1968* .	1388	November 30, 1979	1399
March 19, 1969 .	1389	November 19, 1980*	1400
March 9, 1970 ..	1390		

*Indicates leap year.

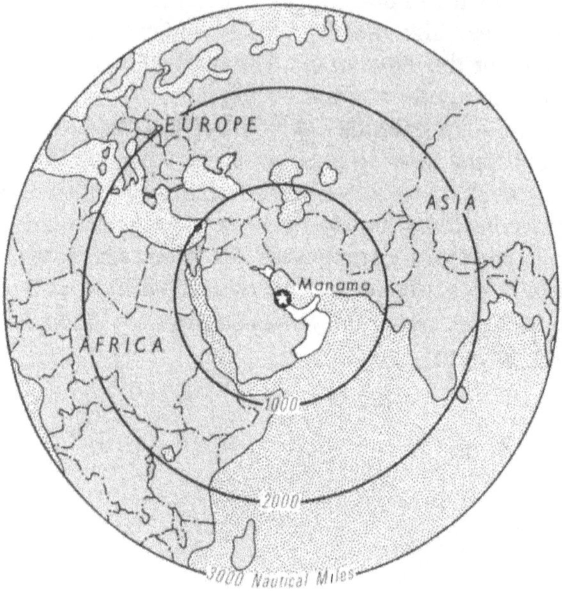

Persian Gulf in its World Setting

COUNTRY PROFILES (DATA SHEET)

	State of Bahrain	State of Kuwait	Sultanate of Oman	State of Qatar	United Arab Emirates
Capital city	Manama	Kuwait	Muscat	Doha	Abu Dhabi
Size (in square miles)	255	7,780	82,000 to 100,000	4,247	30,000 to 33,000
Population[1]	275,000	1,000,000	750,000 to 1,500,000	100,000 to 180,000	685,000
Population growth rate[1] (in percent)	3.6	10	3.1	over 10	20
Literacy[1] (in percent)	40 to 50	60 to 65	5 to 6	10 to 20	20 to 25
Administrative divisions	0	3	1 province; 37 districts	0	7
Name and symbol of currency	dinar BD	dinar KD	rial OR	riyal QR	dirham UD or Dh
Exchange rate (October 1976)	BD1= US$2.53	KD1= US$3.46	OR1= US$2.90	QR1= US$0.26	UD1= US$0.25
Per capita gross national product (in United States dollars)	2,250[2]	11,365[3]	1,250[2]	10,000[3]	8,800[3]
Value of exports (in millions)	BD438[4]	KD2,507[4]	n.a.	QR7,034[4]	UD28,637[5]
Value of imports (in millions)	BD470[4]	KD656[4]	OR231[5]	QR1,622[4]	UD10,910[5]
Armed forces	1,500	15,000	14,000[6]	2,200	15,550
Army	1,500	8,000	13,200	1,600	14,150
Navy	—	200	400	—	200
Air force	—	1,000[6]	400[6]	—	1,200
Other	—	5,000	2,000	600	n.a.
Major seaports	Mina Salman	Kuwait	Mina Qaboos	Doha and Musayid	Abu Dhabi and Dubai

n.a. means not available; — means none.
[1]Estimate as of mid-1976.
[2]1974 estimate.
[3]1975 estimate.
[4]1975 data.
[5]1975 data and estimates.
[6]Excluding expatriate personnel.

THE PERSIAN GULF STATES

TABLE OF CONTENTS

		Page
FOREWORD		iii
PREFACE		v
COUNTRY PROFILE		ix
CHAPTER	1. General Overview of the Societies	1
	2. Historical Setting Pre-Islamic Arabia—The Islamic Period—The Entrance of the Europeans into the Gulf—The Founding of the Modern Gulf Polities: Gulf States in the Eighteenth and Nineteenth Centuries—The Twentieth Century	9
	3. Religious Life Tenets of Islam—Development of Islam—Religion in the Persian Gulf States	41
	4. Social Structure Basic Elements of Social Structure—Social Systems of the Gulf Countries—The Individual, the Family, and the Sexes	55
	5. The Oil Industry in the Persian Gulf States Oil: The International Setting—Oil Industry of Kuwait—Oil Industry of Bahrain—Oil Industry of Qatar—Oil Industry of the United Arab Emirates—Oil Industry of Oman	75
	6. Kuwait Geographic and Demographic Setting—The Economy—Government and Politics—National Defense and Internal Security	121
	7. Bahrain Geographic and Demographic Setting—The Economy—Government and Politics—National Defense and Internal Security	207
	8. Qatar Geographic and Demographic Setting—The Economy—Political Dynamics—Foreign Relations—Mass Communications—National Defense and Internal Security	237
	9. United Arab Emirates Government and Politics—Geographic and Demographic Setting—The Economy—National Defense and Internal Security	275
	10. Oman Historical Background—Government and Politics—Geographic and Demographic Setting—The Economy—National Defense and Internal Security	341
BIBLIOGRAPHY		411
GLOSSARY		429
INDEX		433

LIST OF ILLUSTRATIONS

Figure		Page
1	The Persian Gulf States of Bahrain, Kuwait, Oman, Qatar, and the United Arab Emirates	xiv
2	The Arabian Peninsula in the Ancient Period	13
3	The Persian Gulf in the Medieval Period	19
4	Persian Gulf Oil Fields	74
5	World Oil, 1974	80
6	Kuwait, 1976	120
7	Kuwait, Percent of Population by Nationality, 1970	127
8	Kuwait, Population Pyramids of Kuwaiti and Non-Kuwaiti Residents, 1970	128
9	Kuwait, Al Sabah Family, November 1976	167
10	Kuwait, Government Structure, November 1976	173
11	Bahrain, Population Centers and Transportation Network, 1976	206
12	Bahrain, Average Monthly Family Expenditures, 1975	212
13	Bahrain, Government Structure, November 1976	227
14	Qatar, 1976	236
15	Qatar, Abridged Al Thani Genealogy and Government Ministers, June 1976	257
16	United Arab Emirates, 1976	274
17	UAE, Government Organization, 1976	278
18	UAE, Abridged Genealogy of the Al Nuhayyan Family of Abu Dhabi	289
19	UAE, Abridged Genealogy of the Al Maktum Family of Dubai	291
20	UAE, Abridged Genealogy of the Al Qasimi Family of Sharjah	293
21	UAE, Abridged Genealogy of the Al Qasimi Family of Ras al Khaymah	295
22	Oman, Regions, Roads, and Major Cities, 1976	340
23	Oman, Government Structure, 1976	348

LIST OF TABLES

Table		Page
A	Conversion of Hijra Years to Gregorian Dates	vii
1	Kuwait, Oil Revenues, Crude Oil Production, and Petroleum Exports, Selected Years, 1946–75	95
2	Bahrain, Crude Oil and Refinery Production and Foreign Trade, Selected Years, 1964–75	102
3	Bahrain, Oil Revenues, 1967–74	105
4	Qatar, Crude Oil Production and Oil Revenues, Selected Years, 1950–75	107
5	Abu Dhabi, Crude Oil Production and Oil Revenues, 1962–75	111
6	Oman, Petroleum Statistics, 1967–75	117
7	Kuwait, Public Education, School Years 1966–67, 1970–71, and 1974–75	133
8	Kuwait, Students in Private Schools, School Years 1966–67, 1970–71, and 1974–75	134
9	Kuwait, Daily Per Capita Intake of Food, 1973	137
10	Kuwait, Gross Domestic Product by Sector, Fiscal Years 1969–73	143
11	Kuwait, Summary of Budget Revenues and Expenditures, Fiscal Years 1971–75	146
12	Kuwait, Total Loans Extended by the Kuwait Fund for Arab Economic Development by Project and Country	150
13	Kuwait, Summary of Foreign Trade, Selected Years, 1969–74	159
14	Kuwait, Summary of Balance of Payments, Fiscal Years 1971–74	161
15	Kuwait, Ministry of Defense and Ministry of Interior Expenditures, Fiscal Years 1968–75	192
16	Kuwait, Armed Services Ranks and Insignia, 1972	198

17 Kuwait, Pay Scales of Regular Services, 1972 200
18 Kuwait, National Guard Grades and Pay Scales, 1972 202
19 Kuwait, Non-Kuwaiti Professional Grades and Pay Scales, 1972 203
20 Bahrain, Population by Nationality and Sex, Selected Years, 1959–72 210
21 Bahrain, Summary of Foreign Trade, Selected Years, 1968–74 221
22 Bahrain, Summary of Balance of Payments, 1971–74 222
23 Bahrain, Budget Summary, 1973–76 223
24 Bahrain, Armed Services Ranks and Insignia, 1976 233
25 Qatar, Summary of Foreign Trade, 1969–75 251
26 Qatar, Balance of Payments, 1970–74 251
27 Qatar, Budget Summary, Fiscal Years 1970–75 253
28 UAE, Area and Population by Amirate, 1968 and 1975 300
29 UAE, Summary of Federal Budget, 1972–74 309
30 UAE, Summary of Abu Dhabi Budget, 1971–76 311
31 UAE, Summary of Exports and Imports, Selected Years, 1970–75 320
32 UAE, Summary of Balance of Payments, 1971–74 321
33 UAE, Structure and Composition of Military Forces, 1975 326
34 UAE, Armed Services Ranks and Insignia, 1976 335
35 Oman, Council of Ministers, December 1976 349
36 Oman, Regional Administration by Geographic Location, 1976 350
37 Oman, Government Finances, 1971–75 379
38 Oman, Monetary Developments, 1970–74 382
39 Oman, Foreign Trade, 1970–75 384
40 Oman, Military Expenditures and Forces Strengths, 1973–76 400
41 Oman, Defense Budgetary Authorizations, 1974 401
42 Oman, Armed Forces Ranks and Insignia, 1976 407

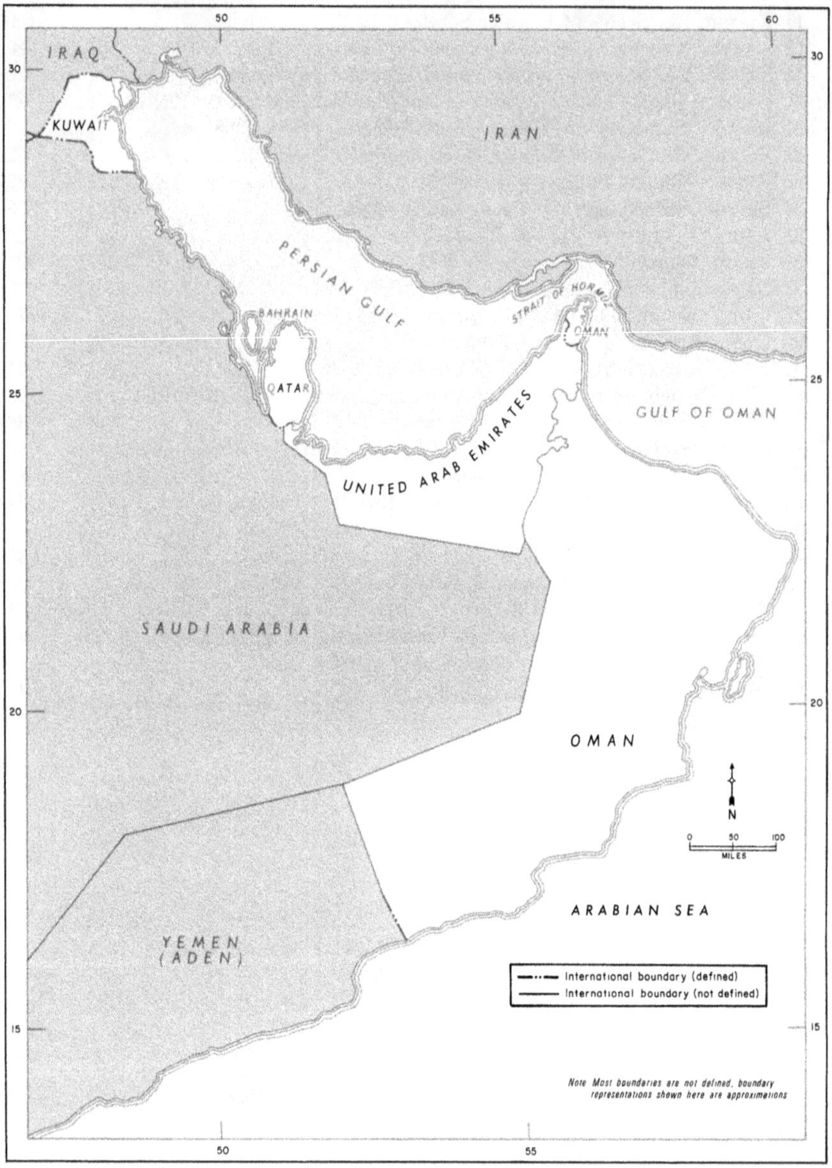

Figure 1. The Persian Gulf States of Bahrain, Kuwait, Oman, Qatar, and the United Arab Emirates

CHAPTER 1

GENERAL OVERVIEW OF THE SOCIETIES

In the aftermath of World War II the several small amirates, shaykhdoms, and sultanates on the Arabian Peninsula side of the Persian Gulf were of marginal concern to most of the world. Only the international oil companies and a few Arabists and archaeologists—mostly British—either knew or cared about the area. Bahrain, Kuwait, Qatar, and the seven shaykhdoms of what was then known as Trucial Oman or the Trucial Coast were under British hegemony as a result of several treaties of protection dating back to the nineteenth century. Although Oman—then known as the Sultanate of Muscat and Oman—had been independent since the seventeenth century, it shared with the other polities on the eastern and southeastern rimland of the peninsula a close connection with and dependence on the United Kingdom and a general obscurity in world affairs.

Within thirty years all of this changed dramatically. In 1961 the State of Kuwait became fully sovereign, ending by treaty its status as a protectorate of Great Britain. In 1968 Great Britain announced that it would terminate its special position and military obligations in the gulf by 1971; in that year, again by a series of treaties, the State of Bahrain, the State of Qatar, and the United Arab Emirates (UAE)—the former Trucial Coast states—became independent. In 1970 Oman became the Sultanate of Oman in the wake of a coup d'etat in which the sultan was deposed by his son Qaboos bin Said bin Taimur Al Bu Said, who in 1976 continued to reign as the Sultan of Oman.

The catalyst of change was not the achievement of independence but a revolution during the early and middle 1970s in the international oil industry (see Oil: The International Setting, ch. 5). When in December 1976 the Organization of Petroleum Exporting Countries (OPEC) met in Qatar for its forty-eighth conference, the deliberations and negotiations were anxiously observed and closely followed by all foreign capitals, the world business community, and the international press. The announcement that the UAE would follow Saudi Arabia's lead by raising the price of oil by 5 percent—whereas the other members of OPEC decided on an increase of 10 percent effective January 1, 1977, and another 5 percent July 1, 1977—was the object of international attention and analysis at a level usually accorded only to the actions of the superpowers. OPEC deployed no military divisions, but by the mid-1970s it had secured and was exercising great power and influence.

As of mid-1976 the petroleum-producing states of the Arabian Peninsula possessed nearly half of the world's proven oil reserves. Saudi Arabia, with over 25 percent of the world's reserves, stood in a class by itself, but the proven reserves of others were impressive. Kuwait had over 11 percent of the proven reserves; the UAE had nearly 5 percent and could reasonably expect more oil discoveries; and Qatar had perhaps 1 percent and hoped for more offshore discoveries. The reserves of Bahrain and Oman, which were not members of OPEC, were small and were expected to be depleted by the late 1980s. Observers thought that additional oil deposits might be found in Oman but doubted that any more oil would be found in Bahrain.

The five political entities shared a number of economic, political, and social characteristics. Except for their oil reserves and for potential fishing industries, which Kuwait in particular was seeking to develop and expand, the economies possessed practically no natural resources. In late 1976 all of the governments were traditional, paternalistic and, in varying degrees, authoritarian monarchies, albeit the UAE was a federation of such monarchies. The indigenous populations were overwhelmingly Arab Muslim, but the governments and economies were heavily dependent on foreign workers and advisers who frequently were neither Arab nor Muslim. In fact, in Kuwait, Qatar, and the UAE noncitizens formed majorities of the populations. Largely because of the new technologies, ideas, and customs introduced by foreign populations and by the social and educational services instituted by all of the governments, the traditional societies either were undergoing or were on the verge of undergoing significant social change. The societies continued to be based on the Islamic, beduin values that had developed over the centuries in the interior deserts of the Arabian Peninsula, and in the mid-1970s it was unclear how rapid or how pervasive the changes would be.

The oil wealth of Kuwait, Qatar, and the UAE has enabled them to provide massive financial loans and grants to other states. At the same time they have also established welfare societies, and Bahrain, with help from its rich neighbors, has very nearly accomplished the same. Medical care—including the services of doctors, dentists, and hospitals—was free, and in the mid-1970s the quality and availability of care were rapidly improving. Kuwait and the UAE were especially active in the construction of hospitals and clinics, which were largely staffed by Indian and Pakistani doctors, nurses, and administrators.

The changes in education were equally impressive and perhaps of greater long-range significance. In 1951, for example, there were no secular schools in the area encompassed by the present-day UAE. By 1975 the largest single item in the UAE budget was for education, and an estimated 80 percent of primary-school-age children were, as required by law, attending school. Except for a few schools for foreigners, there was only one small secular school in Oman before the 1970 coup. Although Oman lacked the vast financial resources of the UAE, by 1976 an estimated 55 percent of Omani children were in school, and expenditures for education were a major item in the national budget. Bahrain instituted free public education earlier than the other states, and by the mid-1970s the literacy level was estimated at 40 to 50 percent. The literacy rate for Kuwait was even higher, estimated at between 60 and 65 percent; for Qatar, 10 to 20 percent; for the UAE, 20 to 25 percent; and for Oman, 5 to 6 percent.

The role of foreigners was critical in the five coastal states. In Kuwait an estimated 53 percent of the 1976 population of slightly over 1 million were foreigners. Although the expatriate population included many Baluchis from Iran and Pakistan, other Pakistanis and Indians, Americans, British, and a large number of Egyptians, the largest foreign group was composed of Palestinians. The government of Kuwait had for years given vast sums of money to support the "confrontation" states (Egypt, Jordan, and Syria) of the several Arab-Israeli wars and had provided financial and diplomatic support for the Palestinian Liberation Organization (PLO). The more aggressive PLO activists in Kuwait felt that Kuwait was not doing enough, however, and they coupled this grievance with complaints that non-Kuwaitis suffered from discrimination in housing, job opportunities, and related matters (see Geographic and Demographic Setting, ch. 6).

By mid-1976 enough members of the Kuwait National Assembly (Assembly) had endorsed one or another of the antigovernment issues to provoke a political impasse. The amir, Shaykh Sabah al Salim Al Sabah, resolved the problem in August 1976 by suspending the Constitution and proroguing the Assembly (see Government and Politics, ch. 6). The prime minister, Shaykh Jabir al Ahmad al Jabir Al Sabah, reorganized the Council of Ministers, instituted new and more restrictive press controls, and increased government surveillance of foreigners and of antigovernment groups. In late 1976 foreign analysts believed that the Al Sabah royal family of Kuwait would not soon resume its fourteen-year experiment with parliamentary politics. The analysts observed that the Al Sabah was attempting to reduce the dependence of the government and the economy on foreigners but that it would be several years before enough trained Kuwaitis were available to staff even a majority of the key technical and managerial positions.

The UAE's population was perhaps even more unbalanced by the presence of foreigners than Kuwait's. The 1968 population had been estimated by the British at about 180,000. Although the data were imprecise, the population in late 1975 was officially estimated at about 656,000, indicating an average increase of over 20 percent a year. Despite significant improvement in health care facilities between 1968 and 1976, the annual rate of natural increase probably did not exceed

3.1 percent. Most of the increase, in other words, resulted from immigration. Most of the teachers were Egyptians; almost all of the medical personnel were Indians or Pakistanis; several thousand Americans and British served as technicians and advisers; and Lebanese and Pakistanis worked as journalists, civil servants, clerks, and accountants. The chief of staff of the armed forces was a Jordanian, and many members of the officer corps also were foreigners. Over one-half, and possibly as many as two-thirds, of the civil servants were foreigners. Foreign observers estimated that the UAE government and economy would continue to rely on expatriate advisers, technicians, managers, and laborers well into the 1990s.

To the extent that there were problems in government in the UAE, the problems resulted not from the presence of foreign agitators but from the structure and the functioning of the federal government. When the union was formed in 1971, a provisional constitution was adopted for a period of five years. The notion was that the 1971–76 period would be one of unification and nation building and that by December 2, 1976, a truly new union—and a new constitution—would have been developed (see Government and Politics, ch. 9).

The prime advocate of unification throughout this period was, and in late 1976 remained, Shaykh Zayid bin Sultan Al Nuhayyan of Abu Dhabi. Most of the UAE's proven oil reserves were in Abu Dhabi; Dubai and Sharjah had much smaller fields, and the discovery of oil in Ras al Khaymah in 1976 had yet to be proved of commercial consequence. The other three members of the union—Umm al Qaywayn, Ajman, and Fujayrah—had no oil and, with a combined population of about 65,000, were described by one analyst as "village-states." As in all the states on the Arabian Peninsula the oil revenues go to the head of the royal family. Shaykh Zayid therefore has had at his disposal vast sums of money, and since 1971 he has devoted massive amounts to development projects in all of the amirates. Schools, hospitals, clinics, new towns, ports, airports, highways, and related projects were constructed everywhere, and foreigners were imported to run the projects. For five years Shaykh Zayid, as president of the UAE, sustained the union by his wealth and his forceful personality.

But progress to actual union was slow, almost nonexistent, through the mid-1970s. The various amirs welcomed Zayid's money, but they were unwilling to relinquish their individual sovereignty. In mid-1976 Zayid announced that he would not serve again as president, an announcement that apparently served the cause of unity. Although a new constitution was not adopted, the existing one was amended. In November, for example, a constitutional provision that enabled each amirate to have its own military force was deleted. Other unification measures were adopted, and Zayid, having secured at least some of the cooperation he deemed essential, accepted another five-year term as president. In late 1976, however, Zayid's money and personality continued to form the glue that held the UAE together.

The population of Qatar also was heavily foreign. In 1976 the population was estimated at 140,000 to 180,000, between one-third and one-half of whom were expatriates. Qatar was socially the most conservative of the coastal states. Its ruling royal family, the Al Thani, and perhaps half of the indigenous population are adherents of Wahhabi (see Glossary) Islam.

In administering its fledgling welfare state the royal family remained dependent on foreign advisers and technicians and on a rising cluster of commoner families. In a pattern followed by the other Arab monarchies, however, Qatar's royal family retained close personal control of the state's fiscal policies and of the key governmental positions. The head of the family, Shaykh Khalifah bin Hamad Al Thani, was the amir, and as such he was also the supreme commander of the armed forces. Unlike most other rulers, Khalifah served as his own prime minister. One son served as defense minister and another son as minister of finance and petroleum. The portfolios of foreign affairs and of interior were held by Khalifah's brothers, and the police force was headed by a nephew (see Political Dynamics, ch. 8).

In Kuwait, Qatar, and the amirates of the UAE succession to the leadership of the royal families was determined by the senior males, who reached a consensus in conformity with family traditions and agreements. In Bahrain, however, the Al Khalifah royal family adopted primogeniture and specified this form of succession in the Constitution. In 1976 the amir was Shaykh Isa bin Salman Al Khalifah. His brother was prime minister, his son and heir apparent was defense minister, and seven other Al Khalifah members held portfolios in the fifteen-member cabinet (see ch. 7).

In late 1976 Bahrain's population was probably about 275,000, although the data on which the estimates were based were questionable. Between 20 and 30 percent were foreigners. About 30 percent of the population was in the working force, and a majority of workers were expatriates, including about 4,500 British and 3,000 Americans. The British and Americans held positions in the petroleum industry and in related fields requiring sophisticated technical skills and managerial experience, and they were also prominent in the banking and commercial center that was developing in Bahrain to serve the Middle East markets (see The Economy, ch. 7).

In late 1976 Oman continued to share many apparent similarities with its fellow Arab monarchies on the gulf, but in some ways it was unique. For example, in each of the royal families that govern the several gulf states there were several adult males who exercised great influence; that is, the ruler usually needed to consult with his brothers, cousins, and uncles in making major decisions. In Oman, as one analyst observed, the sultan *is* the royal family. Members of the Al Bu Said family serve the sultan as advisers and ministers but, because they tend not to have individual bases of power, they quite literally serve at the pleasure of the sultan (see Government and Politics, ch. 10).

Sultan Qaboos' father, Said bin Taimur Al Bu Said, was described by his British advisers as paranoid and reactionary. He sought to seal his dominion off from the world, although he did, grudgingly, grant oil concessions to foreign oil companies; but when he began to receive oil revenues, he spent practically nothing either on social services or on developing an economic infrastructure (see Geographic and Demographic Setting; The Economy, ch. 10). During his first six years as head of government (in essence as *the* government) Qaboos sought to compensate for his father's sins of omission, but his financial resources were not great, and he was starting from an almost zero base. In the mid-1970s practically the only educated Omanis were the sons and daughters of Omani families who had lived abroad during the period of Sultan Said's enforced isolation. Attempts to modernize the society, the economy, and the government therefore depended heavily on expatriates.

Qaboos' efforts to institute social services and to introduce free education were severely hampered by the need to devote large amounts of time, energy, manpower, and money to the eventually successful task of crushing an insurgency in Dhofar Province (see National Defense and Internal Security, ch. 10). Among other things the insurgency—which began in 1963 and remained a minor problem in late 1976—highlighted Oman's dependency on foreigners. In 1976 the chief of staff of the Sultan's Armed Forces (SAF) was British, and the officer corps included many British, Jordanians, and Pakistanis. Moreover an Imperial Iranian Task Force remained on duty in Oman in late 1976 as an apparently essential component of Oman's defense and security forces.

Despite Iran's close relations with Oman and generally cordial relations with the other countries on the gulf, Iran's growing military might has provoked mixed reactions in the Arab states of the gulf. On the one hand Iran's monarch has professed a willingness to use his rapidly expanding military power to combat radical and revolutionary forces seeking to overthrow monarchical systems. On the other hand Iran has several historical claims to various parts of the Arab states on the gulf. Moreover during the mid-1970s Iran was continuing to produce and export its oil at a high volume, and its small gulf neighbors naturally wondered if their oil reserves would be safe when Iran's reserves were exhausted.

Unlike Iran, Iraq was a fellow Arab state and a member of the League of Arab States (Arab League), but the gulf states, Kuwait in particular, reportedly viewed Iraq as potentially a more serious threat. Iraq's ruling Baath Party was a radical, revolutionary group that espoused unyielding hostility to the traditional and conservative Arab monarchies. Since the overthrow of its own monarch in 1958 in a particularly bloody coup, Iraq's succeeding governments have sought to export their regicidal fervor. To Kuwait's consternation old Iraqi claims to Kuwait territory have occasionally been resurrected. In 1976 the Kuwait government reprimanded a newspaper for reporting that Iraqi military units had crossed into Kuwaiti territory, but the reprimand did not deny the newspaper report (see National Defense and Internal Security, ch. 6). An Iraqi intrusion into Kuwait in 1961 was withdrawn in the face of Arab League disapproval and, more important, of a British military detachment that landed in response to an urgent request from the Kuwait government. Iraq subsequently withdrew its territorial claim, but in 1976 Kuwait was not certain if Iraq meant it.

In the mid-1970s there were occasional references by spokesmen of various governments of the area—including Saudi Arabia—to the need for a regional (or gulf) security pact. It was unclear at any given time which governments favored the notion and which opposed it. More to the point, perhaps, were questions of membership: would the pact include only the conservative monarchies, meaning that Iran would be in and Iraq out, or would it be Arab in its composition, meaning that Iran would be excluded and Iraq would be eligible? Although no decision had been announced as of late 1976, foreign analysts speculated that the several Arab monarchies would continue their frequent, ad hoc consultations on security and other matters but would not take the step of formal alliance outside the Arab League.

CHAPTER 2

HISTORICAL SETTING

In 1976, as world attention was focused on the oil-rich and therefore potentially powerful states of the Arabian Peninsula, observers were fascinated by the speed, frequently accompanied by chaos, with which those states were approaching modernization. Images were prevalent of palm-treed shantytowns transformed into metropolitan centers overnight; of Cadillacs parked in front of beduin tents; and of a collection of states, some of which have the highest per capita incomes in the world, ruled anachronistically by patriarchal tribal leaders in a manner reminiscent of the medieval world.

Such images are true perceptions, but it should not be assumed that oil was the first medium to bring life to the Arabian shore of the Persian Gulf. In the century before the discovery of oil and the subsequent use of oil revenues for improving economic strength and for social welfare projects, the Arab states that line the western gulf littoral were at the most somnolent point in their long and turbulent history.

There is evidence that, before human beings had migrated to the European geographic area, indigenous traders of the western coast of the gulf were enriching the first civilizations of the Fertile Crescent both commercially and culturally. Despite periods of decline such trade appears to have continued for millennia. Gulf trade was active during the ancient period but extraordinarily so by medieval times. In the ninth century A.D. gulf traders were traveling to China and returning up gulf waters to Basrah, facilitating the exchange of ideas and goods among the great civilizations of the time. The peoples of the gulf coast, unlike the inhabitants of the interior of the Arabian Peninsula, had always been exposed to outside influences. The cultural and economic area to which the Arab gulf states belonged was not at all limited to Arabia but included the whole of the Persian Gulf area. Arabs and Persians moved with nonchalance from one side of the gulf to the other.

One persistent factor in gulf history is the constant rivalry between merchant states and the desire for hegemony. Another factor, destined to be more fatal to gulf trade, was the interest in the area by outsiders —Europeans, Ottoman Turks, the Wahhabis (see Glossary) of central Arabia, and the Egyptians. After the entrance of the Europeans but before European power reached its apogee in the gulf, Hormuz, located near the Persian coast, was for a time the foremost trading center in the world (see fig. 1). A popular Arab and Persian saying is, "If the world were a ring, Hormuz would be its jewel."

Largely because of events in Europe, the British were able to outstay the Dutch and slowly but effectively build up power in the area. Tribal animosities and rivalries and mutual piracy among gulf states served British interests, and by means of a series of exclusive treaties the British eventually pacified the gulf. Although gulf merchants were still active, they were considerably circumscribed by these treaties. In the early years of the nineteenth century new gulf powers, particularly the states of Kuwait and Bahrain, were emerging and achieving great wealth through trade. Because the states had few natural resources besides pearls and date palms, maritime activity was not only the easiest way for Arab gulf states to achieve wealth but also virtually the only way they could survive. A dramatic change in living conditions occurred, therefore, when European steamships and the associated technology began to appear in the gulf in the 1860s. Although the gulf Arabs were capable of hand-building ships that could carry several tons of merchandise to China and back, they had not developed industrially and could not begin to compete with the European vessels. Fossilized into their political positions and isolated from much foreign influence by British protectorate treaties, the Arab gulf states went into a certain decline and were revived only with the discovery of oil.

In the period before oil gulf Arabs fell back on their natural resources, meager as they were, until the values of these, too, were threatened. In the 1920s the Japanese began producing cultured pearls, and the situations of Bahrain, Qatar, and much of the Trucial Coast area (United Arab Emirates—UAE) were reduced still further. Arabs subsisted largely on minor commerce and other dealings with the Europeans who called there. The worldwide depression of the 1930s reduced traffic through the gulf and consequently reduced income for those who provided services. The people of Dubai, deciding old ways were best, became active in gold smuggling to India.

Oil and the change in British priorities, particularly after World War II, altered the situation. The discovery of oil and the practical utilization of wealth accruing from it were initially slow processes. Gulf Arabs were awakening from a long sleep, and the lack of immediate reform was largely the result of ignorance and underexposure to the kinds of positive social, cultural, and economic change that money could buy. The Arab gulf has never been a monolith, and the individual states moved at different rates depending on custom and income. A curious reversal had taken place. Oman, Ras al Khaymah, and Sharjah, once the most powerful and the oldest established of the gulf states, became the poorest.

Once the states realized their affluent position and what it could do, the traditional welfare system built into the tribal order served them well. In 1976 none of the "welfare" states emerging in the modern gulf was socialistic; yet all that had oil money were extravagant, by any standards, in spending for the benefit of their citizens. This was not an idea introduced by the West; it was rather an expansion of the tribal belief, reinforced by Islam, that no man is rich if a member of his clan is poor.

PRE-ISLAMIC ARABIA

The Civilizations of Prehistoric Arabia

Archaeological investigations of pre-Islamic Arabia are still in an embryonic state, and the results are hypothetical and controversial at best. Through the mid-1900s few people had the physical endurance or the survival techniques crucial to investigate the area. Scarcity of water, the difficulties of desert transport and, until the 1940s, hostile tribes made systematic research a heroic undertaking. Further, very little was known about present-day Arabia beyond its coastal settlements. Initial archaeological discoveries were often the accidental finds of explorers anxious to investigate the isolated land and map its interior.

The oldest evidence of civilized man in northern Arabia is artifacts found sixty miles to the north of Dhahran on the coast of the Persian Gulf. Dated to 5000 B.C., they are identical with those of the Al Ubaid culture of Mesopotamia, the first people to cultivate and settle the Fertile Crescent and the ancestors of the Sumerians, the first known people to develop a high culture. If Al Ubaid culture originated in Mesopotamia, then civilization reached Arabia from the north. If, however, Arabia was the parent site, then the first known agriculturists in the region were migrants from Arabia. This would substantiate the Sumerian myth that agriculture had been brought to Mesopotamia by a fish-man from the Persian Gulf.

From about 4000 to 2000 B.C. the civilization of Dilmun dominated 250 miles of the eastern coast of Arabia from present-day Kuwait to Bahrain and extended sixty miles into the interior to the oasis of Hufuf (see fig. 2). At its zenith in 2000 B.C. Dilmun controlled the route to the Indies and was the trading link between the civilizations of the Indus Valley and those of Mesopotamia. The Mesopotamians regarded Dilmun as a holy place and its people as extraordinarily blessed. In Oman and Abu Dhabi a civilization has been found that might have been related to the one at Dilmun. In Abu Dhabi pre-bronze-age stone buildings and settlements with elaborate tombs suggest a peaceful people and advanced cereal cultivation.

Arabia was only sparsely peopled in the interior. Until about 3000 B.C. the inland was sufficiently verdant to support both cereal agriculturists and herding peoples in the north and hunting and gathering societies in the south. As climatic conditions changed and the desert slowly encroached upon land that had supported both animal and human life, the inhabitants were faced with three choices: to cling to the oases, to move to the coasts, or to leave Arabia. Those who made the third choice and migrated to the north, northeast, and southwest are the only ones who left a historical record in that period.

Ancient Seafaring in the Persian Gulf

Some of the more tenacious and adventurous turned their backs on the inhospitable land and founded thalassocracies, greatly advancing the interchange of commodities and culture in the ancient period. Although the gulf apparently experienced various periods of relative decline in its shipping activities, it was always numbered among the world's great trade routes. At its northernmost point the gulf terminates near the confluence of the Tigris and Euphrates rivers. Goods transported through the gulf from the south and later from East Africa and East Asia reached Mesopotamia and Babylon, states that, though rich in fertile land, lacked the stone, metals, and woods necessary to sustain advanced civilizations. Seafaring trade on the gulf to these areas is documented from the third millennium.

More precise data from the second millennium indicates that Mesopotamia was then importing from three city-states in the direction of the gulf: Dilmun, which had its headquarters on Bahrain; Magan, on the coastal curve of modern Oman lying on the Gulf of Oman; and Maluhha, which in very ancient times was in the region of Sind and the Indus Valley but by the first millennium was identified with Nubia in Ethiopia. Initially Magan functioned as Maluhha's entrepôt. Typical Mesopotamian imports from Magan included copper, diorite, ivory, red ochre (perhaps from the island of Abu Musa), onions, bamboo, wood, and precious stones. "Fisheyes" (pearls) were also listed. Since the best pearl banks are between the tip of Qatar and Sharjah, it is likely that even at this early date pearling was a local industry. Dilmun eventually became the entrepôt for Magan and Maluhha. Dilmun's commercial power began to decline in 1800 B.C., perhaps as a result of the invasion and devastation of the Indus Valley civilization, which disrupted trade in the region for several centuries.

Piracy flourished in the gulf during Dilmun's decline, and Omani seafarers and merchants perforce turned their attention to Dhofar, the southern region of present-day Oman. Dhofar was one of only three producers of the highly valuable aromatic gum resin, frankincense. Frankincense, an essential element in certain ancient Jewish and pagan rituals, was burned as an offering to the gods. It was used lavishly in cremation services and, in Egypt, for embalming. For the funeral of Nero's wife an entire year's harvest was reputedly consumed. Frankincense also has healing properties; it was used as an antidote to poisons and to stop hemorrhages. From Dhofar frankincense was exported by

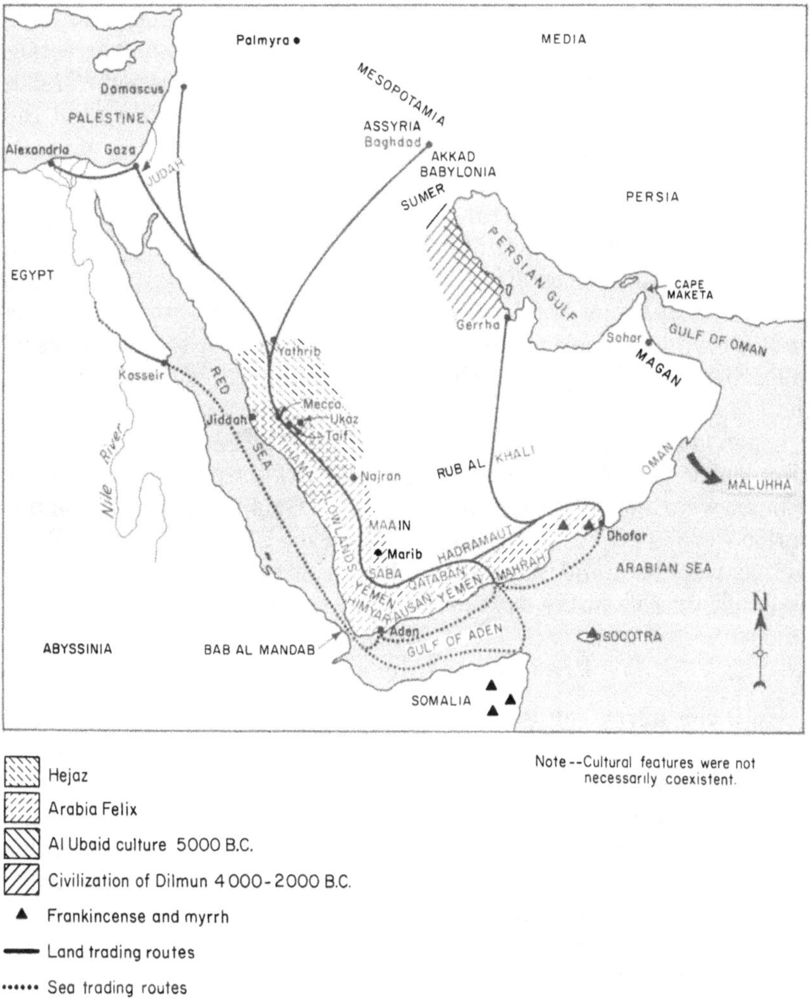

Figure 2. The Arabian Peninsula in the Ancient Period

sea from the port of Sumharam (near present-day Salalah) or transported by camel through the Hadramaut and then usually up the land route through the Hejaz (the western coastal area of present-day Saudi Arabia).

Gulf trade did not cease completely after Dilmun's decline. Alexander the Great's admiral, Nearchos, wrote in his journal of an Arabian cape called Maketa (Ras Musandam) from which cinnamon was exported to the Assyrians. By the end of the third century B.C. gulf trade had revived slightly. Gerrha, opposite Bahrain on the Arabian mainland, had become the most important commercial center in the area and was the local entrepôt for the fabulously wealthy kingdoms of southwest Arabia, among which were Saba (Sheba) and Himyar. At that time

Magan in Oman came to be called Maazun by the Persians, who were increasingly involved in gulf trade. In A.D. 228 the Sassanians established their dynasty in Persia. During the Sassanian period Persian trade in the gulf reached its apogee, and it did not decline until the Arabs conquered Persia in the seventh century.

By the third century A.D. northern Oman's coastal districts were under Persian control. In the fourth century the Persians occupied Bahrain; when King Khosrou I invaded and occupied the Yemen in 520, southern Oman was included in the conquest. The Persians introduced the *qanat* (*falaj* in Arabic), an ingenious irrigation system by which underground channels tap groundwater and carry it to fields several miles away. In 1976 much of Omani agriculture was still dependent on the system.

Oman never acquired the reputation Yemen did because it was virtually denied immediate access to the West by the vast Rub al Khali (Empty Quarter). It is clear, however, that Omanis in the pre-Islamic period were as active in long-distance sea trade between East and West as they were later and that the gulf Arabs, particularly Omanis, played a significant role in the development of ancient trade.

Pre-Islamic Internal History

Until the advent of Islam the dominant political and commercial powers in the peninsula were those of the south—Oman and the declining kingdoms of Yemen. Outside the Omani and Persian trade colonies on the gulf littoral and Hejazi caravan stops turned independent cities, such as Mecca and Yathrib (Medina), much of the population of Arabia was nomadic. In the first three centuries A.D. there were mass tribal relocations throughout the peninsula. Oman and the lower gulf coast with their embryonic local industries and seafaring trade were the main objects of these migrations. The reasons for the migrations are unclear. Certainly the collapse of the dam at Marib, northeast of present-day Yemen (Sana), which had irrigated otherwise untillable land, contributed to it. There was also a great deal of social disruption. The population was increasing, and attacks by beduin on the *hadr* (settled peoples) were becoming more frequent.

Several northern beduin tribes moved south, much to the consternation of the indigenous southerners, who were already pressed for survival because of the paucity of local natural resources. It was about that time that traditions about tribal origins began to take concrete form. The tradition that has had the most serious political consequences for the gulf concerned the difference between northern and southern Arabs. The southern Arabs entered the Islamic period with a sense of an ethnic distinction between themselves and all other Arabs that explained the differences of language, custom, and physiognomy. Popular belief held and still holds that, although all Arabs are descended from Sham ibn Nuh (Shem, son of Noah), the "pure" or southern Arabs (Qahtani) are descended from Qahtan ibn Abir (Joktan ben Eber), or Hud as he is often called, whereas the northern Arabs (called Adnani; in Oman sometimes called Nizari) are descended from Ismail (Ishmael) through Adnan. Although other Arab nationals, particularly those of tribal societies, know or think they know to which group they belong but are little concerned about it, the split is a matter of importance to Omanis and to Arabs of the lower gulf. Many feuds in the eighteenth, nineteenth, and twentieth centuries can be traced to it.

There is very little definite or reliable information about the gulf area from the period of the decline of the great ancient peoples until the advent of Islam. Some scanty information exists for Oman because it was the most developed area of the gulf littoral at the time. In the second century A.D. the Al Azd tribe migrated to Oman. Future imams (see Glossary) and sultans claimed descent from the tribe and usually took "al Azdi" as the final part of their titles. The Al Julanda were rulers of Oman at the time of the Azdite invasion and were vassals of the Persians. The Al Azd, a numerous force, ousted the Persians for a brief period. In a short time, however, the Al Julanda ruled again from the coast in cooperation with the Persians, while the Al Azd moved beyond the mountains, thus creating an internal division in the country that prevailed until 1970.

Also in the second or third century there was a major migration to Tuam (Buraymi Oasis) of two Adnani tribes, the Bani Said and the Bani Abd al Qais. The latter provided the ruling family for Qais, a major medieval gulf port.

Most of the Arabs of the peninsula worshipped an astral triad. In the southern regions—Yemen and Oman—there were also Judaized Arabs, Christians and, as a result of the Persian presence, Zoroastrians. Judaism came to the Yemen, via the Hejaz, when great numbers of Jews fled south after the destruction of the Temple of Jerusalem by Titus in A.D. 70. Theophilus Indus, the most important Christian missionary to southern Arabia, established three churches: at Sana, at Aden, and on the gulf, most probably at Sohar. John, Oman's first bishop, was appointed in the fifth century; Stephen, the last known bishop, was still living in Oman in 676.

THE ISLAMIC PERIOD

Conversion in the Gulf

Unquestionably the event of greatest social, political, and cultural significance for the inhabitants of the gulf and Oman was the conversion of those populations to Islam late in the seventh century. The discovery and production of oil centuries later transformed the marginally productive medieval city-states of the gulf at a speed probably unprecedented in history; Islam has remained the bedrock of the societies, however, and modern economic concerns function to a large extent within the legal and cultural framework that is Islam's legacy.

Because Islam contains many Judeo-Christian ideas—indeed it is considered by Muslims to be the fulfillment of the Judeo-Christian tradition—many of the theological conceptions within Islam were familiar to the inhabitants of the gulf. This was particularly true of the Omanis, who had Christians among them and who had commercial relations with Christians and Jews in the southwest of the peninsula.

The major attraction of Islam for the gulf and Omani Arabs was much the same as for most residents of the peninsula; Islam was a regional phenomenon without foreign overtones. As such it had special appeal for Omanis who had undergone several foreign occupations. There was apparently a relatively rapid conversion of Omanis to Islam. For them a special inducement was that other tribes would join with them to expel the Persians permanently from their lands. There was some recalcitrance on the part of the gulf Arabs to the north; at least they are not cited in Islamic sources as immediately and enthusiastically joining the Islamic ranks.

After an initial period when the peninsular Arabs were only superficially converted to Islam, there were anti-Islamic uprisings throughout the peninsula, even in Oman, where a congregational mosque in Sohar was one of only a few mosques in use in Arabia before the death of the Prophet Muhammad. Among the many self-proclaimed prophets who appeared after the beginning of the Islamic conquest of the peninsula was Dhut Taj Lakit ibn Malik Al Azd of Oman. Abu Bakr, the first orthodox caliph, appointed three generals to suppress the growing revolt led by Dhut. The Al Azd of Sohar and Dibbah (tribesmen of the rebellious Dhut) and the Bani Abd al Qais fought with the generals, and a Muslim victory was achieved. The battle became known as the Day of Dibbah and became a symbol for the defeat of paganism by Islam.

The conquest of the Arabian Peninsula was the most protracted and difficult of all the Muslim conquests. Tribally oriented and wary of strangers, Qahtanis were reluctant to pay zakat to the Adnanis, who provided the Islamic conquest with its raw material. In addition the Qahtanis of Oman and the mixed tribes of the gulf resented the rising power of the Hejaz. Many powerful tribes reverted to their previous situation. Except for Hejazi and Najdi tribes, for the most part only weak ones remained with Islam. Religious conversion of tribes outside the Hejaz had been nominal except for parts of Oman, and Islam had not become an integral part of tribal life during Muhammad's short ministry. Powerful peninsular tribes joined when the conquest showed signs of success. Abu Bakr had to subdue much of Arabia before the conquest of the north began. Muhammad subdued the tribes, but the economic benefits of expansion were necessary to keep their energies directed away from each other.

It is perhaps typical of the traditional factionalism of the Arabian

Peninsula that all three Islamic sects have significant numbers of adherents there, all settled by the eighth century. Because provisions had not been made for the temporal successor of Muhammad, three sects quickly emerged: the Sunni, the Shiat Ali (Shias), and the Kharadjites. Each sect claimed to be following Muhammad's wishes regarding succession and the spiritual direction Islam was to take (see ch. 3). Gulf areas north of Oman adhered to Sunni or orthodox Islam, which holds that the caliph (the successor) be a member of the Quraysh, the tribe of the Prophet Muhammad. Shias believe that only a descendant of the Prophet through the union of his daughter and only issue, Fatima, with his cousin Ali can be the successor. The Shias call the successor an imam (literally, he who sets an example). The Kharadjites believe that any Muslim who meets standards of probity can be imam and that he must be elected to the position.

Early in the Islamic era the Kharadjites broke into subsects, but the Ibadite branch was the only one that survived. Ibadites were harassed by the Sunni Umayyad dynasty and migrated from their center in Iraq to escape Sunni domination. Oman was and is remote from the chief centers of Islamic influence, and many Ibadites migrated there, settling particularly in the Omani interior, an area accessible from the coast only to the hardy and determined. In the eighth century Ibadite Omanis elected their first imam, reinforcing the already established distinction and distrust between the coastal Omanis and those "beyond the mountains."

The Gulf During the Medieval Period

Arabia had become unified, but ironically it had to be virtually emptied to ensure unification. As in the great migrations of the past, Arabia in the immediate post-Muhammad period once again fed its people into the surrounding areas. Peasants were easier to control than refractory tribes, and the social structure and the agricultural assets of the conquered land outside the peninsula encouraged permanent settlements that were necessary in any case to secure Islamic rule. Thus Islamic politics moved from Arabia to Damascus and, in the period between the ministry of Muhammad and the discovery of oil, Arabia became an economic backwater except for trade on the gulf, which increased for the Arabs as the Muslim conquest expanded. Nevertheless, because Arabia was considered the homeland of the Arabs and because Mecca and Medina—Islam's holiest cities—were there, it could not be ignored. Pilgrims who made the haj had to be afforded some protection. Furthermore control of the holy cities gave a ruler instant prestige. The political unit that kept the Hejaz secure was usually Egyptian based, and Egypt was also the source of much of the Hejazi food supply. Islamic rulers in Damascus paid off religious leaders in Arabia who might cause trouble. Arabia became a frontier zone from which political rivals emerged. During the various struggles for religious and political supremacy in Damascus and later in Baghdad, Mecca and Medina became sites for the opposition, and both cities were virtually destroyed in the tumult that ensued. For the most part social organization in Arabia continued as it had been before Muhammad's unification.

During the centuries between the Islamic conquest of the gulf area and the founding of the modern polities in the eighteenth century, the focus of political activity was coastal. Natural resources consisted chiefly of dates, pearls, and camels. Those items were insufficient to supply even the minimal needs of the population, and the people early turned to maritime activities, imitating their Omani neighbors to the south. Favored by the gulf's position at the center of the medieval trading route network, maritime city-states developed and vied for the trade through the gulf.

The golden age of gulf shipping began in the eighth century with the establishment of the Abbasid caliphate (750–1258) in Baghdad. The general economic revival in the Middle East and Europe during the eighth century and the unification of China under the Tang dynasty during the previous century were chiefly responsible for the acceleration of trade. But the gulf was particularly favored, over the Red Sea for instance, because of Baghdad's position due north of the gulf. The intrepid merchant sailors of the area went as far as China. In the middle of the eighth century an Omani, Abu Ubaydah, made the first recorded return voyage from the gulf to Canton. In incredibly flimsy but capacious boats built of local materials, gulf Arabs regularly followed the monsoons and sailed yearly between the major centers of civilization of the medieval world.

The city of Siraf on the Persian side of the gulf was the paramount trading city between 900 and 1100. The accelerated movements of the Seljuk Turks into Iraq and Persia between 1055 and 1155 contributed to Siraf's decline when the Seljuks attacked it from land and sea. Further, by the eleventh century the rise of the Fatimid caliphate in Egypt had siphoned off much of the gulf trade and redirected it through the Red Sea. Between 1100 and 1300 Qais, an island ten miles off the Persian coast, succeeded Siraf as the chief collection and distribution emporium in the gulf (see fig. 3). Finally in the late thirteenth century Hormuz became the chief center; it remained so until 1507, when it was destroyed by the Portuguese. Hormuz' reputation as a place of riches was so legendary that more than a century after its destruction Milton used it in *Paradise Lost:* "High on a Throne of Royal State, which far outshone the wealth of Ormus and of Ind, or where the gorgeous East with richest hand Show'rs on her Kings Barbaric Pearl and Gold, Satan exalted sat."

During the period of Hormuz' floruit, Sohar and Muscat in Oman prospered greatly from their strategic position on the Gulf of Oman near the entrance to the Persian Gulf. During the post-Islamic period,

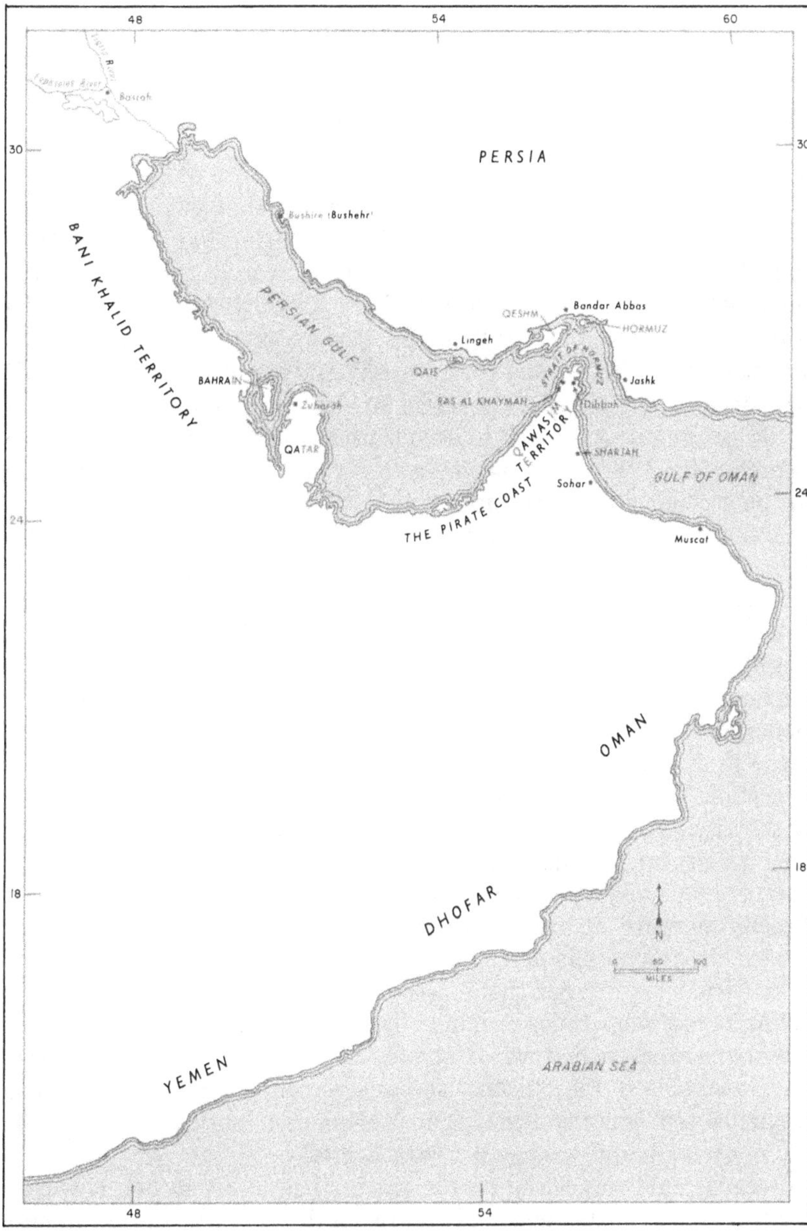

Figure 3. The Persian Gulf in the Medieval Period

although the main trading cities were on the Persian coast, Arabs were usually the towns' rulers. When there was a strong and well-organized Persian government, they paid tribute; otherwise they ruled quite independently. The presence of Persians in Oman and gulf Arabs in Persian territory produced the ethnic mixture still prevailing on the gulf's coasts and led to Iranian claims in the 1970s to islands in the gulf and to parts of its western shore.

Internal Organization

The gulf-state thalassocracies differed from those of other cultures in that political rule did not derive from an originally oligarchic settled merchant class with a long history of urban living and urban institutions. The orientation of the gulf city-states was based on the traditional political model of the desert Arabs. Usually the dominant tribe in the area produced the city's rulers; the shaykh who ruled the tribe was the town's leader and its most affluent merchant. The shaykh's main objects were to secure peaceful internal conditions so that trading would not be disrupted and to secure profits sufficient to maintain his own position. The ruler's income was derived from licenses for trade and pearl fishing, customs duties, and the taxation on such commodities as date palms, all of which were secondary in importance to the profits he made as a result of his own commercial transactions. The shaykh secured his political primacy by payments to his military forces, usually fellow tribesmen; tribute to larger and potentially troublemaking neighboring states; and a variety of subsidies to those in a position to challenge his authority. As has been the case for much of the gulf's modern history, the total income of a city was the shaykh's private purse to dispense privately or in the public interest according to his discretion. The shaykh would attempt to placate nomadic tribes close to the city and to guarantee their cooperation, that is, to encourage their hiring for seasonal activities in the city in order to discourage them from making sporadic attacks on townspeople. City dwellers not directly engaged in trade dove for pearls during the season or cultivated the few arable acres on the city's periphery.

The urban institutions that existed were not the result of the ruler's initiation but were those ordained by sharia, Islamic law. Because of the comprehensive nature of sharia, certain protections and rights were available to the citizens, as were certain restrictions. Common tribal law, urf, was also a conspicuous element in the regulation of daily life but usually did not conflict with sharia.

Because rule was based on the patriarchal desert model, however, the fortunes of a city depended greatly on the strength and astuteness of its ruler. When gulf trade experienced moments of decline, competition between cities was sufficiently fierce to provoke intertribal and intercity violence that usually took the form of piracy on the sea and produced internal confusion at home. If a weak ruler succeeded a strong one, he was soon overthrown; rapid changes in the ruling family became the norm. The flux of political configurations in the gulf city-states was emblematic of the area until the eighteenth-century founding of the modern polities and the eventual protection of their vested interests by the European maritime powers, particularly the British.

THE ENTRANCE OF THE EUROPEANS INTO THE GULF

The Portuguese

Although gulf Arabs, particularly the Omanis, had a virtual monopoly on East-West trade, European powers early entered the competition. The Venetians made a commercial treaty with the Abassid caliph in Baghdad and had business centers in Damascus and Aleppo. When Fatimid Egypt became the predominant Muslim power in the twelfth century, the Venetians established themselves in Alexandria and, in exchange for spices and other eastern delicacies, supplied the Fatimids with munitions, wood, and slaves. The Genoese to a lesser extent also benefited from Arab trading routes established with the East, but they were forced to reduce their trading commitments by the fourteenth century. The Venetians then were the sole corporate link between the Muslim world and Europe until the end of the fifteenth century.

The capture of Constantinople by the Ottomans in 1453 limited Venetian trade to the Red Sea, but the final blow to their Eastern trading activities was delivered by the Portuguese, who entered the Indian Ocean in the last years of the fifteenth century. Vasco da Gama sojourned at many Arab colonies on the east coast of Africa during his voyage in 1497. At Malindi he requested a noted Arab navigator, Ahmad ibn Majid, to sail him to Calicut. Thus familiarized with the vagaries of the monsoons and coastal hazards, the Portuguese were able to divert eastern trade around the Cape of Good Hope to the Mediterranean.

The chief object of the Portuguese venture was to monopolize the trade in spices and other eastern luxuries to which the Portuguese had become accustomed during the Muslim occupations. The crusading mentality also figured in the Portuguese exploration. At least that is one explanation for the wholesale atrocities committed by da Gama on unarmed pilgrims making their way to Mecca. The Portuguese were by no means supported in their ventures by the Christian West. The Venetians, whose commercial interests were also damaged, allied in vain with the Egyptians against the Portuguese. The king of Portugal, bearing the title "Lord of the Conquest, Navigation, and Commerce of India, Ethiopia, Arabia, and Persia", sent Francisco de Almeida to act as governor of the Portuguese settlements in the East. In 1506 Alfonso de Albuquerque started on his journey to India, where he replaced de Almeida as viceroy and governor of India.

Before reaching India Albuquerque secured the Portuguese position in the gulf. He had been given instructions to obstruct Egyptian and Venetian trade in the Red Sea, but he was aware that to blockade the

Red Sea effectively he first had to secure a position in the gulf. He seized the island of Socotra in the Arabian Sea as a convenient midway point. After destroying every Arab ship in his compass he captured Kalhut, Quriyat, Muscat, Sohar, Khor Fakkan, and finally Hormuz, the premier city of gulf trade. The ruler of Hormuz, Shaykh Saif al Din, was forced to become a vassal of the Portuguese. The first of a series of Portuguese forts to be strung along the lower gulf was erected at Hormuz. Albuquerque's initial stay at Hormuz was very short. Troubled by fractious Portuguese crews, he resumed his voyage to India, and Hormuz was temporarily abandoned. In too strategic a position to be neglected for long, it was retaken by the Portuguese in 1515.

The Portuguese set the example to be followed by the English in not interfering with local rule. After a series of local uprisings when the Portuguese attempted to administer the customhouses directly, a treaty between the shaykh of Hormuz and the Portuguese was signed on July 23, 1523, at Minab. The shaykh of Hormuz thereby not only was "protected" by the Portuguese but also was under their direct control. However, the shaykh was confirmed in controlling the internal affairs of Hormuz and the other gulf states under his suzerainty. In 1538 and 1550 the Turks made sporadic attempts, often with local assistance, to dislodge the Portuguese. The Portuguese control remained firm, however, until the early seventeenth century, when a combination of factors made their position untenable.

Dutch and English Involvement

In the late sixteenth century Holland and England began to investigate the possibilities of their countries' involvement with gulf trading activities. John Newberrie of London visited Hormuz in 1580 and returned there three years later with Ralph Fitch. The Dutch decision to raise the price of pepper from three to eight shillings a pound caused great consternation in English mercantile circles. After Fitch published a memoir of his gulf visit, London merchants were sufficiently encouraged to form the English East India Company in 1600.

The English venture fortuitously coincided with the accession to the Persian throne in 1587 of Shah Abbas I, who decided that the Portuguese presence on the coast of his dominions was a severe irritation. The Portuguese had never distinguished themselves by diplomatic niceties in the gulf and had steadfastly refused to pay the relatively small tribute that Persia had always demanded of those who ruled its gulf trading towns. In 1602 Shah Abbas evicted the Portuguese from Bahrain, and between 1608 and 1615 he tried ineffectually to drive them from Hormuz. During those years the English had slowly been acquiring prestige and making friends at the Persian court. Sir Anthony Sherley and his brother Robert had entered the private service of the shah in 1600. When in 1617 Edward Connack carried a letter to the shah from King James I of England asking for support for the English East India Company, he was extremely well received. The shah promised them the port of Jashk from which to trade, granted the English sole control of all silk leaving Persian ports, and provided for the permanent presence of an English ambassador at court.

The Portuguese, who were experiencing increasing difficulties in maintaining their position—in part because of their annexation by Spain in 1581—attempted unsuccessfully to block the ships of the English East India Company from entering Jashk. In 1622 an English-Persian attack on the Portuguese garrison at Hormuz and a post on the island of Qeshm resulted in victory and an agreement that customs collected in Qeshm and Hormuz would be shared by the Persians and the English East India Company. The English established their headquarters at Bandar Abbas, which became the center for political and commercial activities in the gulf for the next 150 years.

Total reclamation of Portuguese-held areas was slow but inevitable. The Persians reclaimed Khor Fakkan and held it for one year; the Portuguese then recaptured it but were ousted by Arab troops under the leadership of the first imam of the Yaruba dynasty in Oman, Nasir bin Murshid. In 1643 Nasir took Sohar, and by 1650 the Portuguese were coerced to abandon Muscat. Only Kunj in southern Persia was left to them.

The Dutch presence in the gulf lasted only 133 years but was extraordinarily lucrative for them. The Dutch East India Company, formed shortly after its English equivalent, was the first public corporation in the world, that is, the first to offer issued shares of ownership in a general public offering, and this undoubtedly contributed to its initial success. Stockholders at home put continued pressure on the Dutch government to support the company's activities, whereas the English East India Company had at first to depend entirely on its own resources. The Dutch were given organizational support in the gulf by the English in return for Dutch assistance in an anti-Portuguese military engagement in 1625. The Dutch, too, used Bandar Abbas as their center. The Dutch had a much wider trading network than the English because of their earlier development of East Asian trade, and consequently they prospered much more. An English-Dutch war in Europe (1653–54) was extended to the gulf, and the English incurred severe losses. Mercantile supremacy passed to the Dutch, and by 1680 their strong presence in both Basrah and Bandar Abbas secured the gulf as a Dutch lake. The English bided their time, and the decline of Holland in the eighteenth century progressively diluted Dutch influence in the gulf. Bandar Abbas was relinquished in 1759 and, when Kharg Island was abandoned in 1765, the Dutch presence was effectively ended in the gulf.

Piracy

From the end of the seventeenth century until 1778 chaos reigned in the gulf. Piracy was a long-established tradition, recorded from Assyrian times. In the modern period the Arabs, particularly the Omanis, were the first pirates to attain general notoriety. It is doubtful whether the word *pirate* can be used as a designation for people using the only means at hand to defend their land from foreign occupation and therefore exploitation, but such was the name given by Europeans who themselves were exceedingly active in piracy on the gulf and elsewhere.

Negative contacts of the English with the Dutch and briefly with the French added to the general instability. By 1688, even though piracy had by no means reached its apogee in the gulf, it was unthinkable that the English could maintain a solely commercial presence there and survive. The English government therefore ceded the island of Bombay to the English East India Company. The English were to protect Portuguese holdings in the East Indies in return for their evacuation of the island. To protect themselves and their commitments to the Portuguese, the company was permitted to organize a civil administration and to acquire troops. The English East India Company merged with "new" companies in 1708 to form the United Company of the Merchants of England Trading to the East Indies.

Residents representing the conglomerate had consular power and rank. Presumably either to increase incentive or to encourage loyalty, residents were required to invest their personal financial resources in some aspect of trade. This caused complications when personal interests deviated from company policy.

Ineffectual as these measures were, they set a precedent for further political and military establishments in the area. By 1763 English merchants were very much in need of such establishments. From the end of the Safavi dynasty in Persia in 1722 a period of almost unbroken internal chaos and anarchy prevailed, and Bandar Abbas was definitely unsafe. With the consent of the shah of the moment, Karim Khan Zand, the company moved its center to Basrah, and a new trading post was formed at Bushire (present-day Bushehr). Karim gave the company freedom from taxation and the sole right to import woolens into Persia and assured it that no other European power would be permitted to establish a port there. The Bushire Residency became autonomous in 1778, and the British Residency in the gulf, dating from that event, remained headquartered at Bushire until it was moved to Bahrain in 1946.

THE FOUNDING OF THE MODERN GULF POLITIES: GULF STATES IN THE EIGHTEENTH AND NINETEENTH CENTURIES

Kuwait and Bahrain

While the British were relocating their trading posts and solidifying their relationship with Persia, the gulf Arabs were involved in a highly turbulent period of mass migrations to the north and central parts of the gulf, local wars, and civil war in Oman. Because of the hitherto unstable political rule in Arabia and because the Arab tribes of the peninsula had never seemed able to remain united long enough to inflict long-term damage to European interests, little attention was paid initially to events that proved to be momentous. An unusually long drought that began in 1722 and the accompanying famine in the Alflaj region of Najd in central Arabia precipitated a major migration of Arab tribes of the Adnani-Anayzah confederation, including the Utub (adjective, Utbi), to the north coastal lands of the gulf. The area was under the domination of the Bani Khalid, whose suzerainty extended from Basrah in the north inland to the eastern part of Najd and south beyond Al Hasa to Qatar.

The Utub had for a long time been under the protection of the Bani Khalid and regularly grazed their flocks in Al Hasa. The Utub were well received, and they settled at a small town they began to call Kuwait, the diminutive of the Arabic *kut,* a fortress built near water. Kuwait had not much to recommend it. It was almost entirely without agricultural resources and lacked a nearby source of potable water. In comparison with the rest of Bani Khalid territory, however, it had a mild climate. Further, Al Hasa, the destination of many of the migrating Anayzah tribes, was quickly becoming overcrowded.

The most influential tribe of the Utub were the Al Sabah, but they were not the only Utub to settle Kuwait. The Al Khalifah and Al Jalahima were the next in importance; at least seven other families or clans emigrated there, though not simultaneously. The town was divided into three sections, and the Al Sabah controlled all from the *wasat,* or central quarters. The exact date of the Utbi settlement is unknown. It was certainly early in the eighteenth century but perhaps later than 1716, a commonly given date, since neither the Al Sabah nor the Al Khalifah were clan chiefs at the time. Until 1752 the Al Sabah exerted mild internal leadership with the blessing of the shaykh of the Bani Khalid, Sulaiman al Hamud. At Sulaiman's death Sabah ibn Jabir, ancestor of the present-day ruling house, was elected as the first recorded Utbi shaykh. There is much discussion about the reason Sabah is remembered as the first of the Utbi shaykhs, since desert Arabs are particularly gifted at remembering genealogy and can usually trace a line for many generations. It is probable that the clans did not come to be called Utub until after their major migration; Utub derives from the

Arabic root *ataba,* meaning to go from place to place.

In 1766 an Utbi settlement was established by the Al Khalifah at the short-lived city of Zubarah in Qatar. At the time of the Al Khalifah settlement on Qatar, nearby Bahrain was under the suzerainty of Shaykh Nasr al Madhkur, an Omani Arab who ruled from Bushire on the Persian coast. Hostilities between Bushire and the Utbi-held areas led to attacks by Nasr on Zubarah in 1778 and 1782. In retaliation a joint attack by the Al Sabah and Al Khalifah netted them Bahrain in 1783, a capture that obliterated Persian influence from the Utbi sphere of influence and gave the Utub access to the richest pearl beds of the gulf while providing a midway point for their increasing mercantile activities. The Al Sabah, particularly, were in an excellent mercantile position because they had access to land as well as sea trade routes. Caravans from Aleppo made regular stops there, and the proceeds enabled them to build a powerful fleet. Because of excellent relations with foreign powers, particularly the British, the Utub with one exception were not forced to engage in piracy. Further the Utbi choice of settling in relatively unoccupied areas ensured unharassed and untaxed access to land and sea trade.

Kuwait's second ruler, Abd Allah al Sabah, succeeded his father in 1762. Abd Allah's long and prosperous reign, which lasted until his death in 1812, set patterns for Kuwait's future political and social development that prevailed until the exploitation of oil and in some cases after it. Abd Allah continued the friendly relationship with the British begun by his father. In 1793 he invited the British to Kuwait, where they remained for two years until danger to their port at Basrah was eliminated. In return the British helped defend Abd Allah against Wahhabi attacks in 1795. Neither the Ottomans in Iraq nor the Persians were strong enough to interfere with Kuwait's rising mercantile power in the eighteenth and the early part of the nineteenth centuries. Shaykh Mubarak al Sabah Al Sabah (reigned 1896–1915) had the prescience to realize that the Ottomans would soon be a substantial threat to his shaykdom and so in 1899 signed an agreement with the British whereby Great Britain assumed responsibility for Kuwait's foreign affairs and for its protection from foreign powers, in exchange for which Mubarak agreed to have no direct relations with foreign powers and not to cede them any land by sale or lease. Until 1946 there were no changes of substance in the agreement. If any change occurred, it was that relations became more intimate as a result of mutual British-Kuwaiti interests.

Bahrain already had an extensive and turbulent history by the time it was captured by the Utub from Kuwait and Zubarah. The Al Khalifah initially ruled Bahrain from Zubarah but established themselves permanently in Bahrain between 1796 and 1798 as a result of devastating Wahhabi attacks, directed particularly at Zubarah.

In the late eighteenth and early nineteenth centuries it appeared that every power in the gulf was claiming Bahrain. Ruled first by the Bani Abd al Qais and then successively by the Umayyads, Abbasids, Persians, Portuguese, and southern Omani Arabs, there were many to dispute Al Khalifah claims. Claims were confused because of changing nomenclature. Until early in the sixteenth century Bahrain was a geographic entity that included not only the present-day archipelago but also the Arab gulf coast from Basrah to the Strait of Hormuz.

Because of the repeated acts of piracy by the Qawasim (adjective, Qasimi), which encouraged other Arab shaykhdoms to join the fray, the British made a concerted attack (their third) against Ras al Khaymah, the headquarters of the Qawasim, in 1819 (see the Trucial Coast: The Qawasim and the Bani Yas, this ch.). The Bahrainis had not entirely abstained from piratical activities but suffered more from them than they gained. Accordingly in 1820 Bahrain signed the General Treaty of Peace with the British, agreeing not to engage in piracy unless they were in a declared state of war. The treaty set the precedent for other states to sign, states that were inclusively referred to as the Trucial Coast until their independence in 1971. Friendly Bahraini-British relations continued, and in 1861 the two parties signed the Perpetual Truce of Peace and Friendship, which included the issues of slavery, British trade with Bahrain, and maritime aggression.

Ottoman involvement in the Arabian Peninsula, usually nominal and concentrated in the Hejaz, became more intense when the Sublime Porte (the government of the Ottoman Empire), fearing the growing strength of the Wahhabi incursions, dispatched its vassal, Muhammad Ali Pasha of Egypt, to subdue the Wahhabis in the eastern part of the peninsula. As the Ottomans increased their activity in the area, they again put forth claims to Bahrain, in 1870 and 1874. Ottoman suzerainty in the gulf could only mean a diminution of British power there. For the benefit of both British and Bahraini interests, treaties were signed in 1880 and 1892. Shaykh Isa bin Ali Al Khalifah agreed, in a treaty similar to that which Britain had signed with Kuwait, not to dispose of Bahraini holdings without British consent and not to establish relationships with foreign powers without British agreement. A British political agent was assigned to Bahrain in 1902. Acting for the Bahrainis, in 1913 the British signed a convention with the Ottomans ensuring Bahrain's independence as a sovereign state. In 1916 a British agreement with Abd al Aziz, future king of Saudi Arabia, ensured that he would not attempt to conquer Bahrain.

Qatar

The withdrawal of the Al Khalifah from Zubarah to Bahrain naturally decreased their power in Qatar, although the Al Khalifah returned for a short period in the nineteenth century and always kept close contact with Qatar. The most important clan in Qatar before the advent of the Al Khalifah were the Al Thani, descendants from Thani ibn Muhammad ibn Thamir ibn Ali of the Bani Tamim, a large Adnani clan. Tradition holds that ancestors of the Al Thani migrated from Najd and settled chiefly in eastern Qatar at the Jibrin Oasis late in the seventeenth century. They eventually moved to Doha (Ad Dawhah), the present-day capital. Apparently the Al Thani were subject to the Al Khalifah until Muhammad bin Thani, shaykh of Doha, began to seek autonomy from them. The Al Thani were, however, powerless against the Al Khalifah until Ottoman influence increased in the eastern gulf. The Ottomans were not concerned with direct rule of the states, realizing that such an attempt would net them little, but they did wish to establish a nominal suzerainty because of the strategic military position of states along gulf waters.

In 1872 the Al Thani became independent of the Al Khalifah when Shaykh Muhammad bin Thani became a *qaim-maqam* (Ottoman provincial ruler). Muhammad was succeeded by his son Qasim (reigned 1876–1913), who had a great vision for Qatar's future and for a time became very influential in the peninsula. Qasim's son Abd Allah (reigned 1913–49) attempted to continue the peninsular policies of his father and also remained under the tutelary direction of the Ottomans, but in 1916 he signed a treaty with Great Britain that was virtually identical with those signed by Kuwait and Bahrain.

The Trucial Coast: The Qawasim and the Bani Yas

The seven Trucial Coast shaykhdoms that became the United Arab Emirates (UAE) in 1971 had for the most part separate histories until the discovery of oil. Historically the five lesser states, that is, those with no oil or much less oil than Abu Dhabi or Dubai, were ruled by a group that had extensive power in the eighteenth and early nineteenth centuries. Their piratical activities were a major instigation for the British to control or at least pacify the gulf by making truces with them and other states.

The Qawasim were the rulers of Ras al Khaymah and Sharjah. For centuries the Qawasim were settled on the Persian coast of the gulf and the coastal area a few miles due north of the Musandam Peninsula (Ras Musandam). Suzerains of Sir, a geographical area that can be thought of as running horizontally across the gulf, the Qawasim were also lords of Lingeh, Qeshm, Kunj, and Luft. Because of the geographic proximity of the Qawasim to Oman and Omani cities on the Persian coast, their histories are intermingled. The Qawasim had made themselves so conspicuous, initially by their trade and later by the magnitude of their piratical activities, that the British in the eighteenth and nineteenth centuries referred indiscriminately to all non-European pirates as the Qawasim, whom they called Joasmees—a misnomer presumably originating from hearing gulf Arabs pronounce the Arabic letter *qaf* with a soft *g*.

It first became apparent that the Qawasim had other interests besides trade when the Qasimi shaykh of Ras al Khaymah assisted the imam of Oman, Sultan ibn Saif II, to seize Bahrain in 1720. A few years later the Qawasim founded a port at Basidu on the island of Qeshm. Because the English East India Company's headquarters were at Bandar Abbas on the mainland behind Qeshm, the British were outraged at the loss of revenue caused by ships' berthing at the closer port of Basidu. Accordingly the company's agent directed a naval foray against Basidu to secure compensation from the Qawasim.

The Qawasim, however, were not long on the defensive. Because of the confusion that prevailed in Persia during most of the eighteenth century, governors were forced to defend their own interests; dependence on the central government usually was futile, and frequently the Persian shah would exploit a governor city.

Mulla Ali Shah, the Persian admiral and governor of Hormuz, therefore sought help from the Qawasim to assist him in defending his governorship against the multitude of demands for tribute made by candidates for the Persian throne. Marriage alliances, a traditional gulf response to political problems, were arranged between the families of Mulla Ali Shah and the Qasimi shaykh, Rashid ibn Mattar ibn Qasim. The Qawasim gained great prestige among the Arabs from the various profits that accrued to them from this coalition. More important, when the Qawasim annoyed the British at Bandar Abbas, the Persians were unable to defend British interests.

The growth of Qasimi power was threatening not only to the British but also to other Arab maritime powers, particularly the Utub of Kuwait and Bahrain and the Omanis. The Omanis and the Qawasim were engaged in almost perpetual warfare during the second half of the eighteenth century; their one truce occurred in 1773 when the Persian shah appeared likely to be a threat to states on the Arabian coast. Conflict between the Qawasim and the Utub came about almost accidentally. Secure in their power, in 1782 the Qawasim attempted to play diplomat between the Al Khalifah and the shaykh of Bushire, who was making claims to Bahrain in the name of Persia. Unfortunately an Utbi vessel captured and killed the crew of a Qasimi boat. The Qawasim, therefore, threw in their lot with Arabs of the Persian coast and attacked the Utbi settlement at Zubarah.

Between 1797 and 1804 there were only three incidents between the Qawasim and the British, and the Qawasim incentive was usually to disrupt Omani trade. Situations in virtually the whole of the Arab coast changed abruptly, however, with the Wahhabi religious and military expansion from central Arabia (part of present-day Saudi Arabia) to the east coast of the Arabian Peninsula in 1800. The Wahhabis are a puritanical sect within Sunni Islam. Since the majority of coastal Arabs were Sunni, with the one exception of Oman, there originally was some sympathy for the movement. At their zenith the Wahhabis controlled the whole of the coast from Basrah to Dibbah. Control of a territory by a religiously motivated force and permanent conversion to its tenets are entirely different matters, however, and only Qatar remained Wahhabi. At the time, however, the Qawasim responded more positively to Wahhabism than did most other gulf Arabs. This was at least partially due to the Qawasim's rivalry with the Omanis for commercial supremacy. The Omanis are of the Ibadite branch of Islam and as such were regarded as infidels by the austere Wahhabis. Presumably the Qawasim hoped that, by aligning with the Wahhabis, they would eventually have access to the spoils that would accrue if Oman fell to the Wahhabis. Because Oman was embroiled in civil war, the Qawasim received help from one of the two Omani warring factions, the Ghafiri, who were Adnani and who, although they did not wish for a Wahhabi takeover in Oman, hoped that the Qawasim would support their cause against their internal enemies.

The Qasimi collaborations with the Wahhabi movement encouraged their wholesale piracies against the British and thus quickened the British desire to make as many firm truces along the coast as possible. Many scholars attribute the increased Qasimi piratical activities not so much to Wahhabi fervor as to British interference in the long-term war between the Qawasim and the Omanis, and particularly to the British support of the Omanis, because wholesale attacks on British vessels did not commence until after 1808 when the war began to acquire some substance. British protection of the Omanis was not due to special allegiance to them or to a particular treaty arrangement. The chief British concern was that the Qawasim, through the Wahhabis, not monopolize gulf trade. The Qawasim felt they had been meanly dealt with, because a British-Qasimi treaty of 1806 had ensured that the Qawasim would respect the English East India Company's flag and that in return the Qawasim would not be harassed during their attempts to recoup part of their Indian trade, which they felt had been usurped by the Omanis.

Anti-British incidents in 1808 prompted a radical change in the British laissez-faire stand on the internal affairs of the Arabs. Lord Minto, the governor general of India, and Rear Admiral William Drury, the naval officer in command of the East India station, decided that the independence of Oman (threatened by Qasimi attacks) was vital to British interests. A British show of support for Oman was also necessary because Sayyid Said, the Omani shaykh, had previously shown sympathy with the French, and the British were still in a state of shock from Napoleon's incursions into Egypt in 1797 and 1798. Accordingly in 1809 the British destroyed Ras al Khaymah, as well as a few small Qasimi holdings elsewhere. The Qawasim recouped their losses, and in

1812 and 1813 they made further attacks on British vessels. Attacks on Ras al Khaymah in 1812 and 1814 were made by the Omani shaykh, Sayyid Said, with assistance from the Bani Yas of Abu Dhabi. Qasimi attacks and British counterattacks continued. A particularly crushing foray was made by the British on Ras al Khaymah in 1820.

Five of the seven shaykhdoms are Qasimi, and all were under the dominance of the shaykhdom of Ras al Khaymah for much of their history: Ajman, Umm al Qaywayn, Sharjah, and Fujayrah. The other two, Abu Dhabi and Dubai, were founded by the previously nomadic Bani Yas. Abu Dhabi, founded in approximately 1761, was originally valued by the Bani Yas for its fresh water and proximity to pearl-bearing oyster beds. The people of Abu Dhabi quickly learned the value of maritime pursuits, and in 1790 the shaykh of Abu Dhabi's most powerful clan, the Al Bu Falah, moved his settlement to Abu Dhabi town. Dubai, in the late eighteenth century, was also inhabited by the Bani Yas, although the port had probably been used by the Omanis for centuries. Initially under the dominance of Abu Dhabi, Dubai, settled by the Al Bu Falasah branch of the Bani Yas, declared its independence in 1834 and soon was a serious rival to the newly mercantile clansmen in Abu Dhabi. The Bani Yas settlements, because of their late founding, did not figure importantly in the gulf until the mid-nineteenth century, but they did ally themselves with the Omanis against the Qawasim. In 1820 the Bani Yas were among the signers of the General Treaty of Peace between Great Britain and the states of the Trucial Coast. Some of the towns whose shaykhs signed the treaty of 1820 have been absorbed by other powers or have undergone name changes, but the original signers were the shaykhs of Abu Dhabi, Dubai, Ajman, Umm al Qaywayn, Jazirat al Hamra, and Hatt and Falna.

The years between 1820 and 1852 continued to be stormy ones for the British and the other signers of the treaty. Abu Dhabi and Dubai began to rise in prominence; Sharjah, formerly second to Ras al Khaymah, became more affluent, and by 1820 Ajman and Umm Al Qaywayn had come into existence. The Qawasim by no means ended their activities, but their potential to build a powerful, independent indigenous state was crushed forever by the British. In 1852 the Perpetual Maritime Truce was signed by the Qawasim and Bani Yas states, and the name of the coast was changed in British records from "Pirate" to "Trucial." The 1852 treaty signaled the beginning of the end of the slave trade—although the British thought it signaled the end. It also fossilized the ruling families of the states and the states themselves, because they had agreed not to combat each other on the open seas. Intertribal and intratribal warfare continued, however, but the fighting was more on land than at sea. The British still refrained from interfering in internal affairs; the unusual result was a situation in which territory was not being taken but deaths and devastation were common.

Although treaties ensured the existence of the states, they did not ensure the conditions under which they would exist, and the resolution of many tribal problems was carried into the twentieth century. In 1892 the Trucial Coast states signed an exclusive agreement stating that Great Britain would be their only foreign and diplomatic contact. The internal histories of the Trucial Coast states were of little interest except to those who lived there. The British were too firmly in control of their destinies.

Oman

Oman, having a continuous history recorded from ancient times and permanent tribal residents recorded from the second century A.D., differed in religion, tribal origins, and world view from the other Arab states of the gulf. While engaged in extensive trade and carrying out frequent forays against some rival states and making alliances with others, Omanis were also concerned with internal divisions caused by geography, tribal origins, and doxological subtleties. Although some of the gulf states have been able to overcome their differences within the virtually all-encompassing embrace of oil, internal Omani problems, particularly those that intensified during the eighteenth century, have consistently been a divisive factor.

Early in the eighteenth century the contest for the succession to the imamate became so violent that it not only solidified the divisions that already existed in Oman but also involved many of the tribes of the future Trucial Coast states. The ulama, the religious leaders, favored the election of Muhanna ibn Sultan al Yaruba, who appeared to fulfill the stringent requirements for an Ibadite imam, whereas tribal leaders wished to elect a young son of the previous imam (see ch. 3). Although Muhanna was eventually elected, the dispute acquired a life of its own. Two factions formed: the Hinawi, named for the Bani Hina tribe and its leader, Khalifa ibn Mubarak Al Hinawi; and the Ghafiri, named for the Bani Ghafir and particularly for Muhammad ibn Nasir Al Ghafiri, who led that faction.

The Hinawi supported the choice of the ulama, and the Ghafiri supported the choice of tribal leaders. Although the choice of imam was certainly an important one to Omanis, the question of who was rightful imam was really only the catalyst in a situation that had long been in the making.

The Hinawis were southerners, that is, Qahtanis; the Ghafiris and their supporters were Adnani. Although both groups were Ibadite, the Hinawis, descendants of original Ibadite tribes, considered themselves the preservers of orthodoxy. Oman was without an imam for various periods when a compromise could not be reached between the two groups.

In the second half of the eighteenth century the imamate was bestowed on a member of the Said clan, an ancestor of the ruler in 1976. A grandson of the first Said ruler and an imam, he made the political error (as far as the tribes were concerned) of moving his capital to Muscat. It was a wise decision considering the commercial advantages and the fact that he could more effectively negotiate with foreign powers, both Arab and European, from a coastal town than from the interior. The more conservative religious leaders of the interior did not, however, see the justification for such a move and began to elect their own imams, who acquired tremendous political power and virtually ruled the imamate as a state separate from the one coalescing around Muscat. Gradually the rulers of Oman, no longer elected as imams, assumed the secular title of sultan.

The sultans of Muscat quickly developed an amicable relationship with the British that was advantageous to both parties. The first treaty, the Agreement of Friendship with England, was signed in 1800. The British felt it protected them from French influence in Oman, and the Omanis felt it greatly enhanced their international prestige and importance. The sultan of Muscat was also the suzerain of Zanzibar, a few other East African coastal towns, and Gwadur on the Makran coast of Baluchistan (in present-day Pakistan).

The British gave what aid they could to Oman during the Wahhabi incursion and, by forcibly subjugating the Qawasim, secured Oman's position. In 1822 the sultan signed the first of three treaties (the others were in 1839 and 1945) with the British to suppress the slave trade in Oman.

Although the British have always been the best Western friends of the Omanis, the United States also made overtures during this period. The Americans, by then able traders, wished to make some arrangement whereby their trade passing through Omani waters would not be so heavily taxed. In 1833 Edmund Roberts, a private merchant, arrived in Muscat having been given the authority to make commercial treaties by the United States Department of State. He effectively negotiated the Treaty of Amity and Commerce, the first American treaty in the area. It remained in effect until 1958, when it was replaced by a similar but updated version.

In 1840 an Omani delegation arrived in New York, the first Arabs to do so officially. The outgoing Sultan Sayyid Said, ruler during the period, had made many gifts to the British, among them the guano-rich Kuria Muria Islands. The British by that time were so closely involved in Omani affairs that the British viceroy in India, Lord Canning, adjudicated between the sultan's quarreling sons after the sultan's death in 1856. One son was given Muscat and the other Zanzibar; Zanzibar, however, was to pay an annual tribute of 400,000 Maria Theresa dollars to Muscat. Zanzibar soon stopped paying the tribute. But because the British had arranged the conditions and because they wished to maintain a stable rule in Oman, which translated into money enough to ensure the sultan's power, British India assumed responsibility for the payment, which it continued to pay until the British Foreign Office assumed the burden in 1947.

The second half of the nineteenth century saw the beginning of Oman's decline and the decline of the rest of the gulf states. Locally built sailing ships were no competition for European steamship lines, and the gulf for the first time in many centuries was becoming an economic backwater. In 1873 Sultan Turki bin Said, who had come to power largely through British support, signed a major agreement aimed at the suppression of the slave trade. The end of slave trading and gunrunning, the last two lines of Omani economic power, dealt the coup de grace to Omani commercial independence. The sultan was hopelessly dependent on the British and had incurred the wrath of the more conservative Ibadites of the interior, because slavery under special conditions is sanctioned by the Quran and because the interior tribes relied on gunrunning.

The British apparently were aware of the difficulties they had caused the sultan. Between 1895 and 1897 Oman was granted two loans financed by the government of India.

Ibadite discontent, fomenting for nearly a century, found a voice in Isa bin Salih, elected imam in 1913. That same year Taimur bin Faisal became sultan in Muscat. By 1915 the imam's warriors had declared jihad (holy war) against the sultan and were beseiging Muscat. Indian infantry assisted the sultan's forces, and Isa bin Salih was driven back. It was a Pyrhhic victory for the sultan, who maintained only the narrow coastal strip near Muscat. In 1920 an agreement was finally reached whereby the imam's forces would not attack the coastal areas under the sultan's aegis and the sultan would not interfere in the internal affairs of the "people of Oman." A new imam was elected immediately after the death of Isa bin Salih. A long-held feeling had reached the treaty table and become fact—Oman was for all purposes two nations. Thus Oman, once indisputable leader of the world's trade and then of the gulf's, entered the twentieth century with depleted resources and a divided kingdom.

THE TWENTIETH CENTURY

British Governmental Transition

As the power of the British in the gulf grew, so did their responsibility and presence there, requiring constant readjustment of their administration. By the middle of the nineteenth century Great Britain was responsible for the affairs of its agents in the gulf but until a century later delegated that responsibility. The English East India Company conducted diplomatic, defensive, commercial, and administrative affairs until 1858. The government of Bombay then assumed responsibility until 1873; the British Indian government then was in charge until its demise in 1947. From 1947 until December 1971 and the independence of the UAE, the Foreign Office in London handled all gulf matters through its political resident in Bahrain and by political agents eventually established in Bahrain, Qatar, Abu Dhabi, and Dubai.

The various treaties that the British made never provided that those states would be under the soverign power of Great Britain—only that Great Britain would regulate and conduct their external affairs. Such conditions brought a measure of security to individual states against the incursions of rivals. But "external affairs" meant the same thing to gulf Arabs as it did to the British—the sea—and any Arab vessel on the gulf was subject to British-drawn guidelines.

In spite of the treaties' being allegedly restricted to external affairs, the internal prerogatives of the Arabs were gradually encroached on, as the natural consequence of the psychological conditioning received by the Arabs, who found themselves having to defer to British restrictions every time they ventured out of port. The conditioning was reinforced by tradition, and the opinion of a political resident on internal affairs could be discounted only at the risk of military or economic sanctions or even exile for the dissident.

Robert Geran Landen, an expert on Oman, writes:

> By the early twentieth century British support had become a more important prerequisite for a ruler's continuance in power than his government's popularity among his own subjects. As late as 1970 such modes of indirect rule remained valid; for instance, the British played a large role in overthrowing the rulers of Sharjah in 1965, Abu Dhabi in 1966, and Oman in 1970.

The adviser system came to be widely used in the gulf. Initiated in 1920 in Oman when the British organized the Muscat Levies, a British-commanded internal security force, the adviser system formed the foundation of many of the administrative practices of the gulf states. Foreigners, at first usually British, were employed by rulers and advised them on political, economic, and social affairs. The discovery of oil increased, sometimes without warrant, the self-confidence of many rulers, who slowly began to make decisions without direct British counsel.

After World War II and particularly after the loss of India in 1947, British priorities in and attitudes toward the gulf changed. The area was no longer necessary as a military frontier from which to protect the empire in India. Because of oil, however, the area was of great economic importance to the British, and a situation reminiscent of the first days of the British in the gulf obtained. Battalions of British soldiers were placed in Oman, Sharjah, and Bahrain. By the 1960s, however, it was clear that the British were losing more in expenditures than they stood to gain by their vigilance. Accordingly Great Britain began to grant full independence to the gulf states. Because, among other things, of the stability of Kuwait's internal rule, it was the first to achieve independence, on June 19, 1961. In 1968 the Labour government in Great Britain announced that its "protection" of the gulf would end in 1971. Bahrain became independent on August 14, 1971, Qatar on September 1, 1971, and the United Arab Emirates (UAE) on December 2, 1971. Although some British troops remained in Oman, the gulf was largely depleted of British forces by the end of 1971. That did not end British interests or influence there, nor did it substantially change the respect and friendship felt by the gulf Arabs for the British.

Gulf Development Before Independence

Although the rate of social and economic change was greatly accelerated by independence, many of the gulf states, with the exception of Oman, were making small starts toward modernization despite the fact that at the end of World War II most of them were nothing more than impoverished fishing villages.

Kuwait's growth after World War II was rapid, although in the late 1940s it was still only a small walled town. Oil had been discovered in 1938 but was not exploited until after the war. Because of Kuwait's early independence most of its social programs were developed after independence. The pace of development was so rapid afterward that Kuwait was able to give substantial aid and advice to the states of the lower gulf that were not yet independent. Many gulf states modeled their social welfare systems, particularly education, on Kuwait's (see Geographic and Demographic Setting, ch. 6).

Initially the greatest postwar change in the gulf was in the development of local military and police forces. The Trucial Oman Levies, formed in 1951 (and later called the Trucial Oman Scouts), was the first. Although a British idea and officered by British volunteers, it functioned as an internal peacekeeping force staffed by local tribesmen. By 1956 the force included approximately 1,000 men. Dubai was the first to install its own police force in 1956; Abu Dhabi was next in 1957; and in the early 1960s the other states of the Trucial Coast followed suit. By 1965 Abu Dhabi had founded its own military defense force; though small, it included army, navy, and air force divisions.

Developmental changes of a social nature in the gulf states were at the initiation of the British, who began to feel that they should undertake some limited responsibility for development schemes. In 1939 the British government opened a clinic staffed by an Indian doctor at Dubai and ten years later began building a hospital there. In 1952 a modest water resources survey and some well-drilling were financed, and in 1955 the first modern school was built at Sharjah. Flushed with success, the political agent in 1955 drew up a five-year development plan. Administrative developments, such as the improvement of the court system and modern buildings for the Trucial States Council, were envisioned, but greater progress was achieved in the social sphere. By 1961 an agricultural school near Ras al Khaymah and a trade school at

Sharjah were established, a teacher training clinic was opened in Bahrain, the Dubai hospital was expanded, four primary-school buildings and a handful of clinics were built, and an antimalaria campaign was begun. The second five-year plan was straitened because of economic difficulties in Great Britain, but in the 1962–63 period a trade school was established at Dubai, and funds were made available in Ras al Khaymah to attack the problems created by severe erosion in the previous decade.

By 1965 both Abu Dhabi and Dubai were affluent enough to assume a major portion of schemes planned by the Development Fund, and in 1966 the ruler of Ras al Khaymah became chairman of the Trucial States Council, which administered the fund. Slowly at first the gulf Arabs, particularly the Kuwaitis, funneled money through the fund and began to make their own decisions about development priorities in the area. The Arab tendency to build on a monumental and grandiose scale was tempered by British practicality in the years before independence, which was important considering the many urgent needs of the region.

As a result of oil produced in commercial quantities by 1932, Bahrain was the only gulf state where modernization was in progress by 1950. Sir Charles Belgrave, resident adviser to the ruler of Bahrain between 1928 and 1956, was largely responsible for the efficient manner in which Bahrain was administered. The educational and economic levels of Bahraini citizens were the highest in the gulf.

Although Bahrainis enjoyed many of the advantages of a welfare state, they were not able to exercise any meaningful political rights. Dissatisfaction with the lack of popular participation led in 1954 to the formation of a nationalist movement, the Committee of National Unicn; it was a pan-Arab alliance of students, intellectuals, oil laborers, and artisans who demanded a number of reforms, including the development of a modern code of law, the formation of a trade union and a legislative council, and the dismissal of Belgrave. A general strike led to investigation of the grievances and authorization to establish an education and health council, half of whose members were to be elected. Although elections were held in 1956, unrest continued until it came to a head during the Suez crisis of November 1956 and widespread rioting broke out. Law and order were eventually restored, and the leaders of the movement were exiled, an action that terminated the first political movement. The outcome of the riots, however, was the removal of Belgrave and the institution of a limited measure of reform.

In an effort to create a more viable political entity, Great Britain in 1951 sponsored the establishment of the Trucial States Council. Members of the council were the seven rulers of the Trucial Coast states. Great Britain announced its intention to remove its forces from the Persian Gulf and other areas east of Suez by the end of 1971, triggering renewed interest among the shaykhs in the possibility of a federation.

With British encouragement the seven Trucial Coast states rulers and their counterparts of Bahrain and Qatar met in Dubai in February 1968 and announced the formation of the Federation of the Arab Amirates. The Supreme Council, composed of the nine rulers, would be established and function on the basis of unanimity. The presidency would rotate among the nine rulers. The council would draw up a charter; would formulate foreign, defense, and economic policies; and would design the federal laws. The agreement would become effective on March 30, 1968.

It quickly became clear, however, that the rulers of the nine states were not yet ready to implement plans to federate by March 30. At the July 1968 meeting a compromise was reached whereby the shaykhs agreed to establish the Provisional Federal Council and a number of ad hoc committees to study such issues as a common currency, a postal system, and a flag and national anthem.

The nine shaykhs continued to meet from time to time, and the various committees made some progress on combining technical services. At a meeting in October 1969 the shaykhs finally decided to appoint Shaykh Zayid bin Sultan Al Nuhayyan of Abu Dhabi as their first president and accepted Abu Dhabi town as the provisional capital. No further agreement was reached concerning the materialization of the federation. Traditional tribal rivalries and distrust and the inability of the rulers to adapt to problems and concepts transcending their tribal origins impeded federation efforts. The disparity in wealth, population, and power among the gulf states compounded the difficulties of forming a working federation.

The attitude of other gulf rulers to Bahrain's inclusion was lukewarm, because Iran was still making claims to Bahrain. As soon as the Iranian claim was settled, Bahrain demanded a representational position based on population within the Provisional Federal Council. When the idea was rejected, Bahrain declared its independence as a state separate from any federation on August 14, 1971. Because the Qataris had not forgotten the Utbi incursions in the eighteenth century and bore a grudge because of it, they refused to be in a position less strong than Bahrain and so declared their independence on September 1, 1971. The probability of a federation appeared to be declining. The people of Ras al Khaymah, furious at their state's being given a minor role in federation affairs—considering its illustrious past—and feeling ill-used because of the lack of support from other gulf states in regard to the Iranian occupation of the Tunb Islands, which Ras al Khaymah claimed, refused to join the federation of the United Arab Emirates (UAE), which was finally announced on December 1, 1971. By the next day the UAE was independent of Great Britain. Ras al Khaymah did not join until February 1972.

Internal Developments in Oman

The twentieth century did not appear to herald great change for Oman until 1938, when Sultan Taimur abdicated in favor of his son, Said bin Taimur. Secularists judged that the direction in which Said moved the sultanate was backward; although he clashed with the Ibadites of the interior, he was basically of their conservative mind. Geographical particularism made it impossible for him to be elected imam by the Ibadites of the interior, but the office of sultan under him became a very close approximation of the imamate. Ibadite restrictions were firmly enforced. Alcohol, tobacco, dancing, singing, and films were strictly prohibited. Women were forbidden to appear unveiled in public and were denied access even to the religious schools.

It is not entirely clear what Said's motivation was. Certainly he did everything in his power to isolate his country from outside influence or even intercourse with foreigners with few minor exceptions: a few British were employed where absolutely necessary in the government and armed forces, and American missionaries ran the country's two hospitals. The missionaries, however, restricted their activities to the purely medical. Entrance visas were denied to virtually all. The few exceptions were made mostly to employees of a British-operated oil firm. The sultan talked a great deal with his British contacts about moderation but increasingly moved away from it. Ultimately he discouraged trade except that necessary to sustain the country at barely subsistence level.

Although the sultan was unwanted in the interior, he nevertheless decided that Muscat was too close to foreign influence and so moved his capital to Salalah in Dhofar Province. Curfews were instituted in all major towns, and city gates were locked a few hours after sunset. Strongly opposed to education, he grudgingly permitted the functioning of three boys' schools. Young men who sought higher education outside the country were routinely denied reentrance visas. Relatively speaking, Said trusted the British more than any other foreigners and also realized the need for military leaders. He therefore permitted his only son, Qaboos bin Said, to attend Sandhurst. When Qaboos completed his education and returned home, however, he was promptly placed under house arrest.

A new imam was elected in the interior in 1954, and he and his supporters attempted unsuccessfully to gain recognition for Oman as an independent state in the League of Arab States (Arab League). Between 1954 and 1959 forces of the sultan and the imam were in a state of constant warfare until the British, at Said's request, helped put down the revolt in 1959. The imam took sanctuary in Saudi Arabia, which had supported him just as the Wahhabis had consistently supported imams against Omani sultans.

Despite Said's paranoia and the enforced isolation of his subjects, modern ideas were spreading, particularly among the young, from the unlikely source of Saudi Arabia. On July 24, 1970, a bloodless coup by the liberals, led by the sultan's son, Qaboos, ousted Said and exiled him to England, where he died in 1972. In 1976 Omanis said, "Before Qaboos, nothing." Certainly this sums up the reign of his father and Oman's position when it finally entered the twentieth century in 1970.

* * *

Information on the period before the founding of the Persian Gulf states in the eighteenth century is scant. Excellent summaries may be found, however, in books that have a later focus. Ahmad Mustafa Abu-Hakima's *History of Eastern Arabia, 1750–1800* is particularly valuable because of his perusal of all Arabic sources and discussion of their validity. Arnold Talbot Wilson's classic, *The Persian Gulf: An Historical Sketch from the Earliest Times to the Beginning of the Twentieth Century,* remains an important and heroic attempt at sorting out the complex dynamics of premodern gulf history. *Oman Since 1856* by Robert G. Landen is the most thorough and scholarly study of the internal and external affairs of Oman and contains an excellent study of Oman's early period. Two key works concerning the British in the gulf are John Barrett Kelly's weighty and meticulous book *Britain and the Persian Gulf, 1795–1880* and Donald Hawley's *The Trucial States.* Hawley's work is a superb one for the general reader because he covers the whole history of the area, including the murky period between World War I and independence.

Three books that are essential reading for the contemporary period are K.G. Fenelon's *The United Arab Emirates: An Economic and Social Survey,* David E. Long's *The Persian Gulf: An Introduction to Its People, Politics, and Economics,* and John Duke Anthony's *Arab States of the Lower Gulf: People, Politics, and Petroleum.* (For further information see Bibliography.)

CHAPTER 3

RELIGIOUS LIFE

From a religious standpoint the west coast of the Persian Gulf is one of the most varied regions in the Muslim world. Almost every form of Islam currently practiced is represented along the gulf; from the puritanical austerity of Ibadism to the ornate mysticism of Persian-style Shiism, virtually the entire theological and liturgical gamut is present. In addition such non-Muslim religions as Hinduism and Christianity count significant numbers of adherents in this heterogeneous cluster of societies.

To state that the gulf area abounds in religious forms is not to state, however, that each individual society contains the entire spectrum. Political and historic forces, some arising during the early centuries of Islam, have distributed the various denominations among the present-day states in such a way that most political entities display a distinctive array of religious groups. In general terms, however, it is safe to say that Islam is the dominant spiritual, cultural and, in many cases, political influence among the societies of the Persian Gulf.

TENETS OF ISLAM

In A.D. 610 Muhammad (later known as the Prophet), a merchant belonging to the Hashimite branch of the ruling Quraysh tribe in the Arabian town of Mecca, began to preach the first of a series of revelations granted him by God through the Angel Gabriel. A fervent monotheist, Muhammad denounced the polytheistic paganism of his fellow Meccans. Because, however, the town's economy was based in large part on the thriving pilgrimage business to the shrine called the Kaabah and to numerous pagan religious sites located there, his vigorous and continuing censure eventually earned him the bitter enmity of the town's leaders. In 622 he and a group of followers were accepted into the town of Yathrib, which came to be known as Medina (from Madinah al Nabi—the Prophet's city) because it became the center of his activities. The move, or hijra (see Glossary), known in the West as the Hegira, marks the beginning of the Islamic era and of Islam as a force in history; the Muslim calendar, based on the lunar year, begins in A.D. 622 (see table A). In Medina Muhammad continued to preach, eventually defeated his detractors in battle, and consolidated both the temporal and the spiritual leadership of all Arabia in his person. He entered Mecca in triumph in 630 and returned there to make the pilgrimage shortly before his death in 632.

After Muhammad's death his followers compiled those of his words regarded as coming directly and literally from God into the Quran, the holy scripture of Islam; others of his sayings and teachings and the precedents of his personal behavior, recalled by those who had known him during his lifetime, became the hadith. Together they form the Sunna, a comprehensive guide to the spiritual, ethical, and social life of the orthodox Muslim.

The *shahada* (literally, testimony or creed) succinctly states the central belief of Islam: "There is no god but God (Allah), and Muhammad is his Prophet." This simple profession of faith is repeated on many ritual occasions, and recital in full and unquestioning sincerity designates one a Muslim. The God preached by Muhammad was not one previously unknown to his countrymen; *Allah* is Arabic for *God,* not a particular name. Rather than introducing a new deity, Muhammad denied the existence of the many minor tribal gods and spirits worshiped before his ministry and declared the omnipotence of the unique creator. God exists on a plane of power and sanctity above any other being; to associate anything with him or to represent him in any visual symbol is a sin. Events in the world flow ineluctably from his will; to resist it is both futile and sinful.

Islam means submission (to God), and one who submits is a Muslim. Muhammad is the "seal of the prophets"; his revelation is said to complete for all time the series of biblical revelations received by the Jews and the Christians. God is believed to have remained one and the same throughout time, but men had strayed from his true teachings until set aright by Muhammad. True monotheists who preceded Islam are known in Quranic tradition as Hanifs; prophets and sages of the biblical tradition, such as Abraham, Moses, and Jesus (known in Arabic as Ibrahim, Musa, and Isa), are recognized as inspired vehicles of God's will. Islam, however, reveres as sacred only the message and rejects Christianity's deification of the messenger. It accepts the concepts of guardian angels, the Day of Judgment (last day), general resurrection, heaven and hell, and the eternal life of the soul.

The duties of the Muslim form the five pillars of the faith. These are the recitation of the *shahada;* daily prayer, or salat; zakat, or almsgiving; fasting *(sawm);* and haj, or pilgrimage. The believer is to pray in a prescribed manner after purification through ritual ablutions each day at dawn, midday, midafternoon, sunset, and nightfall. Prescribed genuflections and prostrations accompany the prayers, which the worshiper is to recite facing toward Mecca. Whenever possible men pray in congregation at the mosque (from *masjid,* place of prostration) under a prayer leader; on Fridays they are obliged to do so. The Friday noon prayers provide the occasion for weekly sermons by religious leaders. Women may also attend public worship at the mosque, where they are segregated from the men, although most frequently those who pray do so at home. A special functionary, the *muadhdhin*, intones a call to prayer to the entire community at the appropriate hour; those out of earshot determine the proper time from the sun. Daily prayer consists of specified glorifications of God. Prayers seeking aid or guidance in personal difficulties must be offered separately.

In the early days of Islam the authorities imposed zakat as a tax on personal property proportionate to one's wealth; the proceeds were distributed to the mosques and to the needy. In addition freewill gifts *(sadaka)* were made. Many properties contributed by pious individuals to support religious and charitable activities or institutions have traditionally been administered as inalienable religious foundations (*awqaf;* sing., waqf). Such endowments support various charitable activities.

The ninth month of the Muslim calendar is Ramadan, a period of obligatory fasting in commemoration of Muhammad's receipt of God's revelation, the Quran. Throughout the month all but the sick, the weak, pregnant women, soldiers on duty, travelers on necessary journeys, and young children are enjoined from eating, drinking, smoking, and sexual intercourse during the daylight hours. Those excused are obliged to endure an equivalent fast at their earliest opportunity. A festive meal breaks the daily fast and inaugurates a night of feasting and celebration. The pious well-to-do usually do little or no work during this period, and some businesses close for all or part of the day. Because the months of the lunar calendar revolve through the solar year, Ramadan falls at various seasons in different years. Although fasting is a considerable test of discipline at any time of year, a fast that falls in summertime imposes severe hardships on those who must do physical work or travel in the desert. Frayed tempers and poor work performances are annual concomitants of the fast.

Finally, all Muslims at least once in their lifetime should, if possible, make the haj to the holy city of Mecca to participate in special rites held there during the twelfth month of the lunar calendar. Those who have completed the haj merit the honorific haji. The Prophet instituted this requirement, modifying pre-Islamic custom to emphasize sites associated with Allah and Abraham, founder of monotheism and father of the northern Arabs through his son Ishmael (Ismail). In Islamic belief Abraham offered to sacrifice Ishmael, son of the servant woman Hagar, rather than Isaac, son of Sarah, as described in the Bible.

The jihad—the permanent struggle for the triumph of the law of God on earth—is an additional general duty of all Muslims, construed by some as the sixth pillar. In addition to specific duties Islam imposes a code of ethical conduct encouraging generosity, fairness, honesty, and respect and forbidding adultery, gambling, usury, and the consumption of carrion, blood, pork, and alcohol.

A Muslim stands in a personal relationship to God; there is neither intermediary nor clergy in orthodox Islam. Those who lead prayers, preach sermons, and interpret the law do so by virtue of their superior knowledge and scholarship rather than because of any special powers or prerogatives conferred by ordination.

DEVELOPMENT OF ISLAM

The Early Period

During his lifetime Muhammad held both spiritual and temporal leadership of the Muslim community; he established the concept of Islam as a total and all-encompassing way of life for man and society. Islam teaches that Allah revealed to Muhammad the immutable principles governing decent behavior; it is therefore incumbent on the individual to live in the manner prescribed by revealed law and on the community to perfect human society on earth according to the holy injunctions. Islam traditionally recognized no distinction between religion and state; religious and secular life merged, as did religious and secular law. Muhammad, however, left no mechanism for selection of subsequent leaders.

After Muhammad's death in 632 the leaders of the Muslim community consensually chose Abu Bakr, the father of the Prophet's favorite living wife, Aisha, and one of his earliest followers, to succeed him. At that time some persons favored Ali, who, besides being a member of the Hashimite lineage, was the Prophet's cousin and the husband of his favorite daughter, Fatima. Ali and Fatima, parents of the Prophet's grandsons, believed the leadership of Islam belonged to them by inheritance. Ali and his supporters (called the Shiat Ali, or party of Ali) eventually recognized the community's choice but only after Fatima's death. The next two caliphs (from *khalifa,* successor)—Umar, who succeeded in 634, and Uthman, who took power in 646—enjoyed the recognition of the entire community.

Dissatisfaction with the rule of Uthman began to mount in various parts of the Islamic empire, however (see ch. 2). For example, the codification of the Quran, which took place under Uthman, hurt the interests of the professional Quran reciters. Some, such as those at Kufah in present-day Iraq, refused to go along with this reform. Others accused Uthman of nepotism. Though himself an early Muslim, Uthman came from the Banu Umayyah lineage of the Quraysh, whose members had been Muhammad's main detractors in Mecca and had resisted him for a long time. The appointment of many members of this house to official posts caused resentment among those who had claims based on earlier loyalty. Still others objected to corruption in financial arrangements under Uthman's caliphate.

Ali, his claim to the caliphate frustrated, became a perfect focus for dissatisfaction. In 656 disgruntled soldiers killed Uthman. After the ensuing five years of civil war, known in Islamic history as *fitnah* (trials), the caliphate finally devolved on Ali. But Aisha, who had long been a bitter foe of Fatima and Ali, objected, demanding that Uthman's killing be avenged and his killers punished by the Hashimites. She helped rally opposition to Ali's caliphate.

The killers insisted that Uthman, by ruling unjustly, had relinquished his right to be caliph and deserved to die. Ali, whose political position depended on their action and their support, was forced to side with them. From his capital at Kufah he refused to reprimand the killers.

At this point Muawiyah, the governor of Syria and a member of the Banu Umayyah, refused to recognize Ali's authority and called for revenge for his murdered kinsman, Uthman. Ali attacked, but the battle of Siffin was inconclusive. Muawiyah's soldiers advanced with copies of the Quran on their spears, thus calling symbolically for God to decide or for the question to be submitted to arbitration. Ali agreed to this settlement, and each side selected an arbitrator.

Some of Ali's supporters, however, rejected the notion that the caliph, the Prophet's successor and head of the community, should submit to the authority of others. By so doing, they reasoned, he effectively relinquished his authority as caliph. They argued that, according to Quranic teaching, rebels must be brought to obedience by force; arbitrating the dispute with the rebellious Muawiyah was therefore wrong. They further argued that the question of Uthman's right to rule had been settled by war during the *fitnah*. When Ali insisted on his course, this group, which came to be known as the Kharadjites (those who seceded), withdrew to Harura near Kufah and chose their own leader.

The arbitration went against Ali in 658. He refused to accept the decision but did not renounce the principle of arbitration. At this point the Kharadjites became convinced that personal interest, not principle, motivated Ali. His support dwindled among all elements of his followers, and he tried unsuccessfully to attack Syria. Muawiyah gained in battle. Ali also engaged in numerous battles with the Kharadjites, including the massacre at Nahrawan in which most of them were killed. In revenge for the slaying of his wife's family in this raid, the Kharadjite Abd al Rahman ibn Muljam al Muradi murdered Ali in 661.

Ali's death ended the last of the so-called four orothodox caliphates, and the period when the entire community of Islam recognized a single head. Muawiyah then proclaimed himself caliph from Damascus. The Shiat Ali, however, refused to recognize Muawiyah or the Umayyad line. They withdrew and in the first great schism of Islam proclaimed Hassan, Ali's son, the caliph. Hassan, however, eventually relinquished his claim in favor of Muawiyah and went to live in Medina, supported by wealth apparently supplied by Muawiyah.

The claims of the Alid line and its supporters did not end here, however. In 680 Yazid succeeded to the caliphate while his father, Muawiyah, was still alive. Ali's younger son, Husayn, refused to recognize the succession and revolted at Kufah. He was unable to gain widespread support, however, and was killed, along with a small band of his soldiers, at Karbala in present-day Iraq in 680. To the Shiites Husayn then became a martyred hero, the tragic reminder of the lost glories of the Alid line, and the repository of the Prophet's family's special claim of presumptive right to the caliphate. The political victor of this second period of *fitnah* was Marwar of the Umayyad line, but Husayn's death aroused increased interest among his supporters, enhanced by feelings of guilt and remorse and a desire for revenge.

Although they did not gain political preeminence in the world Muslim community, supporters of the Alid cause became if anything more fervent in their beliefs. As they had since the earliest days of the caliphate, Ali and his family served as a lightning rod for discontent. With continued conquests the number of Muslims from non-Arab cultural backgrounds grew. It was inevitable that their religious values, quite foreign to the austere faith born in the Arabian desert, should demand some outlet. In the words of Israel Friedlaender, an authority on Shiite history, the Alid cause permitted conquered peoples to "smuggle into Islam some of their most cherished ideas which were essentially un-Islamic and for the most part even anti-Islamic." By this he means that they contradicted the strict, rationalist monotheism of early Islam.

The Shiites founded their objections to the Umayyad and later non-Shiite caliphs on the notion that members of the house of Muhammad, through Ali, were most appropriate successors to his position as both political leader and, more important, imam or prayer leader. Many believed that Ali, as a close associate, early had a special insight into the Prophet's teachings and habits. In addition many felt that he deserved the post because of his personal merits and, indeed, that the Prophet had expressed a wish that Ali succeed him. In time these views became transformed for many Shiites into an almost mystical reverence for the spiritual superiority of Ali's line. Some Shiites also believe that Muhammad had left a written will naming Ali as his successor but that it had been destroyed by Ali's enemies, who then usurped leadership.

Because the correct selection of the imam was the crucial issue over which the Shiites departed from the main body of Sunni Islam, the choice of later successors also became a matter of conflict. Disagreements over which of several pretenders had the truer claim to the mystical power of Ali precipitated further schisms.

The early political rivalry remained active as well. Shiism eventually gained political dominance in Iraq and Persia (present-day Iran), as well as in Yemen (present-day Yemen (Sana)); Shiites are also numerous in Syria and are found in small numbers in most present-day Muslim countries. They constitute about half the Muslim population of Bahrain.

The early Islamic polity was intensely expansionist, fueled both by fervor for the new religion and by economic and social factors. Conquering armies and migrating tribes swept out of Arabia, spreading Islam with the sword as much as with suasion, and by the end of Islam's first century Islamic armies had reached far into North Africa and eastward and northward into Asia. According to legend Ali visited south Arabia during the Prophet's lifetime. Muhammad is also said to have sent Muadh to convert the Yemenis. In local belief this ministry resulted in the construction of the mosque at Al Tanab in present-day Yemen (Sana), which is said to be the first built outside the Hejaz, Muhammad's native region. Nevertheless authorities doubt that in many areas this early Islamic period amounted to much more than political submission to Muhammad's growing state.

Although Muhammad had enjoined the Muslim community to convert the infidel, he had also recognized the special status of the People of the Book, Jews and Christians, whose revealed scriptures he considered perversions of God's true word that were nevertheless in some sense contributory to Islam. These peoples, approaching but not having achieved the perfection of Islam, were spared the choice offered the pagan—conversion or death. Jews and Christians in Muslim territories could live according to their own religious law and in their own communities if they accepted the position of dhimmis, or tolerated subject peoples. This status entailed recognition of Muslim authority, special taxes, prohibition of proselytism among Muslims, and certain restrictions on political rights.

The first centuries of Islam saw the community grow from a small and despised cult to a powerful empire ruling vast domains. This time also saw the evolution of the sharia, a comprehensive system of religious law to regulate life in the community. Derived from the Quran and the hadith by various systems of reasoning, four schools of religious law—the Hanafi, Shafii, Maliki, and Hanbali—are generally recognized; each orthodox Muslim theoretically acknowledges the authority of one of them.

With the passage of centuries Islam gradually absorbed influences from sources other than the prophetic revelation. Pre-Islamic practices reappeared; people in the Arabian Peninsula resumed veneration of trees and stones though maintaining their identification with the Muslim community. Various holy men, especially those who claimed descent from the Prophet, achieved reputations for exceptional spiritual or magical powers. Pre-Islamic beliefs in the inheritance of special spiritual powers in certain family lines blended smoothly into popular Islam. Stories of holy men circulated, and people began visiting these individuals or their graves to seek cures, fulfillment of wishes, or other favors.

Bands of mystics, or Sufis (from *suf,* wool, referring to their rough clothing), sprang up in various countries, claiming to achieve communion with God through various ecstatic or irregular means. Sufi orders gradually arose among both Sunnis and Shiites, and recognized leaders taught particular mystic ways to God. Sufism gained acceptance in large parts of the Islamic world. In most regions people fell away from the singularly austere cult preached by the Prophet Muhammad and adopted practices that softened it and made it more personal and emotional.

Sufi religious life generally centers on orders, or brotherhoods, who follow a leader, or shaykh, who teaches a mystical discipline known as a tariqa (way). Such aids to achieving ecstasy as body gyrations, whirling, dancing, and music figure prominently in many Sufi orders.

Other Developments

Within a century of its founding Islam had spread to the far corners of the known world and had become a significant force in world history. Cultural, political, theological, and social evolution continued in various directions in different countries. Two particular developments, however, bear significantly on states of the Persian Gulf: the founding of Ibadism, the politically dominant sect in Oman, and of Wahhabism, the faith of most natives of Qatar.

The Ibadi Movement

Present-day Ibadis continue the tradition of the moderate wing of the Kharadjite movement. Found in North and East Africa and Iran as well as in Oman, they are the world's only remaining Kharadjites.

Soon after their withdrawal to Harura the Kharadjites elected Abd Allah ibn Wahb al Rasib as their leader. Others joined the group in fair numbers, especially after the results of the arbitration became known. A substantial body of people left Kufah, where Ali's army had camped during the arbitration truce, and joined the Kharadjite camp, now on the Nahrawan canal. The Kharadjites denounced both Ali and Uthman and staged fanatical attacks on their enemies, which often included the killing of women. After Ali's devastating raid on Nahrawan in 658 the Kharadjite opposition continued in scattered uprisings.

In the face of continued violence and fanaticism among their fellows some Kharadjites adopted a quietist position, opposing the indiscriminate continuation of holy war. Abu Bilal Midras ibn Udaiya al Tamimi was leader of this group by 670. After his murder in 681 leadership of the moderates eventually passed to Abd Allah ibn Ibad al Murri al Tamimi. Little is known of the life of the man from whom Ibadism took its name except that in 683 he and his father came from Najd in present-day Saudi Arabia to help defend Mecca against a general of Caliph Yazid. Late in that year Ibn Ibad traveled to Basrah in present-day Iraq, where a sizable body of Kharadjites had settled. Basran followers of the Kharadjite leader Nafi ibn Azraq, who were known as Azraqis, left Basrah to help in the defense of Mecca, but Ibn Ibad, his adherents, and other moderates remained in the town. A split in the moderate ranks occurred when Abd Allah ibn al Saffar, leader of a group known as the Sufris, asserted that non-Kharadjite Muslims should be viewed as polytheists. Ibn Ibad rejected this notion, and the Sufris left Basrah.

The Kharadjites lost political control of Basrah, and the Ibadis entered a state of *kitman* (concealment) during which Ibn Ibad headed what authorities have described as a shadow government. By the end of Islam's first century, however, the growing radicalism of the Ibadis of Basrah had so alienated the governor that he banished the sect, now headed by Ibn Ibad's successor, the Omani scholar Djabir ibn Zaid al Azdi, to Oman on the Persian Gulf. Even in the early days of Kharadjism followers of Abu Bilal had lived in Oman. The arrival of Djabir ibn Zaid and other important scholars greatly increased Ibadi influence in Oman, and Nizwa became a religious center.

Little is known of the doctrines of the early Ibadis. Contemporary Ibadis, like other Kharadjites, feel obliged by Quranic doctrine to elect their leader, or imam. Unlike other Muslims, they do not restrict the choice of leaders to members of the Quraysh or descendants of Ali but believe that any worthy Muslim may be elevated to leadership, regardless of ancestry. They also feel an obligation to overthrow an imam who acts improperly. As a consequence they accept as legitimate the caliphates of Abu Bakr and Umar, the first seven years of the rule of Uthman, and the rule of Ali only until the arbitration. They accept none of the Umayyad caliphs.

Selection of the imam, who acts as religious and military leader and judge, is the responsibility of the community's scholars, as is the power to depose a leader who rules improperly. So long as he follows the Quran and Sunna, however, an imam has absolute power. Removal of an imam accused of improper behavior requires the supposed wrongdoer to meet with the ulama (scholars; sing., alim), who determine whether grounds for the accusations exist. If the imam is found to be at fault, he is offered the opportunity to repent. Failure to do so results in removal from office.

In their relations with other Muslims, Ibadis adopt one of two attitudes: *wilaya* (friendship and cooperation) with those they regard as true believers and *baraa* (hostility) toward all others. Unlike other Kharadjites, however, they regard non-Kharadjite Muslims as *kuffar* (infidels) rather than an polytheists *(mushrikun).* Consequently they do not permit killing of other Muslims on solely religious grounds. Ibadi doctrines even permit marriage with non-Ibadi Muslims.

The Ibadi community can exist in one of several states determined by the ulama. The state of *kitman* occurs during periods of persecution, when individual Ibadis are free to practice *taqiyya* (religious dissimulation) to hide their true religion and thus save their lives. In the state of *difa* (danger) a special *imam al difa* is appointed to lead the defense. When the Ibadis count themselves half as strong as their adversaries in men, arms, supplies, and so forth, they may pass into the state of *zuhur* (manifestation), at which time an imam is openly elected.

The Ibadis are also distinguished by a strict and puritanical ethical code that rejects frivolity and all innovations in worship. Though ascetic, they nevertheless accept mysticism. Commission of a capital sin removes one from the community of good Mulsims and places him among the despised *kuffar*.

The Wahhabi Movement

Muhammad ibn Abd al Wahhab was born in 1703 in the town of Unaynah in present-day Saudi Arabia. A descendant of generations of Hanbali qadis (religious judges), he followed the family tradition of religious studies and traveled to Medina, Basrah, and Damascus, among other places. Even as a young man he showed signs of unusually extreme orthodoxy. Having studied widely and having reportedly dabbled in Sufism, he returned to his native town, where his family was prominent. There he began to preach his own views based on the teachings of the controversial and extremely conservative Hanbali legal scholar Taki al Din Ahmad ibn Taimiya, whose ideas went far beyond the Hanbali norm in strictness and literalness of interpretation. Like his spiritual mentor, Wahhab attracted unfavorable reaction because of the unusual severity of his teachings. He eventually left Unaynah with his considerable household and property and was received into the village of Dariyah, which was ruled by Muhammad ibn Saud.

Wahhab quickly put into effect Taimiya's teaching that the ulama should combine with the *umara* (powerholders; sing., amir) to create a true Muslim society. In 1744 he and Saud concluded a pact stating that the Al Saud (House of Saud) would adopt, fight for, and propagate the Wahhabi doctrines and that in all conquered territory the Al Saud would hold political power and the Al ash Shaykh would hold religious power. (In this usage *Al* means "house of" or "family of.") Thus came into being the partnership that harnessed Wahhabi religious fervor to Saud's dynastic expansionism and that resulted immediately in the creation of a kingdom in Najd and eventually in the development of the modern state of Saudi Arabia.

Wahhab took a literalist view of the Quran and, like Taimiya before him, believed that the absolute and incomparable unity of God was the core of true Islam. Thus the worst sin was *shirk*, the association of anything with God or the worship of anything besides him; such practices and beliefs denied Allah's basic nature. An important cause of *shirk*, Wahhab believed, was the adoption of *bida* (innovations), practices not sanctioned by the Prophet or his followers earlier than the third century of Islam. Some of these supposed innovations in fact predated Islam, but Wahhab denounced them as ungodly later accretions. Prominent among them were the customs of visiting saints and tombs and the veneration of trees and rocks. Forbidden *bida* also included rendering improper honor to the Prophet Muhammad, including the custom of celebrating his birthday. Particularly forbidden was the invocation of the names of the Prophet or of saints in prayer in the hope that they would intercede with God. The Wahhabis believed that, although Muhammad can intercede with God, he does so not in answer to specific requests but only in the case of an exemplary believer.

All special relationships with God are absolutely rejected, whether they involve Sufis, saints, or Shiite imams, as are all ecstatic practices believed to foster such relationships. Access to God is provided only through prayer and obedience to duty, and that access is equal for all true believers. Wahhab accepted such Hanbali beliefs as the literal truth of the Quran, the ungodliness of knowledge not derived from it or the hadith, and the predestination of human events. He went beyond Hanbali teachings, however, in rejecting smoking, shaving, and strong language; requiring rather than merely encouraging attendance at public prayer; enforcing payment of zakat on so-called secret profits; demanding evidence of good character in addition to acceptance of beliefs for admission to the Muslim community; and forbidding minarets, embellishment of tombs, and such aids to prayer as the rosary. He also insisted on strict enforcement of the sumptuary laws established by the Prophet and banned activities he deemed godless frivolity, such as music and dancing.

Unity for Wahhab implied not only the absolute oneness of God but also the absolute singularity of the believer's devotion. The sole duty of human life is to serve and obey God with a resignation that accepts what comes as God's will. Service, furthermore, must be rendered strictly according to God's law. The believer's spiritual zeal thus combines with punctilious obedience to form total submission to God. In the words of Henri Laost, an authority on Hanbali law, Wahhabism implies "the most perfect sincerity placed at the service of the law."

For Wahhab any other approach to God constituted *shirk*. Worse even than non-Muslims were Muslims of other sects or schools; by claiming to follow the true religion of Allah while practicing polytheism, Wahhab believed, they practiced hypocrisy. The duty of the good Muslim was to stamp out such mockeries of true religion. As they gained power, therefore, the Wahhabis proved to be iconclasts. In 1801 they shocked much of the Muslim world, for example, by defacing the tomb of the martyr Husayn at Karbala, a particularly holy shrine to Shiites.

A Wahhabi does not practice his religion only as an individual. Like other Muslims, Wahhab viewed the community as the ideal vehicle for enforcement of God's law. Law is the acting out of faith; Islam is the fulfillment of law. Wahhabis interpret the concept of jihad to include the obligation of the community to enforce the law and the reciprocal obligation of the believer to obey the constituted authority of the Wahhabi imam except where such obedience would lead to straying from the law. The leader therefore has the right and duty to prevent sin, frivolity, and indecency and to enforce such ritual acts as fasting, attendance at prayer, and payment of zakat.

Within their communities Wahhabis believe in the social equality of believers. The only permissible social distinction is between "true" Muslims and others. A Wahhabi ideally strives for the perfection of the example of the Prophet and his contemporaries, who are viewed as the noblest and most perfect of human beings. The followers of Wahhab regard his teachings so highly that they refer to the period before him as *jahaliyah* (ignorance), the term generally used within the broader Muslim community to refer to the period before Islam.

This should not be taken to imply that Wahhab's followers in any way equate him with the Prophet. Rather than preaching anything new, he sought to return Muslims to the true religion of Allah and to the *al salaf al salih* (ways of the pious ancestors). No serious scholar of Islam has accused Wahhab of heresy. Like Muhammad, however, Wahhab provided religious justification for a political movement that sought to sweep away tribal distinctions and bind all Arabia into a unity based on religion.

After the initial successes of Saud and his immediate successors, the Wahhabi movement went into political decline until the beginning of the twentieth century and the rise of Saud's descendant, Abd al Aziz ibn Saud, sometimes referred to as Ibn Saud. A gifted leader, he harnessed religious fervor to political ambition and forged the modern nation of Saudi Arabia under the banner of the Wahhabi cause.

RELIGION IN THE PERSIAN GULF STATES

Bahrain

The native population is about evenly split between Sunni and Shiite Muslims, although the Shiites appear to have a slight numerical edge. The Sunni group, however, includes the Khalifah family—which includes the ruler and his relatives—and therefore exerts a disproportionate social and political influence. The wealthy merchants are mainly Sunnis as well, whereas the Shiites are mostly farmers and fishermen.

Islam is the official religion, and sharia is proclaimed in the Constitution to be the source of legislation. Sunnis and Shiites have their own law courts; Sunnis are further split among the Maliki school, which provides most government officials, and the Shafii school. Both Sunnis and Shiites are highly self-conscious and organized, and for this reason the ulama of both groups play important roles in community life.

Oman

Although the ruling house and the majority of ordinary Omanis are Ibadi Muslims, the population also includes Indian Muslims and several tribes who profess at least nominal loyalty to the Sunni persuasion. These include the Shihuh and Qara of the interior and the Kathiri and Mahrah of Dhofar. The merchant class contains a number of Hindus as well.

Until the mid-eighteenth century the Ibadi imamate was the dominant political institution of Oman. A series of civil wars, however, awarded the imamate by force of arms rather than through traditional election according to theological criteria and brought the office into disrepute. In the late eighteenth century the capital was moved to the coastal town of Muscat by Hamad ibn Said, the imam's son, who unseated his father as political ruler but allowed him to remain in the traditional capital of Rustaq as religious leader. Hamad styled himself sultan rather than imam because he claimed political rather than spiritual authority and established a principle of hereditary rulership that contradicts basic Ibadi teachings. He founded the dynasty called Al Bu Said, which continued to rule Oman in the mid-1970s.

Despite its loss of political power the imamate under the elder Said remained influential among the tradition-oriented elements of the population, especially the tribesmen of the interior. With his death in 1821, however, the community passed into the state of *kitman*, and no successor was chosen. An unsuccessful attempt to revive the imamate occurred in the mid-nineteenth century, but not until 1913 did anyone seriously assert the traditional authority of the imamate. In that year Abd Allah ibn Humayd al Salmi, a conservative Ibadi scholar, led a revolt against the increasingly foreign-influenced royal house. This led to the establishment of an independent state under the imamate of Salmi's son-in-law, Salim ibn Rashid al Khursi, with its capital at Nizwa, the traditional Ibadi center. British intervention on the side of the Al Bu Said, however, ensured the survival of the regime at Muscat (see ch. 2). After the discovery of oil gave rise to greater interest in the interior, the sultan's forces took Nizwa in 1955, and the imamate passed once again into abeyance.

Kuwait, Qatar, and the United Arab Emirates

Specific information on religious life and development of the remaining states of the Persian Gulf, during both the mid-twentieth century and earlier periods, was very scanty and difficult to find in mid-1976. Except in Qatar the native peoples of the lower gulf generally adhere to Sunni Islam. Their religion appears to resemble closely the practices of other parts of the Islamic world. Foreigners represent several other religions, but the bulk of them appear to be Sunnis as well.

Well over 90 percent of native Kuwaitis are Sunni, as are most non-Kuwaiti Muslims in the country. The royal family and the courts generally follow the Maliki rite. Other religions tolerated in the state include the Shiite Islam of Persian traders and the Protestant and Roman Catholic Christianity practiced by Europeans employed in the country. In Qatar the native-born half of the population adheres to Wahhabism. The Al Thani, in fact, is the only Wahhabi ruling house outside Saudi Arabia. Attacks by the Wahhabis in the eighteenth century appear to have dislodged the Khalifah family, which formerly ruled Qatar, and to have forced a move to Bahrain, where they now rule. This helped prepare the way for the rise of the Al Thani.

Wahhabi influence also dominates the Buraymi Oasis, which lies on the border between Oman and Abu Dhabi, one of the United Arab Emirates; this influence is a relic of the great Wahhabi expansion of the eighteenth century. Apart from the Wahhabis, who follow the Hanbali rite, and a concentration of adherents of the Shafii school in Fujayrah, the people of the United Arab Emirates are Sunnis of the Maliki school.

* * *

Sources on the religious life of the Persian Gulf are uneven in both quantity and quality. An adequate, if hardly ample, quantity of material exists on the Ibadis and the Wahhabis, but information on other, more common forms of Islam as practiced in the gulf is very scanty. For Oman and the Ibadis, T. Lewicki's article in the *Encyclopedia of Islam* and Roberto Rubinacci's "The Ibadis" provide good introductions . George Rentz' article "The Wahhabis" serves a similar purpose. Very brief summaries of the religious situation in the Persian Gulf are available in John Duke Anthony's *Arab States of the Lower Gulf* and David E. Long's *The Persian Gulf.* Philip K. Hitti provides a clear discussion of the basic tenets and values of Islam in *Islam: A Way of Life.* For a comprehensive but challenging history of the founding of Islam, the *fitnah* period, and later developments, the first volume of Marshall G.S. Hodgson's magisterial *The Venture of Islam* is highly recommended.

CHAPTER 4

SOCIAL STRUCTURE

The states of the Persian Gulf represent a series of variations on a theme: the adaptation of basically similar tribally oriented Arabian societies to the varying demands and opportunities of the burgeoning petroleum economy. At the beginning of the twentieth century all of the societies, despite some local differences, displayed similar cultural tendencies and social organization. By midcentury, however, several of the gulf peoples had come into one of the greatest windfalls in human history. The discovery that a substantial proportion of the entire world's known petroleum reserves lay along the gulf littoral transformed several of the states from impoverished, backward shaykhdoms into world financial powers in a span of time that may be called, in historical terms, overnight. Petroleum reserves and the resulting wealth are not evenly distributed in the gulf, however, and in some instances the discovery of oil was as recent as the 1970s. Some states, such as Kuwait, have enormous resources and revenues from the exploitation of oil that became available in the 1940s. Others, such as Bahrain, possess only small revenues. In yet other states oil revenues began to have socioeconomic impact only in the late 1960s and early 1970s. The pace and direction of social change in the gulf states thus varies with their geologic fortunes.

For the states affected, the tidal wave of oil wealth brought with it a tidal wave of social change that has swept away much of the traditional social structure and even, in Kuwait, the physical structure of the capital city. Governments have established comprehensive educational programs that for the first time permit upward mobility through achievement rather than personal connections. Greatly expanded employment and commercial opportunities have drawn citizens from rural areas and family-dominated endeavors into the city and the modern sector. A sudden and enormous need for both skilled and unskilled labor has attracted workers from surrounding Middle Eastern countries, from the Indo-Pakistani subcontinent, from Europe, and from North America in such numbers that in the more prosperous states foreigners outnumber native-born citizens. And the vast wealth that came unexpectedly into the control of the states' rulers has greatly enhanced the influence and status of the ruling houses at the expense of the previously highly influential class of wealthy merchants.

Those states not blessed with oil reserves have experienced much less social change, although they naturally are somewhat influenced by their newly rich neighbors. The poorer states retain many more characteristics of the traditional, tribal-based society that has been typical of the Persian Gulf for centuries. Unfortunately, however, neither the ethnography of the traditional societies nor the social structure of the emerging society is well documented.

BASIC ELEMENTS OF SOCIAL STRUCTURE

Apart from its size probably the most important aspect of the oil windfall from a social point of view is its suddenness. The opportunities, pressures, and rewards of exploiting a commodity of unprecedented value fell to societies in no way prepared to deal with them; the gulf states' new prosperity arises not out of any features of their culture or society but merely out of an accident of geography and a fortuitous international situation. The adjustments required have therefore been demanding and in some cases difficult.

Before the advent of oil the typical gulf society had consisted of a port town and its surrounding hinterland, which may have included scattered farming settlements and nomadic herding groups. In both settled and nomadic society tribalism provided the framework for social organization. Each major tribe was associated with a territory in which it enjoyed use rights to agricultural land or pasturage and water. Settled tribes dominated villages, and specific lineages within each nomadic tribe owned particular wells and oases where lineage members gathered with their herds during the scorching summer heat. Substantial segments of most lineages lived permanently on the oases, tending the date trees and other food-bearing plants that grew there. Because village farmers also traced ties to tribes, the line between nomads and settled tribes is unclear in many cases.

In addition to having a territorial structure a tribe is a complex social unit. Tribal social structure is based on the ramification of patrilineal ties between men. A tribe is a group of related families claiming descent from a supposed founding ancestor. Within this overall loyalty, however, descent from intermediate ancestors defines several levels of smaller groups. In cases of conflict, groups of kinsmen mass at the appropriate level of opposition. For example, the grandsons of brothers form two groups in opposition to each other, but they form one unit in opposition to the descendants of the brother of their common great-grandfather.

Tribal societies consist of larger and larger groups bound by weaker and weaker ties of loyalty. The basic unit of organization is the household. A number of households supposedly descended from an ancestor about five generations in the past form the lineage. Four to six lineages combine into a clan, claiming descent from a more distant ancestor. The clan figures most prominently in tribal politics. The tribe, which consists of about seven clans, was formerly the military unit. Each of the fixed descent groups goes by the name of its presumed founder preceded by *al* (people of) or *bani* (sons of).

The positions of shaykh and amir, though the prerogatives of a particular lineage, are not, strictly speaking, hereditary. The choice of the particular individual who will assume leadership in case of a vacancy is made consensually by the heads of influential families from among candidates who have the requisite descent position. In the past the choice tended to go to a man known for his courage, leadership qualities and, when relevant, his luck in battle. By the mid-twentieth century astuteness in business and in negotiating with foreign governments and companies had come to be of great importance. Thus the decision combines elements of personal ability with elements of heredity. No one who lacks the requisite ancestry can aspire to tribal leadership, but ancestry alone does not determine the outcome.

Although the vast majority of the population in the typical gulf society was sedentary, tribalism remained extremely important in the mid-1970s. Most people, whether nomads, villagers, or townsmen, claimed tribal affiliations or at least recognized the validity of such affiliations as the framework for society. The cities grew out of tribal agglomerations, and tribal ties carried considerable weight in social life even in cities. Many towns were organized into tribal quarters, so that town geography represented social organization.

All ruling families of the gulf states base the legitimacy of their regimes on their descent from a leading tribe or lineage long known for its nobility of ancestry. Even in the midst of tremendous social change ruling families attempt to maintain and strengthen the ties to the strong tribes of their countries with whom they claim kinship.

Before the oil industry became the major focus of economic life, the livelihood of the gulf towns came mainly from fishing, pearling, and trading. Essentially lacking usable natural resources, the shaykhdoms exploited their location between the northern countries of the Middle East, Persia, and the Indo-Pakistani subcontinent to develop widespread trading connections based on their skill at navigating the gulf and using the monsoon winds to cross the Arabian Sea. Thus those who controlled the major sources of wealth were the large merchants and shippers, many of them of non-Arab ethnic origin.

Unlike most other Arab societies, the shaykdoms along the gulf have long been dominated by towns. More like city-states than nation-states, several are among the most urbanized polities in the world. The advent of the oil industry has increased the dominance of city occupations, city people, and city ways. The power and influence of the government in the oil states have grown at the expense of the family and the tribe. Rapidly growing educational systems have removed the socialization of children from the exclusive control of the family. Employment opportunities and government subsidies have made younger men more independent of their elders. Governments have taken on responsibility for welfare needs that formerly fell on the family. Social mobility has freed people from the bonds of tribalism and has encouraged the beginnings of self-conscious social classes.

Probably the most striking social change in the oil shaykhdoms has been the influx of large numbers of foreigners to supply needed labor. Nowhere in the gulf states do governments encourage foreigners to remain permanently, largely for fear that they could quickly come to dominate the societies at the expense of the native populations. Naturalization rules are strict in all the states, and in many of them gaining citizenship is virtually impossible. The states also strictly control entrance of foreigners according to labor needs. Separated from local society by their alien nationalities, the foreigners are also kept separate in ordinary social life. In some cases they inhabit special districts occupied almost entirely by foreigners. Compared with the family-oriented local populations, the foreign populations contain disproportionate numbers of young, single men. Both the native populations and the foreigners themselves view the foreign population as temporary, although some foreign workers have spent more than twenty years in the gulf states. The situation is particularly poignant for the many Palestinians who have no hope of gaining citizenship in the countries where they live and no real homeland to return to at retirement. Some states forbid foreigners to own businesses or houses; so even long-term migrants must remain employees and renters.

Because the immigrants come from different countries, they represent a variety of skill levels and cultural and religious orientations. Europeans, especially British, and Americans rank at the top of the hierarchy of training and responsibility; they mainly hold positions as high-level military, technical, or business advisers. A small group, they are almost totally cut off from the local societies. The largest group of foreign residents, and the next in the job stratification, are the Arabs. Though similar in language and basic culture to the host population, the group is internally stratified by occupation. People from northern Arab countries tend to bring a relatively high level of education and professional skills. Palestinians are active in a number of white-collar fields, including communications, government, business, education, and medicine, as are Lebanese, Syrians, and Iraqis. Egyptians belong overwhelmingly to the teaching profession and constitute a high proportion of the teachers in the public school systems, and many work at white-collar government jobs. Jordanians hold many important posts in the military. Southern Arabians, such as Yemenis, come from countries much poorer in educational resources and therefore tend to be less skilled. They are found throughout the gulf states as laborers and mercenary soldiers.

Iranians and people of Persian descent are also present in large numbers. Those from the Arabic and Farsi-speaking regions of southern Iran, some of whom have lived along the gulf for generations, include a number of the wealthiest merchants. Farsi speakers from northern and central Iran arrived more recently in the gulf area and have been active in commerce on a smaller scale. Small and marginal merchants and many laborers have come from the Baluchi-speaking regions of Iran and Pakistan. Most Iranian immigrants, however, are laborers and, except for the Baluchis, who are Sunni Muslims, most are Shiites.

The Indo-Pakistani subcontinent also provides immigrants from a variety of cultural backgrounds. Speakers of Urdu, Sindhi, and Punjabi from Pakistan are generally skilled workers, earning their living as merchants, white-collar and professional workers, and skilled tradesmen. Keralans from southern India and Pathans and Baluchis from Pakistan are generally unskilled, earning their living as menials. Many Pathans also serve in the armed forces.

Despite the social chasm that separates the foreigners from the native populations and despite the resentment many foreigners feel at the barriers erected against their full participation in society, the foreigners are absolutely essential to the countries of the gulf and will remain so until the unlikely event that they can meet their own manpower needs. John Duke Anthony, an analyst of Persian Gulf affairs, quotes an official of a gulf country as saying:

> Were the Egyptians to be removed, many of the school systems would have to close; were the Palestinians forced to leave, the media would cease to function; were the British, Jordanians, Pakistanis, Baluch, Yemeni, and Omani soldiers to be expelled, the defense and internal security network would collapse; were the Iranians, Baluchis, and Pathans who make up the bulk of the labor force to be sent back to their homelands, progress on such vital development projects underway as the building of roads, ports, irrigation schemes, housing projects, schools and medical clinics would all come to an abrupt halt.

SOCIAL SYSTEMS OF THE GULF COUNTRIES

Kuwait

Kuwait probably represents the highest development of the oil economy and therefore the greatest degree of social change. The trends found in the oil countries generally are most visible there. The obliteration of many of the social groupings of an earlier time resulted not only from the waves of immigrants, who account for more than 50 percent of the present-day population, but from deliberate government policy (see ch. 6). During the early 1950s the government purchased the bulk of the real estate in Kuwait City at highly inflated prices. It simultaneously constructed new, planned suburbs on the city's periphery and permitted the construction of unplanned suburbs as well. The policy of assigning families to new homes has resulted in segregation of Kuwaiti from non-Kuwaiti elements of the population. The former mostly inhabit the well-equipped, well-served planned suburbs; the latter, less desirable suburbs and shantytowns.

The ruling Al Sabah dynasty traces its descent from the eighteenth-century members of the powerful Anayzah tribe who migrated from Najd in central Arabia to found present-day Kuwait. About 180 other prominent families, called *asilin* (sing., *asil*), also claim social prominence because of their descent from noble tribes of Najd. These families use the prefix *al* before their names.

Despite this tribal background, however, social change is the hallmark of contemporary life in Kuwait. Rapid economic growth has broken down many of the social relationships that once supported a largely tribalized society. The growth of the Kuwait City complex and the industrial towns of the southeast has reinforced the country's traditional urban character. The bonds of tribe and family have been greatly eroded in the towns, and only a fraction of the population continues to live in the deserts. Class relationships have also changed; the merchants have lost much of their preeminence, and the native-born workers have begun to disappear as a class, their place being taken by foreigners.

Class and class consciousness have emerged as important new elements in Kuwaiti society. The development of a comprehensive, free educational system has removed all social barriers to white-collar occupations for Kuwaiti citizens. Government policy of giving citizens preference over noncitizens in competitions for jobs has produced a situation in which no qualified citizen faces any difficulty in finding desirable employment in which he is better paid than a non-Kuwaiti doing similar work. Some observers have stated that the favorable economic situation has destroyed the willingness of Kuwaitis to accept undesirable jobs or even to remain in school if it proves difficult or unpleasant.

Because they enjoy a virtual monopoly on business opportunities, Kuwaitis have profited enormously in a number of endeavors subsidiary to the main source of wealth. Fortunes have been made in importing, real estate, and a number of other fields. The government's land purchase scheme of the 1950s had as one of its goals placing substantial amounts of capital in the hands of owners of buildings. The families of noble merchants that dominated the economy in the era before petroleum were the chief beneficiaries of the program and of the unprecedented business opportunities that developed soon afterward. One of the most successful has been the Al Ghanim, who transformed a prosperous but traditional trading company into a modern multinational corporation with holdings in the United States and projects in a number of other countries.

Although such families have grown markedly wealthier and more powerful on the international scene, they have lost considerable ground relative to the royal family, which, because it controls the country's enormous oil income, has seen its own power increase astronomically at home and abroad. In addition to these wellborn beneficiaries, numerous ambitious Kuwaitis of more modest social origin have also been able to achieve prosperity. Although personal connections still count heavily in business life, personal ability has increased in importance in the bureaucratic atmosphere promoted by the Kuwaiti economy.

One important change reflecting the country's growing modernization is the new attitude toward women, who are increasingly being liberated from the restraints of custom. Although they do not yet have the vote, they are being educated and trained for such careers as teaching and nursing. The younger Kuwaitis are the beneficiaries of modern education and have a better knowledge of the outside world than their parents, but they have never known anything but affluence and are less appreciative of the government's sweeping paternalism. Throughout the country isolation and provincialism are breaking down because of automobiles, radios, and television sets.

Bahrain

Geographically separated from Kuwait and somewhat less developed, Bahrain is the most advanced of the states of the lower gulf. The outward signs of traditional life have largely disappeared. Bahrain had a well-defined social hierarchy, extending downward from the leading shaykhs through the leading merchants and small shopkeepers to the cultivators and to the ex-slaves and emigrants who served as pearl divers and fishermen. But with the gradual extension of education that has accompanied modernization, a new middle class has developed. Counting bureaucrats, teachers, and students among its ranks, it is in the forefront of the drive for social and political change.

In the mid-1970s the Al Khalifah family, the ruling dynasty, stood at the top of the hierarchy of power and influence, followed by the leading commoners. Before World War II this group had consisted primarily of pearl merchants, but it has since come to include merchants in a broad spectrum of pursuits. Because Bahrain, unlike other states in the gulf, is split about evenly between Sunni and Shiite populations, the leaders of organized religious groups occupy a more prominent position than elsewhere (see ch. 3). The ruling family is Sunni of the Maliki school, and most judges and other government officials have traditionally come from this group, although authorities believe it to be a minority of a minority. The Shiites, formerly mainly farmers and fishermen, have also begun to join the urban working class in substantial numbers.

The population of Bahrain is nearly 100 percent settled and nearly 80 percent urban; over 50 percent of the population is in the capital city. In the mid-1970s foreigners composed about 20 percent of the population and, except for relatively small numbers of Americans and British in high-level technical and management positions in the oil industry and elsewhere, foreigners were concentrated in laboring jobs and the services sector. Because of its strong commitment to education Bahrain has for some years produced a surplus of aspirants to white-collar jobs (see Geographic and Demographic Setting, ch. 7). To reduce unemployment among young Bahrainis entering the white-collar labor market for the first time, the government has adopted the policy of replacing foreigners in white-collar positions with Bahraini citizens.

Qatar

The ruling Al Thani family dominates Qatar not only by its power and social prominence but also by its size. One authority characterized it as "vast"; estimates of family membership range as high as 20,000 people, of whom 500 rate the title shaykh, out of a population numbering between 100,000 and 180,000 in the 1970s. The Al Thani family also supports a large number of retainers, many of whom occupy positions of social and political prominence.

Unlike the cities of Bahrain, which had been centers of trade, pearling, and boatbuilding before the era of petroleum, Doha (Ad Dawhah), the capital and largest city, was little more than a sleepy fishing village, with no deepwater facilities and little in the way of a commercial tradition. Crude oil production in Qatar began in 1949, and since that time Doha has been transformed into a modern city, boasting new roads and houses, water and electrical services, and other modern amenities. Commercial activity has expanded phenomenally.

Although no tradition of commerce existed before the discovery of oil, a number of Qatari nationals have emerged as substantial merchants, but the vast majority of the small and medium-sized merchants are foreign, mainly of Iranian origin. In general foreigners compose a majority of the work force. Qataris, who in past generations earned their living in pearling, fishing, a small amount of pastoral nomadism, and an even smaller amount of farming at the subsistence level, are rapidly disappearing from the manual labor force. The advent of oil wealth has allowed them to hire foreigners to do work that citizens consider unpleasant or undignified. Iranians, Pakistanis, and Indians make up the majority of foreigners, although substantial numbers of Palestinians and somewhat fewer Egyptians are also present.

United Arab Emirates

The United Arab Emirates (UAE) is a federation of seven small shaykhdoms (see ch. 9). Two of them, Abu Dhabi and Dubai, experienced relatively rapid development in the early 1970s; although not as advanced as the states of Kuwait and Bahrain, they are nonetheless well along in the process of social change. Oil production began in

Sharjah in mid-1974, and the influx of oil revenues initiated modest social change. The other four shaykhdoms—Ras al Khaymah, Ajman, Umm al Qaywayn, and Fujayrah—have little hope of discovering oil and therefore of experiencing significant economic growth or social change. The last three, moreover, are so small that one authority characterized them as "village states."

The populations of Ajman, Dubai, and Umm al Qaywayn are nearly entirely settled; between 85 and 95 percent of the populations of the remaining shaykhdoms are settled. Beduin constitute the remainder. Traditionally the coastal people earned their living as sailors, pearlers, and traders; the inland peoples, as camel herders and small-scale farmers. Although the population of the UAE as a whole has grown dramatically as a result of immigration, the foreign workers are not evenly distributed among the shaykdoms.

Abu Dhabi

The passing of traditional society is clearly seen in Abu Dhabi. As late as the early 1960s there were essentially three kinds of inhabitants, each engaged in economic activities of long standing: coastal and island settlers, who were engaged principally in fishing, pearling, and trade; oasis dwellers, who depended on subsistence cultivation; and beduin pastoralists, who roamed the mainland in search of water and grazing land. The development of the petroleum industry, after the discoveries in 1960, led to critical shifts in the social pattern. Abu Dhabi's inhabitants have gradually been drawn into wage-earning jobs in petroleum exploitation, construction, and services. Tribal life is breaking down in the face of increased social mobility, the growth of a money economy, and state programs in health, education, and housing.

Nevertheless Abu Dhabi has more tribal groups than any other shaykhdom in the UAE. Although mainly settled at oases in the interior, they include some nomads. The most important tribal grouping is the Bani Yas, a confederation of fifteen tribes, one of which includes the ruling Al Nuhayyan dynasty. The Al Nuhayyan originated in the Liwa Oasis (Al Jiwa) and is also associated with the settlement of Al Ayn in the Buraymi Oasis, where some members moved in the 1800s. A section of the tribe migrated from Liwa to the coast in the 1700s, although most members remained behind. When oil production began in the 1960s, however, members began to move from Liwa to Abu Dhabi, where the capital had long been located. The move ultimately resulted in the highly unusual phenomenon of a major tribe's changing its primary geographic identification. Although Al Nuhayyan influence remains strong in the interior, the lineage is concentrated on the coast.

Other important tribal groups are the Dhawahir, a confederation of fifteen tribes that is important at Al Ayn, and the Awamir, who are found west of the Buraymi Oasis and south of Al Dafrah (Dafir).

Government policy favors the settlement of the country's remaining nomads; a new town has been constructed for this purpose near Liwa.

In comparison with the royal family the merchants of Abu Dhabi are not a very powerful class. They and the Abu Dhabians employed in the oil industry are, however, the only citizens working in the modern sector. Development of a new middle class is not far advanced. The handful of university-trained men in the country have been assimilated into the more traditionally based elite. The substantial working class at the bottom of society consists of foreigners.

Dubai

Dubai is the commercial hub and largest town in the UAE, a center for entrepôt trade, a thriving port, and a market for tribesmen from Oman. Oil was discovered as recently as June 1966. Dubai owes its existence as a permanent town and its development as a commercial center to the silting up of the Sharjah creek. In 1976 it was the only deepwater port along the coast—a fact that explains its natural growth and development.

The ruling Al Maktum dynasty, which belongs to the Al Bu Falasah segment of the Bani Yas, has taken full advantage of the state's commercial dominance. The ruler is the leading merchant, the largest shareholder in the telephone and electric companies, and the owner of shipping and land interests. Nonroyal merchants also figure prominently in society; many more of the immigrants are merchants than in the other shaykhdoms.

Other Amirates

The other, smaller amirates have experienced the least social change. Because of its past prominence as a British headquarters Sharjah has the highest per capita rate of university graduates in the UAE. In the early 1970s it had difficulty in employing its educated population, but by 1976 the booming economy reportedly provided nearly full employment.

Tribalism is important in these shaykhdoms; in the late 1960s 18,000 of the 28,000 residents of Ras al Khaymah claimed a tribal affiliation. The smaller amirates have experienced no basic social change at all and retain their tribal character. Tribal units are less tied to modern political entities than elsewhere in the gulf. Different branches of the Qawasim tribe rule in Sharjah and Ras Al Khaymah, for example, although in Sharjah they are a large tribe and in Ras al Khaymah a relatively small one. The rulers of the other amirates belong to the most important tribes in their respective states, but the tribes are to some extent also scattered among the other political entities.

Oman

The Sultanate of Oman straddles the corner of Arabia, lying along both the Persian Gulf and the Arabian Sea. Thus it shares features with the amirates to the north and with the Yemens to the west. The country has a long tradition of isolation and localism, and the ethnography of its hundreds of tribes is little known. Traditionally the family and the tribe were the dominant social institutions, and no higher institutions or organizations bound the disparate groups into a single people. The territory's inhabitants range from residents of cosmopolitan port cities to isolated tribesmen speaking aboriginal tongues.

Although the population is estimated to be nearly seven-eighths Arab, there are significant concentrations of non-Arabs, particularly in the neighboring coastal towns of Muscat and Matrah, where non-Arabs predominate. The Baluchis, originally inhabitants of the sultan's former possession of Gwadur on the coast of Pakistan, were introduced into the country as mercenaries. Many are still in the sultan's service, but others are found in various seafaring occupations along the coast. A very few have settled in the interior and have been assimilated by the Arabs. Muscat and Matrah also have large numbers of Indians, both Hindus and Muslims, descended from the early merchants and traders who emigrated to the sultanate during its heyday as a maritime empire. Probably the largest minority group is Negro. For several centuries Muscat was an important entrepôt for the slave trade; although slavery has declined, many Negroes still serve as unpaid retainers for the sultan, tribal shaykhs, and other dignitaries. Those Negroes who are free are employed most frequently as fishermen and pearl divers.

According to the traditions of the country, the Arab majority divides into two principal tribal factions: the descendants of the first Arab settlers, identified by several designations including Yamaniyah, Azdi, Qatitani and, in the eighteenth century, Hinawi; and the scions of the second wave of Arab settlers, known as Nizari, Adnani and, in the eighteenth century, Ghafiri (see The Founding of the Modern Gulf Politics, ch. 2). The first Omanis were of southern or Yemeni origin; the later arrivals, who were from the north and central portions of Arabia, began to push into Oman sometime in the fifth century. For nearly 1,500 years the two groups have been feuding, and the feud is still an important factor in internal politics.

The contemporary appellations Hinawi and Ghafiri date from the eighteenth century. At that time serious internal strife involved every tribe in Oman and developed into a full-scale civil war. The faction led by Muhammad ibn Nasir Al Ghafiri, a leader of the Bani Ghafir tribe, attracted most of the Nizari or northern Arab tribes, whereas the Yemeni or southern Arab tribes joined the standard of Khalifa ibn

Mubarak Al Hinawi of the Bani Hina tribe. For two centuries the two factions, popularly called the Ghafiri and Hinawi, have honored those two chieftains.

Generally the Ghafiri predominate in the northwestern districts, and the Hinawi are more powerful in the southeast. Over the centuries, however, a great deal of intermingling has taken place, so that often a village is split between subgroups of tribes affiliated with the warring factions.

It should be emphasized that the Hinawi-Ghafiri split, in spite of its ethnic antecedents, is primarily a religious division in present-day Oman. The Hinawi have been closely associated with the Ibadi version of Islam, which has been the pillar of Omani particularism over the centuries (see ch. 3). The Ghafiri, though often affiliated with Ibadi Islam, have traditionally been more receptive to outside influences. There are several Ghafiri tribes whose religion is Sunni Islam, and a few others have adapted the Wahhabi (see Glossary) practices traditionally connected with Saudi Arabia (see ch. 2).

Tribalism remains the most important source of social identity for the majority of the people. Of the many scores of tribes and subtribes found in the country, the most important are the Al Bu Said, the royal tribe; the Bani Ghafir, Bani Amr, and Hawasin of inner Oman; the Shihuh and Habus of the Musandam Peninsula; the Janabah of Masirah Island; the Dura of Dhahirah in the north; the Al Kathir, Mahrah, and Qara of Dhofar near the Yemen (Aden) border; and the Al Hirth of Sharqiyya (see fig. 22).

Of considerable ethnographic interest are the Shihuh and Qara, who speak non-Arabic Semitic languages and apparently are descendants of an indigenous population displaced by Arab invaders. The Shihuh are seminomadic, moving each year between winter lodgings in the mountains and summer lodgings on the coast. Different social groupings come into play in the two kinds of settlements; winter settlements are inhabited by one extended family each, summer settlements by several related extended families.

The Qara and other peoples of the Dhofar region are more oriented to local groupings or to tribal fellows in Yemen (Aden) than to the regions to the east. Only in the immediate area of the coastal towns do ties with the rest of the country gain importance.

The tribes formerly exercised substantial political leverage over the towns throughout the country. Tribal leaders received large subsidies to refrain from raiding certain towns or caravans originating from them. Travelers and merchants paid tribute to local shaykhs when crossing tribal lands. The revenue accruing to tribal leaders enabled them to exercise control over the far-flung branches of their tribes. They could afford to be generous and thus commanded widespread allegiance. With the advent of modern transportation in the 1970s, however, the countryside has become more accessible. As a result the towns became more influential and the tribes' power declined.

The advent of the oil industry and the growing strength of the central government has hastened the decline of the tribes' power. As development projects and bureaucratic organizations grow, the tribes are being drawn into a more national orbit. A significant migration from tribal areas to cities has begun, and the sharp cultural differences between the interior and the coast have begun to lessen. The power of tribal leaders in matters beyond their immediate locality has declined, although they remain important in local affairs.

Education is providing a natural vehicle for the development of loyalties to groups beyond the family and the tribe. By the mid-1970s politicized youth clubs supervised by the Ministry of Labor and Social Affairs constituted one of the first examples of voluntary associations in the country.

At the national level the most important social stratum is the ruling family (see The Founding of the Modern Gulf Politics, ch. 2; Government and Politics, ch. 10). A few expatriates, mainly British, occupy positions of importance in the military, the oil industry, and the development program and enjoy the sultan's favor. About twelve merchant houses constitute the commercial elite that forms the next stratum. These families control import franchises and are active in various enterprises other than oil. Nearly all these families are of foreign extraction, mainly Persian, Indian, or Pakistani. Not one of the ten most important mercantile houses adheres to Ibadi Islam; most are Shiites, but the group includes Sunnis and Hindus.

A small urban group of growing importance is made up of the modern-educated intellectuals; in the early 1970s about a dozen were graduates of foreign universities, and a number of others had studied at secondary schools in Africa, Aden, and elsewhere. Such persons often find themselves in cultural conflict with the traditional Ibadi religious leaders. Former slaves of the sultan, who number about 2,000, also form a group of some importance. Many remain in his service and enjoy close ties to the palace.

Foreigners and minority groups are concentrated in the port cities of Muscat and Matrah. Indians, Pakistanis, and Baluchis constitute the most important minorities. Indians and Pakistanis tend to work in commerce, artisan crafts, the military, and the government. Baluchis mainly work as soldiers, farmers, laborers, and fishermen. The Indian community includes Christians from Goa and Kerala who mainly hold white-collar jobs; Sikhs from the Punjab who work as electricians, as carpenters, and in other skilled trades; and Hindu merchants.

THE INDIVIDUAL, THE FAMILY, AND THE SEXES

Social life in the Persian Gulf centers in the family; even in the more developed states no other institution of even roughly comparable strength exists. Although the family is losing ground where social change is occurring most rapidly, family loyalty and particularism still pervade all aspects of life. The household is composed of kinsmen, and among the tribes family ties ramify into tribal structure. The individual's loyalty to his family overrides most other obligations. Ascribed status generally outweighs personal achievements in regulating social relations. One's honor and dignity are tied to the good repute of one's kin group and especially to that of its women.

Sexual segregation has traditionally been absolutely basic to social life and remains significant everywhere. In many places women veil outside their homes. In Kuwait sexual segregation holds sway even in the absence of veiling. The car is useful in this regard; it permits women to move around but prevents contact with strangers. Mixed social gatherings occur only among foreigners and the most highly sophisticated gulf Arabs. Sex is one of the most important determinants of social status. Because of the systematic lifelong segregation of the sexes, men and women constitute largely separate subsocieties, each with its own values, attitudes, and perceptions of the other.

Family and Household

Arabs generally reckon kinship patrilineally, and the household is based on blood ties among men. Ideally it consists of a man, his wife or wives, his married sons with their wives and children, his unmarried sons and daughters, and possibly other relatives, such as a widowed or divorced mother or sister. At the death of the father each married son ideally establishes his own household to begin the cycle again. Because of the centrality of family life, it is assumed that all persons will marry when they reach the appropriate age; many divorced and widowed persons remarry for the same reason. In most areas adult status is bestowed only on married men and often only on fathers.

Although statistics on households were not available in late 1976, the extended family household was quite common in many places, and social values decidedly favored it. Only in such rapidly changing societies as Kuwait does the nuclear family household have significant appeal. Even nuclear households often form parts of extended families living in adjacent houses rather than in the same house. The new houses in the suburbs of Kuwait have been a significant factor in the dispersal of large family groups. Designed for the small family, they encourage greater involvement with neighbors and the surrounding community regardless of kinship ties.

Whatever combination of kin composes it, a household may be viewed as a unit of consumption. The men contribute their earnings to the senior male, who spends them in the interest of the group. Traditionally the individual subordinated his personal interests to those of his family and considered himself a member of a group whose importance outweighed his own. It is not common for persons other than male foreigners to live apart from a family group; even they prefer to live with kin or at least fellow tribesmen or countrymen. Grown children ordinarily live with parents or relatives until marriage; for a girl of respectable family to do otherwise would be unthinkable. Child-rearing practices train the young to be docile members of a family group rather than individualists.

Marriage is a family rather than a personal affair. Because the sexes do not ordinarily mix socially, young men and women have few or no acquaintances among the opposite sex, although among beduin a limited courtship is permitted. Parents arrange marriages for their children, finding a mate through either their own social contacts or a professional matchmaker. Among both villagers and those with tribal ties the preferred marriage partner is the child of the father's brother or someone similarly related, and such marriages are apparently common. In most areas, in fact, a man has a right to forbid his father's brother's daughter to marry an outsider if he wishes to exercise his right to her hand. If the ideal cousin marriage is not possible, marriage within the patrilineal kin group is the next choice; tribes, especially those of noble stature, traditionally practiced virtually total endogamy. The only common exception was a marriage with an outsider contracted for political reasons. Endogamous marriages in traditional society produce several advantages for all parties: the bridewealth payments demanded for the bridegroom's kin tend to be smaller; the family resources are conserved; little danger exists of an unsuitable match; and the bride need not go as a stranger to her husband's house. Although some men educated abroad have married foreign women, most governments officially discourage such matches.

In societies where an emerging class system threatens the traditional tribal domination, such personal achievements as education and job prospects have begun to outweigh descent as criteria in marriage arrangements. In Kuwait, for example, the 180 families who formerly claimed pure descent from noble Arab settlers and married strictly among themselves have begun to accept spouses more distinguished for their personal attainments than for their tribal origins.

In Islam marriage is a civil contract rather than a sacrament. Consequently representatives of the bride's interests negotiate a marriage agreement with the groom's representatives. Although the future husband and wife must, according to law, give their consent, they usually take no part in the arrangements. A young man might suggest to his parents whom he would like to marry; girls usually have no such privileges. Men expect virginity of their brides, but no such expectation exists for bridegrooms. The marriage is registered by the bridegroom and the bride's male representative rather than by the bride. The contract establishes the terms of the union and outlines appropriate recourse in the event they are broken. Special provisions inserted into the standard contract become binding on both parties.

Islam gives far greater leeway to the husband than to the wife. For example, he may take up to four wives at one time, provided he can treat them equally; in the 1970s few men had more than one wife. A man can divorce his wife by simply repeating "I divorce thee" three times before witnesses; a woman can instigate divorce only with extreme difficulty. Any children of the union belong to the husband's family and stay with him in case of divorce. Although economic factors discourage divorce, the one-sidedness of the law symbolizes the woman's subservient position. Men of course exercise authority in the home; but women, particularly when they are mothers, mothers-in-law, and grandmothers, also possess considerable authority.

Men and Women

The social milieu in which the family lives affects the circumstances of the wife to a great extent. In towns and villages male social life goes on outside the home, usually in the shops, cafés, or majlis (see Glossary), where friends meet to chat, where politics is discussed, and where much business is transacted. In the desert, however, women fulfill important economic functions without which the family could not exist. As a result tribal women occupy a position of relative importance and have more freedom. They do not veil, for example. Although casual social contact between the sexes of the sort common in the West is not known, segregation of the sexes is much less pronounced in the desert than in settled communities. Artisan and merchant families are supported by the skills of the men. Women make little financial contribution; their responsibilities are often limited to the household.

In such circumstances it is more likely that women are confined to the home and their social contacts and interests limited to an exclusively feminine sphere. The houses of wealthier settled families traditionally contain distinct areas for men and women—the reception room, where the man of the family entertains male guests, and the women's quarters, from which adult males other than relatives and servants are excluded. Unlike their rural sisters, who move more freely in the fields, villages, and desert, urban women walk in the street veiled and chaperoned, discreetly avoiding coffeehouses, markets, and other public gathering places, as well as any social contact with men.

Arabs assume, and often explicitly say, that men and women are different kinds of creatures. Women are thought to be weaker than men in mind, body, and spirit, more sensual, less disciplined, and in need of protection both from their own impulses and from the excesses of strange men. In courts of law the testimony of one man equals that of two women. The honor of the men of a family, which is easily damaged and nearly irreparable, depends on the conduct of the women, particularly of sisters and daughters; consequently women are expected to be circumspect, modest, and decorous and their virtue to be above reproach. The slightest implication of unavenged impropriety, especially if publicly acknowledged, could irreparably destroy the family's honor. Although family honor resides in the person of a woman, it is actually the property of the men of her natal family. An unfaithful wife, for example, shames her father and brothers far more than her husband. Enforcement of honor, even at the cost of a sister's or daughter's life, is the obligation of the men of a family. Any social standing whatsoever, and even minimal respect, is impossible without it.

Arab societies generally value men more highly than women, and both sexes concur in that estimation. Their upbringing quickly impresses on girls that they are inferior to men and must cater to them and on boys that they are entitled to demand the care and solicitude of women. The birth of a boy occasions greater celebration than the birth of a girl. Failure to produce sons can be grounds for divorcing a wife or taking a second. Barren women, therefore, are often desperately eager to bear sons.

In some areas women marry in their middle teens but, where education has become widespread, the age of marriage has risen. The young bride goes to the household, village, or neighborhood of the bridegroom's family, where she may be a stranger and where she lives under the constant critical surveillance of her mother-in-law. A great deal of familial friction centers on the difficult relationship between mother-in-law and daughter-in-law.

A woman begins to gain status, security, and satisfaction in her husband's family only after she produces a boy. Therefore mothers love and favor their sons, ordinarily nursing them longer than daughters. In later life the relationship between mother and son often remains very warm and intimate, whereas the father is a more distant figure. Observers suggest that women compensate for the emotional lacks in their often rather impersonal marriages and submerged adult lives through their relationships with their sons, who as adults often remain in or near the parental household. The wife who enters such a home finds herself in a distinctly secondary position. Furthermore a girl's own parents are traditionally eager for her to marry as soon as she reaches puberty to forestall any mishap to her virginity; she is therefore not encouraged to remain in the family home.

Although women's power is based largely on their relationships with male relatives, it is far from negligible. Respected women with wide social contacts in female society often act as important sources of information for their menfolk; women can gather a good deal of news about what is being thought and done in the homes they visit, whereas men are forbidden by etiquette even to discuss the female relatives of their friends. Women play crucial roles in marriage arrangements, which often carry important political overtones; their support or opposition often determines which matches come about. Their visits to their natal households serve to strengthen ties between allied men. Their labor makes male hospitality possible. Within the privacy of the family circle women's opinions often carry weight in men's decisions. A prominent example of an influential female is Shaykha Hussa bint al Murr, mother of the ruler of Dubai; according to authorities she exercised considerable political power during the early decades of the twentieth century.

The development of the educational system has presented women for the first time, with options other than traditional marriage and household life. Women of all ages have reportedly responded enthusiastically to those new opportunities, both because of a desire for self-improvement and because the highly desirable modern-educated man prefers an educated wife. It is not at all clear, however, to what extent education has changed or will change the basic nature of relations between the sexes in the gulf states.

* * *

Information on the social organization of the Persian Gulf states is scattered and incomplete. Useful sources, though somewhat limited in their approach to social structure, are John Duke Anthony's *Arab States of the Lower Gulf: People, Politics, and Petroleum* and *Political Dynamics of the Sultanate of Oman.* Also useful are Hassan A. Al-Ebraheem's *Kuwait: A Political Study,* Emile A. Nakhleh's *Bahrain: Political Development in a Modernizing Society,* and Cora Vreede-De Stuers' "Girl Students in Kuwait." Louise E. Sweet presents a summary of ethnographic knowledge through the late 1960s in "The Arabian Peninsula," in *The Central Middle East.* (For further information see Bibliography.)

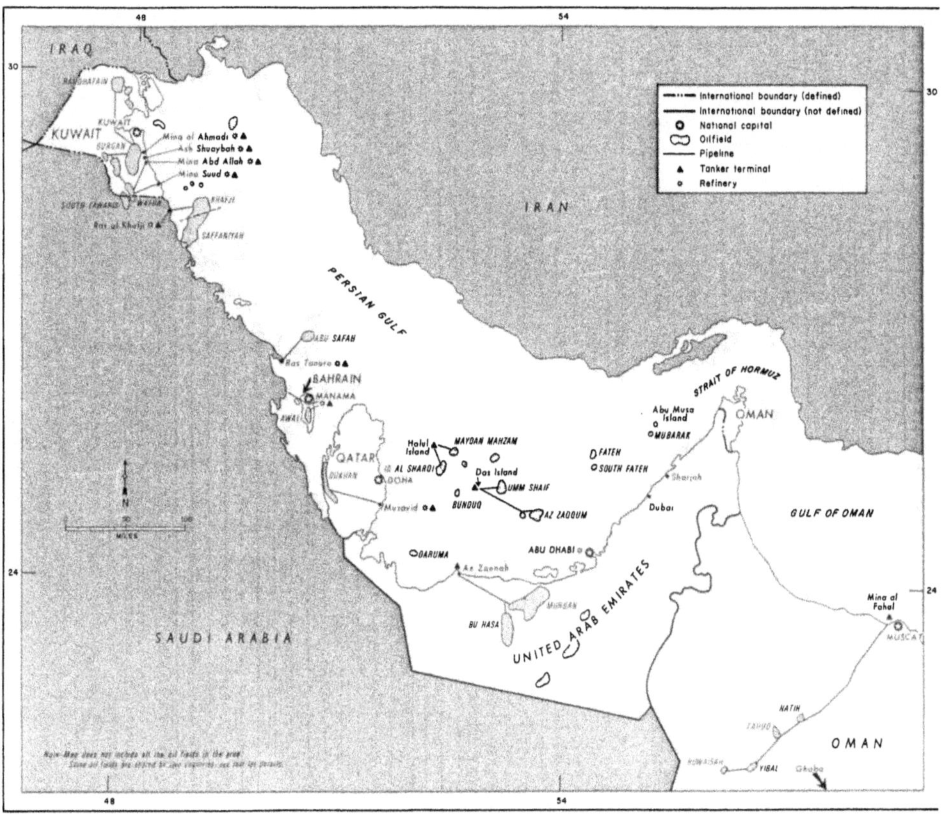

Figure 4. Persian Gulf Oil Fields

CHAPTER 5

THE OIL INDUSTRY IN THE PERSIAN GULF STATES

The discovery of oil introduced change into the poor, backward Persian Gulf shaykhdoms of Kuwait, Bahrain, Qatar, the United Arab Emirates (UAE), and Oman. The amount of change evident in the mid-1970s depended on the size of their crude oil deposits and how long ago the deposits were found. These shaykhdoms also contributed to the profound changes that have occurred in the international oil industry, changes that have tremendously increased the oil revenues of the shaykhdoms and, among other things, made Kuwait the richest country in the world in per capita income. Qatar and the UAE were not far behind.

The first oil discovery in these states was in Bahrain in 1932. The field turned out to be small, and production will last until about 1995 at best. Nevertheless the Bahrain discovery was important because it proved that oil existed in the western and southern reaches of the Persian Gulf and it increased the interest of oil companies in Kuwait, Qatar, and Saudi Arabia (see fig. 4). An American company, the Standard Oil Company of California (Socal), had the concession in Bahrain, the first significant success in penetrating the strong political and economic hold the British then enjoyed throughout the gulf. Encouraged by the Bahrain find, Socal went on to gain the concession in Saudi Arabia, potentially the world's greatest oil producer.

Oil was discovered in Kuwait by a joint Anglo-American company in 1938. The second well drilled hit one of the world's largest oil fields. Subsequent drilling found oil under much of the tiny shaykhdom's territory. Kuwait had 11 percent of the world's crude oil reserves, enough to last seventy-five years at least, and was the fifth largest producer in 1974. When the oil revenues first began to mount, there was a period of lavish spending—for example, on such things as villas and cars—but then control was established and funds were channeled into more constructive endeavors (see The Economy, ch. 6). In order to have sources of income after the oil ran out, Kuwait established a competent investment company to place funds in foreign countries. Kuwait was also instrumental in establishing the Organization of Petroleum Exporting Countries (OPEC) and contributed significantly to its strength in the mid-1970s. The country was the first to negotiate 100-percent ownership of the main oil-producing company, and it has started on a large program for further development of its hydrocarbon resources.

Oil reserves and production were much more modest in Qatar, Oman, and the UAE. Concessions were granted in all three areas in the 1930s to the Iraq Petroleum Company, owned jointly by several of the largest international oil companies. They were not particularly in need of crude oil sources, and development went slowly. World War II and the difficulty of finding oil also contributed to the length of time before production started in the 1950s and 1960s. Among the amirates (see Glossary) of the UAE only Abu Dhabi, Dubai, and Sharjah were oil producers, the last since July 1974. Petroleum exploration was continuing in the other amirates. They followed OPEC guidelines and worked with other Arab oil states in the gulf in petroleum matters.

Total crude oil production from all the Persian Gulf states amounted to 1.86 billion barrels in 1974, approximately 9 percent of world production. Almost all of this oil was exported and accounted for 16 percent of world oil exports in 1974. Crude oil production of each of the shaykhdoms in 1974 was: Kuwait, 929 million barrels; UAE, 613 million barrels; Qatar, 189 million barrels; Oman, 106 million barrels; and Bahrain, 25 million barrels. Production declined in 1975 to about 1.69 billion barrels as worldwide economic recession reduced demand for petroleum products. Economic recovery in the industrialized countries increased the demand for crude oil in the early months of 1976, and preliminary indications suggested that production in the Persian Gulf states would increase during the year.

The bulk of Persian Gulf production was exported as crude oil. Western Europe and Japan were the principal markets. The dependence of West European countries and Japan on Persian Gulf oil, including that from Saudi Arabia, Iraq, and Iran, was particularly high and influenced commercial relations and foreign policies.

Two factors were common to each of the shaykhdoms. The oil sector dominated the economy and provided the bulk of government revenues and foreign exchange earnings. In addition the crude oil fields had high gas pressure that caused the oil to flow to the surface. Few wells required pumping. The crude oil released gases when it reached the surface. These gases were excellent energy sources and provided feedstock for petrochemical plants. Because pipelines and processing facilities were costly and transportation to foreign markets was difficult, it had been too risky a commercial venture to try to make use of the associated gases before the price of energy was so dramatically increased in 1974. Except for some limited uses the gases associated with crude oil production had been burned in the open air. These flared gases presented a spectacular scene at night; the huge waste of energy troubled all of the producing countries.

After OPEC had tremendously increased the price of energy in 1974 and after the increased oil revenues provided the funds for the necessary investments, each of the states developed plans to use the gases being flared. Work was under way in 1976 to collect the gases and build plants to use them. If these efforts are successful, it will substantially diversify the narrow economies of the Persian Gulf states, which are based almost exclusively on crude oil production. The projects will also add considerable capacity to the worldwide petrochemical industry by the early 1980s, which if not adequately coordinated could create troubles for the industry and its workers. Of even greater significance to world energy consumers, the collection and processing of associated gases will set lower limits on crude production in the shaykhdoms, easing the threat of such cutbacks in crude oil production in pursuit of foreign policy objectives as occurred in 1973 and 1974 after the October 1973 War.

OIL: THE INTERNATIONAL SETTING

The era of cheap energy ended in 1973 after OPEC took control of pricing from the international oil companies. The change shocked most of the world. For the small Persian Gulf states, however, it brought revenues by which the rulers hoped to develop their economies to provide for their people after the oil was gone.

One factor contributing to the sharp change in the price of energy was the steeply rising world demand caused by the mechanization of so many activities during the previous 100 years. One estimate indicated that the fuel and power consumed by an average American amounted to the energy equivalent of 200 full-time servants. Between 1962 and 1971 the estimated world demand for energy increased at an annual rate of 4.8 percent, a rate that would result in a doubling in fifteen years. Energy requirements were accelerating so rapidly that observers predicted critical energy shortages in the twenty-first century and the exhaustion of known oil deposits long before that. The primary energy sources—coal, oil, and natural gas—were being depleted at an astonishing rate.

Accompanying the growth in energy demand was a radical shift in the source of energy. In 1859, when the world's first oil well began producing in Titusville, Pennsylvania, men and animals supplied much of the world's power needs, supplemented by waterwheels, windmills, and steam generated by burning coal and wood. Coal powered an early part of the industrial revolution; in 1910, for example, coal supplied 90 percent of American commercial energy requirements. Coal still made up over half the world's commercial sources of energy in 1960, but oil was rapidly displacing it as the world's primary fuel.

Petroleum was used long before the Christian era. Noah reportedly waterproofed his ark with bitumen, an asphalt of Asia Minor used in ancient time also as a cement and mortar. Early man also used oil for medicinal purposes and as fuel for lamps. Until modern times petroleum was collected from natural seepage. Not until 1859, when a drilling rig was set up on Oil Creek near Titusville, was oil sought commercially.

The oil industry grew rapidly after the first well came in. Within a decade Russia, Romania, Canada, Italy, and the United States, were producing oil. Several more countries began producing soon afterward, and oil was discovered in Iran in 1908.

During the late nineteenth century petroleum was used primarily as a lubricant and as lamp fuel. In 1900 about 58 percent of petroleum consumption in the United States, the largest producer and consumer of oil, was in the form of kerosine for heaters and lamps; most of the rest was used as fuel oil for heating and in power plants.

Development of the internal-combustion engine vastly expanded the demand for petroleum products. World War I and World War II greatly accelerated engine development. Improvements in petroleum refining accompanied diversification and refinement of engines. Research added to the uses for petroleum products: a whole new field of petrochemicals emerged, producing dyes, fertilizers, and other products and increasing the demand for crude oil.

Petroleum was cleaner, more convenient, and cheaper than other fuels. In addition it was the unique fuel for internal-combustion engines, feedstock for petrochemicals, and base for lubricants. As a result the market for oil grew much more rapidly than the total demand for energy. World consumption (excluding communist countries) went from 1 million barrels per day (see Glossary) in 1915 to more than 5 million barrels per day in 1940 and 35 million barrels per day in 1971. Between 1962 and 1971 world consumption of petroleum increased at an annual rate of 7 percent, a doubling in ten years. Of the energy consumed in 1970 by noncommunist countries, a little more than one-half came from oil. Coal accounted for nearly one-fourth, natural gas supplied about one-fifth, and nuclear power provided less than 0.5 percent; hydropower contributed the remainder.

Another factor contributing to the sharp rise in the price of energy was the growing international trade in petroleum products. The Soviet Union surpassed the United States as the largest producer of crude oil in 1974, as a result of the declining output of American fields that set in at the end of the 1960s. Both countries, however, consumed most of their own production. The Soviet Union began exporting relatively small amounts of petroleum in the 1950s, and the United States began importing increasing quantities of oil to meet consumption needs some years before. The main stimulus for the international petroleum trade came from the rapid economic growth and increased oil consumption in Western Europe and Japan after World War II. These countries lacked significant crude oil deposits and needed large and increasing imports to satisfy growing consumption needs. Between 1962 and 1971 the petroleum imports of Western Europe more than doubled, and Japan's increased more than fourfold. Petroleum became the most important commodity in value in international trade, and by the mid-1970s petroleum products accounted for more than one-half of all seaborne commerce.

Geologists have determined the outlines of a large basin extending from the Taurus Mountains in southeast Turkey to the Arabian Sea in the south and underlying western Iran, eastern Saudi Arabia, Iraq, and most of the Persian Gulf. In 1974 this basin held 62 percent of the world's proven reserves. The development of these Middle Eastern fields after World War II made them the world's most important crude oil source, supplying 39 percent of world production in 1974 (see fig. 5). Crude oil production costs have been low because pressure in these prolific fields is generally high; transportation costs have been low because the fields are relatively close to water routes. The countries owning these reserves have relatively small populations, little domestic oil consumption, and an interest in exporting oil. As a result these states became the most important exporters of petroleum products, supplying 63 percent of the oil in international trade in 1974.

Structure of the International Oil Industry

Another factor contributing to the sharp rise in the price of energy was the changing structure of the international oil industry. The industry is divided into several distinct phases—exploring, producing crude oil, refining into usable products, transporting, and marketing—and each phase requires costly investments. Even with modern techniques, for example, only 10 percent of the wells drilled in new fields produce oil or gas, and only 2 percent of the wells are significant producers. The industry as first developed consisted of a few very large companies vertically integrating all phases from exploration to marketing.

The petroleum industry before World War II was dominated by seven or eight major oil companies. Five of these were American; the others were European. These major companies held most of the foreign concession agreements for exploration and development. This gave the companies a degree of horizontal integration; they could adjust the output of crude from various areas to match overall marketing needs.

Before World War II the dominance of these companies, vertically and horizontally integrated, gave them a high degree of influence over supply and price through mutuality of interests if not collusion. The major oil companies were able to exert considerable control for some years after World War II; output from the prolific, low-cost Persian Gulf fields was phased into world markets without excessive disruption to pricing, petroleum investments, and employment in the United States and output and revenues in other high-cost crude oil areas. Coal

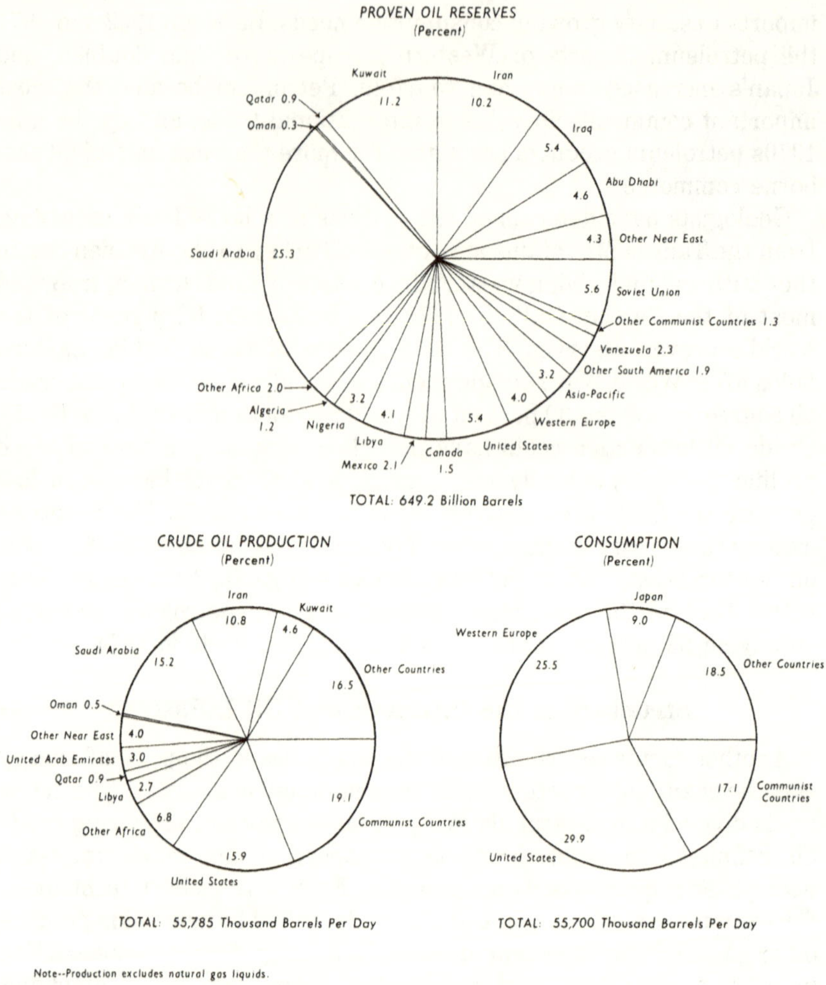

Figure 5. World Oil, 1974

mining throughout most of the world, which had become increasingly more costly as the richer seams were depleted, was less disrupted than it might have been had it been forced to adjust to cheaper oil. The price of energy diminished in relation to other prices but less rapidly than it would have with quicker exploitation of Persian Gulf crude and competitive pricing; the adjustment process was eased for many countries but at the cost of windfall profits to the oil companies.

By the 1950s the dominance of the major oil companies was gradually diminishing. The booming oil business encouraged the entrance of smaller oil companies. Some, such as Getty Oil Company, were private firms; others, such as the French and Italian national petroleum companies, were state-owned businesses. The smaller companies won concessions in oil-producing countries by offering the host governments more favorable terms than those of the major oil companies. The Soviet Union also began to export oil over which the major companies had no control. These developments occurred within the broader framework of the bipolar relations between the leading communist and noncommunist countries. A group of developing countries calling themselves the third world were asserting more control over their own destiny and natural resources at the insistence of domestic nationalistic groups. Diminishing dominance by the major oil companies meant an increasing inability to match petroleum supplies to consumption needs. By the end of the 1950s a growing supply of oil started to push prices downward. The downward pressure on petroleum prices was favored by the consumer, but it reduced revenues for the oil-producing states.

Oil Pricing

Oil pricing is a vast and controversial subject complicated by the vertical integration of much of the industry and the secretiveness of the companies involved. Price can simply match supply and demand, or it can incorporate political decisions to achieve some possibly noneconomic goal. The United States, although an avowed advocate of free enterprise and free trade, had long interfered in the pricing of some commodities—as in former price supports for major agricultural crops—and has regulated phases of the domestic oil and gas industry for particular goals that to a degree affected international prices of petroleum. Other oil-producing countries have also sought to exert various controls over their petroleum resources, with the result that oil pricing and trade have reflected forces other than just supply and demand.

The major international oil companies are largely housed in the United States and have domestic fields, refineries, and outlets. There have been many United States congressional investigations of the oil companies, and their operations in foreign countries have been watched, threatened, and in some instances taken over by the host government. Contrary to their popular image the oil companies have not been free agents concerned only with maximizing profits. This is not to say, however, that they have not wielded considerable domestic as well as international power, done unsavory things, or secured huge profits.

Early in the 1900s the United States was the world's major oil exporter, and international crude oil pricing by the major oil companies was based on the United States price plus transportation costs. This was a base-point pricing system in which the price to the buyer was equivalent to the price of oil shipped to him from the Gulf of Mexico, regardless of actual source. There were other oil-exporting countries at the time, but the oil companies' price was the same to India, for example, even if the oil was shipped from Iran, with transportation costs substantially less than from the Gulf of Mexico. This system worked when there were only a few oil companies with mutual interests; it protected their investments in the United States and other areas and provided a high return for developing fields, such as those in the Persian Gulf. How much the oil company kept and how much the government owning the field received depended on the concession agreement, but there was not a downward pressure on the price of crude oil produced in the new foreign fields in the Caribbean and Middle East.

Between 1945 and 1950 the increased supply of crude from foreign fields, the emergence of independent oil companies, and United States pressure applied through its help in rebuilding postwar Europe brought about major modification of the single base-point price system. The major oil companies set crude oil prices in conjunction with transportation costs from actual point of shipment to establish a series of equalization points—points where the landed price of Middle East oil was the same as crude from the Gulf of Mexico and Venezuela. The equalization point was first southern Europe, then London, and eventually New York. The shifting of the equalization point northward and westward increased transportation costs, thereby requiring lower quotations for Middle East crude although providing it with a competitive advantage in the European markets formerly held by United States, Mexican, and Venezuelan crude.

The oil companies were caught in conflicting squeezes. The cost of producing oil in the Middle East was a fraction of that in Venezuela, which in turn was substantially less than that in the United States. If prices were based on cost of production, Middle East crude would have forced other countries out of the market. The oil companies, American politicians, and American businesses did not want this. In fact Morris A. Adelman, in his authoritative study *The World Petroleum Market,* concluded that an effective coalition in the United States limited imports of foreign oil for the large United States market for at least ten years before official import quotas were established in 1959. The oil companies would have profited from more imports, and the Middle Eastern oil-producing countries would have been happy, at least for a while, if the companies had taken more oil, because their revenues would have gone up. However, the Venezuelans did not want to be forced out of crude production. The oil companies faced a no-win situation. Each producing country wanted to remain in the oil business and earn more money but did not want to suffer competition from cheap suppliers or have the price of its oil go down. The dominance of the major companies over the industry reduced for all producers the pain of adjustment to the large supply of low-cost Middle East crude entering the market.

From about 1950 to the early 1970s there was a system of posted prices for each country—prices at which oil companies would sell crude oil to anybody. Transportation costs were not included. Because "anybody" generally meant independent oil companies that the major oil companies did not want to encourage, posted prices were higher than the actual price at which the bulk of crude oil sales took place. Moreover a short-term oversupply of crude developed in the late 1950s as newly discovered fields began to produce, causing a downward pressure on crude oil prices. By the 1960s, if not earlier, the posted price became only a reference point for calculating the taxes and royalty payments due oil-producing states under the concession agreements.

Economists have had difficulty in analyzing the pricing system because of inadequate data to establish what actual market prices were; perhaps four-fifths of crude oil sales were bookkeeping transactions within integrated companies. There is widespread agreement, however, that actual market prices of crude oil had a substantial downward trend between 1950 and 1970, although posted prices remained fixed for prolonged periods.

The posted price per barrel of Saudi Arabian light (34-degree) marker crude oil was US$2.05 in June 1948, US$1.70 in November 1950, US$2.08 in June 1958, US$1.80 in August 1960, and US$2.18 in February 1971. The market price is the basic price for a typical regional crude —in this case Arabian light used for the Persian Gulf—from which regional prices are derived.

Although the major oil companies exerted considerable influence on crude oil prices, the oil industry was not a closed system, and market forces had an impact. Before 1973 crude oil was priced largely in the context of supply and demand; after 1973 crude oil pricing took into consideration the costs of other energy sources and the long-term supply of petroleum. The frame of reference for pricing decisions was vastly broader and less favorable to consumers than the one that had previously prevailed.

Organization of Petroleum Exporting Countries

The sheer size and integrated nature of the major international oil companies afforded them considerable advantages when negotiating concession agreements with countries having known or probable oil deposits. Moreover many of the concession agreements were arranged with developing countries generally lacking in sophistication and usually hard pressed for funds. The governments of these countries often failed to appreciate the value of their resources and were frequently so desperate for funds that they lacked bargaining power even if they understood the value of their oil. Kuwait, Bahrain, Qatar, Abu Dhabi, and Oman were certainly in this position in the 1930s, when the original concession agreements were let.

Efforts by oil-producing countries to exercise control over their resources without causing excessive domestic economic and financial disruption were difficult because of the pressures the oil companies brought to bear. Mexico nationalized its oil industry in 1938, but it was a costly and prolonged experience. When Iran nationalized its oil industry in 1951, the oil companies refused to buy or transport the oil, causing a severe financial and political crisis in the country for some years.

OPEC was more than a decade in gestation. Part of the initial impetus came from Venezuela, which in 1943 began to press the international oil companies for more control over and a better return on its oil resources. Conscious of Mexico's difficult experience, Venezuela established contacts with Persian Gulf oil states in an attempt to develop coordinated efforts among oil exporters. Some analysts have interpreted the rapid development of the Persian Gulf oil fields after World War II by the major oil companies as both a means of forcing Venezuela to temper its demands and an expansion of an alternative source of oil should the Venezuelan government's demands become unreasonable. Other observers stressed the low cost and other advantages inherent in Persian Gulf oil as a natural cause for development of Middle East fields.

The Middle East countries recognized the usefulness of concerted action to protect their interests. Many of the countries, however, were desperately in need of revenues and independently pushed the oil companies to take more of their oil to increase their income. Not until the oil companies unilaterally cut the posted price in the spring of 1959 to reflect the glut of oil on the market and the growing discounts that they had been giving on posted prices did a movement begin toward joint action. In response to the price cuts the League of Arab States (Arab League) sponsored the first Arab oil conference in the summer of 1959; Venezuela and Iran attended as observers.

As a result of the contacts developed during the 1959 conference, Saudi Arabia and Venezuela issued a joint statement in May 1960 urging all oil-producing countries to adopt common policies to safeguard their economic interests. In August 1960 the oil companies further cut the posted price of Middle East crude oil. The price reduction lopped US$30 million from Saudi Arabia's estimated income for the year and thus severely hampered efforts to balance the budget and execute development programs. When the Saudi Arabian minister of petroleum attended an emergency meeting of authorities from oil-exporting states in Baghdad after the price reduction, he argued vigorously and successfully for the founding of OPEC.

OPEC was formed in September 1960 by Iran, Iraq, Kuwait, Saudi Arabia, and Venezuela to develop some bargaining power among oil-exporting countries vis-à-vis the international oil companies. Other countries joined subsequently. By late 1975 Algeria, Ecuador, Gabon, Indonesia, Libya, Nigeria, Qatar, and the UAE were members. Applications for membership by other oil-exporting countries were under consideration in 1976.

The foundation document of the organization noted the dependence of its members on revenue from oil exports, the wasting nature of the petroleum resources, and the interdependence of all countries. The first objective of OPEC was to restore crude oil prices to the 1958 level and to stabilize prices through mutual cooperation with the oil companies. OPEC acknowledged the need for oil prices that balanced the interests of producing and consuming nations and provided a fair return on investments of the oil companies. The founding members agreed to stick together against efforts by the oil companies to woo any member country or apply sanctions against one that was following a unanimous decision of OPEC.

OPEC policy is officially set almost entirely at its semiannual or more frequent conferences of high-level representatives from each member government. Unanimously adopted resolutions of these conferences become oil policy for the member governments. In 1974 OPEC countries controlled about 70 percent of the world's crude oil reserves, 65 percent of world oil production, and 90 percent of petroleum exports; OPEC resolutions therefore had considerable impact.

Conferences are supported by a full-time staff, including a secretariat, an extensive petroleum library, a legal department versed in petroleum matters, and a technical department that follows technical developments in the oil industry. OPEC selected Vienna as its headquarters site. An economic commission was established in 1964 with liaison to the top levels of the oil departments of the member governments. The economic commission collects data, including apparently very good petroleum statistics, and prepares studies. In early 1976 the studies were not available to the public but appeared to have been the basis for many of OPEC's resolutions since 1970.

OPEC was unsuccessful during the 1960s in restoring crude oil prices to the level preceding the cuts of 1959 and 1960, but it won half the battle in that the oil companies did not make further cuts. OPEC also fostered negotiations with the oil companies to increase the revenues of the oil-exporting states. The negotiations concerned methods of calculating and increasing royalty payments and taxes due the host government that raised the costs of producing crude oil. The added costs at least partly explained the absence of further price cuts in a situation tending toward oversupply in the international crude oil markets.

The payments due the host governments were spelled out in the concession agreements granted the oil companies. The more important concessions had been granted before World War II, when the host governments had little bargaining power. As a result the financial return to the host governments was usually quite low. Venezuela started major adjustments in concession agreements in 1943 by legislating a 50-percent tax on the net income earned by oil companies in the country. The so-called fifty-fifty profit-sharing arrangement, in reality a tax imposed on each barrel of oil produced, was negotiated by Saudi Arabia with the Arabian American Oil Company (Aramco) in 1950, raising Saudi per barrel revenues substantially. Fifty-fifty profit sharing became standard in the Persian Gulf by 1952. OPEC helped refine the calculations of taxes and royalties to the advantage of the host governments during the 1960s. OPEC also helped the oil-exporting governments negotiate substantially better terms in concession agreements granted after its founding.

The June 1967 War between Israel and several Arab states started a chain of events that led to a transformation in the world energy market. The war closed the Suez Canal, and several Arab oil states embargoed oil exports briefly. The canal remained closed until June 1975. The transportation problems of the much longer haul around Africa to Europe were partly offset by the use of new supertankers and by the increasing oil production in Africa, particularly in Algeria and Libya, which were close to southern Europe. Dependence on North African oil became even greater in 1970 as a result of civil war in Nigeria and disruption of operations of the pipeline from Saudi Arabia to the Mediterranean coast in Lebanon. The new military regime in Libya seized the opportunity of its enhanced position to cut back crude oil production and to secure higher posted prices and tax rates from the oil companies under the threat of cessation of exports if its demands were not met. An independent oil company, Occidental, agreed to the much more favorable terms for Libya, and other oil companies acceded to Libyan demands in September 1970.

Meanwhile, in a July 1970 speech, the Algerian representative to OPEC requested that other members seize control of pricing their internationally traded crude oil. OPEC followed with a unanimous resolution in December 1970 calling for substantial revisions and increases in crude oil pricing. Negotiations were held in Tehran in February 1971 for the Persian Gulf region. The posted price of Persian Gulf oil was raised by more than one-third, provision was made for yearly price increases of about 5 percent a year until 1975 to compensate for inflation, and the tax rate was raised to 55 percent. Libya negotiated an agreement in April 1971 whereby it won a substantially larger increase in the posted price plus an increase of the tax rate to 55 percent. Algeria pressed for an increase in crude oil prices above that won by the Persian Gulf but less than that won by Libya. Venezuela legislated rather than negotiated price and tax increases in March 1971.

These assorted actions effectively transferred control over crude oil prices from the oil companies to the producing states. When the United States dollar was devalued in 1971, a clause in the Persian Gulf and Libyan agreements was invoked for new negotiations. An 8.5-percent increase in posted prices was arrived at in Geneva in January 1972 to compensate for the devaluation, and further negotiations took place after another devaluation in 1973.

OPEC members had long discussed having a hand in oil operations, but the oil companies made only token gestures. Discussion switched to taking control of the oil industries in their countries, and it was evident that some members would eventually act. Action came suddenly. Algeria and Libya seized properties of some concessionaires in 1971, and Iraq nationalized some of its fields in 1972. In January 1972 OPEC members from the Persian Gulf began negotiations with the oil companies for a gradual takeover of company operations through participation. Participation—essentially buying part ownership of the operating company—was far less drastic than nationalization, and it had the particular advantage that the services, technical skills, and marketing chains of the oil companies were retained during the takeover. Participation avoided a problem faced by nearly all of the Arab oil states on the Persian Gulf—an acute shortage of highly trained personnel capable of managing their oil fields.

The Saudi oil minister, Shaykh Ahmad Zaki Yamani, led OPEC participation negotiations during several months of hard bargaining in 1972. The result was an immediate 25-percent participation in ownership by the oil states with 51-percent participation by 1983 through annual 5-percent increments starting in 1979. Individual Persian Gulf oil states negotiated actual arrangements, such as compensation to the oil companies for the equity they yielded, as well as the process of phasing the country into marketing so that the companies could continue to meet their sales contracts. Some of the arrangements were not completed until 1974, but they were backdated to January 1, 1973. Participation was generally applied only to the major producing companies; several of the minor companies did not have government participation even by mid-1976 (see specific countries, this ch.).

The Kuwait National Assembly upset the schedule for a gradual increase in ownership when it refused to ratify the 25-percent participation agreement. Some members demanded 100-percent Kuwaiti ownership. Kuwait's negotiators arranged for 60-percent participation in 1974, effective January 1. Other OPEC members and most oil-producing states also negotiated 60-percent ownership effective in 1974. In early 1975 the Kuwait government announced its intention of buying 100-percent ownership in the main producing company, and negotiations were completed in December but backdated to March 1975. Saudi Arabia arranged 100-percent participation in the main producing company in early 1976. Other gulf oil states were negotiating full ownership in late 1976, but some stated their intention of remaining at 60-percent participation, largely because they lacked the expertise and personnel to manage their oil fields.

The October 1973 War between Israel and Arab states created further disarray in oil markets. A selective embargo and phased cutback in crude production was imposed by several Arab countries. No oil-consuming country was seriously hurt by these actions, because the oil companies juggled supplies to keep oil flowing to all countries, but all consumers felt the threat and the supply pinch. Bidding for oil from Iran reportedly reached US$17 a barrel.

OPEC, and particularly the six member states in the gulf, took the opportunity to raise the price of crude oil almost twelve times in 1973 and then raised it by more than 100 percent at the beginning of 1974. The posted price for Arabian marker crude, on which other gulf prices were based, was US$5.12 a barrel in October 1973; the price became US$11.25 effective January 1, 1974. The posted price remained a reference price for assessing taxes and royalty payments. The actual selling price of crude was 93 percent of the posted price, or US$10.46 per barrel. A 10-percent increase decreed by OPEC in September 1975 raised Arabian marker crude to US$12.38 posted price and US$11.51 actual price per barrel. OPEC justified the more than fourfold increase in prices of crude between 1971 and 1975 as necessary to approximate the true cost of energy, compensate the oil producers adequately for depletion of their limited resources, and make up for the long period when oil prices were kept low while prices for manufactured goods continuously increased.

Several times during 1974 the Arab states of the Persian Gulf also increased the royalty rates on each barrel of oil produced and the tax rates on foreign oil companies. The purpose was to diminish the wide differential between the costs to the oil companies of equity oil and oil bought back from the host governments. Equity oil was much cheaper than buy-back oil. As a result of a meeting in Abu Dhabi, the UAE, Qatar, and Saudi Arabia increased the royalty rate to 20 percent and the tax rate on foreign oil companies to 85 percent while dropping the posted price for Arabian marker crude by US$0.40 per barrel effective November 1, 1974. The changes raised the cost of equity oil close to the price of buy-back oil and resulted in an average return to the governments of just over US$10 per barrel. In a December 1974 meeting OPEC members agreed to move toward such a unitary price system. OPEC members and most other oil-producing countries adopted the same royalty and tax rates, which were still in effect in late 1976.

The impact of all these changes on oil revenues of the producing countries was tremendous. Saudi Arabia illustrated the situation for the oil states on the Persian Gulf. Saudi revenues per barrel increased from US$0.22 in 1948 to US$0.89 in 1970. By 1973 they had reached US$1.56 per barrel, and in 1974 they were above US$10.10 per barrel. By the beginning of 1976 revenues were US$11.15 per barrel because of the OPEC price increase in October 1975. Persian Gulf shaykhdoms had a four- to five-fold increase in their total oil revenues between 1973 and 1974 in spite of declining production in several countries. The shaykhdoms suddenly had real wealth on which to base development programs for their economies.

By early 1976 OPEC members had control of crude oil pricing and a large degree of control over production. The members had not given OPEC authority to allocate production among producers, however, and OPEC had no formal means of matching supply to demand. This need arose immediately; world oil consumption declined in 1974 and declined further through much of 1975 because of a recession in the industrialized countries and conservation measures by all oil-consuming nations. Strains were apparent among OPEC members in 1975 because of lower production and declining revenues.

In 1976 OPEC remained a loose confederation of diverse nations that had united to combat the power of the international oil companies. Its diversity may be its salvation in the difficult task of controlling supply to maintain price. There are two main groups within OPEC. The first is made up of oil exporters with large reserves and small populations, such as Saudi Arabia and Kuwait. This group might be called the low-price group, because they were earning more oil revenues in 1975 than they could profitably invest. Immediate higher prices were not very attractive because they would probably intensify development by industrialized countries of new technology and energy sources. The countries' interest was in a moderate price and a slow rate of exploitation so that their reserves would continue to finance their economic development long into the future.

The second group might be called the high-price group. It was made up of such countries as Iran and Indonesia, which had relatively small oil reserves, large populations, and economic potentials other than oil. Because their reserves would not last long, they wanted to maximize current earnings on their limited reserves through high prices. They would then invest the earnings primarily in their own country to develop other resources and industries for self-sustaining growth. Some of these countries pushed for substantially higher oil prices in 1974 and 1975. Iran, for one, wanted OPEC to raise crude oil prices in September 1975 by 20 percent or more.

In 1974 and 1975 OPEC members maintained their united front and even increased export prices in the face of reduced demand for crude oil by importing countries. A few countries cut back crude production to avoid an oversupply that might lead to price cutting by members pressed for funds. Saudi Arabia, Kuwait, and other gulf states shouldered much of the burden of lower crude production and revenues to avoid pressure on OPEC prices in 1975. The Saudi oil minister had said that his country would accept quite low production to support OPEC's prices and a united front but would not lower production so that another oil-exporting country could increase its share of the market.

Saudi Arabia's huge reserves and production potential make it pivotal in OPEC decisions and probably provide veto power in most OPEC discussions, although support from another major exporter would probably be needed to push through Saudi proposals against strong opposition. Saudi officials finally agreed to the September 1975 OPEC price increase of 10 percent, although originally advocating a maximum of 5 percent. The Persian Gulf states and Saudi Arabia supported OPEC during 1975 and 1976, helping to overcome strains in the organization; however, the Saudis continued to argue against higher prices in the immediate future.

Another source of tension in OPEC appeared in late 1975 as a result of the September price increase. It was left to regional groups of OPEC members to determine the actual application of the price increase to the premiums and discounts on each country's crude oil, which varies because of sulfur content, gravity, shipping costs, and other factors. Kuwait tried to arrange a regional meeting for the gulf states, but it fell through, whereupon the country posted its own adjustments to the overall price increase. Iraq took a strong stand against Kuwait's prices, implying that they were undercutting Iraq's prices. The issue of premiums had not been resolved by late 1976 and posed a destructive force to OPEC. Even minor price cutting could destroy the unity of the organization, leaving some oil producers worse off than they were before. The low production costs in much of the gulf gave these countries a distinct advantage over producers with high costs. Price competition could be ruinous for some countries and costly for all.

Organization of Arab Petroleum Exporting Countries

The Organization of Arab Petroleum Exporting Countries (OAPEC) was formed in 1968, largely on the initiative of Saudi Arabia, to provide a supplemental organization for the major Arab oil exporters. Saudi Arabia, Kuwait, and Libya were the founding members. Membership was originally limited to Arab states whose "principal and basic source of national income" was oil. The OAPEC constitution was subsequently amended to permit Egypt and Algeria to join. By 1976 Bahrain, Iraq, Qatar, Syria, and the UAE had joined, bringing the membership to ten countries.

OAPEC has two functions. One is to develop cooperation and promote Arab interests, such as the fund of US$80 million established in 1975 for Arab oil-importing states experiencing foreign exchange difficulties. The second function is to promote joint ventures among interested members in various phases of the oil business. The organization was a juridical entity that could participate in commercial ventures. By early 1976 OAPEC had formed the Arab Maritime Petroleum Transport Company; the Arab Shipbuilding and Repair Company, its first project a dry dock in Bahrain; and the Arab Petroleum Investments Company, to finance petrochemical plants in the Arab world. OAPEC representatives decided in 1975 to set up an Arab energy institute to study alternative energy sources and the best use of oil and to establish a petroleum services company in Libya to study downstream (see Glossary) investments. OAPEC was also working on the question of a single Arab currency.

OIL INDUSTRY OF KUWAIT

First Kuwait Concession

Kuwait is the classic example of the transformation of an oil shaykhdom. Formerly a sleepy maritime village whose population eked out a bare existence in a harsh environment, it emerged in little more than a generation as the wealthiest country on earth on a per capita basis. For centuries oil seepages in the desert indicated oil below, but the amount and quality could only be determined by drilling. The natural seepages in the desert interested oilmen, and in 1911 a company—renamed British Petroleum (BP) in 1954—that was developing oil fields in Iran requested permission to negotiate a concession in Kuwait. The British government refused the request, saying that the area was too troubled. In 1913 the British government sent men to inspect the seepages and make a geological survey. The amir of Kuwait also reaffirmed a previous stipulation that he would grant a concession only to a group recommended by the British government. World War I interrupted another effort by BP to negotiate a concession. By this time the British government had purchased 51-percent ownership in BP as part of an effort to ensure oil supplies for the British navy.

In the early 1920s an unusual character from New Zealand, Major Frank Holmes, appeared in the Middle East seeking oil concessions. He was not an oilman but represented a syndicate that attempted to obtain oil concessions for resale to the highest bidder. Holmes got along well with Arab rulers and obtained several concessions. In 1923 Holmes, competing with BP for a concession in Kuwait, disclosed to the amir of Kuwait the favorable terms he had offered Saudi Arabia to obtain a concession there. The amir then rejected the terms BP had offered, and for several years Holmes and BP negotiated separately for a Kuwait concession. Meanwhile Holmes had won concession rights in the Kuwait-Saudi Arabia Neutral Zone (later partitioned) and in Bahrain. Holmes' syndicate was unable to interest oil companies in the Saudi and the Neutral Zone concessions, and they lapsed. The Bahrain concession, however, resulted in an association between Holmes and the Gulf Oil Company of the United States.

Gulf Oil was short of crude oil sources in the 1920s and sought fields in the Persian Gulf. The company thought prospects for oil good in Bahrain and Kuwait. British treaties with most rulers in the gulf—including Bahrain, Qatar, and Kuwait—had previously kept non-British companies out. Pressure by the United States government, however, helped open an opportunity for American oil companies to participate in a consortium being formed in Iraq in which Gulf Oil had a small share for a short time. The rules prohibited any company in the consortium from undertaking independent action in any part of the old Ottoman Empire. The Red Line Agreement of 1928, which was part of the legal documents setting up the operating company in Iraq, defined the former Ottoman Empire; Kuwait was outside that boundary.

In 1927 Gulf Oil arranged an option agreement with Holmes for his Bahrain concession, but the members of the Iraq consortium were not interested. Gulf Oil then joined with Holmes in seeking a concession in Kuwait where it could act independently. By the early 1930s the United States government was making representations to the British government to give equal treatment to American oil companies trying to establish themselves in the Persian Gulf. The negotiations were long and complicated and probably vigorously pursued by the United States ambassador in London, Andrew Mellon, who had been the active director of Gulf Oil before his appointment. The British resisted until Gulf Oil and BP began to discuss a joint company in 1932. The British approved this solution, opening Kuwait to an American oil company at about the same time another American company discovered oil in Bahrain after getting around the British in a more devious manner (see Oil Industry of Bahrain, this ch.).

Most versions of the original Kuwait concession negotiations portray the amir as stimulating competition between Gulf Oil (allied with Holmes) and BP in order to obtain better terms until the two oil companies joined together and forced much less favorable terms on the amir. In 1975 the final negotiator for BP wrote a history of the negotiations giving a somewhat different view. He said that the amir wanted both American and British participation—affording Kuwait better protection—and that at the time the concession terms were favorable to Kuwait.

The concession agreement was finally signed on December 23, 1934. The concessionaire was the Kuwait Oil Company (KOC), owned equally by Gulf Oil and BP. The KOC was granted exclusive rights to explore, produce, and market oil in Kuwait—including territorial waters extending six nautical miles offshore—for a period of seventy-five years. In return Kuwait received an initial payment of about US$175,000 and a yearly rental of US$35,000 during exploration. The royalty was to be about US$1.10 per long ton of oil produced. Payments were to be in Indian rupees, not gold as stipulated in concession agreements granted by Saudi Arabia, Iraq, and Iran. The amir's demand that a Kuwaiti be on the board of directors of the KOC was rejected. The terms were low in view of the amount of oil subsequently discovered, and the agreement was altered many times over the years to increase the return to Kuwait. The concession was terminated in 1975 as far as it concerned the original companies.

The KOC began operations in 1935 with preliminary surveys. Drilling started in 1936 on the north shore of Kuwait Bay but was unsuccessful. The second attempt, made in 1938 in the desert, hit what was subsequently called the Burgan (Al Burqan) field, one of the largest and most productive fields in the world (see fig. 4). Eight more wells were drilled by 1942 to determine the outlines and capacity of Burgan and to permit construction of ancillary facilities. Because of World War II operations were suspended and finished wells plugged. At the end of the war operations resumed, and pipelines and other facilities were completed to handle 30,000 barrels of crude oil per day. Commercial export of crude oil began in June 1946, and production amounted to 5.9 million barrels, rising to 16.2 million barrels in 1947. Production continued to increase until the 1970s.

The success of the Burgan field encouraged and financed KOC exploration. A number of additional fields were discovered and developed in the eastern part of the country during the 1950s and 1960s. In 1962 the KOC relinquished more than half of its original concession area, primarily in the western half of the country. The bulk of Kuwait's crude production has come from the KOC's eight fields with about 700 producing wells.

Other Concessions

Subsequent concessions contained progressively better terms for Kuwait, partly because of the entrance of small oil companies anxious to acquire crude oil sources and partly because of the activities and exchange of information among oil-producing states. Monetary payments were substantially higher, the duration of the concessions was shorter, schedules for relinquishing underdeveloped areas were established, and opportunities for Kuwaiti participation in the companies were increased.

The American Independent Oil Company (Aminoil) was the successful bidder for Kuwait's rights in the Neutral Zone, receiving a sixty-year concession for exploration and production on June 28, 1948. Aminoil, which was owned by a number of small American oil companies, had a joint operation with the Getty Oil Company, which held the Saudi rights in the Neutral Zone. (Aramco reportedly viewed the terms given Kuwait by Aminoil as excessive and relinquished its concession in the Neutral Zone, which Getty won.) Aminoil started exploratory drilling in 1949 but did not strike oil until March 1953; production started in 1954. Production from the Neutral Zone was shared between the two countries; Aminoil paid royalties and taxes to Kuwait, and Getty paid them to Saudi Arabia. The Neutral Zone was partitioned in 1969 and does not appear on present-day maps, but the partitioning did not affect the concession arrangements.

A group of Japanese companies formed the Arabian Oil Company (AOC), which obtained the concession from both Saudi Arabia (1957) and Kuwait (1958) for exploration and production in the offshore area of the Neutral Zone. An exploitation lease of forty years was granted when oil in commercial quantities was discovered. The AOC started drilling in 1959, and production began in 1961. Some AOC production is from the northern tip of Saudi Arabia's Saffaniyah field, the world's largest offshore field. Saudi Arabia and Kuwait each subsequently bought 10-percent ownership in the AOC.

A concession agreement was granted to the Kuwait Shell Petroleum Development Company in January 1961 in the area offshore of Kuwait proper (excluding the Neutral Zone). Exploration was suspended in 1963 until settlement of conflicting territorial claims by Kuwait, Saudi Arabia, and Iran. In 1962 the Kuwait National Petroleum Company (KNPC) was awarded the concession for an area, largely in western Kuwait, relinquished by the KOC that year. After inviting bids from foreign companies, the KNPC formed the Kuwait-Spanish Petroleum Company in 1967 with a group of Spanish investors known as Hispanoil. Their explorations were unsuccessful.

Production and Facilities

Once production started, it rose rapidly from 5.9 million barrels in 1946 to 125.7 million barrels in 1950 (see table 1). When Iran nationalized its oil industry and international oil companies would not buy Iranian oil in the early 1950s, production in Kuwait was increased to cover much of the demand. By 1955 production had reached 402.8 million barrels a year, more than triple the production in 1950. Some oil from the Aminoil concession in the Neutral Zone became part of annual production in 1954.

Between 1955 and 1970 production increased each year, but the increase varied considerably from year to year. When there was a shortage in the global system, production in Kuwait increased substantially, but in some years the increase was equivalent to only a few days production. Production from Kuwait was eased by the oil companies into world markets without disrupting the system. By 1970 production amounted to 1.1 billion barrels a year. Production from the offshore concession of the AOC began in 1961, but the KOC has always been the principal producer, accounting for over 90 percent of total crude output.

Production continued to rise through 1972, when 1.2 billion barrels were extracted, an average of 3.3 million barrels per day. Meanwhile conservationists in the National Assembly (Assembly) raised questions about the estimates of the country's reserves; some claimed they were about one-third of KOC estimates and that the country needed to place limits on production of crude oil. In 1972 the Assembly placed a limit of 3 million barrels per day on KOC production, which was just met. Kuwait crude is heavy, producing more heating and industrial fuels. KOC production for 1973 was scheduled to meet the 3 million barrels

Table 1. Kuwait, Oil Revenues, Crude Oil Production, and Petroleum Exports, Selected Years, 1946–75

Year[1]	Oil Revenues (in millions of Kuwaiti dinars)[2]	Crude Oil Production (in millions of barrels)[3]	Exports (in thousands of barrels per day)[4]
1946	n.a.	6	n.a.
1947	n.a.	16	n.a.
1950	5	126	n.a.
1955	101	403	n.a.
1960	159	619	n.a.
1965	216	862	n.a.
1970	298	1,091	2,912
1971	354	1,167	3,093
1972	507	1,202	3,214
1973	531	1,102	2,924
1974	2,403[5]	929	2,452
1975	1,687[5]	763	1,800[5]

n.a. indicates not available in sources cited.
[1]Calendar years until 1959. Fiscal years beginning April 1 for 1960 onward.
[2]For value of the Kuwaiti dinar—see Glossary.
[3]Includes Kuwait's share of production from the partitioned Kuwait-Saudi Arabia Neutral Zone.
[4]Exports of crude and products including bunkers.
[5]Preliminary.

Source: Based on information from *Annual Statistical Bulletin, 1974*, Vienna, June 1975, tables 16, 71, 102; and "Kuwait," *Financial Times*, London, February 25, 1976, pp. 16, 19.

per day average by lower production during the summer and increased production in the late fall. Kuwait advocated and imposed the production cutbacks after the October 1973 War, and KOC production averaged 2.8 million barrels in 1973. Total Kuwait crude production amounted to 1.1 billion barrels that year.

Although production cutbacks and embargoes were dropped in early 1974, economic recession and conservation measures in the industrialized countries that buy most Kuwait oil caused crude production to drop to 929 million barrels in 1974, a daily average of 2.6 million barrels from all companies. Preliminary reports indicated that production continued to decline in 1975 to about 763 million barrels for the year, about 45 percent of capacity. Cumulative production from Kuwait's fields had reached about 17 billion barrels by the end of 1975.

Oil fields in Kuwait, as in much of the Middle East, were under considerable pressure, and none of Kuwait's wells required pumping to bring the oil out. When the crude oil reached the surface, considerable quantities of gas were available. The amount of gas varied with the level of crude oil production. In the peak year of crude production, 1972, 648 billion cubic feet of gas was produced, about 10 percent of which was reinjected into wells to maintain pressure. About 28 percent was used to fuel power stations, desalination plants, petrochemical plants, a refinery using the hydrogenation process, and other facilities (see The Economy, ch. 6). Over 60 percent of the gas available was flared because there was no commercial use for it. In 1974, at a lower level of crude production, 467 billion cubic feet of gas was produced, but only 46 percent was flared because the other uses remained relatively constant. Exports of liquefied natural gases amounted to more than 1 million tons.

The bulk of petroleum production was exported as crude oil. In 1974 when production of crude amounted to 2.6 million barrels a day, exports amounted to 2.5 million barrels a day. Western Europe, largely Great Britain and France, bought about 48 percent. Asia was the other major purchaser, taking 42 percent. Japan was by far the most important buyer in Asia, importing 568,000 barrels per day. Singapore was the next most important Asian importer. The remaining exports of crude were shipped throughout the world. The United States bought just under 6,000 barrels a day from Kuwait in 1974. Exports of crude were a higher proportion of total production in 1972, when production peaked and exports amounted to 3.2 million barrels per day.

Each of the producing companies as well as the KNPC has a refinery. Three of the refineries are located in Kuwait proper, and each is on the coast. The largest, owned by the KOC and located at Mina al Ahmadi, had a capacity of 285,000 barrels per day. Production amounted to 130,000 barrels per day in 1974, and peak production since 1968 was 237,000 barrels per day in 1970. Aminoil's refinery was located at Mina Abd Allah. Capacity was 105,000 barrels per day. Production amounted to 85,000 barrels per day in 1974, and peak production since 1968 was 102,000 barrels per day in 1971. The KNPC refinery, located at Ash Shuaybah, had a capacity of 180,000 barrels per day in 1976. Production amounted to 125,000 barrels per day in 1974, slightly less than peak production of 129,000 barrels per day in 1973. Between its completion in 1968 and 1972 the refinery had problems that affected production and profitability. These were corrected by 1973.

Two additional refineries were located in the partitioned Neutral Zone. The AOC refinery at Ras al Khafji had a capacity of 30,000 barrels per day. Production amounted to 18,000 barrels per day in 1974, all of which went to Japan. The Getty Oil Company had a 50,000-barrels-per-day refinery at Mina Suud. The Getty activities were part of Saudi Arabia's petroleum industry, handling that country's share of onshore Neutral Zone production. Offshore sea terminals existed at all the refineries and were capable of handling crude as well as products.

Exports of refined products (including bunkers) amounted to 249,000 barrels per day in 1974 compared with the peak year (since 1968) of 332,000 barrels per day in 1970. Asia bought about half the refined products during the 1970s, and Japan was by far the largest single customer. Ships' bunkers were another important way of exporting refined products. Western Europe, largely the United Kingdom, was a minor market for refined products.

Oil revenues increased more rapidly than crude oil production. Under the original concession agreement the government only received a small royalty for each barrel produced. Government income tripled between 1952 and 1953, largely because of the introduction of a tax on oil company incomes. Some additional improvements in the government's take per barrel of oil produced were negotiated before the 1970s, but they were minor. In 1971 the OPEC nations began rapidly raising the price of crude oil, causing oil revenues to increase substantially. The big change occurred in fiscal year (FY) 1974 when a combination of the very large price increase and the government purchase of 60 percent of the KOC jumped the government's oil revenues from about US$2 billion to more than US$8 billion. Royalties and taxes were no longer the primary source of oil revenues; the main revenue source had become the sale of oil to the former concessionaires and direct to foreign customers.

Government Policy

From the very beginning of the oil industry Kuwait's leaders had wanted to participate in policy and management. BP and Gulf Oil turned aside the amir's demand for a Kuwaiti on the board of directors of the KOC. Kuwait obtained some participation in the AOC concession agreement, but its importance was more symbolic than real.

Frozen out of oil operations by the major oil companies, Kuwait started on its own to develop proficiency in the petroleum industry. The KNPC was formed in October 1960 with the expressed intent that it should become an integrated oil company. Its founding charter allowed it to engage in almost any activity involving petroleum at home or abroad. It began with 60-percent government ownership, the remaining shares being held by private Kuwaiti investors. The government bought out the private investors in 1975.

The KNCP started operations on a small scale, partly because of Kuwait's acute shortage of skilled manpower. It bought out the KOC's local petroleum distribution facilities and became the sole supplier of petroleum products in Kuwait. It participated in foreign refinery operations and established subsidiaries and facilities abroad for marketing petroleum products. Departments for exploration and other aspects of field operations were established within the KNPC to work with foreign companies in the concession area the KNPC had received from the government.

The KNCP also built—with foreign expertise and equipment—a very modern refinery, to use gas that would otherwise have been flared in the Burgan field, employing a hydrogenation process that converts crude into products with sulfur as a useful by-product. The refinery was suited to Kuwait's crude, which is heavy and contains considerable sulfur, and it produced a superior product. The refinery at Ash Shuaybah was completed in 1968, but technical problems caused an unprofitable mix of products for a while. Between cost overruns during construction and a poor range of products, the KNPC lost money for several years until the problems were corrected. Nonetheless the KNPC provided important training in oil company operations for Kuwaitis in upper levels of management.

The Kuwaiti goal of real participation in and control over its oil industry was achieved in 1974 and 1975. In 1974 the government bought 60 percent of the KOC, including the refinery and other installations, reportedly for US$112 million—the net book value of the assets involved. In December 1975—but backdated to March 5, 1975—the government bought the remaining 40 percent of the KOC, reportedly for US$50.5 million. The 1975 purchase arrangements included provisions for the sale of crude to the former owners of the KOC in order that they might honor their long-term contracts. Gulf Oil would receive an average of 500,000 barrels a day and BP 450,000 barrels a day from January 1, 1976, through March 31, 1980, at a discount of US$0.15 per barrel from the prevailing price. BP and Gulf Oil would continue to provide technical services and personnel; some reports indicated that a fee would be charged while others indicated that the price discount would cover services and personnel. The government had also indicated an intention to negotiate complete takeover, possibly in 1976, of the firms operating in the partitioned Neutral Zone, but those negotiations would have to be coordinated with Saudi Arabia.

By early 1976 Kuwait appeared to have major sales contracts for about 2 million barrels of oil a day. Some were with the large international oil companies, and some were with such countries as the Philippines, the Republic of China (Nationalist China), Italy, and Romania. With 950,000 barrels a day to be sold to BP and Gulf Oil, Kuwait appeared to have markets for a rate of crude oil production close to the limit of 3 million barrels per day.

In 1974 the government began a reorganization to handle its growing petroleum business activities. Some false starts were made, such as the formation in 1974 of the Kuwait Oil, Gas, and Energy Company, which was subsequently dissolved. Additional changes were possible, but by late 1976 the government plan appeared to be the following: at the top was a supreme council for oil policy headed by the prime minister; a technical advisory body informed the council and the oil minister of problems and possible solutions and coordinated developments in the industry. The KNPC would be enlarged to act as a holding company over subsidiaries operating in such individual sectors as production, marketing, transportation, and petrochemicals.

In late 1976 the name and form of the subsidiaries under the KNPC were uncertain. There was discussion of retaining the KOC to be responsible for crude production. The KNPC took over responsibility for the KOC refinery at Mina al Ahmadi, however. The government had bought out private interests in the Petrochemical Industries Company (PIC) in early 1976, and the PIC was rumored to be the organization that would take over several large petrochemical plants under construction. In the agreement on the takeover of the KOC, Gulf Oil and BP pledged to use Kuwaiti tankers as much as possible, and an expansion of the tanker fleet was planned. There were reports that the government was about to buy into a private Kuwaiti tanker company and that the KNPC also was planning to buy tankers, leaving uncertain how transportation would fit into the new government petroleum organization.

In 1976 Kuwait petroleum officials continued to express optimism that they would be able to manage their newly acquired assets well. It was true that Kuwait's oil fields were concentrated and operations easier than in many areas of the Middle East. Before it was prorogued in August 1976, the Assembly posed problems, however. It refused to ratify the original agreement for 25-percent participation in the KOC negotiated in 1973, which forced Kuwaiti negotiators back to the table and resulted in the agreement for 60-percent government participation. Some elements in the Assembly were even dissatisfied with 60-percent ownership and demanded 100-percent Kuwaiti ownership, which the government negotiated in 1975.

The Assembly also was responsible for setting a limit of 3 million barrels per day on crude production in 1973 after members reported that recoverable reserves were overestimated. Some assemblymen claimed that reserves were as low as 22 billion barrels, enough for only fifteen to twenty years of production. The government called in foreign experts to estimate reserves, and they supported KOC estimates. Official reserve figures were not available in late 1976, but unofficial estimates of recoverable reserves ranged between 60 and 75 billion barrels, sufficient to last from seventy-five to 100 years at the production rates of the mid-1970s. There were calls in the Assembly in the mid-1970s to limit production to 1.5 million barrels per day, but there is a limit below which production can not drop.

The Kuwaitis have made an effort to use the natural gas associated with crude oil production. Unofficial estimates of natural gas reserves were nearly 32 trillion cubic feet in the mid-1970s, most of which was associated with crude oil. The power plants, desalination plants, the KNPC refinery, and some petrochemical plants ran on the gas produced with oil. Some observers thought crude production needed to be in the neighborhood of 1.5 million barrels per day to run the facilities dependent on gas, but the oil minister indicated in 1975 that crude production needed to be only about 500,000 barrels per day to derive the gas necessary to operate the facilities. The government has planned heavy expenditures on gas-using plants in the late 1970s.

Kuwait had the highest utilization rate of gas (54 percent in 1974) of any major Middle East oil producer, but officials planned much greater use of gas. One project for the late 1970s is to complete a gas collection system, costing the equivalent of about US$1 billion, to bring the gas being flared to industrial centers. A complementary project consists of a large liquefied petroleum gas (LPG) plant to produce 5 million tons a year of butane, propane, and natural gasoline. The LPG plant is scheduled for completion in 1978 at a cost of about US$875 million. Part of the LPG would be exported and part used for feedstock for further processing in additional petrochemical industries to be constructed. The cost for the series of gas projects will probably exceed US$3 billion. The rationale behind the large investment in the gas projects was that it would take advantage of a valuable natural resource that was being wasted, the capital-intensive nature of the production process on the projects would minimize labor requirements in a country having a small work force, and it would broaden the range of exports.

With petrochemicals the government's chosen area for development, large amounts of additional gas will be necessary to feed the plants. The LPG plant will require about 1.7 billion cubic feet of gas per day, far more than was being flared in the mid-1970s. One possibility of meeting future gas requirements would be developing the natural gas field discovered by the AOC offshore of the Neutral Zone. The reserves in the field were not known but were believed large. Because its location was in an area shared with Saudi Arabia, a joint effort by the two countries would be necessary for development. The other possibility for obtaining additional gas would be by increasing crude oil production.

A policy on crude oil production levels had not been necessary since 1973 because world demand had decreased, largely because of depressed economic activity in major oil-consuming nations. Kuwait's crude production and exports in 1975 were more than 40 percent lower than the peak levels of 1972. By 1976, however, worldwide economic recovery was increasing petroleum consumption. The day might not be far off when the demand for Kuwait's oil would push production up against the limit of 3 million barrels per day imposed by the Assembly. Some Kuwaiti petroleum officials, aware of the world's interdependence, favored production above the limit if needed by consumers. Kuwait needs a stable world, partly because of its overseas investments, as much as the world might soon need Kuwait's considerable reserves of oil.

OIL INDUSTRY OF BAHRAIN

In the mid-1970s the oil industry of Bahrain was extremely important to the country's economy, contributing over 70 percent of gross domestic product (GDP), over 80 percent of government revenues, almost 60 percent of export earnings, and more than 4,000 jobs. But to the rest of the world its production, which was small and declining, was of little significance. It was the smallest producer in the Persian Gulf. When oil was first discovered in Bahrain in 1932, however, the discovery had considerable international significance. British treaties with shaykhs throughout the Persian Gulf had kept American oil companies out, but the British oil companies had surplus oil from their operations in Iran and Iraq, which made them less interested in finding additional fields to develop. Moreover seepages in Iran, Iraq, and Kuwait pointed to those countries as more promising areas for exploration than countries lower in the gulf.

Oil was discovered in Bahrain by a convoluted set of circumstances. Holmes' oil concessions in Saudi Arabia and the Neutral Zone had gained him little. In 1925 he obtained a concession in Bahrain, and in 1927 he interested Gulf Oil in the concession. Gulf Oil at that time was a participant in a consortium attempting to exploit oil in Iraq, and the terms of the consortium forbade its members to explore for oil independently in other parts of the former Ottoman Empire. The prohibitions were the legal forerunners of the Red Line Agreement of 1928. When Gulf Oil could not interest other members of the Iraq consortium in the possibilities of the Bahrain concession, Gulf Oil sold its option in Bahrain to Socal in 1927.

It was several years before Socal could begin operations in Bahrain because of the British treaty with the amir. Socal established a company in Canada that subsequently opened an office in Great Britain. In 1930 Socal's option was assigned to its Canadian company, Bahrain Petroleum Company (Bapco), and drilling started. Oil was found in the southern part of the island in 1932. The discovery soon led to concessions—let to other oil companies seeking oil—in Kuwait, Saudi Arabia, and Qatar and in the late 1930s in Oman and Abu Dhabi. In 1933 Socal also gained the concession in Saudi Arabia. In 1936, in order to obtain funds to develop the Bahrain and Saudi concessions, Socal accepted Texaco as an equal partner in a firm subsequently called California-Texas Oil Company (Caltex). Caltex became the owner of Bapco, which operated the Bahrain concession.

A single field, Awali (sometimes called the Bahrayn field) was discovered on Bahrain (see fig. 4). Crude production from this field began in 1935 and increased slowly to a peak of nearly 28 million barrels in 1970, modest production compared with Kuwait's 1 billion barrels the same year. Since 1970 production from the Awali field has been declining and will continue to decline at about 3 percent a year, although improved gas injection techniques slowed the rate of decline after 1972. Production amounted to 24.6 million barrels in 1974 (see table 2). By 1976 reserves were about 300 million barrels, and commercial production from the Awali field was expected to last twenty years at most. In the

Table 2. Bahrain, Crude Oil and Refinery Production and Foreign Trade, Selected Years, 1964–75

Year	Production (in millions of barrels)		Imports of Crude Oil	Exports of Refined Products
	Crude Oil (Awali field)	Refined Products	(in millions of United States dollars)	
1964	18	71	n.a.	n.a.
1965	21	68	n.a.	n.a.
1967	25	84	n.a.	n.a.
1969	28	83	87	184
1970	28	88	95	192
1971	27	90	82	195
1972	26	83	133	221
1973	25	86	192	303
1974	25	90	681	988
1975	n.a.	n.a.	609	895

n.a. indicates not available in sources consulted.

mid-1970s geologists had little hope of finding additional fields on the island, although surveying continued by Bapco and an American company was exploring offshore.

Aramco discovered an offshore field, Abu Safah, in the early 1960s, and production started in 1966. The field extended under waters claimed by Bahrain. The rulers of Saudi Arabia and Bahrain agreed to share production. Aramco operated the field and sent the crude into its own collection system. The Saudi government made payments to Bahrain for its one-half share of production. Production rose rapidly and had reached 34.2 million barrels in 1972. Total production was about 44 million barrels in 1975, almost 85 percent higher than production from Bahrain's field. Production from Abu Safah probably peaked in 1976 and will start declining in a couple of years. Bahrain's income from the Abu Safah field amounted to nearly 50 percent of its total oil revenues in 1974.

Bapco began constructing a refinery soon after oil was discovered, and the refinery began production in 1936. It was enlarged and had a capacity of about 270,000 barrels per day in 1976. Bahrain was unique in the Persian Gulf because it exported no crude oil, only petroleum products. Refinery capacity far exceeded Bahrain's crude production, and imported crude supplied about 72 percent of the input for the refinery in 1974. Nearly all the crude was imported by pipeline from Aramco fields in Saudi Arabia; the refining charge was US$0.06 per barrel in 1976. Small stocks of crude from other areas, notably Iran and Indonesia, were processed in Bapco's refinery. Annual output from the refinery was around 70 to 80 million barrels in the 1960s with a peak of 84 million barrels in 1967. Production remained above 80 million barrels between 1970 and 1974 and peaked at 90 million barrels in 1971 and 1974. Diesel and fuel oils accounted for more than 60 percent of refinery output. Much of the output of the refinery goes to Japan. In 1973 Bapco, with joint financing from a Japanese firm, installed a large unit to remove sulfur from fuel oil, considerably improving its marketability.

Bahrain has large natural gas deposits estimated at 16 trillion cubic feet, most of which is in a large natural gas field not associated with crude oil. The gas is high quality and contains little sulfur. Gas production amounted to about 73 million cubic feet per day in 1968. Most of the gas produced in 1968 was reinjected into the Awali field, and the rest was used at the refinery and the electric power station at Manama, the capital. Gas production increased slowly until 1971, when an aluminum smelter run by electric power cheaply generated using natural gas was completed to process imported materials (see The Economy, ch. 7). Production and consumption of gas increased 75 percent in 1972 because of smelter operations. Average daily gas production reached 274 million cubic feet in 1974.

Government oil revenues in the 1930s and 1940s increased with production because they were based on a royalty of BDO.35 per ton (for value of the Bahraini dinar—see Glossary). The royalty was increased to BD1 per ton in 1951. In 1952 a 50-percent income tax was negotiated with Bapco, following a similar arrangement between Saudi Arabia and Aramco negotiated a short while before. This income tax substantially boosted oil revenues, far exceeding royalties as a source of funds. Bahrain had not joined OPEC as of 1976 (although it was a member of OAPEC), but in 1964 Bapco had agreed to provide Bahrain any benefits that Bapco paid to any other producer in the area. Because Bapco's owners also owned part of Aramco, Bahrain obtained any benefits that Saudi Arabia negotiated. Thus in the 1970s Bahrain's petroleum prices increased with those of other area producers, and the government raised income tax and royalty rates in accordance with OPEC agreements.

The government did not follow the example of Saudi Arabia and other OPEC countries when they negotiated 25-percent participation in the operating oil companies in 1973 because of the country's modest reserves and declining production. The 60-percent participation agreements that followed Kuwait's example were more attractive. The government reached a participation agreement with Bapco in September 1974. The terms included 25-percent government participation in Bapco retroactive for all of 1973 and a 60-percent government share effective January 1, 1974. The government's participation applied only to crude oil production and gas wells, excluding refinery operations. New gas wells were to be 100 percent government owned, however. The government has announced its intention to buy out 100 percent of Bapco production operations, possibly in 1976. Price increases and other OPEC agreements produced an almost sixfold increase in the country's oil revenues between 1970 and 1974 (see table 3).

OIL INDUSTRY OF QATAR

The discovery of oil in Bahrain by Socal prompted British oil companies to head off further inroads into an area they considered their own. The company that became BP sent a negotiator to Qatar to request an oil concession from the amir. At the same time the British Colonial Office urged the amir to make the award to a British company. Pressure from the Colonial Office was thought necessary because Socal and Holmes were attempting to arrange a concession offer. Even though the British shut Americans out, thus avoiding the competitive bidding that took place in Kuwait, BP did not obtain the concession until May 1935. BP immediately assigned the concession to the Iraq Petroleum Company (IPC) for development.

The concession was for seventy-five years and included most of the land area of Qatar. An initial payment of the equivalent of approximately US$175,000 was due on signature. For the first five years the annual rental payment was the equivalent of US$55,000 and thereafter US$110,000. The royalty fee was US$1.10 per ton, although payments were to be made in Indian rupees.

The IPC set up a subsidiary, the Qatar Petroleum Company (QPC), to develop the Qatar concession. The IPC was owned in equal shares of 23.75 percent each by BP, Royal Dutch Shell Oil Company (also a British Company), the Petroleum Company of France (Compagnie Française des Petroles—CFP), and Standard Oil of New Jersey and Mobil Oil Company combined; the remaining 5 percent was held by Participations and Explorations (Partex—a Gulbenkian Foundation). The IPC was not in need of crude oil, and the first well was not started until 1938. Oil was discovered in 1940, but World War II and the IPC's lack of need for additional crude held up the start of production until 1949.

The IPC's wells hit the large Dukhan field on the midwestern side of the peninsula (see fig. 4). The only onshore field discovered as of late 1976, Dukhan is about thirty-four miles long by about five miles wide. It has accounted for most of Qatar's output since production started and in 1974 was producing 223,000 barrels per day, 43 percent of total production. The IPC constructed a pipeline extending across the peninsula to Musayid (Umm Said), on the coast about twenty miles south of the capital, Doha (Ad Dawhah). A deepwater terminal and refinery was constructed at Musayid and a small refinery to produce petroleum products for the local market.

In 1952 Royal Dutch Shell obtained a seventy-five-year concession for offshore exploration and production in the vicinity of Halul Island,

Table 3. Bahrain, Oil Revenues, 1967–74
(in millions of Bahraini dinars)[1]

Year	Bahrain Production			Revenue from Saudi Arabia[2]	Total Oil Revenues
	Royalties	Income Taxes	Total for Local Activities		
1967	3.4	8.8	12.2	1.9	14.1
1968	2.7	9.9	12.6	2.2	14.8
1969	2.8	11.7	14.5	3.0	17.5
1970	2.8	11.6	14.4	2.9	17.3
1971	3.2	13.2	16.4	5.1	21.5
1972	3.3	13.2	16.5	9.5	26.0
1973	—[3]	—[3]	17.6	11.8	29.4
1974	—[3]	—[3]	51.0	48.0	99.0

[1]For value of the Bahraini dinar—see Glossary.

[2]Payments made by the Saudis for Bahrain's share of production from the Abu Safah field.

[3]Not available. The participation agreement negotiated in late 1974 and ratified in 1975 meant that royalties and income taxes were greatly diminished, and most of the government's revenues were produced from sales to the Bahrain Petroleum Company.

Source: Based on information from "Bahrain Enhances Role as a Persian Gulf Center of Communications and Commerce," *IMF Survey*, Washington, December 15, 1975, pp. 372–374.

sixty miles east of Doha. The Shell Oil Company of Qatar (SCQ) was the operating company for the concession. (An Italian firm may subsequently have bought into the SCQ.) The first two wells were unsuccessful. A storm in 1956 wrecked the drilling rig, and drilling did not recommence until December 1959. A small field was discovered in 1960 and a larger field in 1963. Some shipments began from temporary facilities in 1964, but regular commercial production did not start until 1966 after pipelines, storage facilities, and a sea terminal were completed on Halul Island. In July 1974 production from three fields was about 296,000 barrels a day of high-quality oil averaging 36.5 degrees on the American Petroleum Institute (API) gravity scale. The higher the degree, the more such light and expensive products as gasoline are obtained from the crude; 34-degree crude was the standard in the Persian Gulf.

In the late 1960s an offshore field, Al Bunduq, was discovered that extended into the territorial waters of Qatar and Abu Dhabi. A separate company owned by British, French, and Japanese oil concerns was formed to develop the concession. Production started in 1976 and was shared between Qatar and Abu Dhabi. As of late 1976 the size of the field was not publicly known, but the crude was very light, 41 degrees API, and contained relatively little sulfur, making it very marketable.

Additional onshore and offshore concessions were let to other oil groups in 1963 and in subsequent years. The concessions were primarily areas relinquished by the two major concessionaries, IPC and SCQ. Oil in commercial quantities had not been discovered, but in 1976 a group of European and American companies found promising signs offshore. Substantial flows of both gas and oil were found, but further drilling was required to determine the extent of the petroleum deposits.

Although crude production started in 1949, output amounted to only 730,000 barrels. Production in 1950, the first full year, amounted to 12 million barrels (see table 4). Output rose rapidly until 1959, when production declined by about 3 percent. It remained at this level until 1962, when the upward trend resumed. By 1970 crude production amounted to 132 million barrels, an average annual increase of nearly 13 percent between 1950 and 1970. Production increased substantially in the next three years, and 1973 output amounted to 208 million barrels, an average of 570,000 barrels per day. Output declined by 9 percent in 1974 to 189 million barrels (518,000 barrels per day) because of a ceiling imposed by the government. Capacity was about 600,000 barrels per day. About 212 billion barrels of oil had been extracted from Qatar's fields since 1949, nearly all of which was exported as crude oil. Remaining reserves were estimated at about 5.8 billion barrels, enough to last from thirty to forty years.

Qatar had exported no refined products by 1976. Local consumption was small; so almost all of the crude oil produced was exported. Western Europe was the biggest market, taking 57 percent in 1974, and Great Britain was the most important buyer. France, Italy, and the Netherlands were also important purchasers. Thailand was the only significant market in Asia, accounting for about 10 percent of exports in 1974. Exports to the rest of the world were small. The United States imported little oil from Qatar.

Qatar joined OPEC in 1961 and OAPEC in 1970, and the government implemented OPEC decisions on prices, taxes, and royalty rates. In January 1973 the government signed agreements to buy 25-percent participation in each of the two producing companies. In February 1974 agreements were signed for the government immediately to acquire 60 percent of the operating companies. The companies agreed later in 1974 to buy back 60 percent of the government's share of crude production at an average of 93 percent of the posted price. The government could sell its remaining 40 percent of crude production back to the operating companies on the same terms if it did not find markets elsewhere. The participation agreements with the buy-back provisions, along with the sharp increase in prices, tremendously increased government revenues, from the equivalent of US$463 million in 1973 to US$1.8 billion in 1974.

In December 1974 the government announced its intention of purchasing the remaining 40 percent of the two operating companies. Negotiations were completed with the QPC for complete government ownership on September 16, 1976. The terms reportedly included provisions for a US$0.15 per barrel fee to be paid to the former shareholders of the QPC for continuing assistance in operating the fields. Journalists

Table 4. Qatar, Crude Oil Production and Oil Revenues, Selected Years, 1950–75

Year	Crude Oil Production (in millions of barrels)	Oil Revenues (in millions of United States dollars)*
1950	12	1.1
1955	42	34.2
1960	64	54.6
1965	85	69.2
1967	118	106.7
1969	130	117.6
1970	132	112.4
1971	157	199.3
1972	176	255.0
1973	208	463.0
1974	189	1,802.7
1975	159	n.a.

n.a. indicates not available in source cited.
*Converted from British pounds sterling at International Monetary Fund average yearly exchange rates.

Source: Based on information from *Annual Statistical Bulletin, 1974,* Vienna, June 1975, pp. 25, 150.

reported that the government wanted the agreement backdated to December 1974, which would yield it perhaps US$50 million additional revenue. Negotiations were expected to be completed with the SCQ before the end of 1976. The country's extreme shortage of highly trained personnel probably would require provisions for the SCQ's former owners to continue to help in operations.

In 1972 the amir signed the law to establish the Qatar General Petroleum Company (QGPC), but it was not formed until 1974. Its charter empowers it to undertake almost any activity involving gas or oil. The QGPC holds and manages the government's share in the operating oil companies. Presumably the QGPC also has the responsibility for marketing the government's share of oil, but it will take time to develop the staff and contacts to market more than 100 million barrels of oil annually.

The government has made an effort to emerge from the status of a primitive oil-producing country. A small refinery with a capacity of 6,000 barrels per day was completed at Musayid in 1975 to produce products for the domestic market. Its capacity is expected to triple in the future. An export refinery with a capacity of 150,000 barrels per day was under construction in 1976.

A much more costly effort was under way to use the country's natural gas reserves. These reserves were unofficially estimated at 7.5 trillion cubic feet, mostly associated with crude oil. The four oil fields gave up about 560 million cubic feet of gas per day. In 1972 less than 20 percent was used and the rest flared. The proportions probably remained basically the same in 1976. Like other producers in the region the government wanted to use rather than waste this energy resource.

A liquid petroleum gas plant was completed in 1975 to use a part of the gas associated with crude oil production from the onshore field. The initial input was 40 million cubic feet of gas per day, less than 15 percent of the gas the field produced. Output at full production is supposed to be 400,000 tons of propane, 150,000 tons of butane, and 150,000 tons of natural condensates. An agreement was reached with a Shell subsidiary in 1975 to form the Qatar Gas Company (70-percent government ownership and 30-percent Shell) to construct a gas collection system from the offshore fields. Agreements were signed with French firms for two petrochemical plants. One that would produce ethylene from local feedstocks was scheduled for completion at Musayid in 1978. The government's share of the plant was 80 percent. A second plant was under construction in France with minority Qatar equity. An associated tanker company (with 60-percent government equity) was to transport Qatar's petrochemical exports.

Other projects under discussion were more tenuous and costly. The SCQ had found an extremely large natural gas field in its offshore concession area in 1974. As of late 1976 the size had not been announced, but it was described as one of the world's largest by the government, which planned to begin production in 1981. Plans also included two large plants to liquefy gases for export from the natural gas field and from the petroleum-associated gases from offshore oil fields. There also were plans to develop industries to use the gases, such as electric power and desalination plants. Preliminary work and some contracting for equipment was under way by mid-1976 for an integrated iron and steel complex with an output of 400,000 tons a year; the plant would use natural gas in a direct reduction process. There might be slippages in completing schedules, problems with inadequately trained personnel, and other hitches in the plans, but the government was serious about making use of its hydrocarbon resources —particularly the gas that was still being flared.

OIL INDUSTRY OF THE UNITED ARAB EMIRATES

The main developments in the petroleum industry in the UAE predated the federation of shaykhdoms in 1971, and even in 1976 oil policy and revenues remained in the hands of the individual amirates (see ch.2; The Economy, ch. 9). The federation has a petroleum ministry, but it has been largely a paper organization that was expected to become effective only when and if the federation concept takes hold. Meanwhile the industry had to be viewed as a separate structure in each amirate, although crude production in 1974 was almost exclusively (98 percent) from Abu Dhabi and Dubai. Oil production in Sharjah did not start until July 1974. In 1975 UAE reserves were about 30 million barrels—primarily in Abu Dhabi—sufficient to last about sixty years at the 1975 rate of production.

Abu Dhabi

The IPC negotiated a seventy-five-year concession agreement with Abu Dhabi in 1939, four years after obtaining the concession in Qatar (see Oil Industry of Qatar, this ch.). The IPC established the Abu Dhabi Petroleum Company (ADPC) as its wholly owned subsidiary to develop the concession. The ADPC's concession covered much of Abu Dhabi's onshore area, but as of the mid-1970s few details of the concession agreement had been published. Oil was discovered in 1960, and production and export started in 1963. The fields lie inland, and the crude oil is transported by pipeline some seventy miles to tank farms and a tanker terminal at Az Zannah, about 140 miles west of Abu Dhabi town, sometimes referred to as Abu Zaby (see fig. 4). The ADPC was the major producer in the UAE, averaging 918,000 barrels per day in 1974, 65 percent of Abu Dhabi's total output.

In 1953 a concession was granted to the Abu Dhabi Marine Areas Company (ADMA) for a large offshore area for fifty-five years. ADMA was owned by BP (67 percent) and the CFP (33 percent). A group of Japanese firms bought part of BP's equity in 1972. ADMA made its first commercial strike in 1958, and production and export started in 1962. Das Island became the center for offshore operations. Submarine pipelines led from fields, located some distance away, to the island, where storage facilities and a shipping terminal were constructed. ADMA was Abu Dhabi's second major producer, averaging 462,000 barrels per day in 1974.

In the mid-1960s the ADPC and ADMA relinquished portions of their concessions. The amirate then granted concessions to four additional companies, largely from the relinquished areas. The concessionaires represented many industrialized countries. The Abu Dhabi Oil Company (ADOC), owned by Japanese interests, discovered oil offshore in 1969 and commenced production and export in June 1973. Daily production averaged 5,600 barrels in 1973 and 14,000 barrels in 1974, although production of about 100,000 barrels had been expected. By early 1976 production was about 20,000 barrels a day. The Al Bu Koosh Oil Company was formed in 1973—with 51-percent French ownership and several English and American minority owners—to develop an offshore field discovered by ADMA in 1969. Production started in June 1974 and averaged 16,000 barrels a day. Production in the early months of 1976 averaged between 70,000 and 80,000 barrels per day.

The Abu Dhabi National Oil Company (ADNOC) was formed in 1971 as an amirate-owned company for petroleum-related business. Since 1973 it has supplied and marketed petroleum products in Abu Dhabi through a subsidiary. A small refinery with a capacity of 15,000 barrels a day was completed near the capital in April 1976 to produce products for the local market, replacing the need for imports from Kuwait. ADNOC also established an oil tanker company that bought its first tanker in 1975. ADNOC invested in foreign refineries, petrochemical plants, and the Egyptian pipeline from the Red Sea to the Mediterranean. ADNOC handles the amirate's interests in the various equity arrangements in concession agreements.

Production of crude oil rose rapidly through 1974 (see table 5). The average rate of increase was nearly 25 percent per year between 1964 and 1974. Abu Dhabi cut back production in the last two months of 1973, joining other Arab countries in the use of oil as a weapon after the October 1973 War, but production had been so high earlier in the year that output over the full year showed a substantial increase. The rate of growth was modest in 1974, partly because the authorities restricted the operating companies to a ceiling of an average of 1.3 million barrels per day during the latter part of the year. Technical problems in the fields were the main reasons for the ceiling, and in 1975 outside experts were brought in to conduct reservoir studies.

The world recession reduced demand for oil in 1975. The amirate produces a very light crude with low sulfur content, which therefore commands a premium price. The combination of world recession and the tremendous rise in the price of crude oil in the 1973–74 period prompted buyers to seek cheaper crudes. As a result production in February 1975 averaged only 728,000 barrels per day, less than half the average daily production in 1974. Abu Dhabi obtained permission from OPEC to cut its premiums and thus reduce the price by US$0.50 a barrel. This helped sales, and the ceiling on production was raised to 1.5 million barrels per day for the March–December 1975 period. The production limit in early 1976 was 1.6 million barrels per day, probably about 85 percent of existing capacity. The limit on companies was 1 million barrels per day for the ADPC, 450,000 barrels per day for ADMA, and 150,000 barrels a day combined for the two smaller producers. Until early 1976 Abu Dhabi had no refinery capacity, and all crude production was exported.

Although over one-half of Abu Dhabi's crude oil exports went to Western Europe, Japan was the largest single customer. In 1974 Japan bought 36 percent of Abu Dhabi's crude oil exports, France 22 percent, the United States 9 percent, and Great Britain 8 percent.

Abu Dhabi became a member of OPEC in 1966. In 1971 it became a part of the newly formed UAE, and the federation assumed the membership in OPEC, but Abu Dhabi remained the most active of the amirates in OPEC affairs. Abu Dhabi was thought to have amended older concessions to match OPEC guidelines and to have incorporated them in new concession agreements. The amirate has implemented the price changes and increased taxes and royalty rates that have applied in the Persian Gulf region since 1970.

Table 5. Abu Dhabi, Crude Oil Production and Oil Revenues, 1962–75

Year	Crude Oil Production (in millions of barrels)	Oil Revenues (in millions of United States dollars)[1]
1962	5	2.0
1963	18	6.4
1964	68	12.3
1965	103	33.3
1966	131	99.7
1967	139	110.9
1968	182	153.4
1969	229	190.8
1970	284	229.2
1971	387	436.4
1972	440	551.0
1973	559	692.0
1974	613	3,480.0[2]
1975	511	3,800.0[2]

[1]Converted from British pounds sterling at International Monetary Fund average yearly exchange rates.
[2]Unofficial figures reported by journalists.

Source: Based on information from *Annual Statistical Bulletin, 1974*, Vienna, pp. 27, 152; and James Buxton, "United Arab Emirates: Seeking a Formula for Unity," *Financial Times*, London, May 10, 1976.

Abu Dhabi acquired 25-percent ownership in the two major producing companies, the ADPC and ADMA, effective January 1, 1973. In September 1974 Abu Dhabi concluded 60-percent participation agreements with the two main producing companies effective for all of 1974. ADNOC was responsible for the amirate's equity shares. In 1973 and 1974 ADNOC was able to market only a small share of the government's participation oil, and most was sold back to the operating companies under buy-back provisions. Additions to ADNOC's marketing staff of Algerians from that country's national oil company, plus a more aggressive outlook and the reduction of quality premiums, improved ADNOC marketing in late 1975. The company was reported to be ready to request 40 percent of its allocation in 1976, and some officials thought they could market the amirate's full 60 percent of production.

In late 1974 Abu Dhabi announced its intention of taking over 100-percent participation in the two major producing companies. The amirate moved slowly, however, and some observers thought that Abu Dhabi would move toward greater participation only when forced to by events. The amirate had few people trained in petroleum operation, had established a national oil company only a short time before, had considerable oil prospecting yet to be accomplished, and had substantially more technical problems to cope with in its field operations than such other Persian Gulf oil states as Kuwait. Experts suggested that the amirate needed the oil companies not only for management but also as investment partners, particularly in the offshore fields where heavy expenditures were scheduled for the late 1970s. Although there were good reasons for not taking over the operating companies, it would be difficult for Abu Dhabi to hold out after other area producers assume 100-percent ownership.

By early 1976 Abu Dhabi had no participation in the two newest oil-producing companies. The concession agreement with ADOC permitted 51-percent government equity. The company was still operating at a loss in 1975, and the amirate had not exercised its option. Officials had expressed no interest in participation in the Al Bu Koosh Oil Company, although its production was more respectable.

As for future developments, a feasibility study was completed for a 250,000-barrel-per-day export refinery, and in 1976 contracting was awaiting official approval. Construction would require about four years. ADNOC was charged with planning an industrial complex for development over a fifteen-year period, built around a large plant processing gas associated with crude oil production.

Like many other oil fields in the area, most of Abu Dhabi's fields had sufficient gas pressure to bring the crude oil to the surface. Only a few wells required pumps. UAE gas reserves in 1975 were unofficially estimated at 15 trillion cubic feet, the bulk of which was in Abu Dhabi and was associated with crude production. In March 1976 Abu Dhabi proclaimed all gas—present and future—the property of the government. ADNOC was given the task of exploiting the gases, either alone or as a majority partner with foreign companies. The announcement merely formalized what had been apparent earlier.

Construction of a gas liquefaction plant was started in 1973 on Das Island to use the gases being flared from offshore crude production. The plant was designed to produce about 2 million tons per year of liquid natural gas, about 1 million tons of liquid petroleum gas, 220,000 tons of light distillates, and 230,000 tons of sulfur. Cost of the plant probably exceeded the equivalent of US$500 million. Commissioning was scheduled for late 1976 with the first shipment by the end of the year or early 1977. ADNOC had 51-percent ownership of the company that would operate the plant. The minority partners were the foreign firms in ADMA, and BP supervised the project. A Tokyo power company had contracted for the output for a twenty-year period at a floor price of slightly below US$2 per 1 million British thermal units (BTUs), although the actual price would be related to the price of crude oil. The pricing arrangements were renegotiated in 1975 to bring them into line with present-day energy prices because the first pricing arrangements had been made in the early 1970s. ADNOC joined with Japanese investors to form a tanker company (ADNOC had 51-percent ownership) that would operate a fleet of tankers in the future to transport the plant's output to Japanese consumers.

Agreement was reached in April 1976 between shareholders of

ADPC and the government to start a project to collect and process the gases from onshore crude production. At capacity production of 1.3 million barrels of crude per day, these fields would probably yield over 1 billion cubic feet of gas per day, enough for an estimated 185,000 barrels per day of natural gas liquids. The project probably would cost in the neighborhood of US$775 million. Negotiations had been prolonged and tense as the government threatened to go ahead on its own. Two major problems existed. The foreign oil companies insisted that the bulk of the funds would have to be borrowed, an attitude the government interpreted as a lack of commitment on the part of the companies. Eventually the government accepted the idea of borrowing, partly because it needed the expertise and personnel of the companies and partly because the companies apparently convinced the government that they lacked the funds for the project. The second problem related to the government's price for the gas to be sold to the plant; this issue was reportedly resolved in late 1976. This plant will be the hub of an industrial complex for which ADNOC had earlier been instructed to draw plans.

Dubai

The Dubai Petroleum Company (DPC), owned by a multinational group, obtained a fifty-six-year concession in 1963 for offshore exploration and production. Oil in commercial quantities was discovered in 1966 in the Fateh field, and production and export started in 1969. A second field, southwest of Fateh, began producing in 1972. Annual crude production amounted to 4 million barrels in 1969, 31 million barrels in 1970, 46 million barrels in 1971, 56 million barrels in 1972, 80 million barrels in 1973, and 88 million barrels in 1974. Production in 1975 was about 93 million barrels (254,000 barrels per day) in spite of a blowout and fire that hampered production for part of the year. Production capacity was about 330,000 barrels per day by 1976. Dubai had no refinery facilities, and all production was exported as crude oil. The DPC was the only producing company in Dubai. Another concession for onshore and offshore exploration was granted in the early 1970s to a group of companies, but they had not discovered oil by late 1976.

Dubai did not publish information about its concession arrangements or its oil revenues, but foreign observers assumed that the ruler was paid on the basis of OPEC rates. If that assumption was correct, Dubai's oil revenues probably were the equivalent of about US$140 million in 1973, about US$690 million in 1974, and approaching US$1 billion in 1975.

Dubai startled its neighbors in 1975 when it announced that it was taking 100-percent control of the DPC. It would have been the first area government to assume full control of an oil company. The facts were not clear by mid-1976, but the announcement appeared to be an overstatement. From the information available Dubai did not have any participation (equity) in the DPC but had paid or lent the company the equivalent of US$110 million—equal to past company investments—because the company was short of funds. The loan was needed because of a series of mishaps that had kept crude oil production below profitable levels. According to press reports the government did not acquire any equity for its funds, and the DPC's owners were expected to continue making investments for development without a similar responsibility on the part of the government. The arrangements included some provisions for rapid amortization by the DPC, which may have helped the company's earnings per barrel.

Dubai had about 130 million cubic feet of associated gas from its offshore fields. In 1976 a collection system and a liquid petroleum gas plant were under construction; completion was scheduled for late 1978. An aluminum smelter, also under construction, would absorb almost all the gas produced.

Other Emirates

Sharjah was the only other oil-producing amirate in the UAE. In 1969 the amir granted a concession for offshore exploration and production for a period of forty years. Oil was discovered off Abu Musa Island in 1973, and production began on July 18, 1974. The concession was held and operations conducted by the Crescent Petroleum Company, owned by a group of minor American oil companies. In 1974 production from the Mubarak field was 8.5 million barrels (a daily average of 27,500 barrels). It was expected that production would be about 60,000 barrels per day in 1975, but preliminary reports indicated that it was closer to 40,000 barrels per day because of technical problems in the fields. A new well completed in 1975 should have raised production to 50,000 barrels per day or more by early 1976. Production was shared with Iran because of conflicting territorial claims (see Geographic and Demographic Setting, ch. 9).

Sharjah's oil receipts were estimated at the equivalent of US$28 million in 1974. Journalists reported that the ruler of Sharjah paid 30 percent of his revenues to Umm al Qaywayn, another amirate, for its claims in the Abu Musa area.

About ten additional concessions had been granted by Sharjah and the remaining amirates for onshore and offshore activities, but as of 1976 oil in commercial quantities had not been found. Umm al Qaywayn announced the discovery of oil in July 1976, but its commercial value had not yet been assessed. Oil had been found in Ras al Khaymah in 1971 but not in commercial quantities, and encouraging signs were encountered during drilling in mid-1976, creating considerable optimism.

OIL INDUSTRY OF OMAN

Discovery of oil was late in Oman, and proven reserves were small, but the petroleum industry dominated the economy (see The Economy, ch. 10). The oil sector contributed about 68 percent of the GDP, about 99 percent of foreign exchange earnings, and about 97 percent of government revenues in the mid-1970s. Oil production was expected to peak in 1976, however, and begin a gradual decline thereafter unless new fields were discovered. Proven reserves were little more than 2 billion barrels and were not expected to last more than twenty years. Thus it was imperative that the country use its oil reserves to develop other sectors of the economy before this source of development funds was gone.

Oil prospecting began in 1937 when the first concession agreement was signed with Petroleum Concessions, a company formed by the owners of the IPC. In 1951 the concessionaire's name was changed to Petroleum Development (Oman) (PDO). After several years of costly and unsuccessful exploratory drilling, most IPC partners wanted to get out. In 1960 Royal Dutch Shell bought 85 percent of the PDO, and Partex held the remaining 15 percent. In 1967 the CFP bought 10 percent of the PDO from Partex. In December 1973 the Oman government, following the participation agreements negotiated by several Persian Gulf OPEC countries, acquired a 25-percent share of the PDO. In July 1974 the government's share was raised to 60 percent effective January 1, 1974; in mid-1976 the government's participation remained at 60 percent. In 1976 a management committee consisting of two government and four oil company representatives operated the PDO.

Most of Oman's land area was included in the original 1937 concession. In 1951 the PDO relinquished its rights in Dhofar Province, and the sultan then assigned the rights to a friend, an American archaeologist unfamiliar with the oil industry, who nevertheless was able to interest several American oil companies in the concession. Oil was discovered in 1957 in eastern Dhofar, but it was too heavy for fuel and was primarily suited for road construction. In isolated Dhofar it had no commercial value. By the late 1960s more than US$50 million had been invested in this concession without any return. In 1969 the PDO again obtained concession rights in eastern Dhofar Province and subsequently discovered good quality oil but not in commercial quantities. The PDO was continuing exploration in 1976.

Prospecting in western Dhofar had been hampered by tribal unrest and rebellion, but two concessions were granted in 1976 after the rebellion was reportedly suppressed (see National Defense and Internal Security, ch. 10). The competition for the two concessions in western Dhofar was strong, indicating that international oil companies thought the prospects good.

Concessions were granted to several other international oil companies in the 1960s and 1970s. A French group made a promising find in 1975 in Iranian waters in the Strait of Hormuz; the well was one of the deepest in the Persian Gulf (12,000 feet). Additional drilling was scheduled to start in nearby waters in 1976, and the company expected to find oil. Production would have to be shared with Iran if the field extended across the demarcation line.

The Sun Oil Company, an American firm operating on behalf of a consortium of companies in a concession area offshore south of Masirah Island, found sufficiently encouraging geological signs to continue drilling in 1976. Another group of companies had been unsuccessful in drilling offshore of the Batinah coast. A French-Japanese group were conducting preliminary surveys of the Musandam Peninsula in 1976. Oman was not a member of OPEC or OAPEC but closely followed OPEC decisions concerning concession terms and participation arrangements.

Oil in commercial quantities was discovered by the PDO in February 1964 about 150 miles southwest of Muscat, the capital. The main fields were located at Fahud, Natih, Yibal, and Huwaisah behind the Hajar Mountains in the Rub al Khali (Empty Quarter). With the development of the fields, the PDO constructed a pipeline to Mina al Fahal, where a tank farm and a deepwater offshore loading terminal were established. In 1974 the pipeline capacity was 385,000 barrels per day, the tank farm capacity was 3 million barrels of crude oil, and the sea terminal consisted of three offshore single-buoy moorings with a capacity of 8,000 tons per hour. Capacities were probably unchanged by mid-1976, although storage tanks were being constructed to hold imported refined products for bunkering ships or reexport.

Production and regular export of crude oil began in August 1967. Production and export amounted to 20.9 million barrels in 1967 and 87.9 million barrels in 1968 (see table 6). Production of crude oil climbed to 121.3 million barrels in 1970. Production declined in 1971 and 1972 because of a blowout in the Yibal field that proved difficult to control, declines in reservoir pressures in other fields, and other technical problems. Not until 1975 did production and export of crude oil slightly exceed the 1970 level as a result of bringing new fields at Ghaba on stream. Production in 1976 was expected to increase to about 139 million barrels, all by the PDO, the only producing company.

Japan and the Netherlands were the main purchasers of Oman's crude oil, accounting for 38 percent and 20 percent respectively in 1975. Trinidad, France, Great Britain, and Singapore each took from 6 to 7 percent of exports in 1975. United States imports amounted to just under 6 percent in 1975. Because all petroleum exports were crude oil, exports to refining centers such as the Netherlands, Trinidad, Singapore, and Italy were not necessarily for use in those countries.

Although Oman was not an OPEC member, the government implemented most OPEC decisions regarding prices, royalties, income taxes, and participation. For example, Oman's posted price was changed nine times in 1973 alone, reflecting the changes made by Persian Gulf OPEC members. In mid-1976 Oman's income tax rate on oil companies was 80 percent, the rate for royalties was 20 percent, and government participation in the PDO was 60 percent, the same as in many OPEC countries.

Income tax payments by the PDO had been the main source of oil revenues until 1974. Participation agreements with the PDO gave the government a 60-percent share effective January 1, 1974. In 1974 the government paid the PDO the equivalent of US$110 million for its share and paid an additional US$60 million for company operations during that year. In return the government received 60 percent of the crude oil produced, the bulk of which was sold back to the PDO for marketing abroad, forming the main source of oil revenue. The government's direct sales of crude oil to foreign countries amounted to less than 6 million barrels in 1974, but sales more than doubled to 14.2 million barrels in 1975. The participation agreements sharply diminished the importance of royalties and taxes as the main sources of oil revenues.

Omani crude oil is good quality. API gravity ranged between 30 degrees and 34 degrees with less than 1-percent sulfur content and little wax. Shipping distances from Oman to major markets were shorter than from other oil exporters in the Persian Gulf. Omani fields required considerable gas and water injection to maintain reservoir pressure, however. This factor, in conjunction with small fields and extensive pipeline requirements, made Oman's crude among the most costly to produce in the Middle East.

In the early 1970s government officials began to think about major industrial projects based on hydrocarbon resources. Plans were pre-

Table 6. Oman, Petroleum Statistics, 1967–75

Year	Crude Oil Production	Crude Oil Exports	Oil Revenues (in millions of United States dollars)
	(in millions of barrels)		
1967*	21	21	5
1968	88	88	61
1969	120	119	93
1970	121	121	107
1971	107	106	115
1972	103	103	128
1973	107	107	173
1974	106	106	844
1975	125	125	1,070

*Production began August 1967.

Source: Based on information from "Oman: A MEED Special Report," *Middle East Economic Digest*, London, June 1976, pp. 7, 9.

pared for a refinery as a logical development of the Mina al Fahal oil port. The prospect of declining crude oil production (unless additional petroleum were discovered) and budget constraints that appeared in 1974 held up construction. The refinery was under study in 1976 and its future uncertain.

Additional industrial projects were proposed, largely based on natural gas. More than 130 million cubic feet of gas per day was associated with crude oil production. The gas had a high calorific value, was about 69 percent methane, and contained no hydrogen sulfide. Except for some power generation in the oil fields, the gas was flared. Plans had been developed and pipe bought to bring the gas over the mountains to the coast when the budgetary problems of 1974 postponed further industrial development. The scheduled projects called for the gas to be used in a urea fertilizer plant, a gas liquefaction plant, a cement mill, and similar plants. By mid-1976 negotiations were reportedly under way for construction of the pipeline with financial help from Abu Dhabi. Construction of the cement mill appeared likely; the kiln could be fired by oil if the natural gas pipeline project were not undertaken. A desalination and electric power plant will also be powered by natural gas if the pipeline is constructed. Other projects were under discussion in 1976, but none appeared imminent.

* * *

The amount of information on the oil industry in general and on specific areas is quite large and is growing rapidly. Morris A. Adelman's *The World Petroleum Market* is a basic study of the industry, although somewhat overtaken by events since 1973. *Power Play* by Leonard Mosley presents a very readable account of the intrigues and issues in the initial concessions granted by the Persian Gulf oil states. Archibald Chisholm's *The First Kuwait Oil Concession Agreement* provides a detailed description of the negotiations for that important concession. OPEC's *Annual Statistical Bulletin* provides a considerable amount of data on the industry of members in the Persian Gulf. The *IMF Survey* and the *Financial Times* (London) periodically publish review articles on individual Persian Gulf states that contain information on the whole economy, including recent developments in the oil sector. Most of the oil states publish statistical yearbooks that present data on the oil industry, although often after a considerable time lag. (For further information see Bibliography.)

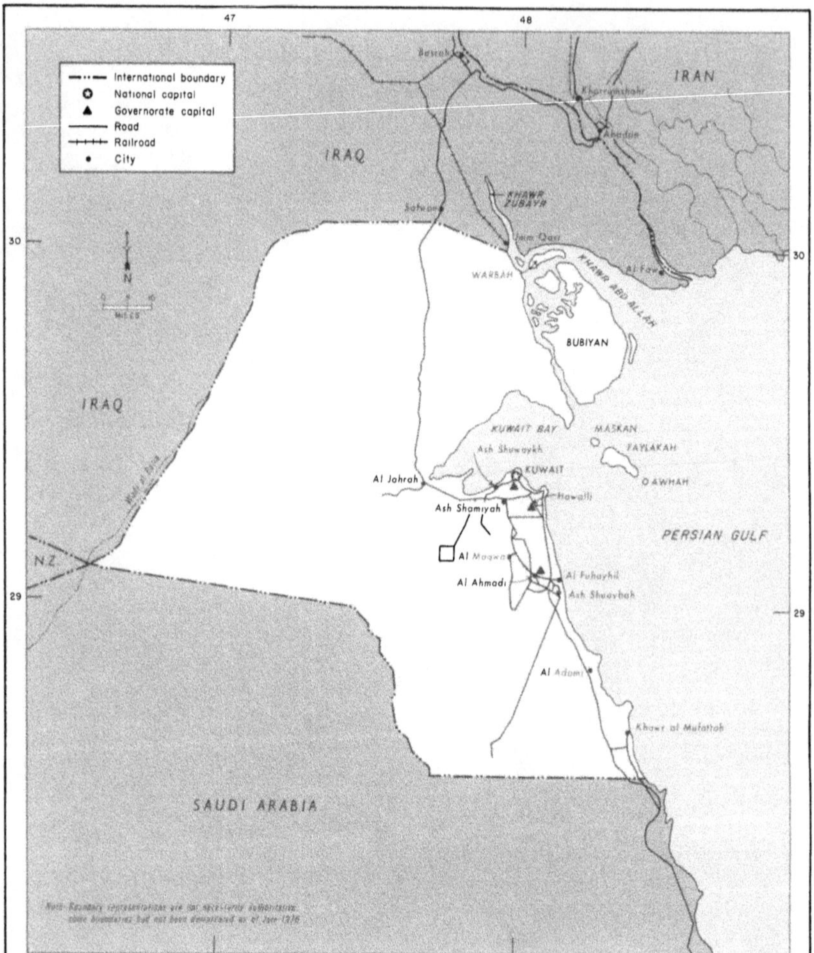

Figure 6. Kuwait, 1976

CHAPTER 6

KUWAIT

At the end of World War II Kuwait was a poor, traditional kingdom whose people earned their living from fishing, pearling, and trading with Persian Gulf neighbors. By the 1970s Kuwait boasted the world's highest per capita income and a system of social services—including public education, free medical care, government housing, and pensions—that placed it among the most advanced welfare states in the world. A number of government programs circulated oil revenues to the people in the form of social services. Kuwait's population had become one of the world's fastest growing, and the country annually attracted large numbers of immigrants from the Persian Gulf, elsewhere in the Middle East, South Asia, and beyond.

The enormous wealth derived from Kuwait's single valuable natural resource—oil—has financed not only the remodeling of government services and the armed forces but also the physical modernization of the state's infrastructure as well as stimulating massive foreign investments, loans, and gifts. In addition to new roads, ports, schools, hospitals, and industrial facilities, the government has supervised the reconstruction of much of Kuwait City and the construction of large, modern residential suburbs. The most modern equipment money can buy has been purchased for the state health facilities and for the new University of Kuwait. The chronic water shortage that plagued Kuwait has ended; construction of the world's largest system of distillation plants has ensured a continuous supply from the Persian Gulf.

This dramatic transformation has not been without problems, however. Foremost among them in the mid-1970s was the challenge of absorbing foreign workers, drawn mainly from surrounding Arab countries, who constituted more than half the population. Strict naturalization laws, intended to retain control of the economy in Kuwaiti hands, prevented foreigners from establishing themselves permanently in the country and taking equal advantage of the services and opportunities made available by the vast oil wealth. The discrepancy between citizens and noncitizens bred resentment, a feeling intensified by the fact that the skills and manpower of the foreigners remained absolutely essential for management of the economy and for many aspects of the government as well.

In late 1976 Kuwait was a constitutional monarchy in the sense that the head of state was a monarch and the system of government was based on the 1962 Constitution. In August 1976, however, the amir (ruler or, in this instance, king) suspended the Constitution and prorogued the National Assembly (Assembly). Since the attainment of full independence in 1961 the most significant political innovation had been the increasing power and prestige of the popularly elected Assembly. Moreover Kuwait was the only state on the Arabian Peninsula that had established a modern legal system.

The monarchy is hereditary within the Al Sabah family, and the reigning monarch is designated by the Constitution as the Amir of Kuwait. In late 1976 the amir was Shaykh Sabah al Salim Al Sabah, who had ruled since the death in 1965 of the first amir of independent Kuwait, Shaykh Abd Allah al Salim Al Sabah. The designated successor was the prime minister, Shaykh Jabir al Ahmad al Jabir Al Sabah. Within the Al Sabah family only the male descendants of Shaykh Mubarak al Sabah Al Sabah, who reigned from 1896 to 1915, are eligible to become amir. In practice only two branches of Shaykh Mubarak's lineage were important in this selection process: Al Salim and Al Jabir. Amir Sabah is of the Al Salim branch (as was his predecessor), and Shaykh Jabir is of the Al Jabir branch.

In late 1976 the amir and the other leaders of the Al Sabah family appeared to command the loyalty of most Kuwaitis and in particular of the principal interest groups within the society. These groups included the armed forces, the tribal and religious leaders, and the merchant class or business leaders. The non-Kuwaiti majority, reported to be at least 53 percent of the population, contributed to the situation that prompted the amir to prorogue the Assembly.

GEOGRAPHIC AND DEMOGRAPHIC SETTING

Physical Features

Located at the northwestern corner of the Persian Gulf, Kuwait is bounded on the east by the gulf, on the north and west by Iraq, and on the south and southwest by Saudi Arabia (see fig. 1). The area of the state is about 7,780 square miles, something less than the size of New Jersey. It is about 130 miles east to west, with a maximum north-south dimension of about 115 miles (see fig. 6). Included in this territory are a number of large offshore islands, the largest of which is Bubiyan, separated from the mainland by a narrow waterway. Of these islands, only Faylakah at the mouth of Kuwait Bay is inhabited. This island is believed to have been a center of civilization in antiquity and is the site of an ancient Greek temple built by the forces of Alexander the Great. The country's only prominent geographic feature is Kuwait Bay, which indents the shoreline for about twenty-five miles, providing natural protection for the port of Kuwait and accounting for nearly half the state's some 120 miles of shoreline.

The northern border with Iraq dates from an agreement with Turkey in 1913 that, though never formally ratified, was accepted by Iraq when that country became independent in 1932; in the 1960s and again in 1976 Iraqi governments made claims to Kuwaiti territory. The boundary with Saudi Arabia was set by the Treaty of Uqair in 1922, which also established the Kuwait-Saudi Arabia Neutral Zone. In 1966 Kuwait and Saudi Arabia agreed in principle to divide the Neutral Zone; the partitioning agreement making each country responsible for administration in its portion was signed in December 1969. The resources in the area, essentially oil, were not affected by the agreement, and the oil from onshore and offshore fields continued to be shared equally between the two countries (see Oil Industry of Kuwait, ch. 5).

Most of Kuwait and the Neutral Zone consists of waterless desert. There is one small oasis at Al Jahrah at the western end of Kuwait Bay and a few wells in the coastal villages. There are no permanent streams, but a few wadis are filled by winter rain. Notable among these is the Wadi al Batin, the broad shallow valley forming the western boundary of the country. At Raudhatain in the north several shallow wadis converge and provide a temporary repository for winter floodwater, which quickly evaporates or sinks into the porous subsoil.

The land's level character is explained in part by the absence of direct drainage to the sea, but fundamentally the monotony is caused by the area's geographic structure. Unfolded Miocene and Pleistocene sedimentary rocks lie almost horizontally on gently folded Cretaceous and Jurassic rocks. There is a continual redistribution of the land surface by flood runoff and erosion.

The Kuwaiti desert is undulating and gravelly, with a few low hills or ridges. The only trees aside from those in date gardens at Al Jahrah are tamarisks in Kuwait City and in a few villages. The most common desert shrub is the *arfaj*, which grows to a maximum height of two and one-half feet and is used for firewood by nomadic tribesmen in the vicinity of the Wadi al Batin. For a brief period in the spring, if winter rains have been adequate, the wadi beds are covered with grass and flowering desert shrubs. The spring flush, however, is very brief because the May sun dries the grass and withers the shrubs and the June winds envelop everything in driving sand.

The climate is somewhat less severe than in other parts of the Persian Gulf. The intense humidity that is characteristic of the region lasts only a few weeks of the year in Kuwait—usually in August—at which time temperatures regularly exceed 110°F and have been known to exceed 130°F. Sand and dust storms are frequent in the summer when the shamal, a strong northeast wind, blows down the gulf from Iraq. Winters are generally pleasant. There is abundant sunshine, and daytime temperatures range between 45°F and 60°F.

Rainfall varies from three to six inches a year in different parts of the country, but actual falls have ranged from one inch in a year to as much as thirteen inches. Sea temperatures in the summer rise to over 90°F in August, increasing the humidity on the coastal lowland. Farther inland the climate is more favorable owing to stronger winds and lower humidity.

Geologically the land is a recently emerged and youthful terrain. In the south limestone has been raised in a long, north-oriented dome that lies beneath the surface debris. It is within and below this formation that the principal oil fields of Kuwait are found. In the west and north layers of sand, gravel, silt, and clay overlie the limestone to a depth of 700 feet. The upper portions of these beds are part of a mass of sediment deposited by a great wadi whose most recent channel was the Wadi al Batin. There are at least two structural highs in the north; one of them, the Raudhatain uplift, has formed a groundwater trap on its western side. Kuwait's principal source of fresh groundwater was discovered there in 1960, but the supply, though considerable, is not sufficient to support extensive irrigation. It is tapped, however, to supplement the distilled water supply that fills most of the needs of Kuwait City. The only other exploited source of groundwater is the permeable zone in the top of the limestone of the Ash Shuaybah field south and west of Kuwait City. Unlike the Raudhatain deposit, the Ash Shuaybah water is saline. About 6.5 million gallons a day of this brackish water were produced for nondrinking purposes, and in the mid-1970s a vast supply remained.

Until the early 1950s Kuwait suffered from an acute and chronic water shortage. Individual households had to hoard winter rainfall carefully because few Kuwaitis had access to ground wells. But as the population began to expand in the twentieth century, it became necessary to import fresh water into Kuwait City from the Shatt al Arab (confluence of the Tigris and Euphrates in Iraq, 100 miles to the north). The city became dependent on a fleet of boats bringing water stored in casks and goatskin water bags. Not only was this water expensive, but the service was often erratic. In the summer months, especially during sandstorms, life in the capital became almost unbearable when weather conditions prevented the arrival of the boats from the Shatt al Arab. The peak year for the carrier fleet was 1947, when it supplied Kuwait City with an average of 80,000 gallons of drinking water per day.

The first step toward increasing the water supply for both consumption and irrigation was taken by the Kuwait Oil Company (KOC) in 1950, when it built the country's first seawater desalination plant, capable of producing 600,000 gallons of fresh water each day. The government, then in the first stages of its program to develop water resources, arranged for 80,000 gallons of the distilled water to be pumped into reservoirs in Kuwait City daily. Government desalination plants continued in the mid-1970s to be an important construction item. As progress in distillation continued, the government began to exploit reserves of brackish groundwater at Ash Shuaybah for use in sanitation, industry, and certain kinds of irrigation. The water carrier fleet had ceased operations in 1951.

By the early 1970s Kuwait had the largest group of desalination plants in the world. Together they were capable of producing about 12 million gallons a day; facilities capable of producing another 17 million gallons a day were under construction and would raise national capacity to 50 million gallons a day. In the early 1970s about 75 percent of all potable water was produced through desalination. The remainder was accounted for by the freshwater reserves at Raudhatain. Because distilled water is tasteless, it was mixed with a small amount of brackish water to make it potable. Chlorine and other chemicals were also added for public health reasons.

The oil-producing sands of Kuwait are extraordinarily rich; they are almost 1,000 feet thick in some places. The porosity and permeability of Kuwaiti sands are the highest in the Middle East, contributing to the ease of petroleum extraction (see ch. 5). The producing fields, moreover, are close to shipping points—a factor influential in the high profitability of Kuwaiti production. In addition the flatness of the terrain makes possible a drilling technique known as skidding, whereby entire derricks and engine houses are shifted from old to new locations. Skidding also includes a procedure for moving the unitized remainder of the rig and equipment.

No metallic minerals have been discovered. A considerable amount of limestone is available in the Neutral Zone, which could be combined with clay deposits along Kuwait Bay to produce cement. Investigations in the early 1960s indicated the presence of some sands that might be exploited for the commercial production of glass.

Overlooking Kuwait Bay is Kuwait City—capital, chief port, and commercial center of the country. One-half the population lives in the city and its suburbs. The broad, tree-lined boulevards, automobile traffic, housing developments, schools, and central business district afford a remarkable contrast with the surrounding desert. Largely rebuilt since 1950, the city extends fifteen miles along the south side of Kuwait Bay and six miles inland and has a radial street plan that forms a concentric pattern outward from the old town.

By the late 1970s the old city had been demolished and rebuilt; only the main gate was left. The business district of the new city resembles that of modern urban areas elsewhere in the world, and the car-to-population ratio is one of the world's highest. Hotels and restaurants are plentiful, and in the mid-1970s the rates were higher than those in New York. The beachfront along Kuwait Bay is excellent, and water-skiing has become a popular sport. Trees and shrubs have been planted in the city and the surrounding suburbs as part of a massive afforestation program that seeks not only to beautify but to screen the metropolitan area against summer sandstorms as well.

The other towns of Kuwait are dwarfed by the Kuwait City complex, which includes such suburbs as Hawalli, with a population of about 130,000. One notable town is Al Ahmadi, the headquarters of the KOC. Al Ahmadi was built to accommodate employees of the KOC and has its own fire department, water desalination plant, mosques, motion picture theaters, and clubs. Nearby is the giant KOC refinery.

Population and Labor Force

According to the April 1975 census, the population of Kuwait was 990,389. Of that number about 47 percent were Kuwaitis, and the remainder were resident foreigners. Those figures represented a 36-percent rise in population since 1971. These statistics present the most salient demographic facts about Kuwait—the numerical dominance of foreigners and the 10-percent annual rate of growth from both natural increase and immigration.

Although Kuwait's rate of natural increase is one of the highest in the world, immigration was mainly responsible for the demographic situation and had been for decades. Kuwait City's population was estimated at 35,000 in 1907, with about another 2,000 people in the territory. The country's first census in 1967 counted 206,473 people, a figure experts believe too low; in 1975 the census counted 467,339.

The great bulk of immigration that brought about this growth occurred within a few decades, especially after 1945. Surveys made in 1957, for example, when immigrants constituted 45 percent of the population, indicated that less than one-eighth had come to Kuwait before 1947; about one-half had come within the preceding three years. Although Kuwait had always experienced some immigration, it was not perceived as a threat to the dominance of the native population until the late 1940s. Until that time the country had no formal code concerning nationality. Between 1948 and 1956, however, a series of laws and decrees sharply differentiated citizens from immigrants. Kuwaitis were defined as those present in 1920 or before and their descendants; all others fell into the category of foreigners. The demarcation between these groups is sufficiently strict for the government's *Annual Statistical Abstract* to issue information under the categories "Kuwaitis" and "NonKuwaitis." The law permits only fifty foreigners a year to gain Kuwaiti citizenship and then only after ten years' residence for Arabs and fifteen years for non-Arabs.

Before the oil explorations of the 1930s about one-half the immigrants came from Iran and the rest from Iraq, Saudi Arabia, and Oman. By the late 1950s the most important immigrant groups were Jordanians, Iraqis, and Iranians (see fig. 7). Most of the Jordanians were Palestinians displaced in the creation of and continuing conflict with Israel. The relative significance of the Jordanians and Palestinians grew with the evolution of Arab-Israeli conflicts. In 1957 Jordanians and Palestinians accounted for about 16 percent of the immigrant population, in 1965 and in 1970 for 20 percent.

Despite their large numbers, or perhaps because of them, foreigners have not been able to become fully integrated into Kuwaiti life. The Kuwaitis and foreigners constitute, in the words of A.G. Hill, an authority on Kuwait, "two separate societies." The distinctions arise in part from such government policies as officially encouraged housing segregation and also from the demographic differences between the two populations.

In distribution by age and by sex Kuwaitis and foreigners differ sharply. The Kuwaiti population pyramid shows almost perfect natural distribution. That of the foreigners, however, represents a population disproportionately composed of young adult men (see fig. 8). The ratio of Kuwaiti females to Kuwaiti males has remained close to ninety-eight to 100 for a number of years; it stood at 97.9 in 1970 and at 98.1 in 1974. The ratio of foreign females to foreign males was only 60.1 to 100 in 1970; it rose to 75.0 by 1974.

This discrepancy occurred because for many years males have migrated to Kuwait either as bachelors or as married men who have left their wives and children at home. Traditionally the Iranians, Syrians,

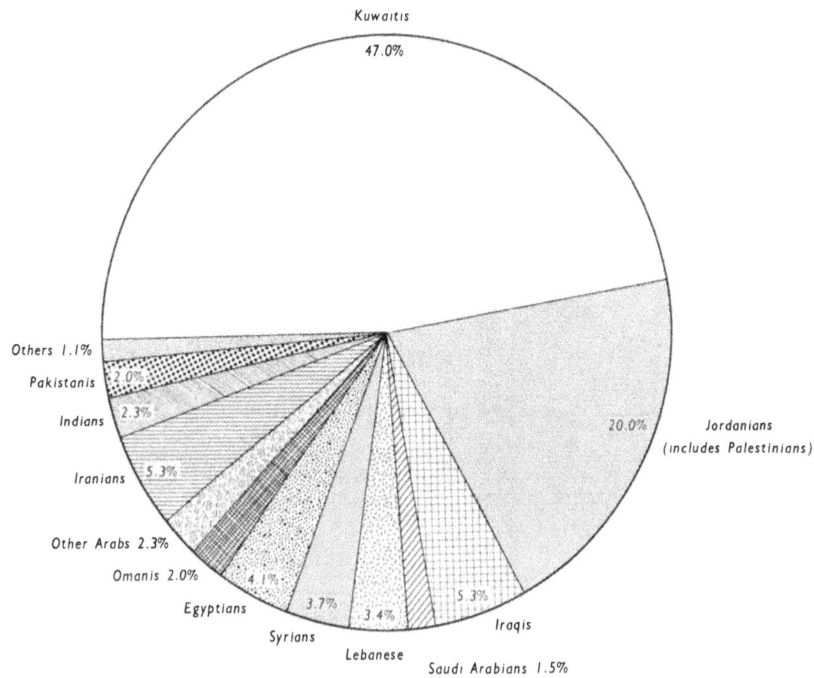

Source: Based on information from *Annual Statistical Abstract, 1975,* Kuwait, 1975, p. 31.

Figure 7. Kuwait, Percent of Population by Nationality, 1970

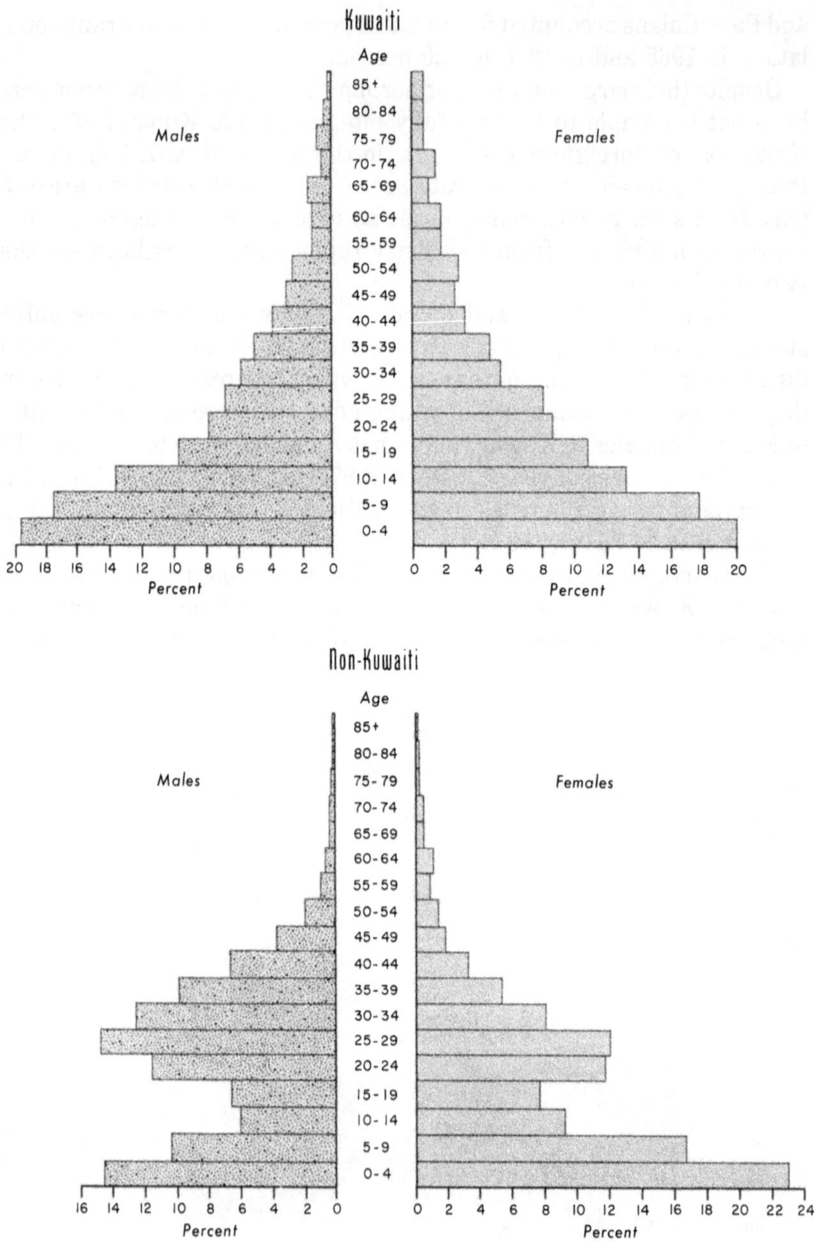

Source: Based on information from *Annual Statistical Abstract, 1975*, Kuwait, 1975.

Figure 8. Kuwait, Population Pyramids of Kuwaiti and Non-Kuwaiti Residents, 1970

Omanis, and people of the Persian Gulf were the groups most disproportionately male; their populations were 70 percent or more male in the mid-1960s. Arabs from such countries as Jordan, Lebanon, and Iraq generally have somewhat higher proportions of females, but the only groups that have traditionally had relatively close sex ratios are Europeans, Americans, and Egyptians. The last group has included a large number of female teachers. Since the late 1950s a trend has been discernible toward a more nearly equal sex ratio among foreigners. More women have been immigrating, which in turn has produced a rise in the number of children.

Foreigners provide services vital to the economy. Non-Kuwaitis appear prominently throughout the employment spectrum filling professional, technical, managerial, and clerical posts—for which there are not enough qualified Kuwaitis—and industrial and menial jobs that Kuwaitis will not accept. In 1970 the number of workers employed was 234,360, of whom 174,720 were foreigners. Non-Kuwaiti professional and technical workers outnumbered Kuwaitis in those fields by more than five to one. Foreign production workers outnumbered Kuwaitis by more than six to one, and foreign administrators and managers outnumbered Kuwaitis by nearly two to one. In 1972 even the civil service was over 60 percent foreign. Because of the number of young adult males, foreigners participate in the labor force in higher percentages than Kuwaitis. In 1970 some 53.7 percent of male foreigners were in the labor force as opposed to 36.1 percent of Kuwaiti males. And 9.9 percent of foreign females were in the labor force in contrast to 1.2 percent of Kuwaiti females.

Despite the importance of foreigners in the demography of Kuwait, their presence does not account for all the spectacular population increase the country has experienced. The rate of natural increase among Kuwaitis is one of the highest in the world, estimated at over 4.5 percent in 1973. Hill attributes this to a combination of an "expectancy of life at birth characteristic of the poor countries of Europe in the 1960s (Spain, Ireland, Hungary) and a fertility level typical of the Arab countries of the Middle East—one of the highest in the world." Further rapid growth is presaged by the country's very young population; in 1970 just over 50 percent were under fifteen.

Authorities attribute the high birthrate to the wealth produced by oil and the rapidity of its availability. Government educational subsidies and free medical care have removed much of the financial burden of childrearing from Kuwaiti parents. Rapidly improving medical facilities have ensured greater safety for both mother and baby. These changes have taken place so swiftly, however, that cultural values have remained traditional, therefore favoring large families. The birthrate of non-Kuwaitis was 37.6 per 1,000 in 1973, somewhat lower than the Kuwaiti rate of 52.4 per 1,000. Non-Kuwaiti birthrates have been rising, however, in part because of the increasing presence of foreign women.

One factor that may somewhat reduce the Kuwaiti birthrate in the future is the gradual increase in the age of women at marriage. Because of greater educational opportunities women in the 1970s were marrying somewhat later than in the past, which might contribute to a decrease in the birthrate. In 1965, for example, 40.5 percent of women between fifteen and nineteen were married as opposed to 33 percent in 1970. Authorities also believe that birth control is practiced, especially among couples who have had a number of children. One authority, for example, estimated that at least one-third of the couples between thirty-five and forty-nine practice birth control.

Education

Prominent among Kuwait's efforts to transform a traditional society into a modern, industrial country has been an ambitious, expensive program to develop universal education. In 1975 education was the largest single item in the government budget, accounting for KD86.7 million (for value of the Kuwaiti dinar—see Glossary). In only a few decades expenditures on this scale have produced a modern, totally subsidized system that is one of the best in the Middle East.

Public education is free to Kuwaiti citizens and many foreigners from kindergarten through university. The government absorbs the costs of books, clothing, meals, and transportation and pays parents an allowance that increases when a child reaches the secondary school level. The government also subsidizes hundreds of students in universities abroad. In addition to providing education to its own residents, Kuwait attracts university students from the Persian Gulf area and elsewhere in the Middle East and has become a leading center of education.

As a result of these efforts the literacy of the population is increasing. Including all residents over ten, the literacy rate was 61 percent in 1970, or 69 percent of the men and 50 percent of the women. Foreigners were somewhat more literate than citizens; 68 percent of non-Kuwaiti men and 65 percent of non-Kuwaiti women were literate, compared with 68 percent of Kuwaiti men and 37 percent of Kuwaiti women. Because persons who claimed only an ability to read were included among literates, it is likely that the actual level of functional illiteracy was somewhat higher than the figures indicate. When the population is broken down by age, however, the effects of the new educational system become apparent. In 1970 about 87 percent of the population between the ages of ten and fourteen was literate, as compared with about 59 percent between thirty and thirty-four and 34 percent over fifty.

The difference is particularly marked in the case of women, who traditionally received little or no education. Whereas 96.5 percent of Kuwaiti women over fifty and 82.4 percent between thirty and thirty-four were illiterate, only 22.3 percent between ten and fourteen were illiterate. The development of women's education in Kuwait has been so rapid that even age differences of five or ten years make substantial changes in literacy levels. About 57 percent of Kuwaiti women between fifteen and nineteen were literate in 1970, as opposed to about 33 percent between twenty-five and twenty-nine.

Until recently the educational attainments of Kuwaitis have generally not been high, in large measure because of lack of facilities. In 1970, for example, only eighty-seven Kuwaitis—ten women and seventy-seven men—reported holding postgraduate degrees, and only 1,260—263 women and 997 men—reported having university baccalaureate degrees. These figures represented a substantial improvement over earlier times. In 1957, for example, there were fifty-one university graduates, all male, and only 197 Kuwaiti secondary school graduates, forty-five female and 152 male. For this reason Kuwait has had to depend on skilled foreigners during the period of initial development. Of the 13,297 holders of bachelor's degrees in 1970, 12,037 were foreigners; so were 986 of the 1,073 holders of graduate degrees.

Development of Education

Kuwait's first modern educational institution, Al Mubarakiyya, was founded by a group of merchants under the patronage of the ruler in 1912. Until that time instruction was available only under religious tutors who gave lessons to individuals or to groups. Training generally consisted of reading, writing, memorizing the Quran, and other elementary studies in religion. The Al Mubarakiyya school, however, was intended to supply clerks trained in modern commercial skills. In addition to religious studies, therefore, it originally offered arithmetic and letter writing. Teachers brought from Palestine added history, geography, and drawing to the curriculum. Supported by contributions from local merchants, the school had scanty facilities and often used texts written by its teachers. In 1921 Al Ahmadia school, the first to offer English, was founded, and in 1927 the first girls' school was founded, offering Arabic, home economics, and Quranic studies. When the worldwide depression caused a drop in the pearl market, then the mainstay of the economy, the merchants found it impossible to continue, and Al Mubarakiyya closed.

The modern period in Kuwaiti education began in the 1930s when an educational mission from Palestine arrived and Kuwaitis began sending boys abroad for schooling. By the mid-1930s the economy again required a steady supply of clerks, and in the 1936–37 school year four primary schools—three with a total of 600 boys and one with 140 girls—were established. The new tax-supported department of education was founded in 1936 to supervise the state schools. In that year teachers from Egypt, Lebanon, Iraq, Syria, and Palestine introduced secondary education using a syllabus from Palestine.

From this small base education developed gradually until the 1950s. In 1945 there were 3,600 students. Serious expansion got under way when the proceeds from oil began flowing into the national treasury. Various special educational facilities developed during the 1950s and 1960s. The first kindergartens and the first technical college were established in the 1954–55 school year. The college began with eighty students studying such fields as auto mechanics and air conditioning and grew to over twelve fields, including chemical engineering. Special education began the following year when thirty-six children enrolled in the Institute for the Blind; in 1973 there were eleven institutes with 1,644 students for deaf or otherwise disabled children. Adult education began for men in 1958 and for women in 1963. In 1962 the department of education was upgraded to a ministry. As of the mid-1970s the most dramatic achievement in Kuwaiti education had been the opening of a university.

Both facilities and enrollments increased markedly in the 1960s. During the 1960–66 period 109 schools employing 2,267 teachers had an enrollment of 44,474 students; by 1970 there were 203 schools, 9,085 teachers, and 150,675 students. While this expansion went on, the number of students per teacher and students per classroom declined substantially.

The small number of educated Kuwaitis and the rapid growth in curriculum and enrollment have required continuing dependence on foreign teachers—mainly Egyptians, Jordanians, and Palestinians. In 1974 only 4,439 of the country's 14,035 teachers were Kuwaitis; they taught mainly in the primary grades, where they constituted over half the faculty. Government policy favors the gradual replacement of foreign teachers by Kuwaitis, and the 1974 figures represent an improvement over previous years; in 1968, for example, about 10 percent of the teachers were Kuwaitis. Nevertheless achievement of the government's goal is far in the future.

The Educational System

Since 1957 Kuwait has followed a program of two years of kindergarten, four years of primary, four years of intermediate, and four years of secondary school. Since 1965 education has been compulsory for Kuwaitis between six and fourteen years of age. Paralleling the intermediate and secondary segments of the secular curriculum is a religious institute for boys. Private schools, teaching in Arabic and foreign languages, exist at all levels as well. Enrollment and facilities in all institutions have been growing rapidly (see table 7; table 8).

A number of options exist for specialization within the curriculum.

Table 7. Kuwait, Public Education, School Years 1966–67, 1970–71, and 1974–75

	Students	Teachers	Schools
Primary:			
1966–67:			
Male	29,368	1,294	37
Female	22,619	1,052	36
Total	51,987	2,346	73
1970–71:			
Male	32,114	1,534	44
Female	25,300	1,279	39
Total	57,414	2,813	83
1974–75:			
Male	45,610	2,490	61
Female	37,971	2,320	50
Total	83,581	4,810	111
Intermediate:			
1966–67:			
Male	16,388	916	24
Female	11,686	754	22
Total	28,074	1,670	46
1970–71:			
Male	27,199	1,627	34
Female	19,866	1,366	30
Total	47,065	2,993	64
1974–75:			
Male	31,014	2,183	48
Female	24,224	2,012	42
Total	55,238	4,195	90
Secondary:			
1966–67:			
Male	4,537	390	5
Female	2,404	222	3
Total	6,941	612	8
1970–71:			
Male	9,214	849	11
Female	6,783	672	8
Total	15,997	1,521	19
1974–75:			
Male	13,692	1,477	17
Female	12,828	1,535	18
Total	26,520	3,012	35

Source: Based on information from *Annual Statistical Abstract, 1975,* Kuwait, 1975, p. 289.

In addition to technical schools commercial high schools are available to boys and girls. Within the general secondary curriculum students specialize in either science or arts in the third year. According to government information student achievement at all levels appears relatively high. Government examinations are given at the end of each curricular cycle. Between 1970 and 1973, for example, 84 percent of

Table 8. Kuwait, Students in Private Schools, School Years 1966–67, 1970–71, and 1974–75

	Arabic Schools			Foreign Schools		
	Primary	Intermediate	Secondary	Primary	Intermediate	Secondary
1966–67:						
Male	2,262	350	--	608	418	52
Female ...	1,310	157	--	591	454	47
Total ...	3,572	507	--	1,199	872	99
1970–71:						
Male	7,947	1,111	88	1,655	585	116
Female ...	6,950	944	123	1,547	545	162
Total ...	14,897	2,055	211	3,202	1,130	278
1974–75:						
Male	6,707	4,871	698	3,226	1,636	401
Female ...	5,628	4,077	578	2,734	1,437	414
Total ...	12,335	8,948	1,276	5,960	3,073	815

-- means number of private schools negligible.

Source: Based on information from *Annual Statistical Abstract, 1975*, Kuwait, 1975, p. 297.

those taking the primary examination passed, 71 percent passed the intermediate examination, 75 percent the secondary sciences examination, and 83 percent the secondary arts examination.

Because essentially all social and economic barriers to education have been removed, all Kuwaitis may attend school; for non-Kuwaitis, however, it is more difficult. Although the government encourages the education of foreign children and admits them to government schools when space is available, in the 1973–74 school year only 61 percent of non-Kuwaiti students were in government schools. They do, however, figure prominently at all grade levels. In the 1975–76 period they accounted for under 4 percent of kindergarten students, about 45 percent of elementary students, about 30 percent of intermediate students, and about 43 percent of the secondary students. Although both lack of facilities and language problems prevent the government from providing education to all foreign students, it does offer subsidies and technical assistance to private schools, sells them books at cost, and inspects them for adherence to government standards.

At the apex of the educational system is the University of Kuwait, opened in 1966 under government control. In its first school year the university had 242 male students studying liberal arts, science, and education and 176 female students in a women's college. By the mid-1970s it had over 4,000 students and 200 teachers in faculties of science; arts and education; law and sharia (Islamic law); and commerce, economics, and political science. Faculties of medicine and engineering were planned. Modern facilities included a library of more than 210,000 volumes; the university published five journals.

Health and Welfare

During the twentieth century the improvement in the country's health conditions has been dramatic. A poor, traditional society almost devoid of medical facilities in the early 1900s, Kuwait had by the late 1960s developed a free, comprehensive modern medical care system and had achieved a standard of health that rivaled that of some European countries. In the mid-1970s the state-controlled health system provided free care to all residents and boasted excellent modern facilities and techniques, including such sophisticated procedures as open-heart surgery.

Budgeted expenditures for health in 1975 were KD45.1 million. In 1975 the country had twelve hospitals and sanitoriums—providing more than 3,500 beds—that included specialized institutions for obstetrics, pulmonary diseases, nervous and mental disorders, tuberculosis, leprosy, and infectious diseases. There were also 115 health units, maternal, child health, and dental clinics, and 268 school health clinics. The government had announced that it intended a 50-percent increase in medical facilities within several years. In addition to the massive establishment run by the Ministry of Health there were eight private hospitals with a total of over 400 beds.

In the mid-1970s there were fewer than 1,000 residents to every physician, and medical and dental specialists practiced in the country. In 1970 some 76 percent of all recorded Kuwaiti live births and 93 percent of all non-Kuwaiti live births were attended by a trained physician or midwife; 62 percent of Kuwaiti women and 82 percent of non-Kuwaiti women gave birth in hospitals. A shortage of Kuwaiti nurses, caused in part by cultural values that discourage young people from entering what is often viewed as a manual occupation, required dependence on large numbers of foreign nurses. Although this caused some problems in communication and assignments, it did not appear to hamper care. The government was taking steps to improve the professional status of nursing.

The development of the health care system and resultant improvement in health conditions paralleled the country's developing wealth and contact with the outside world. In the first decade of the twentieth century the country remained what G.E. Ffrench, an authority on Kuwait, terms "premedical"; it had essentially no effective health care and suffered an infant mortality rate of over 120 per 1,000 live births and a general mortality rate of over twenty-five per 1,000. Natural increase was close to zero. In 1909 an American physician representing the Arabian American Mission of the Dutch Reformed Church arrived in Kuwait. By 1911 he and a small staff had organized a hospital for men and by 1919 a small hospital for women. In 1934 the thirty-four-bed Olcott Memorial Hospital was opened. Between 1909 and 1946 Kuwait experienced some improvement in both health conditions and natural increase. General mortality stood between twenty and twenty-five per 1,000 and infant mortality between 100 and 120 per 1,000 live births.

In 1946 a state health plan supplemented the efforts of the mission facilities and the KOC, and between then and 1950 general mortality fell to between seventeen and twenty-three per 1,000 and infant mortality to between eighty and 100 per 1,000 live births. During that period the country had approximately one doctor and one dentist per 25,000 people.

Between 1950 and 1955 a national health system developed rapidly. Hospital beds numbered more than 1,000, and the effects of the expansion of facilities were visible in a drop in general mortality, which fell to between twelve and eighteen per 1,000, and in infant mortality, which fell to between fifty and seventy per 1,000 live births. As the health care system grew in size and comprehensiveness throughout the 1950s and 1960s, mortality rates approached those experienced in developed countries. The infant mortality rate of thirty-five to forty-five per 1,000 in the late 1960s resembled that of the United Kingdom after World War II, and the general mortality rate stood between ten and fifteen per 1,000.

By the mid-1970s some major problems that had previously plagued the country had been brought under control. Smallpox was no longer an important threat; tuberculosis, however, has proved somewhat more difficult to control. Until a 100-bed sanitorium was opened in 1952, the country had no tuberculosis facilities. In 1956 the World Health Organization (WHO) joined the government in an antituberculosis campaign, and in 1959 a much larger sanitorium was opened. Incidence of tuberculosis dropped steadily because of these measures; in 1965 there were 25.9 cases per 10,000 people, and in 1974 there were 8.6. Because of rapid population growth, however, the 1965 figure represented 1,210 reported cases and the 1974 figure, 817.

In the mid-1970s the major causes of death were not unlike those of European countries. First in numerical importance were infectious and parasitic diseases, followed in order by diseases of the circulatory system (including the heart), accidents and poisonings, and respiratory diseases. In order of incidence the leading diseases in 1974 were measles, mumps, salmonella, hepatitis, chicken pox, and dysentery. Life expectancy at birth for a boy born in 1970 was 66.7 years and for a girl, 70.6 years.

The major historical threats to health—poor sanitation and poor nutrition—had been eradicated by the mid-1970s, although according to authorities intestinal parasites remained a problem. Affluence provided Kuwaitis with a varied diet that gave them essential nutrients in adequate amounts (see table 9). Sanitation standards were generally high. Food handlers underwent inspections, insecticides were widely used, and waste disposal was efficient. The country's extreme dryness, furthermore, prevents the spread of waterborne and flyborne diseases.

Table 9. Kuwait, Daily Per Capita Intake of Food, 1973

Food	Energy	Protein (in percent)	All Foods	Energy (in calories)	Protein	All Foods (in grams per day)
Cereal products	52.6	51.7	36.8	1,646	46.4	467.6
Pulses	3.4	7.9	2.3	106	7.1	29.6
Meat	4.9	14.7	5.9	153	13.2	74.7
Fish	1.5	6.6	2.5	46	5.9	31.1
Eggs	1.4	3.9	2.1	43	3.5	27.1
Dairy products	7.0	8.9	7.0	218	8.0	88.8
Fats and oils	10.2	0.1	3.2	320	0.1	40.8
Vegetables	2.1	1.8	15.1	65	1.6	192.0
Potatoes	1.0	1.1	3.3	30	1.0	41.6
Fruit	8.8	3.3	16.0	275	3.0	203.3
Sugar and sugar products	7.2	-	5.8	225	-	73.6
TOTAL	100.0*	100.0	100.0	3,127	89.8	1,270.2

- means negligible.
*Figures do not add to 100.0 because of rounding.

Source: Based on information from *Annual Statistical Abstract, 1975,* Kuwait, 1975, p. 224.

Some observers believe that the new water and sewerage systems may offer opportunities for the spread of waterborne or ratborne diseases, but this fear had not been realized as of 1976. Because of Kuwait's traditional seaborne trade with India and elsewhere and because of the continuing influx of immigrants, diseases brought into Kuwait sometimes present problems. Ratborne bubonic plague, for example, was a traditional problem, and cases of schistosomiasis and other diseases and parasites requiring waterborne vectors appear in foreigners who have contracted the diseases elsewhere.

In addition to a comprehensive system of health care, the government provided residents with one of the world's most encompassing welfare programs. Payments were made to the disabled, the senile, students' families, widows, unmarried women over eighteen, orphans, the poor, and the families of prisoners. In 1974 almost 8,500 families (21,000 people) received a total of over KD5 million.

Housing

Kuwait ranks as one of the most urban countries in the world. In 1975 population density per square mile was about 127 people. The capital city area grew in two decades from a traditional walled Arab city to a modern metropolis. This growth has been orderly because of government planning and control of development.

In the early 1950s Kuwait undertook the Government Land Purchase Scheme, in which the government bought up essentially all the private property in Kuwait City at very inflated prices. This step was designed to assemble large blocks of land suitable for massive redevelopment, to encourage residents to move to the new suburbs constructed outside the city, and to distribute the accumulating oil wealth among the Kuwaiti population. A British consulting firm produced a plan for a new capital region, including a totally rebuilt urban core and modern peripheral residential areas.

Probably the most important result of the plan was the separation of Kuwaitis and non-Kuwaitis into different residential communities. Although the plan did not propose this, government policy relating to admission to new housing areas produced it. In 1970 the newer planned suburbs had populations more than 80 percent Kuwaiti; the old city was only about 27 percent Kuwaiti. Other suburbs, which were allowed to develop with less planning and fewer services, generally were no more that 25 percent Kuwaiti. Because only Kuwaitis were allowed to own land or businesses, foreigners had not benefited from the astronomical increase in real estate values that the city had experienced.

The suburbs consist of concrete houses set on lots ranging in size from over 8,000 to 10,760 square feet, each with a courtyard surrounded by a high wall to ensure the privacy important to traditional family life (see ch. 4). Arranged in planned communities, these houses have access to open space, schools, shopping, and government-subsidized mosques and recreational facilities. The suburbs are also tied to the city by a radial road system. Municipal services are free.

For a number of years the capital city witnessed frenetic construction. By 1974, for example, the government had supplied over 13,500 dwelling units to low-income families, but 18,400 families remained on waiting lists. Moderate-income Kuwaiti families received long-term home mortgages with no interest. As of 1976 the government planned to build 51,000 housing units in the next five to seven years—16,000 for slum clearance and 15,000 for foreign workers—at a cost equivalent to US$3 billion.

THE ECONOMY

Before the discovery of oil the lack of rain and fresh water for farming forced the people of Kuwait to look seaward for an economic means of support. Boatbuilding flourished, and Kuwaiti craft were known throughout coastal Arabia for their quality. These craft with Kuwaiti sailors plied the waters of the Persian Gulf, the Arabian Sea, and the Indian Ocean. Pearling was a large-scale activity until the rise of the cultured pearl industry in Japan in the 1920s and 1930s. The good harbor in Kuwait Bay stimulated other maritime activity and led to the development of a merchant class in Kuwait City. During the eighteenth and nineteenth centuries the city rivaled Basrah in Iraq as an entrepôt for the trade between India and parts of the Middle East. The legacy of seafaring and trading gave the Kuwaiti people a maritime outlook and an aptitude for trade that distinguished them from other people of the region.

Although maritime activities provided employment in Kuwait, incomes were low, and life was harsh, except for a few wealthy merchant families. Even the ruling family had a very limited income. In the mid-1930s the entire budget for the palace and the royal family was the equivalent of about US$7,500 a year and was contributed directly or indirectly by the merchant families. In 1938 average per capita income in the country was estimated at the equivalent of about US$35 a year, and total public revenue, two-thirds of which came from import duties, amounted to US$290,000. Some observers attributed the resourcefulness and independent spirit of the Kuwaiti people to their struggle for survival under adverse conditions.

Conditions had begun to change by the time oil was found in 1938 (see ch. 5). Oil field activities created jobs at much higher pay than many traditional pursuits, attracting Kuwaitis from labor-intensive, low-income sectors of the economy and workers from other countries. World War II held up development of the fields, and commercial production of crude oil began in 1946, ushering in a construction boom. More workers were attracted from abroad, and by the 1950s foreigners outnumbered Kuwaitis in the work force. By 1948 oil revenues had reached about US$6 million, sufficient for the government to begin public projects to improve living conditions. In 1949 construction began on a public hospital, a government building, schools, and roads.

The government was not prepared for the changes that occurred; for there was little government when oil was discovered, and little was needed. Initial public projects were started with little basic planning and often from simple sketches. Public administration developed in response to the increasing flow of oil revenues. But one basic change started with the first oil company payment: the oil revenues were paid to the amir, concentrating in his hands, and subsequently those of other government officials, the distribution of the oil wealth. The position of the amir and the government was enhanced relative to other segments of the society and was made free of dependence on financial support or taxes from the rest of the community.

The oil revenues continued to mount. By the 1960s Kuwait had become the richest country in the world in national income per capital. The rapid escalation of crude oil prices between 1973 and 1974 combined with the government's purchase of part ownership in the oil companies extracting the country's crude oil caused the government's oil revenues to jump from US$1.8 billion in fiscal year (FY) 1973 to US$8.3 billion in FY 1974. By 1975 per capital gross national product (GNP) was the equivalent of US$11,365, the highest in the world according to a study by the Union Bank of Switzerland; another study ranked the United Arab Emirates (UAE) first, however.

The flow of oil revenues created a problem unusual to most governments in the world—what to do with all the money. By the mid-1970s Kuwait had the most extensive welfare program in the world. Citizens received free medicine in advanced facilities, free education through university, and subsidized food, housing, and utilities. Some longtime observers thought the welfare program had sapped the initiative and resourcefulness of the Kuwaiti people. These observers cited the general lack of investment and risk-taking by private investors, often forcing the government to make the initial investments; private capital joined when the risk was nil or after they were assured of profitability. Other observers thought the Kuwaitis retained their old values and cited particular individuals, companies, and institutions that illustrated outstanding initiative and vigor.

The government also channeled large sums to finance economic development in foreign countries, primarily other Arab states. Complete figures on foreign aid were not publicly available, but government officials claimed that it amounted to between 6 and 10 percent of GNP for several years in the early 1970s.

Government officials were keenly aware, while channeling the oil revenues into various activities, that the country's oil resources were a wasting asset; a day would come, although it was eighty to 100 years away in 1976, when there would be no more oil. Officials attempted to invest surplus oil funds to create future sources of income and employment. Investment in the local economy was difficult because of the country's lack of water, natural resources, and skilled manpower, although large investments in education and training were upgrading the work force. A substantial part of investments therefore had to be in foreign countries.

Kuwait's high-income economy was very unusual. In the 1960s, even with unrestricted imports and extensive welfare programs, only between 50 and 60 percent of public and private income went into consumption, and 40 to 50 percent of income was saved for investment. In 1974, after the large jump in oil revenues, public and private consumption amounted to only 25 percent of gross domestic product (GDP). Gross domestic capital formation amounted to another 5 percent, leaving approximately 70 percent of GDP for aid and foreign investments. Most economists expected consumption and domestic investment to rise through the late 1970s, but Kuwait was still expected to have a very large amount of money available for foreign investments.

The option of limiting crude oil production to only the amount needed to finance Kuwait's development was precluded before the 1970s by the foreign oil companies' control of production. In the 1970s the oil-exporting countries generally and Kuwait specifically exerted increasing control over crude oil production and pricing. In 1975 Kuwait bought out all foreign interests in the Kuwait Oil Company (KOC), which produced more than 90 percent of the country's crude oil. By 1976 the government was in a position to establish whatever production level officials thought in the country's best interest (see ch. 5).

Some officials maintained that oil was more valuable in the ground than sold to foreign countries for pieces of paper issued as money, shares of corporations, titles to property, and claims on foreign governments and institutions. The 1970–75 period was characterized by considerable international financial instability resulting in changes in currency values and worldwide inflation along with rapid increases in prices of imported commodities. In addition some countries became fearful of investments in institutions and property by oil exporters such as Kuwait, adding to the ever-present risk that those nations might impose restrictions on investments or on withdrawals of profit and capital. In spite of the risks involved in converting oil safely stored in the ground at no expense into money or similar financial claims, the government in the mid-1970s continued to maintain crude oil production at levels considerably in excess of Kuwait's financial needs.

How long Kuwaiti officials would be willing to accept increasing assets for declining oil reserves depended on a great many factors and was unpredictable. Petroleum officials indicated that crude oil production would at least be maintained at current levels (about 2 to 3 million barrels a day between 1973 and 1975), and they were making investments that would lessen the likelihood of a reduction in oil output. The officials realistically pointed out that the country was as dependent on commodities, technical skills, and financial stability in the rest of the world as other countries were dependent on Kuwait's oil. The government believed it was acting responsibly in handling petroleum and financial affairs, particularly by avoiding disruptive movements of the country's considerable international reserves in pursuit of higher earnings. This suggested that government authorities would continue to produce the amount of oil needed by the rest of the world from Kuwait unless depletion of its reserves became excessively high.

Structure and Growth of the Economy

The discovery of oil in Kuwait produced a rapid and dramatic change in the economy. The small size of the domestic market, the lack of skilled workers, and the shortage of local capital posed no problems to the oil companies because they were linked to the world economy. Thus the oil industry grew much faster than other sectors. High wages first attracted labor from farming, fishing, and boatbuilding to the oil industry and, in a second stage, to government service, construction, and the many service jobs in a high-income economy. Traditional labor-intensive industries could not compete in wages for labor and either receded to a fraction of their former significance or disappeared entirely. Transformation of the economy was well under way after ten years, and after twenty-five years there was little resemblance to the economy of the late 1940s.

One of the most dramatic and traumatic economic changes was the influx of foreign workers. The high wages attracted many workers, primarily from nearby areas. Foreign workers provided skills greatly lacking among Kuwaitis and the labor for menial work considered too demeaning for Kuwaitis. Economic growth was based largely on imported labor. The working force increased from 80,000 in 1957 to 184,000 in 1965, to 242,000 in 1970, and to about 304,600 in 1975. Kuwaitis accounted for about 30 percent of the work force in 1957, about 23 percent in 1965, and about 30 percent in 1975. Even in the mid-1970s most observers thought that importing additional workers was critical to further growth and diversification of the economy.

By the mid-1970s the oil sector dominated the economy, contributing between 70 and 75 percent of GDP. Official statistics included crude oil production and some minor extractive industries in a broad category of mining and quarrying, which accounted for 69 percent of GDP in FY 1973 (see table 10). Oil refining was included in, and was the most important activity of, the manufacturing sector. Total employment in the oil sector was about 7,600 workers (three-quarters of whom were non-Kuwaitis) in 1970 and probably remained unchanged in 1976. The sharp increase in oil revenues in FY 1974 added to the importance of the oil sector, but data were unavailable in late 1976 to measure the gain.

Government services, the second most important sector of the economy, contributed 10 percent of GDP in FY 1973. Expansion of the services provided by the government caused a steady increase in the number of workers employed by the government, from 66,600 in FY 1965 to 112,000 in FY 1974. Approximately one-third of the work force was employed by the government in the mid-1970s—about one civil servant to every four citizens. Some observers thought the ratio was kept high in order to distribute oil revenues among the population; government employment was looked on as a sinecure by many Kuwaitis, resulting in overstaffing and inefficiency in most government offices. At the same time observers noted that the few truly competent Kuwaiti administrators were overworked because of the numerous tasks they were required to administer.

All other sectors of the economy contributed less than 20 percent of GDP in FY 1973. Wholesale and retail trade had always been a significant part of the economy, but its importance declined appreciably in the 1970s as other sectors developed, even though growing oil revenues brought in more imports for distribution and sale each year. Financial services grew rapidly in the 1970s, surpassing trade in importance in FY 1972. Manufacturing was another sector that expanded rapidly in the 1970s as a result of the construction of some large plants. The government's efforts to diversify the economy should expand manufacturing activities through the 1970s. Construction activity slowed during FY 1972 and FY 1973 but picked up after large projects were started. By 1975 agriculture and fishing were a minute part of the economy, contributing a small fraction of 1 percent of GDP.

Because of the dominance of the oil sector, economic growth was largely paced by the growth of oil revenues. Between 1955 and 1970 oil revenues rose at an average rate of 7 percent per year. Although statistics to measure economic growth were not available until the 1960s, the economy probably grew at least as fast. A mission from the International Bank for Reconstruction and Development (IBRD, also known as the World Bank) estimated economic growth at about 8 percent a year between 1959 and 1962. Oil revenues increased more slowly in the late 1960s, which was reflected in the rest of the economy. Published official GDP data showed that the economy stagnated between FY 1967 and FY 1969 and declined in FY 1970, although there was reason to question the downturn; GDP dropped statistically in 1970 because of sharp curtailment of government services, but the drop was not reflected in the budget, suggesting that an error or a change in method affected the official GDP estimate for FY 1970.

The rate of economic growth accelerated in the 1970s as government revenues per barrel of oil produced increased dramatically. GDP amounted to KD3.2 billion (for value of the Kuwaiti dinar—see Glos-

Table 10. Kuwait, Gross Domestic Product by Sector, Fiscal Years 1969–73[1]
(in millions of Kuwaiti dinars)[2]

Sector	1969	1970	1971	1972	1973
Agriculture and fishing	5	4	4	4	4
Mining and quarrying[3]	557	652	902	936	1,447
Manufacturing	36	38	42	60	75
Utilities	37	39	47	36	43
Construction	39	34	40	21	21
Trade	85	81	90	107	113
Transport, storage, and communications	35	36	41	59	66
Government services	177	58[4]	159	216	208
Financial services[5]	18	19	21	123	134
Gross Domestic Product[6]	989	961[4]	1,346	1,562	2,111

[1]Fiscal years begin April 1; these were the latest published figures in late 1976.
[2]For value of the Kuwaiti dinar—see Glossary.
[3]Predominantly crude oil production.
[4]An unexplained aberration in published statistics that does not follow budget data, which suggested an increase in government services.
[5]Primarily finance, insurance, and real estate.
[6]At market prices.

Source: Based on information from *Annual Statistical Abstract, 1975,* Kuwait, 1975, p. 151.

sary) in FY 1974, an average growth rate of 34 percent a year between FY 1971 and FY 1974. This high rate of growth was misleading, however, for it incorporated an unusual shift in the country's terms of trade. If GDP were adjusted for the shift in terms of trade caused by the rapid escalation of crude oil prices and the government purchase of ownership in the oil companies, GDP growth would have been substantially less though still impressive.

Kuwaiti officials have objected to the standard methods of computing GDP for an economy, such as theirs, that is excessively dependent on crude oil production. They noted that GDP was achieved largely by drawing on capital in the form of petroleum reserves, for which an allowance should be made. World Bank economists acknowledged that GDP for Kuwait and some other oil states should be calculated differently than for most economies or perhaps reduced by 30 percent, but by late 1976 no method had been agreed on.

In spite of the problems of measuring Kuwait's economic growth in the 1970s, the very large increases in oil revenues contributed to boom conditions in 1976 that might last into the 1980s. The flow of funds stimulated both the public and the private sectors, contributing to growing diversification of the economy. Changes were under way in the mid-1970s that might make the economy of the 1980s as different from that of the 1960s as the latter was different from that of the 1940s.

Budget and Fiscal Policy

Under the first oil concession, payments were made to the amir. Government in the usual sense did not exist, and the amir and his close advisers decided how much of the oil revenues would be spent and in what way. In time ministries, budgets, financial controls, and other aspects of modern public administration were instituted, but decision-making remained centralized in a small group at the pinnacle of the government (see Government and Politics, this ch.). Certain policies were established soon after oil revenues began to mount, however, that continued to influence fiscal policy in the mid-1970s.

Government officials have been keenly aware that oil is a depleting asset and that, because the country has few other resources, preparations have to be made for the day when there will be no more oil. Until the mid-1970s the government had almost no control over the level of crude oil production; the primary control was over use of the oil revenues. The government received the oil revenues and used budget expenditures to control the amount of money diffused into the economy and in what form. Almost from the beginning government current expenditures were kept below revenues, leaving a surplus for investment. The limited investment opportunities at home meant that much of the surplus had to be invested abroad. World Bank economists estimated that about 25 percent of government revenues were turned into foreign assets during the 1950s.

By the 1970s investment income made up the main source of government revenues other than those from petroleum (see table 11). Earnings from foreign investments were generally reinvested. The usual revenue sources for most countries were not important in Kuwait. The modest tariff on most imports contributed only about KD8 to KD9 million between FY 1970 and FY 1973, less than 2 percent of total revenues. An income tax was applied only to foreign corporations on their earnings inside the country, and the only foreign companies with appreciable earnings in Kuwait were oil companies. Most of the oil revenues were derived from that tax until 1974, when the government began buying out foreign oil companies. The government's equity participation in oil operations entitled it to an appropriate share of crude production. When the government acquired complete ownership of the KOC in 1975, the income tax became minor, and sales of petroleum by the government became the principal source of oil revenues. Fees for such services as electricity, water, and transportation contributed a portion of revenues shown in the budget. Some economists thought that the full costs of many of these services were not passed on to users, however, and that services required net expenditures, that is, a subsidy to consumers.

Government employment was selected quite early as a means of disbursing oil revenues into the economy. A large, overstaffed bureaucracy soon developed. Salaries were high and were raised frequently to keep abreast of price increases; many fringe benefits were provided. As a result Kuwait had one of the best paid civil services in the world, and personnel costs took up a very large portion of current expenditures. In terms of annual budget allocations by ministries for all current expenditures, education consistently received the most, climbing from KD22 million in FY 1967 to KD62 million in 1974 (see Education, this ch.). Defense and the National Guard received the next largest allocation, increasing from KD21 million in FY 1968 to KD42 million in FY 1975. Other ministerial allocations were considerably smaller than those for education and defense. Moreover the budget usually had large allocations for unidentified purposes that presumably were used by ministries in need of funds.

A comprehensive social welfare program was another early way of distributing income from oil. By the early 1970s one-third or more of current expenditures was for the program, at a cost equivalent to more than US$325 per person. Education and public health accounted for about 70 to 75 percent of the costs of social welfare. Other parts of the program included housing, social welfare payments, and religious services. The government's intent in the early days of mounting oil revenues was to improve the poor living conditions of Kuwaiti citizens, who

Table 11. Kuwait, Summary of Budget Revenues and Expenditures, Fiscal Years 1971–75[1]
(in millions of Kuwaiti dinars)[2]

	1971	1972	1973	1974	1975[3]
Revenues:					
Oil receipts	354	506	544	2,404	1,687
Investment income	42	51	89	162	270
Other receipts (net)[4]	28	41	42	39	48
Total Revenues	424	598	675	2,605	2,005
Expenditures:					
Current[5]	263	270	285	442	609
Development	51	60	73	106	249
Domestic land purchases	20	23	25	27	50
Foreign aid	36	44	153	290	n.a.
Total Expenditures	370	397	536	865	908
Surplus[6]	54	201	139	1,740	1,097

n.a.—not available.
[1]Fiscal years begin April 1.
[2]For value of the Kuwaiti dinar—see Glossary.
[3]Proposed budget before actual revenues and expenditures were known.
[4]Includes fees received less some costs for such services as electricity, water, and port use.
[5]Includes expenditures for business activity in local companies.
[6]Transferred to general reserves, much of which was invested abroad.

Source: Based on information from "Kuwait," *Financial Times*, London, February 25, 1976, p. 16; and "Kuwait's Efforts to Pace Development Result in Large Rise in Foreign Assets," *IMF Survey*, Washington, November 24, 1975, p. 351.

then formed the bulk of the population. Even after Kuwaitis became a minority, the initial bias remained, and in 1974 benefits for non-Kuwaitis were substantially less than for Kuwaitis, causing dissatisfaction among resident aliens.

Another fiscal policy adopted in the early days of the oil boom was to use government funds to purchase land. The land was to be used for public purposes or resold. Real estate speculation soon set in, and land prices increased many times. An example cited by World Bank economists was that space to park a car in the center of Kuwait City in the early 1960s cost nearly US$20,000. The government continued to buy land at inflated prices, reselling it at a fraction of its cost. Land purchase was a major government program, accounting for about one-quarter of total government expenditures between 1957 and 1967. The program remained in the mid-1970s, though at a somewhat lower cost and as a much smaller proportion of the budget. The policy effectively disbursed oil funds into the economy, but economists have criticized it as an ineffective way of stimulating the economy. A relatively small group benefited from the large profits and placed part of their earnings abroad instead of creating businesses in Kuwait.

A corollary of the land purchase and resale program was the government's reluctance to undertake development itself—to build a sizable public sector. At the outset the government was too small to undertake large projects, and public works in 1949 were started without planning, cost estimates, or much else aside from sketches. But as the government's competence grew, preference for private initiative remained unless the investment costs were beyond the means of private investors. The government encouraged private investment with loans, equity capital, government contracts, and other incentives. Even in the large-scale, capital-intensive projects related to petroleum the government sought private capital.

By the mid-1970s that policy was being modified, however. Government officials decided that development of industries dealing with petroleum after it reached the surface was a sound investment and should be pursued. Private investors were bought out in two major government-sponsored companies when the government began large-scale investments to develop the base for petrochemical industries. Journalists reported that the government was also investing in large tankers to carry petroleum products to markets. Nevertheless by late 1976 the modification of policy had only affected petroleum-related activities.

The reliance on private investors for economic development and the government's initial lack of administrative organization resulted in rapid but haphazard economic growth. A group was formed in 1960 to guide the country's development. World Bank advisers suggested changes, and a planning board was established in 1962 with broad responsibilities for long-range planning and other economic matters. Several private citizens were on the board, which was headed by the prime minister. A central statistical office was formed under the jurisdiction of the board, substantially improving statistical data on the economy.

The planning board's staff prepared a five-year plan in 1966 and submitted it to the National Assembly (Assembly). It was adopted after a few modifications. The plan, covering FY 1967 through FY 1971, suggested investment allocations by the private sector and the government. The major allocations went to transportation and communications; electric power and water desalination; and housing, education, and health facilities; but actual investments fell far short of the plan. Some observers claimed that actual investments amounted to only 65 percent of the plan, and the government reported that public service investments were only 24 percent of the allocation. For example, housing, which was the major target for government public service investment, received only 17 percent of planned funds.

Many factors contributed to the shortfall in investment expenditures. A slowing in the growth of oil revenues and the necessity of providing aid to Arab states after the June 1967 War diminished the amount of funds available for domestic investment. Many developing countries have had difficulty meeting investment targets, and Kuwait was no exception in its first planning effort. Most ministries lacked the administrative skills to implement the plan. Moreover Kuwaiti businessmen were not overly cooperative, resenting and thwarting government efforts to influence their investment decisions. Perhaps the disappointing experience of the first plan discouraged efforts on another, because a second five-year plan was not drafted to follow the first. In late 1976, however, journalists reported the outline of a second five-year plan.

Total planned investments in the proposed five-year plan (FY 1976–80) amounted to KD4.4 billion (US$15 billion). According to press reports the public sector accounted for over 75 percent of investments and the private sector 24 percent (KD1.1 billion). Major investments included KD1 billion in the petroleum sector and KD703 million in housing. The plan anticipated that 115,000 additional workers would be needed, raising the work force to 412,200 by FY 1980.

Preparing for the future when oil would run out resulted in the formation of organizations to handle the country's growing financial assets. The first was an investment office established in London in 1952. British nationals with experience in London investment circles staffed the office and invested funds for the government. In that year investment relations were established with a large New York bank. Kuwaiti officials were primarily afraid of losses, and instructions to the investment offices reflected this, diminishing the return but protecting the capital. By the end of 1962 the government's foreign assets exceeded US$1 billion, primarily invested in bonds and treasury bills in Great Britain and the United States. By 1976 the number of foreign banks and companies handling the government's investments had expanded considerably, and they were located in many industrialized countries.

The best known Kuwait investment organization was the Kuwait Fund for Arab Economic Development (KFAED), formed in late 1961. In the 1970s it was the source of only a relatively small portion of the country's foreign aid, however. The success of the fund had led to its use as a model for similar investment funds established by individual Arab states and the regional fund, the Arab Fund for Economic and Social Development (AFESD), formed by several Arab countries in 1968 with headquarters in Kuwait.

The KFAED was established in response to recommendations by several Kuwaiti missions that had toured many foreign countries about the time Kuwait achieved full independence. The idea behind the establishment of the fund was to choose a vehicle that would prove the country's place in the international community along with its sense of responsibility toward poorer members by helping to finance their development. The cynical view was that the country sought to buy international acceptance and protection immediately after Iraq had made claims on Kuwait territory (see National Defense and Internal Security, this ch.). The choice of an investment fund as a major dispenser of aid was guided in part by a desire to keep Kuwaiti aid free of politics.

An investment fund had the additional advantage that, if investments were sound, loan repayments would maintain the country's financial resources even after the oil had run out. Moreover an investment fund fitted the Kuwaiti mercantile outlook and aptitude. The investment fund proved to be both profitable to its donors and economically rewarding to its recipients.

The KFAED was established as a public corporation originally capitalized at KD50 million and empowered to borrow money and issue bonds. KFAED capital was increased to KD100 million in 1963, KD200 million in 1966, and KD1 billion (approximately US$3.4 billion) in 1974. The government provided the KFAED's capital from budget reserves at no charge, and the several increases in capitalization freed the KFAED from any need to borrow money or issue bonds. The original charter confined its aid activities to Arab countries, but changes in 1974, when KFAED capital was increased to KD1 billion, expanded potential aid recipients to any developing country in the world.

Control of KFAED policy was vested in a board of directors, all Kuwaiti citizens from influential segments of the community. The board was headed by the minister of finance until 1974, when the prime minister became chairman of the board at the time that capitalization was greatly increased. In actual practice the director general of the fund, Abdlatif Y. al Hamad—because of his competence, dynamism, and proven trustworthiness to the country's rulers over many years—continued in 1976 to exert considerable influence over policy as well as daily operations of the fund. As was typical of Kuwaiti institutions, only Kuwaitis could make decisions concerning disposition of Kuwaiti resources, although the director general relied on the technical expertise of foreign professionals. A small, highly qualified staff of ten to fifteen senior professionals—almost exclusively expatriates—supported by about fifteen junior professionals—largely Kuwaitis—evaluated projects and made recommendations, but decisionmaking was reserved to the director general.

Although its charter permitted other kinds of financial aid, the KFAED had confined itself almost exclusively to financing specific projects, largely because it was patterned after and had been advised by experts from the World Bank. The KFAED's staff had gained recognition for their competence and freedom from political considerations in evaluating loan projects.

From late 1961 through March 1974 the KFAED had committed a total of KD136 million to finance projects in other Arab countries (see table 12). Of this total 37 percent went for transportation projects, 28 percent for agricultural projects, 18 percent for industrial projects, and 17 percent for electric power projects. By early 1976 an additional KD165 million in loans had been approved, bringing the total to KD301 million. India and Pakistan were among the non-Arab countries that had received loans since 1974.

The KFAED made loans only to independent governments or to public corporations when the recipient government guaranteed repayment. By 1976 there had been no default on any KFAED loan. The loans were mostly concessionary with substantial grace periods and an interest rate between 2 and 4 percent. A few loans carried no interest —only a 0.5 percent service charge to cover administrative expenses. KFAED investment income amounted to KD17 million in FY 1974. Investment income became part of reserves from which additional loans were made.

The KFAED had operated throughout its history with an informal structure and a small staff in contrast with almost all other Kuwaiti government organizations. The large expansion of its capital and area of operation in 1974 presumably will eventually require substantial changes in organization and staffing, but by 1976 the staff had been increased by only 40 percent, although the number of loans approved had doubled in less than two years. Project evaluation was minimized on some recent loans by participating with the World Bank in joint financing of projects.

Several additional investment companies, some private and some with partial government ownership, were established after the KFAED. These companies also channeled government funds into commercial investments, but they did not confine their activities to developing countries or economic projects. Investments included property, stocks, and bonds in industrialized countries as well as projects such

Table 12. Kuwait, Total Loans Extended by the Kuwait Fund for Arab Economic Development by Project and Country[1]
(in millions of Kuwaiti dinars)[2]

Country	Industry	Electric Power	Transport	Agriculture	Total[3]
Jordan	3.5	3.3	0.0	6.5	13.2
Iraq	3.8	3.0	0.0	0.0	6.8
Lebanon	0.0	1.8	1.0	0.0	2.8
Syria	2.0	0.0	7.0	0.0	9.0
Egypt	4.5	0.0	23.3	0.0	27.8
Tunisia	0.0	8.4	0.9	9.1	18.4
Algeria	0.0	0.0	10.0	0.0	10.0
Morocco	0.9	0.0	0.0	10.1	10.9
Sudan	6.2	0.0	7.0	10.0	23.1
Yemen (Aden)	0.0	0.0	0.0	0.3	0.3
Yemen (Sana)	1.9	0.0	0.3	2.2	4.4
Bahrain	1.5	7.4	0.5	0.0	9.4
TOTAL[3]	24.2	23.9	50.0	38.1	136.2

[1]Late 1961 through March 1974.
[2]For value of the Kuwaiti dinar—see Glossary.
[3]Figures may not add to total because of rounding.

Source: Based on information from *Annual Statistical Abstract, 1975,* Kuwait, 1975, p. 186.

as the Sumed (Suez to Mediterranean) pipeline in Egypt. The country's long experience with surplus revenues had produced highly diversified foreign investments handled by a wide variety of financial organizations.

The International Monetary Fund (IMF) reported the government's foreign assets at US$2.8 billion at the end of 1972 and US$4.4 billion at the end of March 1974. Estimates published by the *Financial Times* of London gave foreign assets as US$8 billion at the end of March 1975. An additional increase of foreign assets occurred by March 1976, perhaps amounting to US$3 to US$4 billion. The majority of these assets were held in long-term investments, because Kuwait had financial organizations set up before the large surpluses that followed the jump in oil revenues.

Aside from building up foreign assets with surplus funds, Kuwaiti officials were proud of their record of channeling large sums to help other developing countries. They claimed that 6 to 10 percent of GNP went to foreign aid in the early 1970s. An IMF study of aid provided by members of the Organization of Petroleum Exporting Countries (OPEC) showed that only Saudi Arabia exceeded Kuwait in the amount of aid extended. From January 1973 to June 1975 Kuwait committed US$5.4 billion to foreign countries and disbursed US$3.2 billion. Commitments went from US$700 million in 1973 to US$2.0 billion in 1974; US$2.8 billion was committed in the first six months of 1975. Disbursements amounted to US$500 million in 1973, US$1.2 billion in 1974, and US$1.5 billion by June 1975. A significant but unknown part of the aid was extended as outright gifts. Grants to Arab states, primarily Egypt, Syria, and Jordan, probably amounted to about US$450 million in FY 1974 and at least US$670 million in FY 1975.

The bulk of Kuwaiti aid has been in loans. Besides project financing channeled through the KFAED, the government provided developing countries with credits, some of which were the kind that would have had difficulty qualifying even under the concessionary terms of the KFAED or the World Bank. Other Kuwaiti organizations provided loans to or joint investments with organizations in developing countries to start projects. Kuwaiti aid also went to international organizations in the form of donations to relief agencies, purchases of World Bank bonds, and contributions of capital to regional institutions.

Evaluating Kuwaiti claims about the share of GNP channeled into foreign aid was difficult. Confusion existed between the country's investments abroad based on commercial considerations and concessionary credits and investments provided because the recipient could not obtain or afford them from commercial sources. It was questionable, for example, to classify purchases of World Bank bonds, a secure investment, as foreign aid to underdeveloped nations. Some of Kuwait's investments in developing countries resembled corporate investments from industrialized countries. Claims by Kuwaiti officials often implied a definition of foreign aid that, if used universally, would also considerably raise the value of aid programs provided by other countries. In spite of the confusion, however, it was clear that a substantial part of the country's national income went to foreign aid, probably amounting to at least 5 or 6 percent of GNP in the mid-1970s.

Agriculture and Fishing

Agriculture has always been severely limited in Kuwait. The soils are poor, climatic conditions are severe, and water is extremely scarce (see Geographic and Demographic Setting, this ch.). Rainfall is so low that cropping requires irrigation. The small amount of farming before the discovery of oil depended on a few wells along the coast and on Kuwait Bay. Subsequently geologists discovered two major groundwater sources. A large pool of fresh water trapped ages ago northwest of the bay was used for drinking in the urban centers, leaving little if any for irrigation. Large amounts of brackish water were discovered southwest of Kuwait City and used in limited irrigation.

The boom period that followed the discovery of oil attracted many farmers to high-paying jobs in other sectors. Farming declined, and almost all food was imported. In the 1950s the government became interested in the farm situation. An experimental farm was set up relying largely on the brackish water that was available. The combination of knowledge gained from work at the experimental farm and the high incomes of the urban population caused a limited resurgence in farming. The farm population increased by more than 90 percent between 1965 and 1970 and amounted to 2,850 people in 1970. The cropped area increased by 60 percent between 1970 and 1974.

In spite of the resurgence farming still played a very small part in the economy. Agriculture (including fishing) contributed only 0.2 percent to GDP in FY 1973. Almost no cereal crops were grown, and the population depended on imports for food. Domestic production in 1974 supplied only about 13 percent of the vegetables consumed, 6 percent of the milk, 20 percent of the eggs, and 45 percent of the poultry.

The total area considered cultivable amounted to a little more than 4,000 acres. In FY 1973 about 2,000 acres were devoted to crops and about 250 acres to fruit trees. Presumably the remainder was not cultivated because of the lack of irrigation. A much larger area contained coarse vegetation, sustained by sporadic rainfall, that provided limited seasonal forage for livestock.

Vegetables were about the only crop that was economical to grow. Freshness provided local produce an advantage over imports. In FY 1973 about 1,600 acres were planted with vegetables, and production amounted to 20,000 tons. Tomatoes, radishes, and melons accounted for about 60 percent of the output. Clover and other livestock feed were planted on about 400 acres, and according to the Kuwait government production amounted to 22,000 tons. A wide variety of fruits were grown, but dates were by far the most important. There were 17,500 date palms that produced nearly 600 tons of dates in 1974.

The trend had been away from small farms and toward larger ones, which in the Kuwait context meant over twelve acres. A very large farm would be over fifty acres. In the mid-1970s there were probably about 500 farms. In 1970 about 60 percent of the farms were less than twelve acres, and 40 percent were less than 2.5 acres. Most of the cultivated area was in holdings that depended on hired labor, almost completely non-Kuwaitis. In 1974 there were about 2,400 agricultural laborers, more than half of whom were employed by the government.

In 1974 there were about 136 poultry farms and seventy farms for dairy cattle. The estimated livestock population was 1.5 million chickens, 3,800 cattle, 6,900 sheep, and 1,300 goats. Poultry raising on special farms had proved to be profitable and was expanding. Feed for poultry and dairy cattle was partly imported and partly from local supplies. Before the discovery of oil nomads had maintained sheep and goats on the coarse vegetation sustained by rainfall, and presumably most of the sheep and goats were still moved about to different forage areas in the mid-1970s. In value of production meat was far more important than vegetables. In 1974 meat from poultry alone accounted for 43 percent of the value of agricultural production, milk 27 percent, vegetables 22 percent, and eggs, wool, and other meat products the remaining 8 percent.

Traditional fishing resembled agriculture in its reliance on centuries-old techniques and large labor input. Once oil was discovered, a fisherman in a nonmotorized boat operating close to shore using traditional equipment could not earn as much as he could in other employment. The government attempted to protect the local fisherman by prohibiting commercial trawlers from fishing within six miles of the shore, but the measure was insufficient. Fishing had declined sharply by the 1950s, and most of the fish marketed in Kuwait had been caught by foreign trawlers offshore. The census listed 1,200 fishermen in 1970, only 137 of whom were Kuwaitis.

In 1959 a private Kuwaiti fishing company was formed with substantial capital, including three modern trawlers manned by non-Kuwaitis. The primary object was to catch shrimp for export to the principal markets of Japan and the United States, although small shrimp and fish were sold in the local market. The business was profitable, and the company expanded. Two additional Kuwaiti fishing companies were formed and expanded rapidly. Competition among the three companies and with other fishermen in the Persian Gulf grew intense. Shrimp beds became depleted, and the catch declined, reportedly from 17,000 tons in 1968 to 9,000 tons in 1970.

The fish catch declined further in 1971, and the fish companies were in poor financial condition. In 1972 the three companies were merged into the United Fisheries of Kuwait, and the government became the largest single shareholder. A major reorganization of fishing operations was also undertaken. Measures were instituted to protect shrimp in breeding periods and breeding areas. Joint ventures in foreign coastal countries were established, and in lieu of royalty payments United Fisheries furnished the capital, expertise, and boats and the foreign partner the fishing area, crews, and other considerations. The company extended operations far beyond the gulf to the Arabian and Red seas, the coast of southern Africa, and even Australia.

By 1974 United Fisheries was the world's largest shrimp company with 157 trawlers, ten factory ships, and seven mother ships. The boats also caught fish and had a profitable lobster operation with the Yemen (Aden) government. The company, located in Kuwait, had a large, modern freezing plant and other facilities for processing the catch as well as producing such fishing equipment as nets. The fish catch for the local market and shrimp for export had increased by 1974, but figures on the total catch were not available.

By the mid-1970s fish products, primarily shrimp, had become a major nonpetroleum export. Many experts thought there was considerable potential for further development of the fishing industry. Studies had shown that fishing (excluding shrimp) in the Persian Gulf could be increased several times without danger of overfishing through the use of expensive equipment. Kuwaitis, with their fishing tradition and easy access to capital, were in a position to diversify the economy by looking seaward again.

Industry

Industry developed slowly in Kuwait. By FY 1973 the manufacturing sector's share of GDP was just under 4 percent and, although industry had expanded by 1976, the effect on its share of GDP was slight. Employment in industry totaled 32,000 people in 1970, of whom about 1,700 were Kuwaitis. Earlier employment figures reflected the growth of manufacturing. Industrial employment amounted to 6,600 in 1957 and almost 18,000 in 1965.

Major constraints hampered industrial development, however. Other than oil and gas the country has few natural resources. No metallic minerals have been discovered. Limestone from the partitioned Kuwait-Saudi Arabia Neutral Zone and local clay formed the basis for a cement industry, and sand was available to manufacture glass products. The lack of fresh water was another constraint. The small size of the domestic market limited production for local consumption to small-scale operations. Moreover the open economy, which was maintained before and after the discovery of oil, provided little protection from foreign competition. Industrialists interested in large-scale production had to think in terms of foreign markets and established competitors. The lack of a school system until recently meant few skills in the labor force. The commercial tradition in the country predisposed most entrepreneurs to invest in trade rather than manufacturing. After oil was discovered, labor costs escalated, and in a few years wages in Kuwait were higher than in almost any other area in the Middle East.

The earliest industries, boatbuilding and handicrafts, were doomed once large petroleum deposits were found. Traditional boatbuilding was a dying industry by the early 1960s because high wages were pricing wooden boats out of the market, even though the shipwrights retained their skills. The discovery of oil created a demand for new industries, initially satisfied by the oil companies themselves. The oil company operations needed water, electricity, and refined petroleum products—these were the first modern industrial plants built in the country. The government soon took over production of electricity and fresh water.

The demand for electricity expanded rapidly, and supply expanded even more rapidly. Installed capacity increased from 30 megawatts in 1956 to 1,364 megawatts in 1974. Production of electricity went from 87 million kilowatt-hours in 1956 to 4.1 billion kilowatt-hours in 1974, an average increase of 24 percent per year. In 1974 nearly all the generating capacity, 1,204 megawatts, was located near the industrial area of Ash Shuaybah. Two additional stations were constructed, which would raise installed capacity to 2,168 megawatts by the end of 1976, when the last power plant was scheduled for commissioning. Ministry of Electricity and Water officials, who operated the power stations, expected the latest additions to take care of the demand for electricity through the 1970s. Contracts were let in mid-1976 to expand generating capacity to 2,628 megawatts by 1980. Most observers rated the ministry and its operations as competent and efficient. Because of their expertise ministry officials helped other Arab countries solve power problems. The ministry had kept generating capacity ahead of demand, and the electric power base had imposed no constraint on industrial development.

In a country with no streams and few subterranean sources, provision of water was crucial to both inhabitants and industrial development. No alternative existed but distillation of seawater, however costly. The oil companies installed the first desalination plant, and the government's first unit was completed in 1953 with a 1 million gallon per day capacity. By March 1973 desalination capacity was 52 million gallons per day, the largest continuously operating facility in the world. Production of distilled water went from about 700,000 gallons per day in 1954 to 27.5 million gallons per day in 1974. Contracts were let in mid-1976 that would raise desalination capacity to about 80 million gallons per day by 1978.

The desalination plants operated in tandem with electric power generators. Steam, heated by natural gas from the oil fields, passed from the generators to the evaporators of the desalination units to produce distilled water. The discovery of underground water sources in the 1950s contributed to the water supply. The freshwater pool northwest of Kuwait City supplied less than 1 million gallons per day in 1974. About 20 million gallons per day of brackish water from underground sources was pumped in 1974, the bulk of it going to agricultural and industrial users. Some of the brackish water, however, (5 to 7 percent in the early 1960s) was added to the city water supply. Seawater was used for cooling systems in industrial plants, and the expansion of industry in the late 1970s would require installation of considerably greater pumping capacity to provide the seawater needed for cooling. Some 40 percent of the buildings and houses in Kuwait were not yet hooked to the freshwater system; they stored fresh water in tanks filled from tank trucks. Officials expected a large increase in freshwater use when most houses and buildings in urban centers had piped water.

Oil refining was by far the most important manufacturing activity, contributing 74 percent of value added by the sector in 1971. Each oil company operating in Kuwait had its own refinery, and the government built its own in 1968 (see ch. 5). In 1975, when the government bought complete ownership of the KOC, the government also became the operator of the largest refinery in the country.

Privately owned small businesses—the most numerous kind of manufacturing establishment—were engaged in final production of consumer products. The more important products were furniture and fixtures, food and beverages, and paper and printing. Investment in most of these businesses was small, and they survived largely because of the transportation costs of imports.

A group of large-scale businesses gained importance in the 1970s. They were few in number and generally issued shares that were publicly traded. The government owned shares in some of them. The largest was the Kuwait National Industries Company, in which the government owned 51 percent of the shares. The company's assets were valued at KD2.4 million in 1973, and it employed about 1,700 people. It operated plants that manufactured bricks, chemical detergents, asbestos products, and lead-acid batteries. The company had part ownership of three large concerns producing metal pipes, cement, and prefabricated buildings. Production of some products—particularly batteries, metal pipes, and prefabricated buildings—was geared to foreign markets. Bricks, blocks, and cement were for the domestic construction industry. Cement capacity was 300,000 tons a year in 1975, far less than local construction needed. Expansion of the cement plant was under way in 1976 and would raise capacity to 700,000 tons in about 1978. A separate company milled flour and produced an array of products partly for export.

The public sector consisted of petroleum-related industries, operated by two companies in 1976. The Kuwait National Petroleum Company (KNPC) operated the government's two oil refineries (see ch. 5). The other company, the Petrochemical Industries Company (PIC), was formed in 1963 with 80-percent government ownership. It began with modest facilities but acquired additional plants over the years. In 1973 the PIC bought the 40-percent share foreign oil companies held in a chemical fertilizer plant, giving it 100-percent ownership. In 1975 the PIC's larger facilities were three urea plants with a total capacity of 644,000 tons a year, an ammonium sulfate plant with a capacity of 165,000 tons a year, three ammonia plants with a total capacity of 700,000 tons a year, and a sulfuric acid plant with a capacity of 132,000 tons a year. In 1974 urea production amounted to 517,000 tons and exports to 553,000 tons, making Kuwait a major exporter of chemical fertilizers. In 1976 the government bought out the remaining private investors in the PIC, making the company wholly government owned.

In 1976 the government was investing perhaps as much as US$3.5 billion to develop petroleum-related industries; the KNPC and the PIC played key roles. The program included a gas-gathering system to collect the gas being flared, a large processing plant for the gas, and petrochemical plants to produce intermediate products from portions of the gas. Some of the gas would provide the heat source for the planned expansion of capacity in electric power, water desalination, and the cement industry. The program would add to the manufacturing sector of the economy and the variety of exports, but the petrochemical industries could not function without petroleum production and thus did not provide a hedge against the day when oil reserves were gone.

Most of the larger manufacturing plants were located in the Ash Shuaybah Industrial Estate, established in 1964 and operated by a government agency. The government provided such necessary facilities as roads, gas, electricity, and water and rented or leased industrial sites at nominal rates. By 1976 the industrial park was being expanded to accommodate new petrochemical installations and other public and private plants. The Ash Shuaybah port facilities were also being greatly expanded to handle imports of equipment and the anticipated growth of exports. Port facilities would eventually handle the export of natural gas liquids.

The government took other steps to encourage industrialization beyond the industrial park, equity capital, and loans. An industrial law of 1965 provided new industries with exemption from direct taxes and customs duties and tariff protection of up to 15 percent. The law required that local firms have 51-percent Kuwaiti ownership, however, which discouraged some foreign investors. The Investment Bank of Kuwait was created in 1974, with 49-percent government ownership, to provide medium- and long-term industrial financing, filling a financial gap that had hindered development of manufacturing. The bank also prepared feasibility studies, drawing on World Bank personnel for assistance.

There were many observers, Kuwaitis and foreigners, who questioned how far the country could and should go in industrializing. The country's one advantage was cheap energy close to its source. The big disadvantage to many Kuwaitis was that industrialization meant more foreign workers. Few Kuwaiti citizens worked in industrial plants, and few Kuwaiti industrialists had developed, although the international currency fluctuations in the early 1970s spurred the interest of local investors in domestic manufacturing. Even the government moved cautiously. There was not the rush into commitments with the rapid rise of oil revenues that had overtaxed the financial, physical, and human resources of many oil-exporting countries in the mid-1970s. It appeared that government officials might also be uncertain how far to go in industrial development.

Foreign Trade and Balance of Payments

Petroleum products dominated exports, accounting for 95 percent of value in 1974. Oil exports have been valued on the basis of the oil revenues the government received (see ch. 5). This method had the disadvantage that income taxes—the main source of oil revenues before 1974—were paid by the foreign oil companies in the year after production. An alternative method of valuing oil exports applied posted prices to the quantity of oil exported, but this procedure substantially overstated actual foreign exchange earnings because until 1974 Kuwait did not receive payments even close to posted prices. Local purchases and local wage payments by the foreign oil companies created relatively small amounts of additional foreign exchange earnings, but these were service transactions in the balance of payments rather than exports.

Exports other than petroleum increased from KD4 million in 1954 to KD117 million in 1974, an average annual increase of about 18 percent. The growth of nonpetroleum exports was much more rapid after 1968 (an average increase of 40 percent a year between 1968 and 1974), reflecting price increases, greater reexports, and real expansion in the Kuwait economy. Kuwait statistics do not show reexports separately, but the long tradition of trade made reexports probably the most important part of the export trade into the 1970s. Reexports probably accounted for much of the increase in exports of machinery and transport equipment and for some of the increase in manufactured goods and food, beverage, and tobacco products in the 1970s (see table 13). Domestic investments in chemical fertilizer plants, the fishing industry, flour mills and bakeries, and other industries contributed to the growth of nonpetroleum exports, particularly after 1968. Chemical fertilizer exports were the most notable example, increasing from KD4 million

Table 13. Kuwait, Summary of Foreign Trade, Selected Years, 1969–74
(in millions of Kuwaiti dinars)[1]

	1969	1971	1973	1974
Exports[2]				
Petroleum Products[3]	279	354	531	2,403
Nonpetroleum Products:				
Food, beverages, and tobacco products	5	8	12	10
Crude materials	1	1	2	3
Chemicals	5	6	20	47
Manufactured goods	3	4	13	17
Machinery and transport equipment	7	12	17	32
Other	2	3	6	8
Total Nonpetroleum Products	23	34	70	117
Total Exports	302	388	601	2,520
Imports[2]				
Food and live animals	34	41	53	69
Beverages and tobacco	6	6	8	9
Chemicals	11	11	13	19
Manufactured goods	50	51	66	113
Machinery and transport equipment	86	77	107	156
Other	44	46	64	89
Total Imports	231	232	311	455

[1]For value of the Kuwaiti dinar—see Glossary.
[2]Imports and exports free on board (f.o.b.).
[3]Valued in terms of oil revenues received by government and based on fiscal year starting April 1.

Source: Based on information from *Annual Statistical Abstract, 1975,* Kuwait, 1975, p. 228.

in 1969 to KD32 million in 1974. Nonetheless the economy produced very few exportable products other than oil.

Nearby Arab countries bought 50 percent of the nonpetroleum exports in 1974. Asian countries bought an additional 43 percent and the United States and Western Europe most of the remainder. Saudi Arabia was the most important buyer of Kuwait's nonpetroleum exports, accounting for 30 percent in 1964 and 18 percent in 1974; the Persian Gulf states accounted for 17 percent in 1964 and 12 percent in 1974. Japan was the second most important market, accounting for 17 percent of nonpetroleum exports in 1974. Iran had always been an important buyer, and in the early 1970s Japan, the People's Republic of China (PRC), India, and Pakistan became important markets for Kuwait's exports of chemical fertilizer.

Imports increased from KD30 million in 1954 to KD656 million in 1975, an average annual increase of 15.8 percent. The rate of growth was nearly double that of oil revenues and considerably faster than that for the rest of the economy. The high rate of growth reflected greater imports for improved living conditions for most inhabitants and a larger volume of reexports. After 1972 worldwide inflation raised the price of Kuwait's imports very substantially—by nearly 40 percent between 1973 and 1974 alone—contributing to the increased value of imports. Although imports more than doubled between 1973 and 1975 when oil revenues increased dramatically, the increase was more restrained than that of many other oil-exporting countries in the gulf. Nonetheless Kuwait's port facilities were overtaxed, and at any time in mid-1976 more than 100 ships were waiting to unload.

Kuwait's high-income economy and limited domestic production have meant that a large part of imports were for consumption—ranging between 40 and 50 percent from 1965 to 1974. Foods and beverages ranged between 15 and 20 percent of total imports. Various manufactured goods, such as clothing, shoes, watches, and photographic supplies, ranged between 25 and 30 percent of total imports. Imports of passenger cars for individuals were very small (0.01 or 0.02 percent) from 1965 to 1973, but in 1974 they amounted to 6 percent of total imports. Most passenger cars were imported for industrial or government use.

Imports of machinery and equipment usually ranged between 15 and 20 percent of total imports from 1965 to 1973, but in 1974 they dropped to 10 percent. Equipment for direct industrial development amounted to KD22 million in 1974 and KD27 million in 1973—more than double what it had been in the two previous years. Imports of electric power plant equipment increased considerably in 1973 and 1974.

Intermediate goods ranged between 33 and 38 percent of total imports from 1965 to 1974. A wide array of products were included—wheat for milling, cement, auto parts, and lubricants—but the most important goods were yarns and fabrics, iron and steel, and other metal products.

Because of the predominance of manufactured products in imports, the bulk came from industrialized countries. Japan supplied 17 percent of imports in 1974, the United States 14 percent, and West European countries 43 percent. The percentages would be higher if military equipment were included in imports. Arab countries supplied only 6 percent of imports in 1974, mostly fruits, vegetables, eggs, and live animals. The developing countries of Asia supplied about 15 percent of imports in 1974; India and the PRC were the most important, supplying about 3 percent each.

Foreign exchange earnings have been an exceptionally high proportion of national income. In FY 1974, when oil revenues increased dramatically, foreign exchange earnings exceeded 80 percent of GDP; in the 1960s they exceeded 50 percent. The large amount of foreign exchange relative to the rest of the economy has kept the country virtually free of any balance-of-payments constraints and permitted an open economy. Most imports were duty free or assessed at a modest 4 percent of their value. Only a few items, such as some cereal products, carried a 10- to 15-percent ad valorem duty to protect new industries. Commercial importing, however, was restricted to Kuwaiti citizens and firms. The government budget and fiscal policy were more important factors affecting the level of imports than foreign exchange earnings. The country had a surplus in its balance of payments for years.

The huge jump in oil revenues between 1971 and 1974 increased the country's holdings of foreign assets more than fourfold during the period in spite of substantially larger international payments (see table 14). The country's long access to imports and the government's moderation in initiating new development projects caused imports to rise more slowly than the combined receipts from oil revenues, earnings on foreign investments, and expanded exports of nonpetroleum products. The increased number of expatriates working in the economy raised the value of remittances they sent back home, however, and the government expanded its loans and grants to other countries from 99 million special drawing rights (SDR) (see Glossary) in FY 1971 to SDR817 million in FY 1974. Private capital outflows exceeded SDR1.5 billion over the two years of FY 1973 and FY 1974. Nonetheless the country increased its foreign assets by more than US$3 billion in FY 1974 alone (see Budget and Fiscal Policy, this ch.).

Table 14. Kuwait, Summary of Balance of Payments, Fiscal Years 1971–74[1]
(in millions of SDR)[2]

	1971	1972	1973	1974
Transactions of Petroleum Sector (net)[3]	1,478	1,613	1,882	6,705
Other Goods and Services:				
Exports (f.o.b.[4], excluding petroleum)	113	144	221	369
Imports (c.i.f.)[5]	−662	−760	−928	−1,564
Invisible (net)[6]	307	352	523	539
Total Other Goods and Services (net)	−242	−264	−184	−656
Current Account Balance	1,236	1,349	1,698	6,049
Capital and Remittances (net)[7]	−495	−423	−1,109	−2,894
Monetary Movements[8]	−741	−926	−589	−3,155

[1]Fiscal years (FY) begin April 1.

[2]SDR—special drawing rights of the International Monetary Fund (IMF). The average yearly exchange rate was SDR1 equals US$1 in 1971, US$1.09 in 1972, US$1.19 in 1973, and US$1.20 in 1974.

[3]Includes oil revenues, dinars sold to oil companies, and net transactions by the government's oil company. The figure for FY 1971, which reflected balance-of-payments data published by the government, is extremely large relative to government statistics on oil revenues.

[4]f.o.b.—free on board.

[5]c.i.f.—cost, insurance, and freight.

[6]Mainly earnings on foreign assets.

[7]Includes errors and omissions.

[8]A minus sign indicates an increase in foreign assets.

Source: Based on information from "Kuwait's Efforts to Pace Development Result in Large Rise in Foreign Assets," *IMF Survey*, Washington, November 24, 1975, p. 353.

Money and Banking

The Indian rupee was the principal currency circulating in Kuwait until 1959. A higher price for gold in India than the world market price led to the smuggling of gold from Kuwait to India, where rupees were obtained. Subsequent conversion of rupees into foreign exchange by Kuwaiti merchants to buy more gold strained India's foreign exchange reserves. In 1959 a special gulf rupee replaced the Indian rupee in Kuwait, a step taken to halt the gold smuggling. In preparing for full independence Kuwaiti officials decided to issue their own currency for reasons of prestige and also to earn returns on foreign exchange backing for the currency issue. The Kuwaiti dinar was introduced on April 1, 1961.

The initial value of the dinar was established at 2.488 grams of fine gold; this value remained unchanged in late 1976. The corresponding rate was KD1 equaled US$2.80 between 1961 and 1971, when the United States dollar was depreciated. Since then the value of the Kuwaiti dinar has fluctuated in terms of the United States dollar. In June 1976 KD1 equaled US$3.40. On March 18, 1975, Kuwait ceased to link the dinar to the United States dollar and pegged it to a weighted average of several currencies important in Kuwait's foreign trade.

In 1961 a currency board was established to issue dinar notes and to maintain currency reserves behind the note issue. This function and regulation of the banking system became the responsibility of the central bank, which was established in 1969. The commercial banking system consisted of six banks in 1976. The first bank in Kuwait was British, founded in 1941, but subsequent laws prohibited foreign banks from conducting business in the country. When the British bank's concession ended in 1971, the government bought 51-percent ownership. The other five commercial banks were owned by Kuwaitis.

The early focus of commercial banks on foreign trade financing and investment abroad led the government to establish some specialized banks. The Savings and Credit Bank was established in 1961 to channel investments into domestic projects in industry and agriculture. The Industrial Bank of Kuwait (49-percent government ownership), specializing in medium- and long-term loans to industry, was established in 1974 to facilitate the flow of credit to finance industrial expansion. During the 1960s two investment banks were formed (with the government the largest shareholder) to accept deposits for local and overseas investment. Private interests formed a real estate bank in 1973.

The high-income economy placed the commercial banks in the position of having large deposits and few local borrowers. A large part of their business continued to be overseas investment. Several banks joined to form a bank in London as an outlet for funds, and many Kuwaiti banks held shares in foreign banks. The liquidity situation began to change in 1973. International currency instability diminished the attractiveness of foreign investments, and Kuwaiti investors turned to the local economy. Businessmen began to borrow more to expand operations and develop new businesses, but the heaviest borrowing went to finance extensive speculation in real estate and stocks of Kuwaiti companies. Money became tight in early 1974, and the central bank placed deposits with the commercial banks to meet the need for immediate liquidity. Later that year the central bank imposed high liquidity ratios on the commercial banks. When the demand for credit expanded sharply in 1975, liquidity problems in the banking system again emerged. Credit remained tight in 1976.

A 7-percent interest limit passed several years earlier by the Assembly created some of the problems in the banking system in 1976. As the demand for credit increased, banks raised interest rates to attract deposits. Eventually the gap between interest paid depositors and interest earned on loans closed. Service charges and other fees added to the costs of borrowing and eased the banks' position somewhat but had the effect of cutting off loans of more than a few months' duration. With a maximum of 7-percent interest, the commercial banks lost the ability to charge different rates for short- and long-term loans. Consequently long-term loans had become difficult to obtain from commercial banks in the mid-1970s.

The government used many Kuwaiti financial institutions in placing its surplus funds for investment. Several investment banks, both private and partly government owned, became active in the mid-1970s in managing, placing, and underwriting foreign bond sales in Kuwait, including issues of the World Bank. The government invested large sums in bonds in 1975, assuring the local investment firms of substantial profits. The government also channeled some of its foreign real estate activity through Kuwaiti firms.

Banking and investment activities were more highly developed than activities in other financial institutions. Although quite active in the mid-1970s, the stock market was informal. Quotations were restricted to local companies, and business was conducted in banks and coffeehouses. A British consultant studied the stock market in 1976 and reportedly found the value of trading larger than that on the London Stock Exchange. Journalists reported that establishment of a regular stock exchange that would eventually trade some foreign stocks was under discussion. Efforts were under way to develop a secondary market for bonds and other securities. A government bank floated a bond issue in 1975 for the purpose of developing a secondary market. The bonds were sold by commercial banks, and the central bank agreed to accept them as discountable liquid assets. Active trading in the bonds followed, and observers thought the market would continue to develop.

Problems in 1976

The economy was booming in 1976, but beneath the surface were some serious problems. Potentially the most serious was the distinction made between Kuwaitis and non-Kuwaitis in distributing economic benefits. Observers have noted for fifteen years or more the resentment harbored by non-Kuwaitis about their treatment. The problem was most severe in housing.

Housing has lagged behind population growth. In 1976 there were an estimated 141,000 housing units for a population of 1 million. The population of the shantytowns was estimated at more than 120,000 and growing by 6,000 a year. The pressure on housing has existed for years, but presumably the high cost of land and materials held up construction of sufficient units, particularly for low-income families. Government help to Kuwaitis enabled many of them to buy houses, but non-Kuwaitis, prohibited by law from owning land or property, had to rent. As the housing stock dwindled in relation to the population, it was the low-income worker and the renter who were forced to build hovels in shantytowns. The shantytown population was mostly unskilled non-Kuwaiti laborers.

With a projected requirement of more than 100,000 additional workers in the labor force by 1981, the government in 1976 established a large program to improve housing. The plan was to construct about 51,000 housing units over five to seven years on reclaimed land and redeveloped slum areas. The total cost could reach US$3 billion. One contract was awarded to foreign contractors in mid-1976 for a housing estate for 20,000 middle-class residents in 3,300 two-story dwellings. Schools, mosques, and a medical and recreation center were part of the estate. Construction bids were being examined for slum clearance and building projects for 16,500 low-income dwellings in September 1976. Some observers questioned whether the housing program would be greatly underfulfilled, as previous ones had been, and whether this program was enough even if the goals were met.

Another problem was the construction boom under way in all of the oil-exporting countries on the Persian Gulf. All ports were congested, although Kuwait's ports were not as bad as others in the area. Kuwait's ports were being expanded, but completion was some time away. Meanwhile the inflow of materials was improved during 1976 by better administration in the ports and longer working hours. Yet as many as 100 ships were waiting to unload during the summer of 1976. The boom affected other parts of the economy as well. Bus transportation in the urban areas was not able to keep up with the increased number of passengers, for example, and occasional commodity shortages cropped up. These were temporary problems, however, that would disappear when new facilities were completed.

A potentially more serious problem was competition among the gulf oil states for the limited number of highly skilled personnel to install and operate the gasworks and petrochemical plants that these states intended to build. Many phases, particularly in petrochemicals, involved very sophisticated technology and required competent personnel to keep them operating. Some experts questioned whether sufficient trained manpower would be available if all the oil states attempted to build at the same time.

GOVERNMENT AND POLITICS

Political Dynamics

In late 1976 observers believed that the Al Sabah royal family continued to be in complete control of all important aspects of the nation's economic, military, and political affairs. Although the title of amir was formally sanctioned in 1961 when the country became fully independent from the United Kingdom, the Al Sabah dynasty had maintained continuous rule within the country since 1756.

The monarchy in Kuwait is based on the concept of a traditional hereditary amirate. Succession must be accomplished by the selection of a shaykh from the descendants of the seventh ruler of Kuwait, Shaykh Mubarak al Sabah Al Sabah (see fig. 9). Rule is limited by the dictates of sharia—the sacred body of Islamic law—and, in choosing a successor to the amir, the nominee of the amir must be approved by a majority vote of the National Assembly (Assembly). In 1966 the Assembly approved the choice by Amir Sabah al Salim Al Sabah of Shaykh Jabir al Ahmad al Jabir Al Sabah. In the absence of the legislature, as in late 1976, the traditional consensus of the senior members of the royal family, sanctioned by the traditional groups within Kuwait, would probably perform this function.

The foremost groups that must be accommodated in carrying out the decisionmaking process are the royal family, the principal tribal leaders, the important merchant families, the ulama (religious scholars), and the senior military officers. In late 1976 Amir Sabah appeared to enjoy the loyalty of all these elements. The prestige and power of the Al Sabah family line depended not only on the family's noble lineage but also—and perhaps of greater significance—on the fact that from the outset the revenues from oil production have been paid to the head of the family (see The Economy, this ch.) (see Oil Industry of Kuwait, ch. 5).

In the early twentieth century events took place that produced the predominance of the lineage of the present leaders of the royal family. Two sons of Mubarak, Jabir II (reigned 1915–17) and Salim (reigned 1917–21), dominated the affairs of the royal family as well as the domestic affairs of the country. From their relatively stable positions of influence and power within the Al Sabah family they created a situation in which their successors could be chosen only from among their descendants. Although Mubarak had other sons, it seemed unlikely in 1976 that a ruler would be chosen from their descendants.

The royal family of Kuwait is based on these two sons of Mubarak, and the two important branches of the family derive their names from them: the Al Jabir branch and the Al Salim branch. Salim's successor in 1921 was Jabir's son Ahmad, and the intention of the family from that time forward has been to alternate the rule between the two branches. This is what occurred when Salim's son Abd Allah succeeded Ahmad in 1950. These two branches of the royal family are sometimes referred to as "the cousins," and politics in Kuwait revolve around their activities.

When Abd Allah al Salim Al Sabah (also known as Abd Allah III) began his rule in 1950, the country had begun to realize an extended period of prosperity through increasing oil revenues, and his attempts to modernize the country eventually led to the renegotiation of Kuwait's protectorate treaty with the United Kingdom. By 1961 oil revenues almost thirty times greater than they had been in 1950, the basis for modern governmental and legal systems laid by Abd Allah, and a public welfare system based on the British model all provided the foundation for a smooth transition to full independence and further modernization of the country under the guidance of the Al Sabah family.

Full Independence and the Reign of Amir Abd Allah (1961–65)

The first eleven years of Shaykh Abd Allah's reign witnessed a gradual process of growth and modernization, highlighted by rapid progress in the fields of education and social welfare. The amir intended to promote parliamentary development by way of constitutional rule by the monarchy. In 1961 all important government posts were held by members of the Al Sabah family, but it was obvious to the royal family that their participation in government would decrease. Participation by commoners increased, and of the new participants the most important were recruited from the major merchant families, that is, Al Khalid, Al Ghanim, Al Saqr, and Al Salih.

On August 26, 1961, Amir Abd Allah ordered the election of a twenty-member constituent assembly to draft the country's first permanent constitution. This group completed its work within a year; and on November 11, 1962, the amir promulgated the final document. Kuwait is stated to be an independent sovereign Arab state and a part of the Arab nation. The country's official language is Arabic, its state religion is Islam, and the major source of legislation is the Islamic sharia. Freedom of opinion, the press, religion, and demonstrations are mandated in accordance with the dictates of law. Kuwait is described as a hereditary amirate with an indepen-

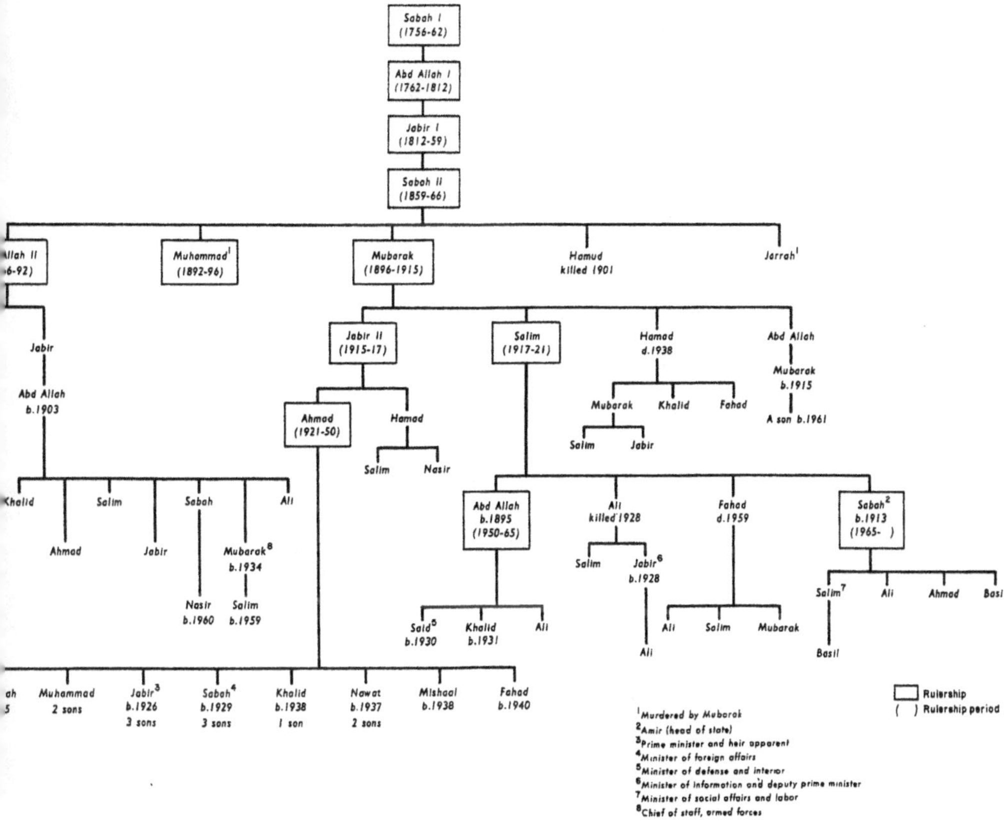
Sabah I (1756-62)
Abd Allah I (1762-1812)
Jabir I (1812-59)
Sabah II (1859-66)
Muhammad[1] (1892-96)
Mubarak (1896-1915)
Hamud killed 1901
Jarrah[1]
Jabir
Abd Allah b.1903
Jabir II (1915-17)
Salim (1917-21)
Hamad d.1938
Abd Allah
Mubarak b.1915
A son b.1961
Mubarak
Khalid
Fahad
Salim
Jabir
Ahmad (1921-50)
Hamad
Salim
Nasir
Khalid
Salim
Sabah
Ali
Ahmad
Jabir
Mubarak[8] b.1934
Nasir b.1960
Salim b.1959
Abd Allah b.1895 (1950-65)
Ali killed 1928
Fahad d.1959
Sabah[2] b.1913 (1965-)
Salim
Jabir[6] b.1928
Ali
Salim[7]
Ali
Ahmad
Basil
Said[5] b.1930
Khalid b.1931
Ali
Ali
Salim
Mubarak
Muhammad 2 sons
Jabir[3] b.1926 3 sons
Sabah[4] b.1929 3 sons
Khalid b.1938 1 son
Nawaf b.1937 2 sons
Mishaal b.1938
Fahad b.1940
Rulership
( ) Rulership period
[1]Murdered by Mubarak
[2]Amir (head of state)
[3]Prime minister and heir apparent
[4]Minister of foreign affairs
[5]Minister of defense and interior
[6]Minister of information and deputy prime minister
[7]Minister of social affairs and labor
[8]Chief of staff, armed forces

Figure 9. Kuwait, Al Sabah Family, November 1976

dent judiciary, an elected national assembly, and a ministerial system.

By January 1963 the first Assembly had been elected, and its fifty members had assumed their responsibilities. Shaykh Sabah, Amir Abd Allah's heir apparent, was named prime minister of the fourteen-member Council of Ministers; eleven ministers were members of the royal family. The Islamic tradition of the ruler's appointing a consultative assembly *(shura)* had been used by Kuwaiti rulers in the past, but the new Assembly was the first actual legislature the country had ever known.

From the beginning of the Assembly's role in government, the amir was confronted by opposition groups within the membership. The best known was the Arab National Movement (ANM), headed by Ahmad al Khatib. As a progressive, leftist organization, the ANM's efforts to promote rapid social change were countered by the traditional, tribal-oriented members. By 1965, Amir Abd Allah's last year as ruler, the government's inability to meet the rising demands of its newly politicized citizenry led to the country's first major constitutional crisis, one that resulted in the naming of a new cabinet for the third time in as many years.

The royal family members who refused to adapt to the country's new democratic institutions or who were ill prepared to assume the burden of public office were replaced by influential and more adaptable commoners. Only five members of the royal family received ministerial posts in the new fifteen-member Council of Ministers. In the decade that followed this ratio of royal family to commoner membership on the Council of Ministers remained the same through five government reorganizations.

The changes inside the country were not primarily a result of the conflict between royal family and commoners but reflected the demands of the growing non-Kuwaiti population, which tended to support the progressive elements of the Assembly. Many of the non-Kuwaitis had for a long time contributed to the development of the country and felt that the differentiations made between the two groups by the government were unjust. Amir Abd Allah, however, insisted that his government provided resident aliens with greater freedom and privilege than anywhere else in the Arab world and noted that non-Kuwaitis were able to protest their situation even though they were not citizens of the state.

Kuwait under Amir Sabah (1965–)

Abd Allah died on November 24, 1965, and was succeeded by his brother, Shaykh Sabah. The alternation of successive rulers between the two major branches of the family was broken. Sabah, the twelfth ruler of Kuwait, was seen as a compromise choice within the family, however, because he was older and was highly respected by the traditional groups in Kuwait. The transition of authority was quickly accomplished, and the Al Jabir branch of the family was placated by the choice of Shaykh Jabir al Ahmad al Jabir Al Sabah as the country's new prime minister.

In early 1966 Amir Sabah chose Jabir to be his heir apparent, and in May 1966 the nomination received unanimous approval in the Assembly. This action completed the process of succession, but the Assembly continued to plague the government and its new leaders.

The first major challenge to the new government occurred during the June 1967 War when labor unions went on strike, accusing the government of not supporting the Arab states and the Palestinians. The crisis was soon resolved, however, when Kuwait took a leadership role at the League of Arab States (Arab League) Summit Conference, held in Khartoum in August and September 1967. Kuwait offered financial assistance to those Arab states involved in the war (called the confrontation or front-line states) to promote economic recovery. The royal family had been a strong supporter of the Palestinian groups and, from 1967 into the mid-1970s, contributed massive amounts of aid to their efforts.

Although the regime was striving to maintain cordial relations with Western nations, during the June 1967 War Kuwait joined the other Arab oil-producing states in stopping the flow of oil to the West. This action somewhat placated demands from the Assembly that the government nationalize the oil sector and end its agreement for military assistance with the United Kingdom. Shortly thereafter, however, the British announced their intention to withdraw from the gulf states by the end of 1971 and specifically noted that the Kuwaiti-British agreement would end on May 13, 1971 (see National Defense and Internal Security, this ch.).

Under Jabir's leadership in the late 1960s the government used excess oil revenues to accommodate both internal and external pressures. These included a massive foreign aid program, an extensive redistribution of income within Kuwait, and a gradual increase in benefits to the resident non-Kuwaitis. After the Assembly elections of early 1971 the government hoped that better cooperation could be achieved between the executive and legislative branches. The new Assembly took an active interest in foreign affairs and in the government's negotiations for greater participation with foreign oil companies in the oil industry.

The right of free speech and the open forum of the Assembly permitted deputies to state their opinions not just to the government but to audiences outside Kuwait as well. Outspoken criticisms of other countries involved Kuwait in disputes between Arab states, and the Assembly's demand to have more information on the actual wealth of the country and on the way oil revenues were to be used led to heightened public criticism of the government. By early 1973 the Assembly was in a strong enough position to take a stand on the issue of Kuwaiti participation in the ownership and operation of the oil-producing companies (see Oil Industry of Kuwait, ch. 5).

During the October 1973 War Kuwait's leaders acted to prevent criticism by the Assembly of nonsupport of the Arab side in the war. Kuwaiti troops fought on the Suez Canal battlefront against Israel, and Amir Sabah called for an emergency meeting of the Organization of Arab Petroleum Exporting Countries (OAPEC) to determine a united policy for the use of the "oil weapon" against Western states supporting Israel. The government's efforts were supported by most Assembly deputies, and by January 1974 the government seemed to be in a stronger position.

Although in early 1974 a participation agreement for a 60-percent share met opposition in the Assembly by those who demanded "100 percent or nothing," the fact that the 60-percent share gave the state controlling interest convinced enough deputies, and the agreement was ratified in May 1974. The government also committed itself to renegotiation of the remaining 40 percent before 1979; in the event, in December 1975 Kuwait became the sole owner of the Kuwait Oil Company (KOC).

The government hoped that the January 1975 elections for the Assembly would change the composition of that body so that it would more frequently support government objectives and policies. The new Assembly was convened on February 9, 1975, and most observers concluded that its membership had moved in a more moderate direction. Twenty-five new members were elected for the first time, and at least twenty of the fifty elected members were considered supporters of the royal family and government. As it turned out, however, many of the new deputies were either leftist or progressive and joined in the growing criticism of the government's foreign policies. The protests were specifically aimed at the government's policy on the civil war that began in Lebanon in mid-1975 and at Kuwait's support for Egypt's second disengagement agreement with Israel in September 1975 (see Patterns of Foreign Policy, this ch.).

In 1975 the Palestinian population among the non-Kuwaitis in the country was estimated to be at least 270,000 and had become a vocal, sometimes hostile group. Amir Sabah in his support of the Palestinians permitted greater flexibility to them to enter government service and play essential roles in the private sector, particularly in the news media. The Lebanese crisis continued into 1976 at even higher levels of violence and brought dissension among Arabs in its wake. The amir's government, standing by its determination not to involve itself in the domestic affairs of other Arab states, became the target of criticism from deputies whose sympathies were with the leftist Muslim and Palestinian groups in Lebanon. Kuwait's position of favoring an effort of all Arabs for a cease-fire was seen by these deputies as a betrayal.

By mid-1976 Amir Sabah had received assurances from the leaders of Palestinian organizations in Kuwait that they were receiving sufficient political and economic support. The adversary relationship between the government and the Assembly continued, however, and through the summer of 1976 little progress was made either in passing new legislation or in permitting the government to play a mediating role in the crisis in Lebanon. Amir Sabah therefore dissolved the Assembly for a period long enough to redefine its role in the political system; the Assembly was closed by decree on August 29, 1976, ending the fourteen years of parliamentary experimentation in Kuwait.

The amir's actions, which included suspending a number of constitutional articles and clamping down on the press, were not within the authority reserved to him in the 1962 Constitution. His actions, however, were reportedly received with enthusiasm by many Kuwaitis, and many of the traditional leaders thought them long overdue. Stating that the measures were only "a pause for reflection on our path to democracy," Prime Minister Jabir established a committee with a mandate to report within six months what amendments to the 1962 Constitution were needed. The committee's proposals were to be endorsed by the Assembly or by a referendum within a period not to exceed four years. Few observers believed that an election for a new assembly would take place before the revision of the Constitution took place.

Amir Sabah also decreed on August 29 that the prime minister and the Council of Ministers should resign immediately and that Shaykh Jabir should form a new government with additional authority to carry out the legislative functions of the state. On September 6, 1976, Shaykh Jabir announced the formation of a new cabinet that included seven members of the royal family out of a total of eighteen—an increase of three new ministerial posts. One of the newly created posts, Ministry of Legal and Legislative Affairs, was filled by Shaykh Salman ibn Duaij Al Sabah. This ministry was seen by observers as the governmental organ responsible for the legislative functions of the dissolved Assembly.

In late 1976 the amir's government seemed to be in complete control of the country's foreign and domestic affairs. The amendments proposed by the constitutional committee were expected no later than January 1, 1977, and reports indicated that elections for a new assembly might be called shortly thereafter. Attempts were being considered to placate the non-Kuwaiti community by extending the right to become a Kuwaiti citizen to more persons and increasing the community's participation in the welfare system. The future direction of political development in Kuwait seemed unclear, but most observers believed that the country would retain a constitutional, parliamentary monarchy and begin another phase of governmental experimentation.

Government Structure

Central and Local Government

The office of the amir of Kuwait, as established by a decree when the country became fully independent in 1961 and as defined by the Constitution, dominates government (see fig. 10). Ministries and civil servants are placed under a rigid system of central control directed by the amir and his heir apparent, who is also the country's prime minister. In 1976 Amir Sabah was head of state and commander in chief of the Kuwait armed forces; Shaykh Jabir was prime minister and head of government; and seven members of the Al Sabah family were cabinet ministers.

Effective separation of legislative and executive functions was the highlight of political developments that began in 1963 with the first elections to the Assembly. The development of a ministerial system was also a major political accomplishment. What had emerged by the mid-1970s was a diverse system of government that was able to respond to most of the needs of the country and its people. The governmental system gradually became more effective and was one of very few political systems in the world providing almost total public welfare

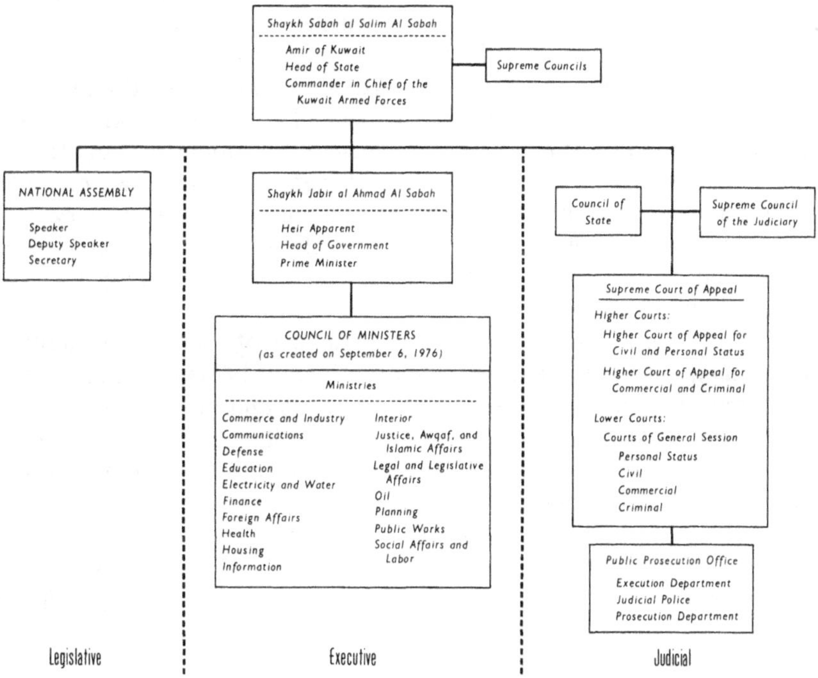

Note On August 29 1976 Amir Sabah dissolved the National Assembly for an indeterminate period

Figure 10. Kuwait, Government Structure, November 1976

programs to its citizens and many resident aliens. There were few financial constraints on the government to appropriate funds for these programs and for the operation of government at all levels; the major deficiency within the government was the lack of trained personnel to put the government's plans into motion.

With few exceptions the most important ministerial posts, such as defense, foreign affairs, information, interior, social affairs and labor, and the newly created Ministry of Legal and Legislative Affairs, were held by members of the royal family in 1976. One notable exception was the post of minister of oil, held by Abd al Muttalib ibn Abd al Husayn al Kazimi, who was also Kuwait's representative to OPEC and OAPEC. He and other nonroyal members of the cabinet had attained prominence by means of technical knowledge that was sometimes complemented by political influence as a member of a leading family.

Under the authority of the Council of Ministers were many autonomous agencies and public corporations, such as the KOC and the Kuwait Fund for Arab Economic Development (KFAED). They were formally organized in the structure of government; the more important agencies were headed by executives who were sometimes more influential than the ministers to whom they were responsible. These agencies and corporations had no independent power, however.

Kuwaiti nationals must always be given preference for employment in the government, and the government has continued a policy of replacing non-Kuwaitis with Kuwaitis whenever possible. In mid-1975 the Civil Service Commission (established in 1960 under the direction of the Council of Ministers) was instructed to fix minimum age and educational qualifications for all government positions and to streamline what the government decided had become an overgrown civil service. The government was the principal employer in the country with approximately 90,000 employees in 1976. All non-Kuwaitis employed, estimated at 60 percent of the total, had been thoroughly checked before being hired.

The government must compete with the private sector for qualified personnel, including managers, and by the mid-1970s this had become a source of conflict between the government and business leaders. A fortunate consequence of the need for qualified professionals, however, has been that few Kuwaiti nationals educated abroad have failed to return after completing their education, a major problem of many developing countries. In an effort to solve problems between the government and the private sector, the Council of Ministers created the Planning Board in November 1975 and, among other things, assigned it the task of identifying the needs of the public and private sectors and coordinating long-term goals.

In 1976 only a few senior civil service positions were filled by persons who possessed adequate educational qualifications. Government service has been a vehicle for the government to distribute oil revenues to its people and to prevent unemployment among Kuwaitis. Some offices were heavily overstaffed, and many Kuwaitis adhered to traditional ways that interfered with the functioning of a modern bureaucratic apparatus. Non-Kuwaitis have reportedly been the backbone of the civil service but were barred from higher posts, permanent job status, and some financial benefits that Kuwaitis were receiving.

Kuwait has been referred to by many observers as a city-state, and the capital, Kuwait City, was the center of all activity. Nevertheless the Ministry of Interior has been formulating plans and establishing institutions for a system of local government. In addition to Kuwait City two other major cities, Al Ahmadi and Hawalli, have been designated to carry out the functions of local administration. The governors, who reside in these three cities, administer jurisdictions that cover the entire country. Although in 1976 it was not clear to outsiders what these areas were, each was called a provincial governorate, and the governors were under the authority of the Ministry of Interior. In late 1976 the three governors, who were appointees of the amir, were members of the royal family.

Each governor administers his jurisdiction through a central municipal council composed of ten elected and six appointed members. The councils elect a president and a deputy president from among their members. The country was divided further into ten administrative divisions as determined under the election law for the Assembly; five deputies were elected from each district. The primary functions of the local governments involved the administration and implementation of such programs as education, housing, and health.

The Legislature

Historically the Al Sabah family had always depended on the traditional concept of community consensus to bolster and justify its position as the ruling family. Often referred to as beduin democracy, this consensus was important before the discovery of oil because of the influence of the merchant families who dominated the economic affairs of the country. By the 1960s the royal family's control of revenues from the oil industry had so altered power in favor of the royal family that the merchant families had become more and more dependent on royal family favors. The tradition of consensus was nonetheless an important aspect of the 1962 Constitution.

According to the Constitution legislative powers were to be shared by the amir and the Assembly, and the executive and legislative branches of government were to be on an equal footing except that the amir could take actions to dissolve the Assembly. If the Assembly achieved a two-thirds majority concerning any piece of legislation, however, its power overrode the amir's. Moreover the Assembly quickly became an institution of national public opinion; it was able to voice its concerns over the government's policies in an open forum with little interference.

Elections were held in 1963, in 1967, in 1971, and in 1975, and each resulted in an assembly made up of numerous factions. The electorate probably never amounted to 20 percent of the eligible voters, and political parties were banned. Voting was restricted to male Kuwaitis who were literate and over twenty-one; servicemen and members of the police force could not participate. Ministers could run for a seat in the Assembly, but those who did not or who were defeated were still considered ex officio members of the Assembly.

In order to be a candidate one must have Kuwaiti citizenship, be able to read and write Arabic, and be at least thirty years of age. The factions that developed in the Assembly, although not members of formal political parties, were supported by such organizations as unions, the ANM, and numerous other economic and cultural associations. In addition many candidates came from important tribal, religious, and merchant families.

The diversity of membership in the Assembly was not conducive to an efficient legislative process, and the functioning of government was often delayed. When the Assembly was able to reach consensus about a piece of legislation, not infrequently their actions were in direct opposition to known government policies. The Assembly probably demonstrated a much greater spirit of independence and exercised more authority than its architects had envisioned. According to the amir the dissolution of the Assembly in August 1976 was more to rethink what position a legislature should fill in Kuwait than to return to absolute monarchical rule. The government gave itself until August 1980 to amend the Constitution and to redefine the legislative functions or, rejecting that possibility, transform the government structure into one in which the legislative and executive functions are discharged by a single branch of government.

Legal System

The 1962 Constitution describes the judiciary as an independent branch of government, noting that judges will not be subject to any authority in administering justice. Although in late 1976 the Constitution was suspended and in the process of being amended, the status and independence of the judiciary will reportedly not be affected. During this interim period royal decrees had the force of law in the absence of a legislature.

The constitutional basis for the legal system, as well as the organization of the judiciary and the codification of law, has been derived from many sources. The frameworks of various other Arab states were important, especially the Egyptian, Iraqi, and Bahraini experiences. Traces of the French legal code were taken from the Egyptian system, and elements of English law were borrowed from the Bahraini legal system. The legal system also relies on the Islamic legal provisions of the Majalla, the Ottoman Civil Code of 1876. The Majalla was the first attempt by Islamic legal scholars to advance the use of sharia through the systematization, codification, and promulgation of extrapolations of sharia by the authority of the state. In 1976 the only countries still maintaining and applying the Majalla as the civil law of the land were Jordan and Kuwait.

In the areas of civil and personal status law the legal system reflects the conservative unifying force of sharia. The basic tenets of sharia are believed to have been revealed by God and govern an individual's activities from the most private to the most public. Sharia is no longer the sole source of public and private law, but according to the Constitution legislation must never conflict with sharia.

Both sharia and the Majalla are open to additional interpretations, and neither would be called exclusive by most Islamic legal scholars. Interpretations of the sources of sharia are accepted or rejected according to the Islamic scholars whose works led to the development of the four schools of Islamic jurisprudence (see ch. 3). These schools began to appear in the second and third Islamic centuries (ninth and tenth centuries A.D.), and for historical reasons each school predominates in a given geographic area.

The legal system of Kuwait as it affects personal status is primarily based on the traditions of the Maliki school, but the civil codes are derived from the Iraqi Civil Code and the Majalla, which were based on the Hanafi school. The Maliki school, based on the practice of judges at Medina (part of present-day Saudi Arabia), predominates in North Africa, West Africa, Upper Egypt, Bahrain, and Kuwait.

In 1976 the Supreme Court of Appeal was the highest judicial authority, although the amir might act as a final court of appeal and as a source of pardon. The court's primary duties were to hear appeals from the higher courts of appeal, and it was also competent to decide on the constitutionality of laws and regulations.

There were two separate higher courts of appeal, or appellate courts, at the next lower level: one dealt with family law and inheritance and with civil rights, the other with commercial matters and with minor crimes and felonies. The function of these courts was to hear appeals from the lower courts.

The lowest level of the court system was based on the courts of general session, divided into four divisions and numerous sections. The four divisions were personal status, civil, commercial, and criminal. Sharia was applied in the divisions of personal status and civil jurisdiction and where applicable elsewhere. The numerous promulgated public laws and regulations applied for the most part in criminal cases and also in commercial and labor disputes.

The unified court system was administered by an independent judicial body called the Council of State. Its functions included rendering legal advice and drafting bills and regulations concerning the judiciary. The ultimate authority in regulating the judiciary, however, was the Supreme Council of the Judiciary. This body was assigned to protect and ensure the independence of judicial authority.

For the most part judges and lawyers were Kuwaitis, although in 1976 there were some non-Kuwaiti lawyers. The law required a foreign lawyer to receive permission from the Ministry of Justice, Awqaf, and Islamic Affairs before practicing in Kuwait. It was unclear if this law were still in effect, however; a ministerial decree issued in September 1975 prohibited the practice of law by non-Kuwaitis.

The Public Prosecution Office is part of the regular legal system as established in the Constitution. Its function is to supervise the affairs of judicial police, enforce penal codes, and pursue offenders. Bringing offenders to prosecution describes only part of its duties; a subdivision of this office, the Execution Department, is equally important. When a prosecution has been successful and the judgment is final, the judgment must be taken to the Execution Department, whose duty it is to carry out the court's decision.

Under the authority of the Supreme Court of Appeal a Kuwaiti decree of September 14, 1975, established the State Security Court with jurisdiction over political cases. Attached to the court is the prosecution office of state security, but in late 1976 it was not clear whether this additional judicial body was actually part of the formal, independent judicial system or was directly under the minister of interior.

Patterns of Foreign Policy

In late 1976 Kuwait's foreign policy continued to focus on the maintenance of its security on the Arabian Peninsula; defense of Arab unity and opposition to Israel; advocacy of shared goals and objectives by the world Islamic community; and expansion of commercial, industrial, and diplomatic relations with such key industrialized, oil-consuming nations as the countries of Western Europe, Japan, and the United States. These countries were the most important customers for Kuwait's oil exports and were the sources of the economic and military goods and services that the Kuwaiti government was purchasing.

Kuwait, primarily in conjunction with Saudi Arabia, Egypt, and the newly established gulf states, continued to have a moderating role in intra-Arab politics. The government's greatest regional apprehensions were over the influence and politics of the radical or Marxist-oriented leaders of some Arab states, particularly Iraq. Moreover Kuwait's concept of Arab unity differed from that of those Arab leaders who espoused alignments between or among states as a prelude to a permanent merger of states and governments. The Kuwaiti view of Arab unity sought consensus and cooperation among independent Arab states.

The major foreign policy decisions were made by Amir Sabah and Prime Minister Jabir in consultation with their senior advisers. These advisers included but were not necessarily limited to the foreign minister, Sabah al Ahmad al Jabir Al Sabah, the defense minister, Said al Abd Allah al Salim Al Sabah, the oil minister, Kazimi, and the under secretary of foreign affairs, Rashid ibn Abd al Aziz al Rashid. Moreover the policy decisions of the government for oil production and exports were made within the framework of the Organization of Petroleum Exporting Countries (OPEC) and in concert with OAPEC. Kuwait's active participation in those bodies constituted a vital part of its diplomatic effort (see ch. 5). Until the dissolution of the Assembly in August 1976 that body also played an active role in formulating legislation concerning many aspects of foreign policy.

From 1899 to 1961 the United Kingdom handled Kuwait's foreign relations, but this arrangement was formally terminated in June 1961 when the British signed a treaty with Amir Abd Allah recognizing the independence of Kuwait and providing for British military assistance if Kuwait were threatened and if the amir requested protection. After consultations with the Kuwait government in May 1968, the United Kingdom announced that it could no longer honor the defense commitment. On May 13, 1971, the British carried out that decision, and the two countries terminated the last clauses of the treaty of 1899.

Kuwait's constantly increasing wealth has provided the government leverage in its relations with countries in the region. Through loans, grants, and investments Kuwait has become a major financial contributor to those Arab governments with few resources for development, to the military effort against Israel, and in particular to the Palestinian refugees and organizations. The government's policy of providing aid and loans to promote the development and stabilization of Arab countries during the 1960s and early 1970s was extended in the mid-1970s to include many other nations of Africa and Asia. The best known vehicle for channeling aid and assistance was the KFAED (see Budget and Fiscal Policy, this ch.).

The divisions among the radical Arab states and the moderate and conservative Arab states grew out of the events of the 1950s and 1960s when military groups in Egypt, Iraq, Libya, and Yemen (Sana) overthrew traditional monarchies and such traditional states as Jordan and Morocco were threatened by similar coup attempts. Beyond the political differences among the Arab states, the Arab-Israeli wars of 1948, 1956, 1967, and 1973 also produced frustrations and tensions within the Arab community of nations. Kuwait's foreign policy within the region has reflected both its continued distrust of the radical Arab regimes and its commitment to the Arab struggle with Israel. Despite the radical politics of the Palestinians, the Kuwaitis generally endorse the goals and activities of the Palestinian Liberation Organization (PLO).

The use of the oil weapon in support of its Arab allies, not only during the October 1973 War but also during the June 1967 War, has had its effects on Kuwait's relations with both the Western world and many parts of the third world that have depended on its oil resources. Various third world nations that either had been neutral in the Arab-Israeli conflict or had supported Israel changed their policy to that of endorsing the Arab position. Kuwait's diplomats have had an active role in the United Nations (UN) and other forums in securing and keeping this support.

In 1976 a major concern of the Kuwaiti government was the civil war in Lebanon; the Kuwaiti position opposing outside intervention or interference in Lebanese domestic affairs was hotly debated in the Assembly. The radical elements, particularly among the non-Kuwaitis in the country, brought to life the tensions in Iraqi-Kuwaiti relations (see National Defense and Internal Security, this ch.). At the regional level Kuwait remained concerned with the events affecting the Persian Gulf and the whole of the Arabian Peninsula. The status of Saudi relations with the Persian Gulf states has been vital to these events.

In 1971 the century-old protectorate relationship between the United Kingdom and the shaykhdoms of the lower Persian Gulf was replaced by full independence for Bahrain, Qatar, and the United Arab Emirates (UAE). Kuwait had maintained a close relationship with the royal family of Bahrain for many decades and, as a result of the British withdrawal, increased diplomatic and other relations with Bahrain. A major goal of Kuwaiti policy has been to find solutions to the many unresolved territorial disputes (see ch. 2). As of late 1976 the government remained cool to a proposal by Iran for a formal collective security arrangement among the states bordering the gulf—a position that is probably based on Kuwait's consistent skepticism about Iran's long-range goals.

During the 1970s Kuwait continued a policy of expanding relations with Western industrialized nations as well as selected communist states. In 1976 Kuwait maintained sixty-six embassies throughout the world and had established diplomatic relations with almost eighty countries. Kuwait was admitted to the Arab League within two months after independence in June 1961 and was admitted to membership in the UN on May 14, 1963, as its 111th member.

Kuwait's relations with the United States and most other Western states were cordial and sometimes, as in the case of the United Kingdom, very close. Most were oil-consuming countries that also could supply industrial and technical expertise and equipment and sophisticated military weapons. In the mid-1970s Kuwait maintained diplomatic relations with most of the oil-producing and non-Western oil-consuming countries and was one of the first countries to promote the idea of interdependence of oil producer and oil consumer. Kuwait has also established close relations with Japan in order to acquire manufactured goods and the technical expertise and manpower needed to modernize the country. Kuwait has vigorously opposed the policy of apartheid and in 1976 had relations with neither South Africa nor Southern Rhodesia.

Mass Communications

In late 1976 mass communications were closely scrutinized by the government after the crackdown on the press as a result of the emergency measures imposed in August 1976. The press was the main target in the government's attempts to lessen domestic turmoil. The amir accused the press of violating formal guidelines relating to criticism of foreign governments and Kuwait's own foreign policy and of making abusive remarks about certain domestic policies. The amir decreed the suspension of constitutional articles guaranteeing freedom of the press and opinion and imposed strict interim guidelines on the media until new amendments to the 1962 Constitution had been prepared and adopted. In the aftermath of these actions the Ministry of Information, under the authority and guidance of the Council of Ministers, became the primary source of information and was in charge of government information policies both at home and abroad.

When the country attained full independence in 1961, there was little in the way of mass communications. Informal channels disseminated news and information, but the government quickly recognized the high amount of distortion and exaggeration produced through these channels and rapidly developed policies that promoted the growth of newspapers and other print media. Throughout the 1960s government aid and grants were allocated to media establishments to encourage their growth, and by the mid-1970s a popular and respected press had emerged, primarily based in Kuwait City.

The process set a precedent for the government to grant concessions before a newspaper or periodical could start printing. By the early 1970s, however, the minister of information determined that the growth of the press in Kuwait had reached its limit, and the last newspaper concession was granted in 1971. The entire financial apparatus of the media had changed by that time, and the need for government aid was no longer necessary because the press establishments were operating as very profitable corporations. Significantly the concessions were confined to Kuwaiti citizens, but by the mid-1970s the majority of editors and writers were non-Kuwaitis.

The freedom offered reporters and editors until August 1976 encouraged the immigration of journalists from throughout the Arab world. The salaries were also attractive, and Kuwait, unlike most of the Arab states, maintained a policy that encouraged the independence of the press. When the government has taken actions against a newspaper or journal, the penalties have been relatively mild; for example, the most drastic action taken against a newspaper during the 1976 crackdown was a three-month closure. No press establishment has ever lost its license over a conflict with the government, although that action was clearly possible under the new press regulations.

The press regulations of September 1976 permit the Council of Ministers to suspend any newspaper license for a period not exceeding two years and the minister of information to suspend a license for a period not exceeding three months. Any newspaper or journal may be permanently cancelled, however, at the direction of the Council of Ministers.

For the most part there was a clear division among the press establishment in Kuwait in 1976. Most were moderate in tone and complied with government guidelines that required journalists to respect the country's national interests and to avoid remarks that might lead to domestic unrest. A substantial faction within the press, however, included more radical journalists who were well known throughout most of the Arabic-speaking world. Many were of Palestinian origin and members of the ANM; others were refugees from the civil war in Lebanon. All represented causes in which Kuwait itself played little part.

In early 1976 fewer than fifty newspapers and periodicals were published, primarily in Kuwait City. The combined circulation of the eight dailies probably did not exceed 50,000, and newspapers were not passed on for others to read as they were in most other Arab countries. Another Kuwaiti practice seemed to be for readers to buy more than one daily newspaper. This was not common in other Arab countries. Of the eight dailies six were in Arabic and two in English. All newspapers in the country were founded after 1960.

In the mid-1970s *Ar Rai al Amm*—dating from 1961—was the leading daily newspaper in Kuwait. Its title means "public opinion," and its publication was subsidized by the government for a time. Originally it was printed in Beirut and at that time was seldom distributed more than three or four times a week. By 1976 *Ar Rai al Amm* had become a major publishing organization, and the circulation of the newspaper had reached 30,000 in a population of about 1 million people. The five other Arabic-language dailies were *Akhbar al Kuwait* (1961), *Al Qabas* (1972), *Al Siyassa* (1965), *Al Watan* (1964), and *Al Anbaa,* which was given a special concession to begin publication in 1976.

Only one of these dailies was closed during the crackdown on the press in September 1976; *Al Watan,* a Marxist-oriented, pro-Palestinian newspaper, was suspended for one month. The government's new policies were aimed primarily at a number of weeklies and periodicals that the government perceived as posing a threat to the country's security. At least four were closed for a three-month period, and one, *Al Talia,* published by a leader of the ANM, would probably remain closed for a much longer period.

In 1976 approximately one of every two residents owned a radio receiver, which gave Kuwait one of the highest proportions of ownership in the world. One of every five Kuwaitis had a television receiver, also a very high proportion. Under the authority of the information minister, the Kuwait Broadcasting Service was in charge of all domestic and foreign radiobroadcasting, and Television of Kuwait controlled and maintained all television broadcasting facilities in the country. As government-owned and -operated organizations their primary roles were to present government policy, foster a sense of national identity, and provide entertainment. In 1969 Kuwait became a charter member of the Arab States Broadcasting Union and in 1976 of the Islamic Radio and Television Organization. Through these organizations Kuwait promoted cooperation in the field of information, particularly among the states of the Arabian Peninsula.

Kuwaiti television broadcasts were in Arabic only, but radiobroadcasts were in English and Urdu in addition to Arabic. Both radio and television facilities were modern; nine radio stations produced four major programs, and the three television transmitters (in Mutlaa, Raudhatain, and Faylakah) had the ability to cover every area of Kuwait and parts of Saudi Arabia, southern Iraq, and southwestern Iran. Because of the growing non-Arabic-speaking population, the English-language broadcasts were complemented by two English-language newspapers—the *Daily News* (founded 1963) and the *Kuwait Times* (founded 1961).

Foreign publications were subject to preentry scrutiny by the Ministry of Information, and certain publications have been banned. In late 1976 it was reported that after almost four years of planning the Kuwait News Agency would soon begin operations under government ownership and control and would most likely handle preentry censorship. In addition the news agency would be responsible for developing a unified information organization to provide domestic and international news agencies with reports and information on all political, economic, and cultural matters. Until 1976 the primary source of information concerning government policies had been the *Kuwait al Yau*, the official government gazette published each Sunday since 1954.

Many national and international news agencies had offices in Kuwait, and broadcasts from other parts of the Arabian Peninsula, Egypt, and other Arab nations could be heard in some parts of the country. The broadcasts of the Voice of America (VOA), the British Broadcasting Corporation (BBC), and Radio Moscow could be picked up most of the time on shortwave receivers. Informational activities were conducted by a number of nations through their diplomatic and cultural representatives stationed in Kuwait. Mass communications from countries unfriendly to Kuwait were officially banned, but little could be done to interfere with their transmission.

NATIONAL DEFENSE AND INTERNAL SECURITY

Background of Military Development

Under the British protectorate formed by Great Britain and the shaykh of Kuwait in the agreements of January 1899 and November 1914, Great Britain guaranteed defense of the country, and the shaykh ruled over internal affairs (see ch. 2). Kuwaiti forces consisted of royal guards and a small domestic police service, or constabulary, under British guidance. Saudi Wahhabi (see Glossary) raiding parties attacked Kuwait in 1920 and again in the 1929–30 period but were driven back; Kuwait remained a protectorate until after World War II.

The need for national armed forces was clear as the time of independence approached. By the mid-1950s Kuwait had become a major oil producer, and the new oil revenues not only obviated the need for any British subsidy but also provided more than ample funds to pay for as much military expansion as the Kuwait government could handle. By June 19, 1961, when Kuwait became fully independent, the 600-man constabulary (dating from 1954) had been expanded with British equipment and training to a combined arms brigade of about 2,500 men. The Kuwaiti air force and navy (basically a coast guard service) were also established in that year under British guidance, which took the form of a military training and assistance mission.

The exchange of diplomatic notes and the mutual support and consultation treaty by which Kuwait became fully independent (abrogating the 1899 agreement) had scarcely been accomplished when Iraqi Prime Minister Abdul Karim Qasim asserted a claim to Kuwait on grounds that it was properly part of Iraq and had been stolen in 1899 by British "colonial imperialists." Kuwait immediately appealed for assistance to the UN, the Arab heads of state, and Great Britain. Under the independence agreements Great Britain was committed to providing assistance if requested, and British troops began landing early in July 1961.

The Iraqi threat subsided, and on July 20 Kuwait was admitted to membership in the Arab League. In October the British force was withdrawn and replaced by detachments from Egypt, Jordan, Saudi Arabia, and Sudan under Arab League aegis to guarantee Kuwait's independence. These troops were in turn withdrawn by the end of February 1963 after a military coup in Iraq resulted in the killing of Qasim and the overthrow of his regime. The successor government in Iraq sought improved relations with Kuwait and in October 1963 formally recognized Kuwait as an independent state.

The new Kuwaiti armed forces were expanded and equipped after the crises of the early 1960s. Kuwait sought to play a moderating role among Arab states, using its vast oil wealth generously in economic and military grants and loans to assist many Arab states and causes, including the Palestinian Liberation Organization (PLO). On May 13, 1968, after Great Britain's declaration of intent to withdraw its forces and defense commitments from the Persian Gulf region by the end of 1971, Great Britain announced that its 1961 commitment to assist Kuwait if requested to do so would terminate on May 13, 1971. This development, preceded by a troublesome border incident with Iraq in 1967 and followed by the negotiated division of the Kuwait-Saudi Arabia Neutral Zone in December 1969, indicated increased responsibilities for the Kuwaiti armed forces.

A new politico-military crisis exploded on March 20, 1973, when Iraqi forces occupied the Kuwait border post of Al Samitah and shelled another post. Each side blamed the other for provoking the incident. Kuwait immediately declared a state of emergency, reinforced the border, and appealed to Arab and world opinion. Under pressure from the Arab League and Egypt, Iraq withdrew from the occupied territory in early April. A major factor occasioning this withdrawal, according to some reports, was a Kuwaiti hint that its financial aid to the Arab world in its conflict with Israel might have to be reconsidered if Iraq did not desist. During the negotiations the Iraqi foreign minister voiced a claim to the Kuwait offshore islands of Warbah and Bubiyan (see fig. 6). The border incident and the claim to the islands reminded Kuwait that the old Iraqi designs on Kuwait were not dead (see ch. 2).

On September 17, 1976, one of the government-controlled newspapers in Kuwait carried a brief news report that an Iraqi military force (size unknown) had crossed the Iraqi-Kuwaiti border and had established a "base" some two to three miles inside Kuwait territory. Radiobroadcasts from Cairo on the same day quoted Kuwaiti sources as stating that Egypt was seeking to mediate the border dispute. On September 19 *Ar Rai al Amm* published a letter it had received from the Ministry of Information; the letter chastised the newspaper for its report on the border issue but neither denied nor confirmed the accuracy of the newspaper's report. The fact remained that neither the boundary problem nor the Iraqi claim to the two islands had been settled by late 1976.

The development of Kuwait's armed forces has thus taken place in the context of the national development of a traditionalist Arab country that has been fully independent for only fifteen years. Kuwait is a country unique in the world because of its small size but enormous, single-source wealth, a country governed as a welfare state and family preserve by a paternalistic, constitutionally established monarchical line and one in which less than half the population are citizens. Because of these circumstances Kuwait's defense and security forces have been and are characterized by defensive orientation, professionalism, high pay and emoluments, high morale, small size, the best equipment that money can buy and that can be used, and close identification with and support of the ruling Al Sabah house. In the mid-1970s the principal external influence on equipment, organization, and training continued to be British, although additional strong influences from the United States and France appeared after 1973.

External and Internal Threats

The country's small size and extremely limited military power base effectively preclude aggressive policies. The nation's vast oil wealth, however, makes it vulnerable to aggression by its much larger neighbors. British protection shielded Kuwait during the first half of the twentieth century and well after World War II as the governments and alignments of the Persian Gulf states became stabilized. Kuwait, in other words, was not virgin territory open to claims by an adventurous power. The grave political consequences of disturbing the recognized power balance in the gulf region and the likely repercussions appeared to be a deterrent against incursion by distant powers or expansionism at Kuwait's expense by such countries as Iraq, Saudi Arabia, and Iran. The small princely states and the United Arab Emirates (UAE) of the lower gulf—with whom Kuwait has cordial relations—do not present conceivable threats.

The principal threat continued to be from Iraq because of its hostile governmental ideology, its much smaller proven oil reserves, and its historical claim to the Kuwait territory that appeared to persist despite the 1961 recognition of Kuwait's independence. As former United States Under Secretary of State Joseph J. Sisco stated in testimony before the House Committee on International Relations in June 1975, "Kuwait's primary concern has been the absence of any acceptance by Iraq of the present boundary between the two countries."

An additional external threat to Kuwait arises from its extensive political and economic involvement in the Arab-Israeli conflict. Since independence Kuwait has been very active in Middle Eastern affairs, using its money to create favorable relations, mediating inter-Arab disputes (with much success in some instances), supporting Arab nationalist causes in general, and contributing billions of Kuwaiti dinars to the "confrontation states" of Egypt, Jordan, and Syria and to the Palestinian movements in the conflict with Israel. Although Kuwait City is far distant from Israel (about 800 airline miles), Kuwaiti military and logistical installations might not be immune to Israeli air attacks in any future Arab-Israeli war.

Internally the principal threat to Kuwait's security appears to lie not in ordinary crime or possible palace coups but in political agitation and potential public disturbances generated by radical activists from among the large Palestininan minority estimated to account for almost half of the noncitizens, who make up about 53 percent of the population. Despite its well-known financial support for the Palestinian political and guerrilla movements and its earlier tolerance of strongly pro-Palestinian factions in the press, Kuwait has not been invulnerable to criticism and has experienced a number of terrorist incidents by radical Arab elements. In late summer 1976 a combination of internal stresses and external factors produced a domestic crisis in which Amir Sabah al Salim Al Sabah dissolved the National Assembly (Assembly) on August 29 and in the next few days closed down opposition newspapers and instituted a period of rule by royal decree (see Mass Communications, this ch.). In this constitutional crisis the ruler and his supporting cabinet family—apprehensive for the security of their regime and fearful that the bitter conflict in Lebanon would spread into Kuwait—confronted the legislature and the PLO and Palestinian elements in the country (see Internal Security, this ch.).

Missions of Defense and Security Forces

The classic missions of defense of the state by land, sea, and air are applicable to the armed services of Kuwait, with adaptations occasioned by the country's particular circumstances. The preservation of national sovereignty and honor requires armed forces to validate a national identity and independence. The small size of Kuwait and its population, however, provides a correspondingly small base for the development of military power. It is clearly not militarily rational to suppose that Kuwait alone could successfully defend itself indefinitely against a sustained and determined attack by—for example—Iraq, whose Soviet-equipped forces were in 1976 at least ten times as large as those of Kuwait. Kuwait's defensive strategy, therefore, must be politically one of developing friendly supporting alignments with larger powers but avoiding disturbance of the status quo in the gulf region and militarily one of imposing costly delay on any attacker. In his 1975 congressional testimony on Kuwaiti purchases of several advanced United States weapons systems, Under Secretary Sisco summed up the situation: "These weapons systems have been purchased by Kuwait for the purpose of reinforcing its defense in order to have sufficient forces to slow down an aggressor long enough for either friendly regional forces or diplomacy to come to its aid and bring an end to the fighting."

Political, military, and geographic factors indicate that a war fought by Kuwait's forces would be short—probably measured in days rather than weeks. During that time, however, well-trained and -led Kuwaiti forces with high-quality equipment could conceivably generate such a high level of violence that the initial losses to a potential attacker might in fact discourage such an attack. Kuwait has ample funds to buy the most powerful and sophisticated nonnuclear weapons systems in existence; the open arms sales policies of European nations make many of the systems available; and Kuwait's money and friendly relations with the United States, the United Kingdom, France, and the more advanced Arab countries make training assistance readily obtainable. Because of this combined diplomatic-military-economic position, Kuwait's prospects for survival as an independent state in face of an external attack appear far better than might be supposed.

The army's mission is to contain or delay any ground attack as far forward as possible and to defend the capital. The mission of the air force, in conjunction with the Hawk missile system being developed in 1976, is air defense. Its additional missions include interdiction of the invading force, close air support of ground forces and reconnaissance and transport functions. Ground and air forces would seek to impose rapid, heavy initial casualties on the attacker, and the development of a credible, recognized capability for so doing is inherent in their peacetime training requirements.

Kuwait's National Guard—a semiautonomous body—has guard and surveillance duties on the border and at the oil fields and public security functions in inland rural areas. It acts as a reserve for the regular forces, reinforces the metropolitan police as directed, and in times of war or emergency would have rear area security responsibilities. Kuwait's navy is, in fact, a police coast guard, and has such usual missions as coastal patrol, surveillance of shipping, and prevention of smuggling and illegal entry. The force has no naval combat capability, and defense against attack by sea would have to be conducted by air and on the beach unless more effective capabilities are developed (see Public Security Forces, this ch.).

Kuwait's national police, operating in the metropolitan capital city complex and urban coastal areas, have the usual police functions of crime prevention and investigation, maintaining order and tranquillity, guarding officials and government offices, traffic control, and public assistance. In emergencies, if police capabilities prove insufficient, regular forces as well as the National Guard may be used to establish public order.

In addition to defending the country the regular armed forces may be assigned other tasks by the ruler either at home or abroad in support of larger Arab causes, as shown by stationing a Kuwaiti detachment with Egyptian forces on the Suez Canal in 1970 and 1973. The ruler's policy of aid to the forces of the Arab confrontation states could include the use of Kuwaiti airfields, ports, communications, and logistic facilities, and such activities would probably involve assistance by the Kuwaiti armed forces.

Constitutional Provisions, Command, and Organizational Relationships

Under the 1962 Constitution "the Amir is the Supreme Commander of the Armed Forces. He appoints and dismisses officers in accordance with law." The Constitution also states that "the Amir shall declare defensive war by decree. Offensive war is prohibited." Constitutionally the amir has power to decree martial law for up to fifteen days; periods in excess of that time require approval by the Assembly. If that body has been dissolved by the amir, the state of martial law must be considered by the successor at its first sitting. The Constitution also states that safeguarding the integrity of the country is a trust for every citizen; that military service shall be regulated by law; that by law only the state may form armed forces and public security bodies; that any mobilization must be regulated by law; and that a supreme defense council shall be set up to conduct defense affairs, to safeguard state integrity, and to supervise the armed forces. The council, established by Law Number 24 of 1963, included the amir, the prime minister, ministers of defense and interior, and others who might be designated.

Because of the size of the army the minister of defense directs the armed forces through the army chief of staff and his headquarters staff. The small air force is responsible to the army chief of staff but has its own commander and a limited functional autonomy. Somewhat similarly the National Guard has its own commander but is under the Ministry of Defense. The public security forces, all under the minister of interior as director of national security, include the police and the coast guard. The coordination of military and security forces has been greatly simplified since January 1965 because the ministries of defense and interior have continuously been headed by the same man: Shaykh Said al Abd Allah al Salim Al Sabah, nephew of the amir (and eldest son of the amir's predecessor) and cousin of the prime minister, Shaykh Jabir al Ahmad al Jabir Al Sabah. The tight grip of the ruling family on the organs of power is further illustrated by the incumbents of the offices of chief of staff and deputy chief of staff who were respectively, in late 1976, Major General Shaykh Mubarak al Abd Allah al Jabir Al Sabah and Brigadier Shaykh Salim al Muhammad al Jabir Al Sabah. Most of the senior posts of the security services were also believed to be held by members of the Al Sabah family.

Expenditures, Procurement, and Foreign Aid

At the time of independence in 1961 Kuwait's expenditure for defense and security was about KD16.4 million (for value of the Kuwaiti dinar—see Glossary). Total forces were about 2,500 in a population of about 220,000. By 1966 the annual expenditure was about KD25.6 million. The first and main surge of personnel expansion had occurred, and total forces numbered 7,000 to 8,000 in a population of about 470,000. During the 1961–66 period arms procurement increased, but the main procurement surges were yet to come.

By a single order costing approximately KD6.4 million in 1968, Kuwait purchased fifty British Vickers thirty-seven-ton medium tanks, to be delivered over the next four years. During this four-year period small numbers of such other major items as jet fighters, combat helicopters, and transport aircraft, including the C–130 Hercules, were secured from Great Britain and the United States.

The major defense procurement action occurred in 1973. In May of that year the government announced plans to order arms from the United States valued at about US$500 million. On July 3, 1973, the Assembly authorized the expenditure of some KD422 million, or about US$1.4 billion, over the next seven years for military equipment. By mid-1976, however, the funds reportedly had been expended or committed and were to be supplemented by further allocations.

The 1973 program was distinguished by five major transactions: two with France, two with the United States, and one with Great Britain. In February 1974 Kuwait announced that it had signed contracts with France for arms purchases totaling about KD25.2 million. This announcement was followed by simultaneous statements in mid-April 1974 from the French and Kuwaiti defense ministers that France had agreed to sell Kuwait its most advanced jet fighter, the Dassault Mirage F-1. Early reports indicated that thirty-two of these aircraft were on order for delivery in 1976; by early 1976, however, it appeared that twenty might be a more accurate number. Also in 1974 Kuwait contracted with France for ten Aerospatiale SA–330 Puma combat helicopters, twenty Aerospatiale SA–341 Gazelle combat helicopters, and a mixture of the best West European antitank missiles to arm the helicopters and for surface use. These missiles included the French SS–11 and Harpon and the new HOT—a Franco-German Euromissile. The total value of these transactions was not accurately known in 1976, but in 1974 the cost had been estimated at KD41.5 million.

The largest single arms deal was that announced on November 7, 1974, to secure thirty-six McDonnell-Douglas A-4M Skyhawk jet fighter-bombers from the United States for KD74 million, about US$250 million, for delivery in 1977. On the same date a second agreement was signed with the United States for the purchase of fifty improved Hawk air defense missile launchers with a total system cost of KD26.6 million, about US$90 million, also for delivery beginning in 1977. These costs, as for other major purchases, included supporting equipment and ammunition and allowances for spare parts and training.

Until the negotiation for these massive arms contracts with France and the United States in the 1973–74 period, Great Britain had been almost the exclusive military supplier. From 1965 through 1974, according to calculations of the United States Arms Control and Disarmament Agency (ACDA), Great Britain provided about 88 percent of Kuwaiti arms procurement, the United States about 9 percent, and all other sources the remainder. After 1975 Great Britain once again claimed a major share of the Kuwaiti market when it was announced in late February 1976 that a contract had been concluded for 165 British Chieftain main battle tanks at a cost of about KD66.0 million, or approximately US$230 million.

From time to time Kuwaiti defense authorities have indicated that arms and equipment might be purchased from the Soviet Union. For example, in the interview of April 17, 1974, in which he announced the contract for the French Mirage F–1, the defense minister, Shaykh Said, also stated that a military delegation had visited the Soviet Union to consider its weapons and terms of availability. This development may have come about as a combination of Kuwaiti desires to check all major possibilities, keep its options open, diversify its sources, and placate the radical Arab states whose arms supplies were chiefly from the Soviet Union.

In September 1975 the deputy prime minister, Shaykh Jabir al Ali Al Sabah, stated that arms would be secured from the Soviet Union under an agreement reached earlier that year. In the next two months another Kuwaiti military delegation and the foreign minister, Shaykh Sabah al Ahmad al Jabir Al Sabah, visited Moscow for talks on sophisticated arms purchases, and it was again announced that agreement had been reached. Reports indicated, however, that only artillery—a major category of weapons neglected in the Kuwaiti inventory—had been specifically provided for. The defense minister in late July 1976 stated that a new decision on Soviet arms had been reached and implied that procurement would be initiated. So far as was known in late 1976, however, Kuwait had neither received nor utilized Soviet matériel or training. Some analysts observed that, if Kuwait did commence arms deals with the Soviet Union, this action might be for the purchase and transfer of arms as aid to such friendly Arab countries as Syria and Egypt, whose forces possess Soviet hardware.

Although Kuwait's wealth is not unlimited, it has been more than ample for the operations and maintenance costs of its small forces and for its large arms procurement expenditures, which may be regarded as extraordinary in view of the size of the forces. In the surplus welfare economy of Kuwait the expression "relative burden," commonly used to mean military expenditures as a percentage of gross national product (GNP), becomes academic. Definitions and calculations of GNP differ among sources. Considering all military expenditures, however, the relative burden in Kuwait in the mid-1970s has been estimated by ACDA to be in the range of 2 to 5 percent of GNP.

Kuwait has been both a source and a recipient of foreign military aid —a source of money, a recipient of training assistance. Egypt, Jordan, and Syria have been the conspicuous beneficiaries of Kuwait's financial aid. Some of the grant or long-term loan aid given to these countries —and to the PLO—has been specifically designated as military; some of it has simply been in the form of undesignated budget support funds, which may or may not have been used wholly or partly for military purposes. The variety and number of transactions thus makes it virtually impossible to determine with accuracy Kuwait's actual total military expenditures.

Training assistance by Great Britain for the armed services and security forces was continued after independence. Many Kuwaiti officers have been trained at the British military academy at Sandhurst and many more at military academies in such Arab countries as Egypt and Jordan. Great Britain also provided flight and technical training. After the negotiation of contracts with France and the United States in 1973 and 1974, however, diversified training assistance for the new equipment became necessary. Although the British connection continued, by early 1976 it was estimated that at least 3,000 Kuwaitis were attending military courses abroad and that the majority were in the United States and France. Among the earliest in the United States were twenty-four pilot trainees who arrived in March 1974. By January 1975 there were 100 Kuwaiti military students in the United States at fourteen army and air force installations. By mid-1975 over 500 had received United States military training in Kuwait or the United States. Since then requirements and numbers have increased. The Hawk missile system, for example, has required training for perhaps 1,000 Kuwaitis, and the first step after selection and screening has necessarily been one of intensive English-language instruction.

Kuwait has no real arms production industry. It has pledged to contribute financial support to the Arab Military Industries Organization, whose formation by Egypt, Saudi Arabia, Qatar, and the UAE was announced in July 1975 for the purpose of reducing Arab dependence on the great powers as arms suppliers. By late 1976 it appeared that this project was evolving slowly and would require at least three or four more years before significant production could be expected. Kuwait's contributions, at least initially, would be classed as development or investment funds rather than military expenditures.

Military expenditures in general include capitalization procurement, military aid, and operation and maintenance of forces. The last consists of all personnel-related costs, upkeep of facilities and parts, maintenance of equipment, and such consumables as food, ammunition, and gasoline. Kuwaiti government figures for expenditures by the defense ministry, including the National Guard, show an increase of 98 percent in fiscal year (FY) 1975 as compared with FY 1968 (see table 15). These

Table 15. Kuwait, Ministry of Defense and Ministry of Interior Expenditures, Fiscal Years 1968–75
(in thousands of Kuwaiti dinars)[1]

Year	Expenditures[2,3] Ministry of Defense	Expenditures[4] Ministry of Interior	Total Defense and Interior	Total Kuwait Government	Defense and Interior as Percent of Government Total
1968	21,082	16,924	38,006	344,037	11.1
1969	24,000	19,085	43,085	285,164	15.1
1970	25,688	19,296	44,984	311,143	14.5
1971	26,000	20,194	46,194	334,210	13.8
1972	30,393	22,091	52,484	382,528	13.7
1973	32,043	28,455	60,498	453,914	13.3
1974	39,049	32,648	71,697	474,253	15.1
1975	41,798	35,876	77,674	452,523	17.2
TOTAL	240,053	194,569	434,622	3,037,772	14.3

[1]For value of the Kuwaiti dinar—see Glossary.
[2]Includes National Guard; does not include military aid.
[3]Fiscal year 1977 to be approximately KD65.8 million.
[4]Includes security services (police and navy) as main item.

Source: Based on information from *Annual Statistical Abstract, 1975,* Kuwait, 1975, p. 41.

figures represent actual expenditure outlays in the years cited and do not include grants-in-aid to other countries. A similar comparison for the Ministry of Interior (including the police as the main item) shows an increase factor of 2.1; expenditures for the two ministries were 2.0 times as great in the latter year. Using different definitions that included annual outlays, obligations incurred, and grants-in-aid given, ACDA estimates indicate that Kuwait's military expenditures in 1975 were about 5.7 times as great as in 1968. The substantial increase reflects not only Kuwaiti expansion and inflation but also magnified military and financial aid to a number of Arab countries after the June 1967 War.

Manpower and Personnel Policies

Although the obligation of military service is legally established, the armed forces and security forces have been manned by volunteer career personnel. All military personnel are Kuwaiti citizens. The Kuwaiti military manpower available in the peculiarly structured population of the country is sufficient for the small forces of somewhat more than 15,000—including police—but is by no means excessive. Personnel strength is supplemented by foreigners in a quasi-military civilian capacity under a category called "Non-Kuwaiti Professionals," organized under five grades. Technicians have been recruited from among the resident Palestinians, Egyptians, Jordanians, and other foreign nationals; they did not have command positions. The number of Non-Kuwaiti Professionals working for the armed forces in late 1976 was not reliably established in public sources.

Some observers have speculated that the Non-Kuwaiti Professionals employed by the services for administrative, technical, and logitical work might subvert or take over the armed forces. Although the chances of sabotage or terrorism could not be entirely ignored, this possibility was considered slight because the command positions were all held by carefully chosen Kuwaitis. Air force pilots had to be Kuwaitis, and the combat arms and the National Guard were almost all manned by Kuwaitis, who received higher pay than non-Kuwaitis. In the mid-1970s observers estimated that up to 80 percent of the men of the combat arms and the National Guard were of beduin background. Among them the tribal traditionalist patterns so thoroughly known to the Kuwait ruling house are particularly strong, and when well paid these men are not particularly susceptible to the inducements of foreign urban radicals.

The total armed forces and security strength was about 2,500 in 1961; it increased to about 8,000 in 1968 and to about 15,000 in 1975. The principal jump was from 10,000 to 14,000 in the 1970–71 period. These figures were believed conservative, but a maximum number of

25,000 may represent as large a force as the system and population can manage.

After more than two years of study the Assembly on January 24, 1976, passed a compulsory induction law under which all Kuwaiti males between the ages of twenty-one and thirty years would be required to have military training for eighteen months, a move indicating that the military expansion program might be encountering manpower problems requiring a draft. Numerous exceptions to the law were provided, however, in the form of deferments and exemptions for several categories of students, dependents, and others, and as of late 1976 the induction law was not yet in operation.

Status of the Armed Forces in the Society

Some commentators have described Kuwait's armed forces as an elite element in the society whose loyalty is secured by traditionalism and professionalism but above all by privileges, high pay, and the best equipment that money can buy. Other observers have noted that, among the commercial, sea-oriented peoples of the coast and the urban foreign concentrations that have come about with the oil wealth, military personnel are not held in high esteem. Both of the views have some validity. The well-kept armed forces do not appear to have been used or to be regarded as an agency of heavy oppression as of late 1976; they are, however, clearly identified with the ruling house and its government.

Organization and Administration

The Army

In 1976 the regular ground forces had a reported strength of 8,000—a figure that had been used in public sources since 1973 and was probably low. The principal combat formations were one armored brigade and two composite brigades each having infantry, armor, and artillery units, supported by a central administration and logistical facilities. Organizational patterns and administrative and logistical systems revealed the legacy of the long and continuing British influence.

Major items of equipment included fifty Vickers and fifty Centurion British medium tanks; a mix of about 250 British Saladin armored cars, Saracen armored personnel carriers, and Ferret scout cars; and a total of about forty British 25-pounder gun-howitzers and United States 155-mm howitzers—both old but proven pieces for light and medium artillery purposes. Since 1962 antitank defenses—other than tanks—have been based on the British Vigilant antitank ground missile of about one-mile range. The addition of the

French SS–11 and Harpon antitank missiles with improved guidance and about a two-mile range has substantially upgraded antitank defense since 1974. The United States improved Hawk air defense missile system being developed in 1976 was expected to include about fifty launchers, possibly organized in five batteries of nine launchers each. This system will provide one of the best national air defense missile coverages in the world but will add, in all probability, at least 1,000 men to the total army strength.

The Kuwaiti tank holdings were also of particular interest. To supplement the mid-1970s inventory of about 100 middle-aged but by no means obsolete Vickers and Centurion medium tanks, the government in early 1976 had ordered about 165 new British Chieftains. On the basis of fifty tanks per battalion this would suggest five battalions, leaving perhaps fifteen tanks as a maintenance reserve. Three Chieftain battalions might then be allocated to the armored brigade and one Vickers or Centurion battalion to each of the composite brigades. In any case an increase in regular personnel of perhaps 1,500 men seemed likely as a result of new tank acquisitions unless the older vehicles were to be retired or otherwise disposed of.

Infantry, armor, and artillery training centers were located on the outskirts of Kuwait City and near the Burgan (Al Burqan) oil field. There was no military academy or staff college in late 1976, but the expansion of forces and new high-technology acquisitions indicated that attention would be given to the formation of academies and advanced technical schools.

The Air Force

The air force, commanded in 1976 by Colonel Marzuq Ajil, had a reported strength—probably low—of 2,000. Before independence Kuwait had about a dozen light transport and communication aircraft and in 1961 immediately began developing the capability to handle high-performance aircraft by procuring six British Aircraft Corporation (BAC) Hunting Jet Provost trainers. Procurement thereafter included a variety of jets and transport and trainer aircraft, including helicopters. Great Britain supplied most of this inventory and also provided training until the major contracts were let with France and the United States in 1974.

In early 1976 Kuwait had thirty-two combat aircraft, organized in three squadrons: one fighter-ground attack squadron of six British FGA-57 Hawker Hunters; one fighter-interceptor squadron with fourteen supersonic F-53 and T 55 BAC Lightnings armed with British Hawker-Siddeley Redtop air-to-air missiles; and one counterinsurgency (COIN) squadron of twelve BAC–167 Strikemaster jets (an improved version of the Hunting Jet Provost). In addition there were a transport squadron with two Canadian DHC-4 Caribou, two United States Lockheed C-130 Hercules, one old Hawker-Siddeley Argosy, and several older and smaller craft and a helicopter squadron with four Bell 205 and two Bell 206 (jet) aircraft made in Italy and one British Westland Whirlwind.

Additional aircraft being delivered beginning in mid-1977 as a result of the 1974 contracts consisted of thirty-six United States A–4M Skyhawk fighter-bombers. Twenty French Gazelle and ten French Puma helicopters armed with the HOT antitank missile, and twenty French Mirage F–1 air superiority fighters possibly armed with the French Matra Magic air-to-air missile were in Kuwait at the close of 1976. The first squadron of Mirage F-1s (believed to be six) was delivered on July 26, 1976, having been flown from France by Kuwaiti pilots trained by the French.

As in the case of the new tanks and Hawk missiles, the addition of the new aircraft seemed certain to result in an increase of air force strength by at least 100 percent unless some aircraft were retired or transferred. Both the International Institute of Strategic Studies in London and the Stockholm International Peace Research Institute (SIPRI) have consistently reported that all or some of the Mirage F–1s are intended for Egypt in one of the multicornered arms transfers that have become common in the Middle East.

Whether or not Kuwait keeps the Mirage F–1s, the increased size of the air force presented operations and training problems in terms of airspace and ground facilities. Analyst John Duke Anthony has noted that Kuwaiti authorities have expressed interest in the use of the small offshore Bahraini island of Al Muharraq "as a site for stationing units of the Kuwayt Air Force to add protection against the possible revival of an Iraqi threat to Kuwayt." On April 17, 1974, Defense Minister Said was quoted by the Kuwait press as saying, "There is an understanding between Kuwait and the neighboring states on this subject. Their airspace is open to our planes for training within the purview of coordination on the subject." Regarding ground facilities, reports in early 1976 indicated that a new airbase was to be constructed at a cost of about KD57.5 million.

Whether the Hawk missile system was to fall under air force command, to be classed as an army ground force element, or to become a separate service (air defense command) under central armed forces command was not public knowledge in 1976. In any case the expansion of the air force suggested that it was likely to become structurally and administratively more autonomous.

The National Guard

There is little known about the Kuwait National Guard—including the strength, which may be several thousands, organized into area or district platoons, companies, and possibly battalions. It does have its own commander, a post requiring political reliability and familiarity with traditional tribal and village conditions. The guard is budgeted under the Ministry of Defense and responds to the directives of the minister and the chief of staff.

The National Guard is supported by the army logistical system for supply and maintenance; its requirements for equipment and facilities are much simpler than those of the regular army and are met at lower priority. Regular army officers are detailed to duty with the National Guard, but the grades of warrant officer and below are specifically established and paid for by the guard. The National Guard's volunteer part-time personnel constitute both an army reserve and a militia for guard and surveillance duty. Personnel are also available for internal security functions in rural areas or as reinforcements for the police in urban concentrations. The small size and power of the National Guard make it far less than a military counterpoise to the regular forces, but it does constitute a semiseparate source of security control and intelligence.

Conditions of Service

Military Justice

The military justice system is embodied in law reflecting British and Egyptian influences (see Government and Politics, this ch.). Military courts, based in constitutional origins of authority, have jurisdiction only over military offenses committed by members of the armed and security forces within the limits specified by law. This limitation is lifted when martial law is in force; at such times martial law courts are formed in addition to the usual courts-martial system applicable to the administration of the military and security services themselves.

Uniforms, Ranks, and Insignia

Uniforms in all services closely resemble those of the corresponding British services, a consequence of the long period of British influence and of British guidance in the formation of Kuwait's national forces. Nevertheless distinctively Arab items of dress and accoutrement are seen in some units.

The system of officer, noncommissioned officer, and other ranks is the British system as modified and generally standardized among the Arab countries of the central Middle East and North Africa (see table 16). The Arabic words for the various ranks are the same for all services including the police and coast guard. Specific distinction may be made for the particular service by adding the appropriate service modifier; for example, *naqib al shurta* is a captain *(naqib)* of police *(shurta)*. Insignia of rank also follow the basic British system in modified form, employing a five-pointed star (corresponding to the British pip), a crown, and the crossed saber and baton.

Pay Scales

In 1976 the pay scales of the regular armed forces and police forces were believed to be the most generous in the Arab world and probably among the highest in the third world. Each grade had an entry pay rate and a maximum rate after a designated number of in-grade steps. Special, fixed additional living expense allocations called social allowances were paid to all grades, varying in amount according to grade and whether the recipient was single or married (see table 17). A minimum period in years of service was required in each grade as a prerequisite for promotion to the next grade. The noncommissioned officer and other grades specifically established for the National Guard differed slightly from the regular services and, being less than full

Table 16. Kuwait, Armed Services Ranks and Insignia, 1972

Rank	Arabic Name	Insignia[1]
Supreme Commander	Al Amir (The Ruler)	(Personal choice)
Field Marshal[2]	Mushir	Crossed saber and baton within a wreath
General[3] / Lieutenant General[3]	Fariq	Crossed saber and baton and one crown
Major General	Liwa	Crossed saber and baton and one star
Brigadier General	Amid[4]	Crown and three stars
Colonel	Aqid	Crown and two stars
Lieutenant Colonel	Muqaddam	Crown and one star
Major	Raid	Crown
Captain	Naqib	Three stars
First Lieutenant	Mulazzim Awwal	Two stars
Second Lieutenant	Mulazzim Thani	One star
Chief Warrant Officer	Wakil	One gold bar
Sergeant Major/Warrant Officer	Raqib Awwal	Crown and three chevrons
Sergeant	Raqib	Three chevrons
Corporal	Arif	Two chevrons
Private First Class	Jundi Awwal	One chevron
Private	Jundi	—[5]

[1]All stars are five pointed.

[2]Grade may be assigned to minister of defense.

[3]No grade of full general listed in Kuwaiti pay tables and no incumbent in grade of lieutenant general in 1976.

[4]May be translated as brigadier or brigadier general and regarded as either depending upon whether British or United States customs, respectively, are being followed.

[5]No insignia.

Source: Based on information from U.S. Department of Commerce, Office of Technical Services, Joint Publications Research Service, "Law Concerning Military Pay and Allowances, Kuwait," 1972.

time, were less well paid (see table 18). The least well paid full-time members of the defense and security establishment, as a group, were the Non-Kuwaiti Professionals (see table 19).

Public Security Forces

The Police

In the mid-1970s the police continued to be organized corresponding to the nation's three governorates of Hawalli, Al Ahmadi, and Kuwait City. In each of these the governor, who was appointed by the amir, was assisted by a police officer known as the director of security, who was appointed by the minister of interior. Under the director of security there were within each governorate a number of area or district police commands, each headed by an area commander who was appointed by the governor.

The Kuwait Governorate was the largest. Its director of security was a full colonel; the nine area commands under his jurisdiction were headed by officers having the grades of captain, major, or lieutenant colonel. These officers supervised a total of twenty-seven police stations, each manned in shifts by three or four officers, three or more noncommissioned officers, and ten or more policemen and police guards.

The central directorate under the Ministry of Interior included administrative services and the departments of public security, general police affairs, criminal investigations, execution of sentences, passports, and traffic; a special branch targeted against subversion, conspiracy, and espionage; and the Kuwaiti Police Academy and Training School. In late 1975 the police service was opened to unmarried Kuwaiti women between twenty and twenty-six years of age who possessed high educational and aptitude qualifications.

The total number of police in 1975 was about 2,000. A five-year plan announced by the Ministry of Interior in September 1975 projected an increase of about 3,000 by the end of the plan. New buildings for the Traffic Department and the police academy were projected, and plans were under way for the development of better transportation and communications. Reportedly literacy among the police had increased from 46 percent in 1967 to 80 percent in 1975, but ability to read and write with fluency was limited to about 10 percent.

Police officer grades were the same as in the regular armed forces, except that in 1976 the highest listed police rank was that of major general. Officer base pay and social allowance scales were also the same. The police, however, did not use the rank of warrant officer but did have five grades of noncommissioned officers and other ranks; in descending order master sergeant, corporal, policeman first class (may be translated as lance corporal), policeman, and police guard. The pay and allowances for a master sergeant were about the same as for a chief warrant officer in the army and for a police guard about the same as an army private. The intermediate police noncommissioned grades were somewhat more highly paid than their army counterparts according to the 1972 scales. The police, like the regular services, employed the category of Non-Kuwaiti Professionals and on the same pay scale.

The Navy (Coast Guard)

The strength of the navy was reported in 1975—as it had been for three previous years—at about 200 men, organized as an element of the police forces under the Ministry of Interior. Its equipment was of British design and origin and included twenty-eight coastal patrol craft varying in length from thirty-five to seventy-eight feet and two eighty-eight-foot landing craft used as logistical vessels for work parties and transport. Some of the patrol boats had speeds up to twenty-four knots; but the heaviest armament carried was the single 20-mm gun, and the force had no naval combat capability.

In 1974 the minister of defense and interior stated that proposals were being studied for the procurement of new and faster boats equipped with long-range sea-to-sea rockets. The reference here was to the high-technology surface-to-surface missile and torpedo boats of various designs usually called fast attack craft, which were becoming popular among Middle Eastern naval forces in the 1970s. No procurement was known to have occurred by late 1976. By February of that year, however, it was reported that Kuwait had made the decision to establish a navy and was entering the world market for combat vessels. At about the same time Kuwait contracted with a British firm for the construction of a naval base on the coast just north of its southern border at a cost of about KD90 million. A navy, if formed, would fall under the Ministry of Defense and might absorb the coast guard and its functions. In any case further substantial expenditures and requirements for additional manpower and training appeared to be on the horizon.

Internal Security

Crime and Prosecution

Under the Law of Criminal Procedure of 1960 offenses against the law are classed as misdemeanors or felonies, according to their severity, with a third category of administrative violations covering infractions, such as minor traffic violations, judged to be less than misdemeanors. According to government statistics for the eleven-year period 1964 through 1974, the number of annual convictions for mis-

Table 17. Kuwait, Pay Scales of Regular Services, 1972
(monthly in Kuwaiti dinars)[1]

Rank	Entry Base Pay	Maximum Base Pay	Fixed Social Allowance	
			Single	Married
Field Marshal	650	650	Allowance for a minister	
Lieutenant General[2]	550	550	Allowance for a minister	
Major General	460	504	65	65
Brigadier General	400	440	65	65
Colonel	328	382	27	65
Lieutenant Colonel	264	312	27	65
Major	222	250	27	65
Captain	186	210	22.5	60
First Lieutenant	156	176	22.5	60
Second Lieutenant	140	148	22.5	60
Chief Warrant Officer	138	154.5	22	48
Sergeant Major/Warrant Officer	103	132.5	22	48
Sergeant	91	100	22	48
Corporal	80	88	22	48
Private First Class	70	78	20	45
Private	62	68	20	45

[1]For value of the Kuwaiti dinar—see Glossary.
[2]No incumbent in 1976.

Source: Based on information from U.S. Department of Commerce, Office of Technical Services, Joint Publications Research Service, "Law Concerning Military Pay and Allowances, Kuwait," 1972.

demeanors varied within a narrow range. This would appear to be remarkable, since the population doubled during the period and the capital city's urban concentrations of foreigners increased. The statistics for felony convictions appear to be more realistic because they show an approximate doubling during the period. Many crimes are believed unreported, including those in which offenders are dealt with under traditional tribal or village and familial customs. Police area commanders in the Kuwait Governorate have seen a need for more "market guards" and in 1974 doubled the number of mobile night patrols. The stresses of urban congestion, inadequate housing, social unrest, and political agitation and the apparent need for 3,000 more police all pointed to the likelihood of increased crime problems in the late 1970s.

The Public Prosecution Office is organized under the attorney general, who is responsible to the Ministry of Justice, Awqaf, and Islamic Affairs (see Government Structure, this ch.). The 1962 Constitution provides that "the Public Prosecution Office shall conduct penal charges on behalf of the society. It shall supervise the affairs of the judicial police, the enforcement of penal laws, the pursuit of offenders,

Table 18. Kuwait, National Guard Grades and Pay Scales, 1972
(monthly in Kuwaiti dinars)[1]

Grade[2]	Base Pay Entry	Base Pay Maximum	Fixed Social Allowance Single	Fixed Social Allowance Married
Warrant Officer	77	102	22	47
Sergeant	62	87	22	47
Corporal	50	75	22	47
Guardsman	40	65	22	47

[1]For value of the Kuwaiti dinar—see Glossary.
[2]Officers of the National Guard are detailed from the regular army.

Source: Based on information from U.S. Department of Commerce, Office of Technical Services, Joint Publications Research Service, "Law Concerning Military Pay and Allowances, Kuwait," 1972.

and the execution of judgments. . . . As an exception, a law may entrust to the public security authorities the conduct of prosecutions for misdemeanors." Under the exception police courts and prosecutors do in fact dispose of most cases of administrative violations and many lesser misdemeanors.

New prison construction was planned under the Ministry of Interior's five-year plan of 1954 to conform to UN resolutions on penal institutions and regulations propounded by the Kuwaiti Council of Ministers in 1972. The number of inmates in Kuwait prisons reportedly dropped from a 1972 average of 5,913 to 5,071 in 1973 and 3,198 in 1974 but was believed to have risen in 1975 and 1976.

Subversion and Political Security

The chief problem of public security in 1976 appeared to be that of social unrest and political security rather than ordinary crime. Despite its heavy contributions of cash and policy support to the PLO and the factional Palestinian nationalist and guerrilla groups, Kuwait has not been immune to extremist terrorism and inter-Arab radical intrigue. For example, eight members of the Popular Front for the Liberation of Palestine (PFLP) were deported in 1972 for bombing a Jordanian embassy vehicle. The government announced in May 1972 that Palestinians would not be permitted to settle permanently because it was expected they would all "return home eventually." Between September 1973 and February 1974 Kuwait was embarrassed on three occasions by terrorist aircraft hijackers and an embassy seizure, and in 1975 authorities prosecuted several Palestinian terrorists in a special state security court for bombings at an American insurance company office and at the British Council.

A confrontation with the PLO developed in August 1976 directly related to the ongoing civil war in Lebanon. Kuwait's policy had been one of seeking a cease-fire and mediation between the leftist Pales-

Table 19. Kuwait, Non-Kuwaiti Professional Grades and Pay Scales, 1972
(monthly in Kuwaiti dinars)*

Grade	Base Pay		Fixed Social Allowance
	Entry	Maximum	
Professional, First Class	73.75	85.00	20
Professional, Second Class	60.75	70.50	20
Professional, Third Class	44.25	58.00	20
Professional, Fourth Class	30.75	42.00	20
Professional, Fifth Class	18.75	28.75	20

*For value of the Kuwaiti dinar—see Glossary.

Source: Based on information from U.S. Department of Commerce, Office of Technical Services, Joint Publications Research Service, "Law Concerning Military Pay and Allowances, Kuwait" 1972.

tinian side in Lebanon and the Phalangist right, supported for various reasons by the "radical" Arab government of Syria. Kuwait refrained from condemning the Syrian intervention in Lebanon and, by continuing its policy of aid to Syria, appeared tacitly to endorse it. This infuriated the Palestinian activists among the large Palestinian element of the population, whose press organs attacked the governments of Syria and, by implication, Kuwait. In response the amir dissolved the Assembly on August 29, 1976, accepted the cabinet's resignation, closed down the offending newspapers, suspended the provisions of the Constitution on press freedom, announced that a constitutional revision would be drawn up in the near future, and formed a new cabinet composed of the former prime minister and ministers plus the new ministers that was believed to portend tighter security measures.

The PLO and associated Palestinian factional groups, excluded from control in Egypt, discredited in Syria, driven out of Jordan, and apparently defeated or fought to a standstill in Lebanon, were thought by some observers to be seeking to establish a new base in Kuwait in 1976. It appeared highly doubtful, however, that a Palestinian-dominated state would be even minimally acceptable to neighboring Saudi Arabia, Iran, or the other smaller gulf states, to say nothing of Egypt, Syria, and Jordan. Given this condition and the apparent continued loyalty to the Al Sabah government of the combat arms of the army and air force, the possibility of a successful Palestinian-backed subversion of government in Kuwait appeared slight, although radical elements might generate public violence and a condition of security repression harmful to the country and—most of all—to these elements themselves.

* * *

Kuwait: Urban and Medical Ecology by G.E. Ffrench and A.G. Hill presents a wealth of information on a variety of subjects. Harry Winstone and Zahra Freeth's *Kuwait: Prospect and Reality* offers a more general but still useful survey. A brief treatment of Kuwaiti education can be found in A.L. Tibawi's *Islamic Education.*

A useful but dated survey of the Kuwait economy was prepared by economists of the International Bank for Reconstruction and Development in *The Economic Development of Kuwait.* The *IMF Survey,* November 24, 1975, contained a brief summary of economic developments. Between 1973 and 1976 the *Financial Times* of London published annual reviews of Kuwaiti developments. The Kuwait government publishes many statistical series; the *Annual Statistical Abstract* contains a broad range of data though not as up-to-date as one would like. Many current statistics are carried in the IMF monthly *International Financial Statistics* and the UN *Monthly Bulletin of Statistics.*

Hassan A. Al-Ebraheem's *Kuwait: A Political Study* is an excellent introduction to the political system of Kuwait. Kuwaiti attempts to develop democratic institutions within a constitutional framework are analyzed in various works by Abdo I. Baaklini, such as "The Legislature in the Kuwaiti Political System." Kuwait's role in regional and international relations is covered in depth in Soliman Demir's *The Kuwait Fund and the Political Economy of Arab Regional Development* and David E. Long's *The Persian Gulf: An Introduction to Its People, Politics, and Economics.*

Among the more reliable sources on Kuwait's military holdings and procurement patterns are such standard reference works as the annual issues of *The Military Balance* by the International Institute for Strategic Studies and the appropriate publications of Jane's yearbooks. The publications of the Stockholm International Peace Research Institute (SIPRI) are particularly valuable, especially *World Armaments and Disarmament: SIPRI Yearbook, 1975* and subsequent editions. Dale R. Tahtinen's *Arms in the Persian Gulf* presents an overview of the region. Although occasionally difficult to obtain, the reports of the hearings by various congressional committees and subcommittees frequently are extremely useful; see, for example, *The Persian Gulf, 1975: The Continuing Debate on Arms Sales* by the Special Subcommittee on Investigations of the Committee on International Relations of the House of Representatives. (For further information see Bibliography.)

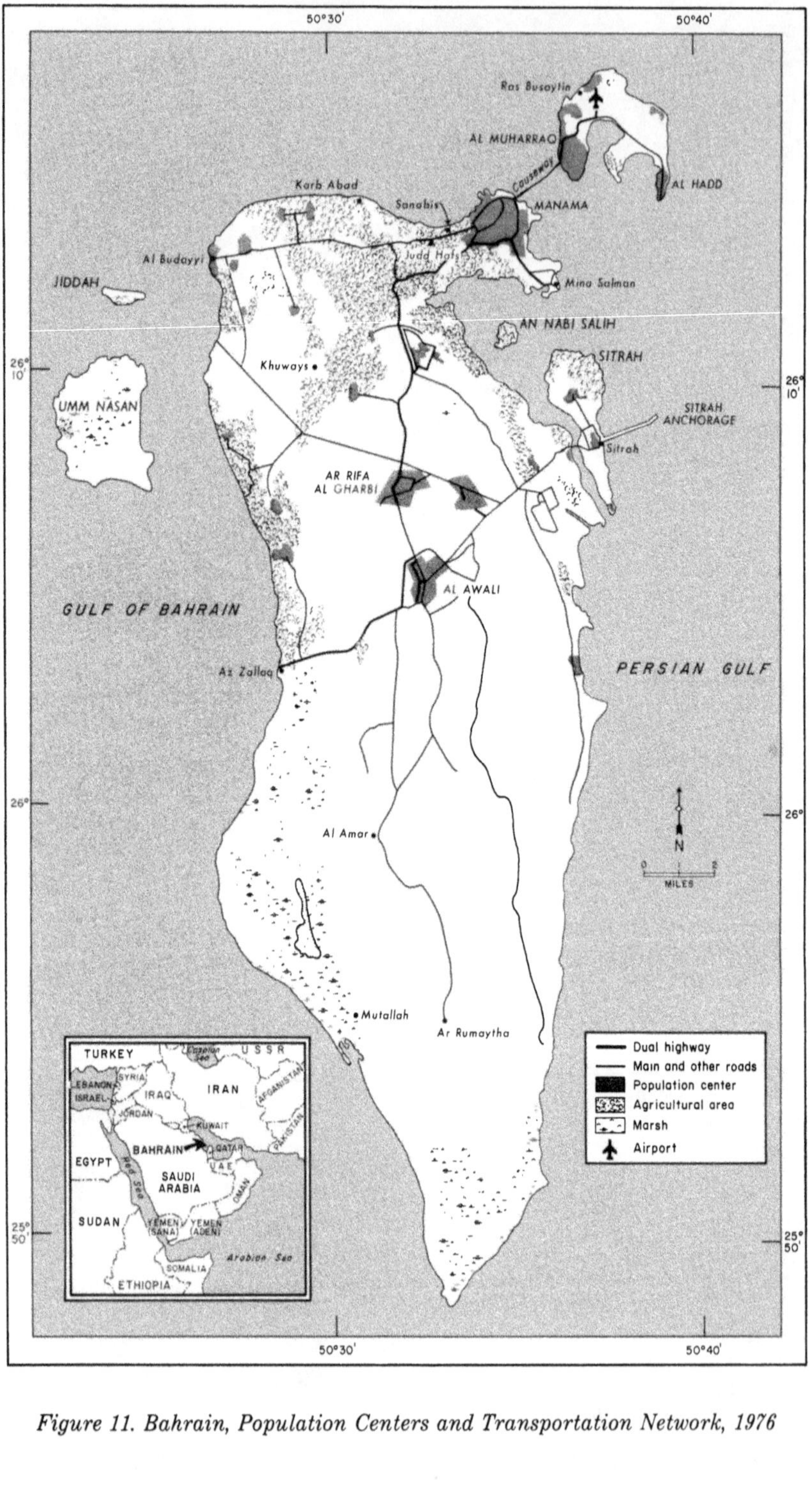

Figure 11. Bahrain, Population Centers and Transportation Network, 1976

CHAPTER 7

BAHRAIN

The country's ruler in 1976 was Shaykh Isa bin Salman Al Khalifah, who had become the head of the Al Khalifah family on the death of his father, Shaykh Salman bin Hamad Al Khalifah, in 1961. Shaykh Isa, who was born in 1933, was the tenth Al Khalifah ruler of the Bahrain archipelago; when Bahrain secured its independence from British protection and suzerainty on August 14, 1971, Isa became the first Amir of Bahrain.

In late 1976 the Al Khalifah continued to dominate the government and the society. Isa's brother, Shaykh Khalifah bin Salman Al Khalifah, was prime minister and head of government; Isa's eldest son, Shaykh Hamad bin Isa Al Khalifah, was defense minister and heir apparent. Seven other members of the royal family served in the fifteen-member cabinet.

The paramount family among the noble beduin tribes from the interior of the Arabian Peninsula who in 1783 expelled the Persians from the islands, the Al Khalifah by the late nineteenth century had adopted a form of hereditary succession. Unlike most Arab monarchies, which select the heir apparent from among the several able males within the royal family, the Al Khalifah succession is based on primogeniture. Since the rule of Shaykh Ali bin Khalifah Al Khalifah (1868–69) each ruler has been succeeded by his eldest son. A royal decree based on the consensus of the family leaders could designate someone other than the eldest son as the heir apparent, but foreign observers believed that the person designated would be of the ruler's immediate family, probably a brother or another son of the ruler. The 1973 Constitution specifies that future rulers must be from the lineage of Amir Isa.

The political stability and relative economic prosperity of the royal family and Bahrain were fixed by the three men who ruled from 1869 to 1961: Shaykh Isa bin Ali Al Khalifah (1869–1932), his son Shaykh Hamad (1932–42), and his grandson Shaykh Salman (1942–61), the father of Amir Isa. During this period the various shaykhs signed treaties of protection with Great Britain, granted oil concessions to foreign companies, and laid the groundwork for the government that was formed after independence. Bahrain was the first Arab state to benefit from the discovery of oil (1932), the first to institute general and free education and public health services (1925), and the first to experience serious domestic unrest (in the 1950s).

By the mid-1970s Bahrain had peaked as an oil producer, and its relatively small production was expected to decline sharply in the 1980s (see Oil Industry of Bahrain, ch. 5). Its economic future and hence its ability to sustain a variety of social welfare programs depend on a continued expansion of refining and shipping capacities and of Bahrain's role as an entrepôt and financial-commercial center (see The Economy, this ch.).

GEOGRAPHIC AND DEMOGRAPHIC SETTING

Bahrain comprises an archipelago of thirty-three islands and a total land area of about 255 square miles. The main island accounts for 85 percent of the total land area; it lies at the entrance of the Gulf of Bahrain, an inlet of the Persian Gulf between the coast of Saudi Arabia and the Qatar Peninsula (see fig. 1). For centuries this position has given the island group regional importance as a trade and transportation center. Only Bahrain, Sitrah, Umm Nasan, and Al Muharraq are of significant size; the remainder are little more than exposed rock and sandbar.

Bahrain, from which the archipelago takes its name, is about thirty miles long and about ten miles wide at its broadest part. Most of the island is desert; low outcroppings of limestone form rolling hills, stubby cliffs, and shallow ravines. The interior contains an escarpment that rises about 450 feet to form Jabal Dukhan, a steep-sided hill where oil was first found in the archipelago (see Oil Industry of Bahrain, ch. 5).

Manama, the capital, is located on the northeastern tip of the island of Bahrain. The main port, Mina Salman, is also located on the island, as are the major petroleum refining facilities and the business and commercial centers (see fig. 11).

Manama is linked with the island of Al Muharraq and its major city of the same name by a 1.5-mile causeway; the international airport is also on Al Muharraq. The island of Sitrah is linked to Bahrain by a bridge spanning a shallow channel. North of Sitrah is An Nabi Salih, where freshwater springs irrigate numberous date groves. Northwest of Bahrain is the rocky islet of Jiddah, which serves as a prison settlement, and south of Jiddah the larger island of Umm Nasan, the personal property of the ruler and his private game preserve. About twelve miles southeast of the island of Bahrain and close to the coast of Qa+ar are the Hawar Islands, the subject of a territorial dispute between Bahrain and Qatar.

By the mid-1970s the major islands were well served by a system of paved roads, some of which were dual highways and most of which were in excellent condition. Between 1966 and 1976 the number of vehicles had more than doubled, and traffic accidents and road congestion had become major problems. The situation was eased somewhat with the introduction of a public transportation system of 120 buses in 1971. Other steps initiated in the mid-1970s included the construction of belt highways around the capital on reclaimed land and additional causeways and bridges linking the three major islands. Travel to other islands, which have only limited land transportation, was by boat. There were no railroads in the country, and none was planned.

Although the climate is relatively pleasant from October to April, it is characterized by intense heat and humidity during the summer months. Between June and September the daytime temperature frequently reaches 106°F. Daily temperatures are fairly uniform throughout the islands. A dry southwest wind, known locally as the *qaws*, periodically blows sand clouds across the barren end of the island of Bahrain toward Manama.

The average annual rainfall is less than four inches; no year-round rivers or streams exist. Rain tends to fall in brief torrential bursts during the summer months, flooding the shallow wadis that are dry the rest of the year and making secondary dirt roads impassable. Little of this water is reserved for irrigation or drinking.

Climatic and geographic conditions of the archipelago offer limited shelter for wildlife. Among commonly found species are gazelles, hare, jerboas, mongooses, and lizards. Migratory patterns increase the bird population during the winter months; typical of those found are ravens, sparrows, bulbuls, pigeons, and flamingos.

Population and Working Force

The 1971 census indicated a total population of 216,078, of which about 18 percent were non-Bahraini. The projection for 1975, which was based on the assumption of no change in vital rates, placed the population at about 266,000. In late 1976 the population was probably about 275,000, but the data were incomplete. Improvements in health and welfare services have been reflected in an increasing annual rate of population growth from about 2.6 percent in 1950 to the extremely high rate of 3.6 percent in 1971. Continued growth at this rate would double the population by 1992. The high cost of providing housing and basic goods and commodities for such an expanded population was a serious concern of the government, and in the mid-1970s the Ministry of Health advocated a family planning program to encourage smaller families.

According to the 1971 census there were about 38,000 expatriates in the country, a figure that had remained fairly constant since 1965 (see table 20). The high percentage of non-Bahraini, which may have increased somewhat during the first half of the 1970s, indicated the attractiveness of the labor market, especially for those with specialized skills. Exact figures were not available, but the majority of foreigners, especially the less skilled, were from such Arab countries as Oman and Saudi Arabia and from Iran, India, and Pakistan. A report in 1976 indicated that there were an estimated 4,500 workers from Great Britain and 3,000 from the United States among the expatriates. Immigration to Bahrain was rigidly controlled by a system of visas and residency and work permits. Acquisition of Bahraini nationality was permitted but very restricted.

The economically active population in 1976 was about 80,000, or only about 30 percent of the population, indicating a high dependency rate in the country. The majority of the working force were employed in manufacturing, oil production, or service occupations. Although the data were imprecise, the "service personnel" category probably meant, for the most part, government employees. A greater percentage of Bahraini than non-Bahraini were employed in clerical, sales, transportation, and communication positions. By industry the service (or government) sector employed the largest number, followed by construction, mining, and manufacturing. Expatriates from Oman and Iran played a major role in the construction industry, and Indians and Pakistanis were important in foreign and domestic trade. British and Americans were active in the petroleum industry, in managerial positions, and in other jobs requiring highly technical skills.

In 1971 unemployment amounted to about 3 percent (1,800 workers). The increase in construction projects during the early 1970s largely absorbed unskilled labor. Unemployment was found primarily among graduates of high schools and liberal arts colleges who were seeking positions for the first time. About 78 percent of the unemployed were Bahrainis. The government was seeking to remedy this situation by insisting that skilled positions be filled by Bahrainis, but a number of expatriates were expected to remain in the labor force to fill unskilled

*Table 20. Bahrain, Population by Nationality and Sex, Selected Years, 1959–72**

	1959	1965	1971	1972
Bahraini				
Males	59,913	72,368	89,772	93,004
Females	58,821	71,446	88,421	91,604
Total	118,734	143,814	178,193	184,608
Non-Bahraini				
Males	17,709	27,016	26,542	27,498
Females	6,692	11,373	11,343	11,751
Total	24,401	38,389	37,885	39,249
Total				
Males	77,622	99,384	116,314	120,502
Females	65,513	82,819	99,764	103,355
TOTAL	143,135	182,203	216,078	223,857

*Month for reporting varies from year to year.

and highly technical jobs. Increased mechanization, specialized training, and greater use of women were planned to reduce reliance on expatriates; the number of working women in fact doubled between 1971 and 1975.

The 1971 census indicated that the population was relatively young, about 44 percent of the total being under fifteen years of age. For expatriates this age-group constituted about 25 percent of the total; for Bahrainis, about 49 percent. For Bahrainis the male-female ratio was 101.5 to 100; for non-Bahrainis, 234 to 100.

Most people lived in the Manama-Muharraq metropolitan area; its percentage of the population had increased from about 55 percent in the early 1940s to about 66 percent in 1971, and it was twelve times the size of the next largest town. Metropolitan growth was somewhat offset in the mid-1970s by the construction of new housing projects outside the capital complex, but government projections indicated that Manama proper was still expected to double in population by 1988. Most expatriates resided in urban areas, and the rural population was almost exclusively Bahraini.

Living Conditions

In many respects the Bahrainis enjoyed a better standard of living than citizens of other gulf states. Families below subsistence levels of income were eligible for support allowances from the Ministry of Labor and Social Affairs. Education was free for all citizens, and medical services were free for the needy. Population centers were relatively modern, and electric service was available to most of the population under a government program. An improved pension plan for government workers was implemented in 1975, and a national social security system was under consideration in 1976. Living standards during the mid-1970s, however, reflected the effects of substantial inflation that had been only partly offset by pay raises and government subsidies (see The Economy, this ch.).

The most serious concern for the average family was the high cost of food. Despite government efforts to reduce food prices, including subsidies for basic food commodities, almost half of the average family budget was spent for food (see fig. 12). Although the country grows various fruits and vegetables—some of which were exported—that provide a wide range of minerals and vitamins, it was not known to what extent the population had a balanced diet. The reliance on imported foodstuffs moreover, exposed the consumer to worldwide increases in the price of food. Fish, the most common and favored food and source of protein, was caught locally. Demand exceeded supply, however, and imports were needed. Experimental programs in poultry and dairy farming could provide new sources of protein (see Agriculture and Fishing, this ch.).

A shortage of housing was one of the more serious problems. Although rents remained low for Bahrainis because of government controls and subsidies in the mid-1970s—averaging less than 15 percent of monthly income—non-Bahrainis faced soaring fees. Most Bahrainis, however, lived in highly overcrowded quarters. Most residences had three rooms or fewer. In some older structures, particularly in small villages, there were as many as nineteen people living in one house. The average number of occupants of new housing was five. The safety of existing structures, moreover, was uncertain since there was no building code. A growing number of houses had running water, but increased saline levels required the purchase of drinking water from various distribution centers; the cost of desalination of all drinking water by 1980 was to be carried by the government. Major urban areas and new housing in other areas had closed sewerage systems, but these facilities were greatly strained. A new system for recycling wastewater for irrigation was under consideration in 1976.

The Ministry of Housing was established in 1974 to conduct a survey of the existing situation, develop a plan for further development, draft

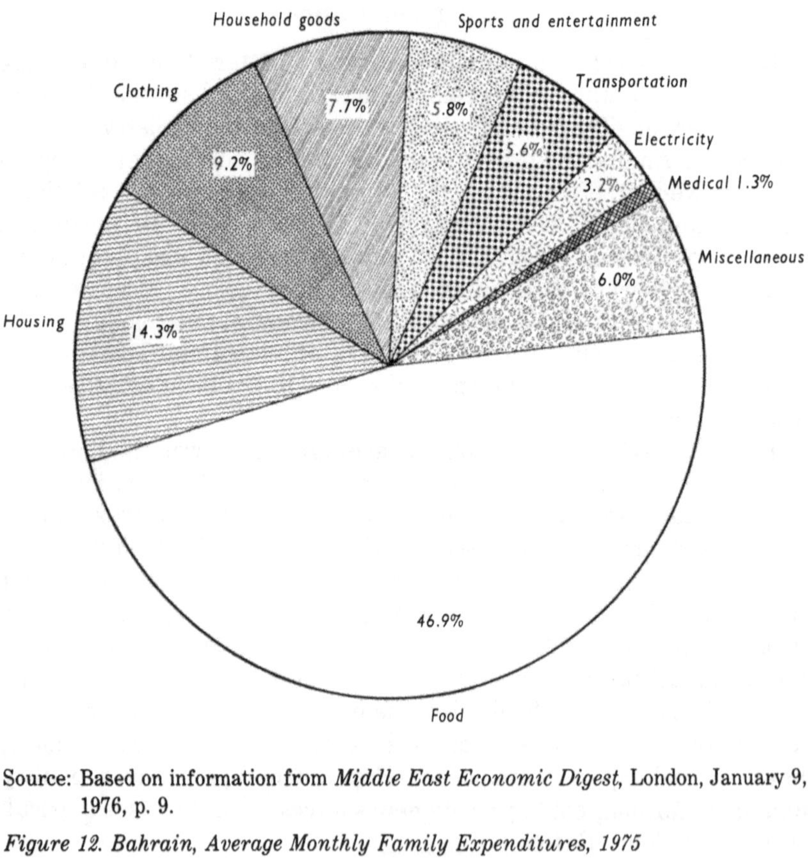

Source: Based on information from *Middle East Economic Digest*, London, January 9, 1976, p. 9.

Figure 12. Bahrain, Average Monthly Family Expenditures, 1975

a building code, and oversee future construction. To be included under the new program were government-constructed housing to be rented or sold at low cost, low-term financing for private construction of houses, and funds for the renovation of existing houses. Efforts to reduce urban congestion, such as the construction that began in the late 1960s of a community called Isa Town about five miles from the edge of the capital complex, were to be given new emphasis. By 1975 the ministry was actively engaged in letting contracts for the construction of precast and prefabricated housing throughout the country under an ambitious plan designed to produce thousands of houses over a five-year period.

Free medical services—including immunization, outpatient treatment, and hospitalization—were established in Bahrain in 1925, and such endemic and infectious diseases as smallpox, trachoma, and dysentery have been significantly reduced. By the mid-1970s about 80 percent of all births took place in hospitals, and infant mortality had been reduced to about twenty-three deaths per 1,000 live births. Tuberculosis remained a serious problem only among those living in crowded quarters. An estimated 8,000 to 10,000 people suffered from alcoholism. Occupational diseases and such hazards as air pollution were of increasing concern.

Medical facilities—including seven hospitals, four school clinics, and twelve other centers—provided a total of about 900 beds in 1971. In 1976 efforts were under way to expand medical facilities under a long-term plan developed in consultation with the World Health Organization (WHO); the plan included major upgrading and expansion of the Sulmaniya General Hospital in Manama. The construction of a 100-bed private hospital to treat patients from upper income families who usually sought treatment abroad was proposed in 1975. Private facilities were also available for military personnel and for employees of certain firms operating in Bahrain.

Although the number of medical personnel—including ninety-nine doctors and 444 nurses in 1971—was gradually expanding to levels considered optimal by the Ministry of Health, the medical service was forced to rely on expatriate personnel. Bahraini women continued to be reluctant to enter nursing, and in the mid-1970s about 85 percent of the nurses were foreigners. Various training programs in the country and abroad had begun to increase the participation of Bahrainis in the medical service by the mid-1970s, however.

Education

Bahrain has been investing heavily in education for over fifty years, and throughout the early 1970s education was one of the largest recurrent expenditures in the national budget. Primary education for boys was initiated in 1919, and separate facilities for girls and various secondary programs were subsequently established. In 1976 students continued to be segregated on the basis of sex. Education was free through the secondary level; students received supplies, uniforms, meals, and transportation to and from school at no charge. About 50 percent of the population over ten years of age in the mid-1970s was literate. About one-fourth of the total population was enrolled in school. The vast majority of these attended government schools, but there were a few private schools.

The educational system was composed of six years of primary education, three years of intermediate education, and three years of secondary education. Secondary education included general, technical, commercial, and religious education for boys and general and commercial education for girls. Higher education consisted of teacher training programs and study abroad. Fellowships were available for study abroad, and in the mid-1970s about 1,400 Bahrainis were engaged in such study. Curriculum was controlled by a government inspectorate and included a wide range of subjects. Adult literacy and vocational training programs were also available. Little was done to educate the physically or mentally handicapped, although in 1974 a school for the blind was established in Al Muharraq.

In 1971 there were about 37,000 students in primary schools, over 7,000 in intermediate schools, and over 5,300 in secondary schools. The percentage of girls enrolled in intermediate and secondary schools had increased significantly since 1965. Ninety percent of all intermediate school students sought and were automatically granted admission to secondary school. Three times as many secondary school students specialized in liberal arts as in the sciences. Upon graduating most students sought clerical and office jobs rather than continue with higher education, although most encountered difficulties in finding such positions. The government was reconsidering its admissions policy for secondary school and seeking means of increasing interest in technical training.

Projections based on growth patterns in the early 1970s indicated that throughout the 1970s enrollment would increasingly exceed the number for which teachers were available and that by school year 1979 there would be a shortage of about 1,800 teachers. By the mid-1970s all primary school teachers were Bahraini, and the government wanted to replace all non-Bahraini teachers in intermediate education. The government had begun a program to increase student interest in teaching, including a monthly stipend to all students enrolled in teacher training programs.

THE ECONOMY

Farming and fishing supported most of the people of Bahrain before oil was found. The islands also served as ports for nearby areas, particularly the Eastern Province of Saudi Arabia. Early trade contacts resulted in a cosmopolitan attitude in Bahrain and the development of a modern educational system earlier than in most of the Persian Gulf area (see Education, this ch.).

Discovery of oil introduced profound economic changes. The oil industry, unhampered by domestic constraints, grew much more rapidly than the rest of the economy. National accounts statistics were unavailable over time, but the revenues from oil production primarily set the pace for economic growth. The infusion of oil money first stimulated trade and construction and, after a lag, the rest of the economy. The faster growing sectors attracted workers from such traditional pursuits as farming, fishing, and boatbuilding because of the higher pay. By the 1970s considerable change had occurred.

Economists from the International Bank for Reconstruction and Development (IBRD, also known as the World Bank) made estimates of Bahrain's gross domestic product (GDP) for 1972 from the available data. According to their calculations the share of GDP controlled by the oil sector (including refining) was 66 percent; agriculture and fishing, 1 percent; manufacturing (including utilities but excluding oil refining), 6 percent; construction, 2 percent; trade, 6 percent; government service, 9 percent; and all other services, 10 percent. Total GDP amounted to BD133 million (for value of the Bahraini dinar—see Glossary), yielding a per capita GDP the equivalent of about US$1,350. Gross national product (GNP) probably was about one-half of GDP, or about US$650 per capita, because of oil company earnings remitted abroad.

Economic growth and change occurred after 1972, but statistics were unavailable to measure them. The increased revenues per barrel of oil produced raised GDP and the oil sector's contribution to it sharply. The increase of oil revenues stimulated other parts of the economy, particularly trade, construction, banking, and real estate. Between 1973 and 1976 there was substantial inflation. The government raised salaries, subsidized basic food costs, cut electric rates, and imposed rent controls. Control of rents was least effective. By mid-1976 prices for land, housing, and office space had reached very high levels. Some observers claimed that it was cheaper to reclaim land from the gulf than to purchase a site on dry land. A construction boom was also under way, and costs escalated because of bottlenecks in the supply of materials.

There was also real progress (apart from inflation) in the economy after 1972. New industrial plants were completed, contributing to the value added by the manufacturing sector. The boom conditions throughout the gulf resulted in greater use of Bahrain's port for entrepôt trade. Many of the business people seeking contracts in the gulf arrived first in Bahrain, and airport and hotel services considerably increased. In 1976 Bahrain became the Middle East terminal for supersonic Concorde flights. Bahraini urbanity, amenities, and a more educated work force attracted many firms seeking a base for operations in the gulf.

The need to import workers was another profound change that accompanied the discovery of oil. By 1971 the work force of about 61,400 was 37 percent non-Bahraini. Bahrain was not as badly off as some gulf states where expatriates outnumbered the local population, but the boom conditions since 1972 were pushing in that direction. Some observers guessed that the work force was approaching 80,000 in 1976 and that expatriates outnumbered Bahraini workers. The influx of foreign workers was a major cause of the housing shortage. Thus the government wanted the number of foreign workers minimized and Bahrainis trained to take over jobs. Vocational and on-the-job training was pushed in the mid-1970s so that fewer expatriates would be needed in the future.

The Bahraini government has long been aware that the country's petroleum reserves were small and would last only a short time (see Oil Industry of Bahrain, ch. 5). A substantial part of the income from oil was spent to develop the schools, ports, airport, communication links, roads, and other facilities that would maintain the country's position as a regional center. When oil revenues begin to decline in the late 1970s, the other economic sectors will become increasingly important as sources of income and employment. Creation of jobs will become increasingly important because of the population's high rate of increase.

Agriculture and Fishing

Agriculture, including fishing, was the major economic activity before the discovery of oil. In the mid-1970s, however, agriculture's contribution to GDP probably was less than 1 percent. In 1971 the sector employed about 4,000 workers, 7 percent of the work force. Farming and livestock production employed about 2,680 workers, of whom 1,960 were Bahrainis. Production from farming and fishing failed to meet demand, and most food was imported.

Cropping depended completely on irrigation because of the lack of streams and the scarcity of rain. From earliest times springs and wells on the northern rims of the islands of Bahrain, Sitrah, and An Nabi Salih supplied the irrigation water. Groundwater sources were diminishing and becoming more brackish in the mid-1970s, however. Indiscriminate drilling of wells had caused seepage of seawater and contamination of freshwater aquifers by brackish water. The increased salinity of irrigation water combined with inadequate drainage brought salts close to the surface, killing plants in some areas and forcing farmers off the land. In 1972 the government prohibited drilling of private wells and took other steps to halt the deterioration, but the scarcity of water and its salt content posed major constraints on agricultural development.

The cultivated area consisted of about 15,000 acres near the wells; about two-thirds of this area contained date palms, and the remainder was used to grow vegetables, fruits, and forage crops. Even though dates were the principal crop, the country did not produce enough for domestic consumption. Statistics were lacking on most phases of agriculture, including livestock production. Foreign experts estimated that in the early 1970s there were about 5,000 head of cattle, 10,000 sheep, 500 camels, and 40,000 poultry.

The productivity of both land and labor was low in agriculture. In the late 1960s the government introduced such measures as subsidized inputs and improved marketing to help farmers. The government also provided tractors and motorized sprayers for plowing, land-leveling, and spraying services at cost to farmers. Farm cooperatives were encouraged. Nonetheless many experts expected the scarcity of water and its salinity, the traditional agricultural practices and primitive marketing, and the attraction of higher paying jobs in other sectors to keep agriculture from any significant development.

Fishing was similar to agriculture in its dependence on primitive practices and its low productivity per worker. In 1971 about 1,300 fishermen (of whom 1,035 were Bahrainis) fished primarily in local waters using traditional baskets and traps. Marketing was largely as it had been for generations. Statistics were unavailable on the catch, but it did not meet local demand in the mid-1970s. Fish were plentiful in the Persian Gulf, however, and fishing had a development potential.

Shrimping was the only part of the fish industry that had been modernized by 1976. The Bahrain Fishing Company (40 percent foreign owned) was formed in 1967, and by the mid-1970s it operated fifteen modern steel trawlers in Bahraini waters and off the nearby Saudi Arabian coast. The company had shore facilities to process, package, and freeze shrimp for foreign markets. The shrimp catch was about 2,000 or 3,000 tons a year, almost all of which was exported. Shrimp were the only significant export product of complete Bahraini origin other than petroleum.

Industry

Modern industry came to the country with the discovery of oil. An oil refinery was completed in 1936 and enlarged over the years (see Oil Industry of Bahrain, ch. 5). By 1972 oil refining contributed 38 percent of GDP, and all other forms of manufacturing (including electric power and water supply) contributed an additional 7 percent. In 1971 manufacturing other than oil refining employed 7 percent of the labor force. By 1976 other manufacturing activities had increased their contribution to GDP slighly, but oil refining still dominated the sector.

Industry other than oil developed slowly. Oil and natural gas were the country's only known commercially exploitable natural resources, although manganese deposits may turn out to be exploitable. The lack of local materials for processing, the small size of the domestic market, a commercial policy that kept tariffs low to encourage the reexport trade, and the inadequate domestic supply of skilled personnel imposed major constraints on the manufacturing sector.

Electric power and water for industrial development have been in short supply at times but generally remained ahead of demand. The first electric power station was constructed in 1929 to supply the capital city, but not until 1953 was a start made on extending electricity to the rest of the country. Generating capacity grew from 1.8 megawatts in 1950 to 220 megawatts in 1976. Production of electricity grew from 2 million kilowatt-hours in 1950 to 330 million kilowatt-hours in 1973, an average annual increase of 25 percent. Consumption of electricity by both households and industry increased substantially in the mid-1970s, requiring acceleration of construction plans. By late 1976 construction was to be completed on power projects that had originally been planned for completion in 1980. A third power station was to be constructed in the late 1970s.

The government in 1972 approved the country's first desalination plant—because of the diminishing groundwater level—with a scheduled capacity of 5 million gallons a day. The plant, part of an electric power station operating on natural gas, was scheduled for commissioning by late 1976; it was partly financed by a Kuwait investment fund. The distilled water was to be mixed with brackish well water to reduce the salt content by more than one-half and to make the most use of the available water supply. Nearly all urban houses and other buildings had piped water.

The bulk of the country's manufacturing was by small, privately owned establishments, many employing fewer than ten people. These establishments included bakeries, repair shops, producers of building materials and of paper, and printing shops. A few larger establishments were engaged in ship repair, electric motor rewinding, processing food and beverages, manufacturing plastic products, production of oil-well-head equipment, and aluminum smelting. In 1971 a census reported 145 manufacturing enterprises (excluding oil refining and utilities) employing about 4,100 workers, of whom 44 percent were Bahrainis.

In 1965 the government established a development bureau to attract industrial investments as a means of diminishing dependence on oil. Incentives offered to private investors included energy and utilities supplied at nominal rates, freedom from income taxes, low import duties and exemptions under certain circumstances, and government equity participation in some projects. An industrial area was set aside near the port of Mina Salman for lease at nominal rates, although the tenant had to undertake reclamation and development of the site. The government encouraged foreign investors but preferred 51-percent local ownership. The various policy measures effectively promoted industrial development.

The most notable achievement in industrial development was the formation of the Aluminum Bahrain Company (ALBA) in 1968. The government (which held 19-percent ownership) joined with six international metal companies to form ALBA. Construction of an aluminum smelter, with a capacity of 120,000 tons of ingots a year, was started in 1969, and production began in 1971. Each partner received part of production in accordance with its equity share.

Since electric power is a major cost in aluminum production, the smelter was located in Bahrain to take advantage of the plentiful, cheap natural gas to power the electric generators. The government supplied the natural gas at a nominal rate under a twenty-year contract. Alumina was imported from Australia under a twenty-year contract. Production figures for the mid-1970s were not available, but production probably was less than capacity because a worldwide recession had sharply reduced foreign markets for aluminum. Several of the original investors had sold their interests to the government. By 1976 the government owned 78 percent of ALBA. The smelter employed about 2,300 workers, more than 85 percent of whom were Bahrainis. Nearly all of the aluminum produced was exported.

ALBA's original shareholders used or marketed their shares of production. One of the companies sold the government's share. As the government's ownership of ALBA increased, the need for a sales department became apparent. In 1976 a complete marketing organization was being formed by ALBA. The government also encouraged development of aluminum processing facilities to expand employment and income. An aluminum atomizer plant with a capacity of 3,000 tons a year, partly government owned, was completed in 1973 to produce aluminum powder for paints and explosives. In 1976 a Swiss company was building an extrusion plant, and a factory to produce aluminum electric power transmission cables was reportedly about to be built to process part of ALBA's production. Construction of an aluminum rolling mill was under consideration.

A flour mill was completed in 1972, with 8-percent government ownership, 20-percent ownership by a Kuwaiti flour company, and 72-percent ownership by Bahraini private investors. The mill imported Australian wheat, and the government banned flour imports to protect the mill's local market. Three ship repair and servicing companies completed the larger manufacturing establishments in 1976.

A major addition to the economy scheduled for completion in July 1977 was a large, modern dry dock for crude oil carriers of up to 450,000 deadweight tons. The project was approved by the Organization of Arab Petroleum Exporting Countries (OAPEC) in 1972, and seven OAPEC members were financing a substantial portion of the costs, which were estimated at nearly US$300 million in 1976, almost three times the original estimate. International companies provided the equipment and were constructing the dry dock. A management contract had been signed with an experienced West European repair yard for large tankers. An extensive training program formed part of the service to be provided. Employment was expected to reach 2,200 by 1980, and OAPEC wanted all operations to be performed by Arabs as soon as possible. Bahrainis expected substantial upgrading throughout the work force from the diffusion of skills by dry-dock employees over the years. The dry dock, located at the center of much of the petroleum trade, might turn out to be profitable but, if Dubai and Iran constructed large dry docks for tankers as planned, the resulting excess of repair facilities in the gulf could create disruptive competition and underutilization.

Foreign Trade and Balance of Payments

Bahrain was a trade center long before oil was found. The country's favorable location and efficient port continued to support a large volume of transit into the mid-1970s, encouraged by an industrial area and transit zone free of customs duties for goods reexported. Most reexports went to nearby areas, particularly Saudi Arabia. Bahraini foreign trade statistics excluded imports and exports of petroleum products, which were published elsewhere (see Oil Industry of Bahrain, ch. 5). Alumina imports for the aluminum smelter and exports of frozen shrimps appeared to be excluded from trade statistics, presumably because the processing operations were located in the free-trade zone.

The country depended on imports for a wide range of products for domestic use and for reexports (see table 21). About one-third or more of imports were reexported in the mid-1970s. Imports increased from BD26 million in 1959 to BD176 million in 1974, an average annual increase of nearly 14 percent. Worldwide inflation raised international prices sharply between 1973 and 1974 and was the major reason the value of imports increased by 38 percent in 1974. The impact of increased oil revenues primarily affected imports in 1975. Partial year data suggest that the level of imports increased about 35 to 40 percent in 1975; the increase caused some congestion and delays at the port. Industrialized countries supplied the bulk of Bahrain's imports; in 1974 the United States accounted for 18 percent, the United Kingdom 15 percent, Japan 13 percent, and all other European countries 19 percent.

Exports other than petroleum were largely reexports. They increased slowly from BD11 million in 1959 to BD32 million in 1973, an average increase of about 8 percent a year. In 1974, however, exports more than doubled to BD72 million, largely because of increased exports of aluminum ingots, which amounted to 98,000 tons valued at BD32 million. Japan purchased most of the aluminum output in 1974 and became Bahrain's major export market. In 1974 Japan imported

Table 21. Bahrain, Summary of Foreign Trade, Selected Years, 1968–74
(in millions of Bahraini dinars)[1]

	1968	1970	1972	1974
Imports[2]				
Food and live animals	10	14	15	25
Beverages and tobacco	1[3]	3	3	5
Chemicals	–[4]	3	6	15
Manufactured goods classified by material	14	23	27	48
Machinery and transport equipment	17	24	30	53
Miscellaneous manufactured goods	7	11	16	23
Other	3	3	3	8
Total Imports[5]	52	80	100	176
Exports[6]				
Food and live animals	2	4	6	8
Beverages and tobacco	1	1	1	2
Manufactured goods classified by material	6	7	21	40
Machinery and transport equipment	3	2	5	9
Miscellaneous manufactured goods	5	5	8	12
Other	1	3	1	2
Total Exports[5]	18	22	43	72

[1]For value of the Bahraini dinar—see Glossary.
[2]Excludes crude oil for refining and precious metals. Imports on basis of cost, insurance and freight (c.i.f.).
[3]Beverages included in Other category.
[4]Included in Other category.
[5]Figures may not add to total because of rounding.
[6]Excludes petroleum products and frozen shrimp but includes reexports. Exports on basis of free on board (f.o.b.).

Bahraini goods worth BD25 million, 35 percent of Bahrain's nonpetroleum exports. Saudi Arabia imported goods worth BD22 million from Bahrain in 1974, some 31 percent of exports. Other nearby areas, such as the United Arab Emirates (UAE), Qatar, Kuwait, and Iran, purchased about 15 percent of Bahrain's nonpetroleum exports, primarily reexports.

A high level of imports (primarily equipment and materials for such development projects as the aluminum smelter, power projects, and the flour mill) exceeded exports and revenues from oil, creating trade deficits between 1970 and 1973 (see table 22). Through 1972 the trade deficits were largely financed by capital inflows, and the country's balance of payments was in surplus. The flow of foreign capital diminished in 1973, requiring a drawdown of foreign assets. In 1974 oil revenues increased nearly threefold, sufficient to pay for a substantially larger import bill and a build up of foreign assets by more than US$100 million. The country's balance of payments probably remained in surplus in 1975. Gross official reserves almost quadrupled, rising from US$72 million at the end of 1973 to US$271 million on June 20, 1975.

Table 22. Bahrain, Summary of Balance of Payments, 1971–74
(in millions of SDR)[1]

	1971	1972	1973	1974[2]
Oil sector revenues (net)	49.2	63.0	84.7	231.0
Other exports	51.1	96.4	124.1	157.5
Non-oil imports	−220.5	−224.3	−268.8	−350.7
Trade Balance	−120.2	−64.9	−60.0	37.8
Services, transfers, and capital amounts (net)[3]	144.8	65.5	40.7	53.3
Monetary movements[4]	−24.6	−0.6	19.3	−91.1

[1]SDR—special drawing rights of the International Monetary Fund (IMF). The average yearly exchange rate was SDR1 equals US$1 in 1971, US$1.09 in 1972, US$1.19 in 1973, and US$1.20 in 1974.

[2]IMF projections based on partial year data.

[3]Bahrain did not publish balance-of-payments data, and this entry was a catchall for the payments and receipts for which data did not exist.

[4]A minus sign indicates an increase of foreign assets.

Source: Based on information from "Bahrain Enhances Role as a Persian Gulf Center of Communications and Commerce," *IMF Survey,* Washington, December 15, 1975, p. 374.

Budget and Fiscal Policy

Oil revenues financed most government activities, providing 72 percent of all revenues in 1976 (see table 23). Another major tax source was import duties, which were expected to yield a little over BD2 million in 1976. About BD15 million in fees were expected to be collected for such government services as the post office, electric power, airport operations, and use of government property. Foreign loans and grants provided BD16 million on the revenue side of the 1976 budget. It was not possible to determine from the available data the extent of budget support grants from other Arab countries and how much of the aid consisted of developmental loans to cover such projects as expansion of electric power and desalination capacity.

Current expenditures had risen rapidly, from BD15 million in 1968 to BD91 million in the proposed 1976 budget. The increase reflected expanded services, such as education and utilities, and higher personnel costs as government salaries were increased to match rising prices. In the 1976 budget, for example, the Ministry of Education received 13 percent of current expenditures, the Ministry of Interior 11 percent, the Ministry of Public Works, Electricity, and Water 10 percent, and the Ministry of Defense 9 percent. Subsidies (primarily for basic foods) to combat the effects of inflation amounted to BD6 million in both the 1975 and 1976 budgets. The amir had received one-third of royalties and income taxes from the Bahrain oil concession in earlier years, but it was uncertain what changes might have been introduced for the head of state's expenses after the government became 60-percent owner of the producing oil company. The 1976 budget allocated BD6 million for the amir's personal and official use.

Table 23. Bahrain, Budget Summary, 1973–76
(in millions of Bahraini dinars)

	Actual		Proposed	
	1973	1974[2]	1975	1976
Receipts				
Oil revenues	29	99	111	131
Other taxes and fees	16	17		34
Foreign loans and grants	0	2	23	16
Total Receipts	45	118	134	181
Expenditures				
Current expenditures	34	46	68[3]	91
Capital expenditures	8	19	61[3]	100
Total Expenditures	42	65	129	191
Surplus or Deficit (–)	3	53	5[3]	–10

[1]For value of the Bahraini dinar—see Glossary.
[2]Estimated actual spending before final accounting.
[3]Preliminary information indicated expenditures were substantially less than planned, resulting in a surplus in the magnitude of BD40 million.

Source: Based on information from "Bahrain Enhances Role as a Persian Gulf Center of Communications and Commerce," *IMF Survey*, Washington, December 15, 1975, p. 373; *Financial Times*, London, February 10, 1976; and U.S. Department of Commerce, Office of Technical Services, Joint Publications Research Service, "1976 Budget Breakdown," 1976.

Although current expenditures had risen rapidly, capital expenditures grew even more quickly, from BD2 million in 1968 to BD100 million in 1976. The increase in oil revenues had been primarily allocated to building up the economy. Development projects touched almost all phases of the society, such as schools, roads, ports, sewers, and public buildings; but housing, electric power, and the supply of fresh water had received priority.

The government was prudent in increasing expenditures as oil revenues increased. The budget has generally shown a modest surplus in most years since the late 1960s. There were extrabudgetary operations, however, that may have created deficits not shown in budget data. The government published little information about the extrabudgetary operations or the state's reserves. Loans provided to the company operating the aluminum smelter when it was starting production in the early 1970s were an example of extrabudgetary operations. The BD10 million deficit in the 1976 budget was to be financed by drawing on state reserves. Perhaps a more significant reflection of past extrabudgetary operations and the size of state reserves was a statement by Bahrain's prime minister in August 1976 that the Kuwait proposal to provide support for Bahrain's budget was being implemented in the magnitude of US$100 million annually. Observers surmised that other Arab oil states, such as Saudi Arabia, Qatar, and the UAE, were contributing to the budget support, presumably in the form of grants.

The announcement of budget support from Arab oil states saved Bahraini officials from painful decisions. Oil revenues may have peaked for Bahrain in 1976, and they were certain to start declining within a very few years in any event (see Oil Industry of Bahrain, ch. 5). The choice before the government was between reducing expenditures and increasing revenues. There were opportunities to reduce expenditures by eliminating such subsidies as the provision of basic foods, water, and electricity at a fraction of cost and decreasing the number and salaries of government employees, but major cuts in expenditures would have hit economic development the hardest. There were many ways revenues could have been increased because the population was almost untaxed, but the impact might have been counterproductive. If an income tax were imposed on businesses or individuals, new businesses might not settle in Bahrain, and individuals, particularly the expatriates who contributed much to the economy, might choose to be employed in nearby tax-free oil states. Budget support from other countries has removed the immediate need to adjust Bahrain's fiscal policies to the country's changing situation.

Money and Banking

The Indian rupee was the primary currency in Bahrain until 1959, when the gulf rupee was introduced to halt gold smuggling into India from the Persian Gulf states. In 1964 the Bahrain Currency Board was established, and in 1965 it issued the Bahraini dinar and maintained the reserve funds behind the dinar notes. (The dinar was also the main currency in circulation in Abu Dhabi until the UAE was formed in December 1971 and issued its own currency.) The Bahrain Currency Board had no central bank functions other than note issue. The Bahrain Monetary Agency was established in 1973 to replace the currency board and to exercise bank functions including control of commercial banks.

For many years there were only three commercial banks (all British) in Bahrain, but after 1968 the number increased rapidly. By early 1976 there were sixteen commercial banks, and two more were scheduled to begin operations shortly. Most of the banks in Bahrain were branches of long-established foreign banks. They were attracted to Bahrain by its central location in a major oil area and its developed facilities. The growth of economic activity that accompanied increasing oil revenues in the early 1970s resulted in an expansion of domestic liquidity, averaging about 25 percent a year, and an increase of credit to the private sector by the banking system of about 50 percent a year. Most lending financed trade and construction activities, but credit was also available for long-term industrial projects, personal consumption, and other needs. The central bank was not in a position to exert much control over the banking system until 1975, but little control was needed because the bankers and borrowers approached banking conservatively. Some reserve requirements were imposed on commercial banks in 1975, but controls remained minimal in late 1976.

A major new development in Arab banking was the government's decision in 1975 authorizing offshore (foreign currency) banking facilities to be established in Bahrain. The offshore banking units would not accept local deposits or operate in the local economy. The theory behind the move was that short-term depositors from the oil-rich states around the gulf would prefer to place foreign currency deposits with banks in the same time zone rather than in New York or London. The large international banks responded to the idea because of the chance to develop their business with gulf depositors and because of Bahrain's lack of taxes and its good communications and other facilities. Bahrain society was less austere than most nearby states, making it easier to recruit the expatriates needed for the banking operations. The decision was announced in September 1975; by mid-1976 thirty-two licenses had been issued, and sixteen offshore units were in operation. The assets of the offshore banks in Bahrain amounted to US$3.5 billion at midyear. A large, London-based international money broker also received permission to open an office in Bahrain, which would provide additional outlets for offshore banking.

If the offshore banks prosper, the effect on Bahrain's economy could be substantial. Office overhead expenditures and salary payments would provide the initial stimulus, but over the longer run other money market operations would be attracted, and Bahrainis could learn the sophisticated skills of foreign exchange dealing. Offshore banking could add a new dimension to the economy.

GOVERNMENT AND POLITICS

The establishment of a formal government structure and a bureaucracy began in the 1920s with the introduction of free, modern health facilities and a secular educational system. The treaties with Great Britain placed only foreign affairs and defense under British control, but British influence and tutelage in domestic matters were extensive. During the late 1960s Shaykh Isa bin Salman Al Khalifah, who at independence in August 1971 assumed the title of Amir of Bahrain, instituted extensive administrative reforms. In 1970 he created the first formal executive authority, the Council of State. This twelve-member body was the first high-level government body without a British member; only four members of the royal family were included in it. Each member of the council was in charge of a government department; these departments became ministries at independence.

In December 1972 the members of a constituent assembly were elected in the country's first national election. In response to Amir Isa's request, the Constituent Assembly prepared the draft of a constitution, which the amir issued on May 26, 1973. Within six months elections for the country's first national legislature were held, and on December 16, 1973, the amir convened the first session of the National Assembly (Assembly). Thirty of the members had been elected for a four-year term; the other members were, as provided in the Constitution, the members of the cabinet.

Within the newly elected Assembly coalitions emerged. At one end of the spectrum was the majority group made up of the country's more traditional elements. This group was divided from the rest of the Assembly, however, by the constant quarreling between its Sunni and Shiite factions. The other end of the spectrum contained the more dissident groups, which also did not always cooperate with each other; these included the labor leaders and the socialist-oriented members, who were often supported by radical groups committed to the overthrow of the monarchy. Some observers asserted that fanatical conservatives and left-wing radicals were overrepresented in the Assembly, leaving the large group of middle-of-the-road Bahrainis underrepresented.

The free elections and the resultant Assembly seemed to lessen the absolute authority of the monarch; but political parties were not permitted to organize within the country, and labor unions, promised by the 1973 Constitution, were still not sanctioned by the amir. The Assembly was in constant dispute with the prime minister and the cabinet. In 1974 new elections for the Assembly were held, but the new members produced even greater friction. The government barred a number of the newly elected radicals from taking their seats, stating that members of the Bahrain National Liberation Front had no place in the Assembly.

By mid-1975 the attempts to develop a constitutional monarchy had become a burden on the government's day-to-day activities, and the sharing of power that the government had hoped would produce stability was actually producing instability. On August 24, 1975, Prime Minister Khalifah bin Salman Al Khalifah resigned along with his cabinet in protest over the Assembly's interference in the running of the government. The following day Khalifah formed a new government, and Amir Isa dissolved the Assembly.

With the dissolution of the Assembly came numerous arrests of dissidents, including at least one former Assembly member. Amir Isa announced in August 1976 that the Assembly would remain dissolved for an indeterminate period. In late 1976 the fifteen-member cabinet continued to perform legislative as well as executive functions under the firm guidance of Amir Isa and Prime Minister Khalifah (see fig. 13). Eight members of the cabinet were members of the royal family.

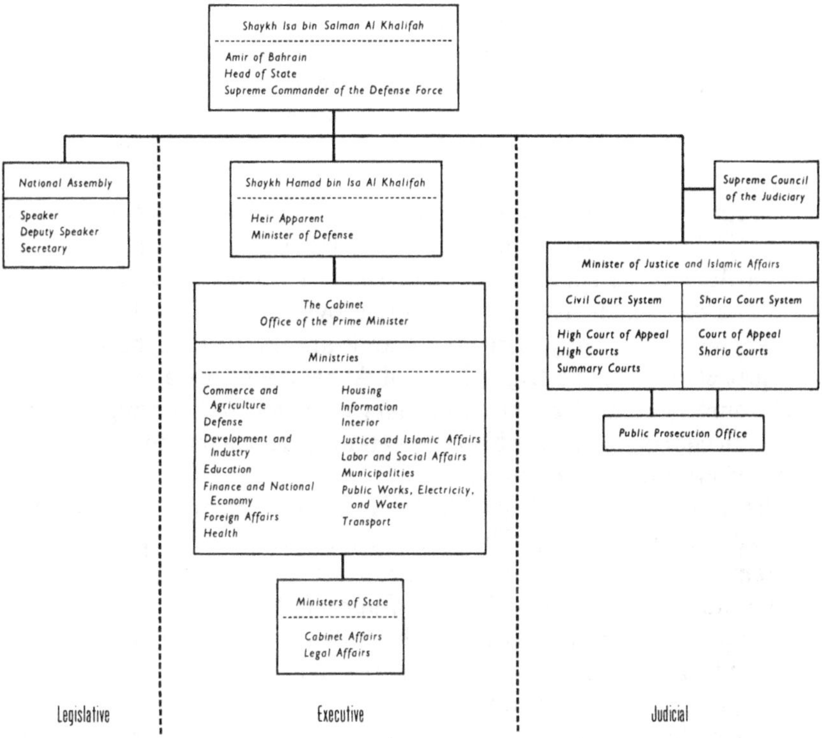

Note The National Assembly was prorogued on August 25 1975, in August 1976 the amir ruled out the possibility of its being reconvened in the near future

Figure 13. Bahrain, Government Structure, November 1976

Legal System

Bahrain was the first gulf state to begin developing a non-Islamic legal system. This development was begun in the mid-1920s under the first of numerous British legal advisers, Sir Charles Belgrave; in the early 1950s a formal position of British judicial adviser was created to accelerate the codification of statutory law. In 1976 the process continued, but there remained remnants of many systems of law, including urf, or customary tribal law, two schools of Sunni sharia law (Maliki and Shafii), Shia sharia law, and civil law as defined in numerous codes and regulations.

At the pinnacle of the judicial system was the amir, who retained the power of pardon and whose office was the final step in the appellate procedure. By the mid-1950s Bahrain had developed a local court system effective enough to replace British jurisdiction at the local level in 1957. Since independence the minister of justice and Islamic affairs, acting for the amir, has exercised the judicial authority.

Theoretically, according to the modified 1973 Constitution, the judiciary is a completely independent and separate branch of government.

The court system is extensive; the nature of an offense determines in which court a claimant appears. The formal dual court system, which has jurisdiction over all other court systems, comprises the civil court system and the sharia court system. Decisions handed down by courts of urf law, for instance, could be appealed to one of the appropriate recognized courts.

Sharia courts were established in response to the Sunni-Shia division within Bahraini society. The courts, found throughout the country, deal primarily with personal status law. There is a sharia Court of Appeal in Manama, which has final authority over the many sharia courts under the minister of justice and Islamic affairs. In 1976, however, the highest court was the High Court of Appeal, which was a part of the civil court system. The court's primary duty was to hear appeals from the high courts, but it was also competent to decide on the constitutionality of laws and regulations.

The high courts heard appeals from the network of summary courts throughout the country. The sharia court system and the civil court system form the formal unified court system, which was administered by an independent body called the Supreme Council of the Judiciary. The supreme council determined the jurisdiction of each court and of the Public Prosecution Office. In practice, however, it was probably the minister of interior who supervised the functions of prosecution. After the Assembly was dissolved in August 1975, a state security law that the Assembly had refused to enact was promulgated. It was unclear in late 1976 whether the law was still in effect and whether separate courts for national security offenses had been established.

Mass Communications System

In 1976 the government continued to maintain rigid control over the country's mass communications, as it had at least since the termination of the Assembly and the suspension of various articles of the 1973 Constitution. Although the press in Bahrain dates only from the 1930s, it played a dramatic role during the 1950s and 1960s in the political dynamics of the country. Before 1957 the press was for the most part an independent institution with few controls on its freedom to print whatever it chose. Because the press supported the riots and strikes of labor groups in the mid-1950s, however, all independent periodicals were suspended by the government.

The government issued a press law in 1965 that allowed for the development of newspapers according to specific guidelines, which among other things proscribed criticism of the royal family, the government, or friendly foreign countries. After 1965 the press began to grow again, and in 1967 the country's first Arabic-language daily, *Akhbar al Khalij*, began operation. By 1976 there were eleven weekly newspapers; two were government owned and operated, two were published by the Bahrain Petroleum Company, and the remainder were privately owned. Although most newspapers were published in Arabic, about one-third of the press run was in English.

In 1976 publishing houses continued to need government licenses. The two foremost publishing houses were the Arab Printing and Publishing Establishment and the Arabian Printing and Publishing House.

In 1955 the government established a government-owned and -operated radio system, the Bahrain Radio Station. Television, introduced as an independently owned system in 1963, was taken over by the government in February 1975 and named Royal Television Bahrain.

The primary channel chosen by the government to reach its people was radio; at least half the population owned radios in 1976, and the remainder had access to them. Unfortunately for the purposes of the government in its attempt to unify the country through mass communications, radio listeners could also hear broadcasts from antigovernment groups outside the country. Radio and television broadcasts were in Arabic only.

The only foreign news agency in the country was Reuters (United Kingdom). Broadcasts from other parts of the Arabian Peninsula could be heard, and those from Saudi Arabia were particularly popular. Many international broadcasts could be picked up by Bahraini radio listeners. In 1976 a coalition of friendly Arab Persian Gulf states agreed to the establishment of the Gulf News Agency, which would be headquartered in Bahrain to promote cooperation and development of better mass communications on the Arabian Peninsula.

Patterns of Foreign Policy

In late 1976 Bahrain's major foreign policy objectives were improving cooperation among Arab states, especially those located on the Persian Gulf; supporting the Arab cause versus Israel; and building bridges through diplomacy to expand the domestic economy, establish reliable trading patterns, secure protection against external attack, and gain foreign economic assistance. Unlike many of the other Persian Gulf countries, Bahrain could not rely solely on oil production as a diplomatic bargaining point.

Bahrain was among the world's resource-poor countries and was forced to import almost all its needs. This dependence on the outside world was a major consideration in foreign policy decisionmaking. In 1976 the friendships the nation's leaders had been able to develop—especially with gulf neighbors, Western nations, and Japan—might very well determine the survival of Bahrain itself.

Since independence the amir has taken an active role in foreign affairs, especially in regional matters. The prime minister and the defense minister frequently travel abroad on diplomatic missions. According to foreign observers the foreign minister, Shaykh Muhammad bin Mubarak Al Khalifah, has developed the government's most efficient bureaucracy. Although the British remained in complete control of Bahrain's foreign relations until independence in August 1971, the Ministry of Foreign Affairs was established under Shaykh Muhammad in 1969, and he was ready to replace the British in this area at independence.

The government's closest bilateral relations have been with Kuwait. The countries' royal families are often referred to as Islamic cousins, and there has been a strong affinity between the rulers of Bahrain and Kuwait for well over a century. Kuwait's financial assistance to Bahrain began before independence and continued in 1976.

For economic and strategic reasons Bahrain also depends heavily on Saudi Arabia. The direction of Bahrain's social and political development has been criticized by Saudi leaders as too progressive, but those same spokesmen have noted that Saudi Arabia feels a responsibility to protect the integrity of Bahrain, its government, and the Al Khalifah family. These three royal families—Al Sabah of Kuwait, Al Saud of Saudi Arabia, and Al Khalifah of Bahrain—consider themselves the most important lineages on the Arabian Peninsula.

After the 1968 British announcement that all British military forces would be withdrawn from the Persian Gulf region, Bahrain joined Qatar and the other Trucial Coast states in plans to form a federation (see The Twentieth Century, ch. 2). Bahrain saw itself as the obvious leader among these states, a view not shared by the several other royal families. In the end, after the Iranian claim to Bahrain was resolved by the assistance of a United Nations (UN) mission and through Kuwaiti and Saudi mediation with the government of Iran, the Al Khalifah family decided to remain outside the federation, as did the Al Thani family of Qatar. The remaining seven Trucial Coast states, however, agreed to plans for a combined state, the UAE.

Bilateral relations with both Qatar and the UAE were cool during the early and mid-1970s, but Bahrain has made moves intended to solve past differences and begin an era of cooperation among the Arab states of the gulf. This pattern of foreign policy had been somewhat successful by late 1976, not in small measure because of Bahrain's urgent need for financial assistance and Arab skepticism over Iran's long-range intentions.

After independence Bahrain continued its close relations with the United Kingdom through a treaty of mutual friendship and expanded relations with most of the West European countries, Japan, Australia, and the United States. In 1976 Bahrain maintained diplomatic relations with almost fifty countries; twelve had embassies in Manama, and many others accredited their ambassadors in Saudi Arabia or Kuwait to Bahrain as well. The country was admitted to the League of Arab States (Arab League) and to the UN in September 1971.

An interesting aspect of the government's actions after independence was the leasing in December 1971 of a section of the evacuated British naval base at Al Jufayr to the United States as its command post for the United States Navy's Middle East Force (MIDEASTFOR). The government has faced much criticism from within and outside its borders, especially during the October 1973 War between Israel and Egypt and Syria. At that time Amir Isa announced that Bahrain was terminating the lease but, after negotiations that resulted in an increase in the rent, he agreed to the continued use of the Al Jufayr facility by the United States until June 1977. In late 1976 negotiations were still in progress to extend the agreement beyond June 1977. (On June 30, 1977 the Bahrain government terminated the lease.)

NATIONAL DEFENSE AND INTERNAL SECURITY

In 1976 Bahrain continued to rely on its close relations with Kuwait, Saudi Arabia, and the other moderate Arab states and with the United Kingdom and the United States as the major deterrent to invasion. Bahrain's small, modestly equipped army would not be able to repel an invading force, and the country possessed neither an air force nor a navy. The nation's defense rested on the assumptions that, because Bahrain was poor in natural resources, it was unlikely to be invaded and that, should it be invaded, its friends and protectors would come to its aid.

Armed Forces

In 1976 the nation's sole military organization was the Defense Force. Its approximately 1,500 men were organized into one infantry battalion, one armored-car squadron, and small administrative and support units. In early 1976 its weapons inventory consisted of eight Saladin armored cars, eight Ferret scout cars, six Mobat 120-mm recoilless rifles, and six 81-mm mortars.

The Constitution designates the amir as supreme commander of the Defense Force. Amir Isa's son and heir apparent, Shaykh Hamad bin Isa Al Khalifah, was minister of defense and commander in chief of the Defense Force.

In 1976 the country had no conscription legislation and, because there have consistently been more volunteers for enlistment than there have been openings, such legislation was not envisaged. The Defense Force recruited primarily within the Sunni community (see ch. 3). Although specific information was not available in late 1976, most members of the small officer corps were members of either the Al Khalifah family or one of its cadet branches.

Pay and compensation were either comparable or superior to those of civilian positions. In the mid-1970s, for example, one of Bahrain's serious domestic problems was inadequate housing, but members of the Defense Force and of the public security forces (police) were given priority in housing allocations and received some monetary assistance.

Military uniforms closely resemble those of the British army, reflecting the long period of British influence and guidance. Rank insignia are modifications of the basic British system, using a five-pointed star in place of the British pip, an eagle (or falcon) instead of the crown, and crossed saber and baton instead of crossed sabers (see table 24).

Public Security Forces

Although official data were not available in late 1976, the combined strength of the public security forces was estimated by observers to equal that of the Defense Force, that is, about 1,500 men. The forces included three units: the Security Force; the Special Branch, which was the intelligence and criminal investigation wing of the Security Force; and the police. All units came under the Ministry of Interior, which was headed by Shaykh Muhammad bin Khalifah bin Hamad Al Khalifah.

The police are stationed in the municipalities and local communities. They perform such standard police functions as controlling traffic, patrolling the local markets, and responding to citizen complaints of petty thefts and minor misdemeanors.

The Security Force is a mobile unit with offices in Manama and five other municipalities. Among its duties are controlling riots, especially those related to labor strikes or work stoppages, and controlling aliens' entrance into the country. In 1976 its equipment included two lightly armed patrol craft and two Scout helicopters.

The Special Branch performs both criminal investigation and political intelligence gathering. Many of the senior officers of the Special Branch and of the Security Force are non-Bahraini. In 1976 British, Jordanians, and Pakistanis were responsible for many operations, particularly technical equipment and the surveillance of resident aliens. Foreign observers described the two units as particularly effective in intelligence gathering.

Internal Security

The Bahraini population is divided almost equally between the two major branches of the Muslim religion: Sunni and Shia. The royal family, the other noble families, and the important merchant families are predominantly Sunni; the working class and small farmers are usually Shia. In addition there are extreme differences between urban and rural areas in the social status and economic well-being of the people. Finally, the politics of the people range from the ultraconservatism of the traditional Sunni tribal and religious leaders to radical Marxism.

Divisions within the Bahraini society have been complicated by the presence of a large number of non-Bahrainis (approximately 15 to 20 percent of the population in 1976). These included those associated with the government or the business community—who for the most part

Table 24. Bahrain, Armed Services Ranks and Insignia, 1976

Rank[1]	Arabic	Insignia[2]
Supreme Commander	Al Amir (The Ruler)	(Personal choice)
Field Marshal	Mushir	Crossed saber and baton within a wreath
Lieutenant General ..	Fariq	Crossed saber and baton and one eagle
Major General	Liwa	Crossed saber and baton and one star
Brigadier General ...	Amid[3]	Eagle and three stars
Colonel	Aqid	Eagle and two stars
Lieutenant Colonel ..	Muqaddam	Eagle and one star
Major	Raid	Eagle
Captain	Naqib	Three stars
First Lieutenant	Mulazzim Awwal	Two stars
Second Lieutenant ...	Mulazzim Thani	One star
Chief Warrant Officer	Wakil	One gold bar
Sergeant Major or Warrant Officer ...	Raqib Awwal	Eagle and three chevrons
Sergeant	Raqib	Three chevrons
Corporal	Arif	Two chevrons
Private First Class ..	Jundi Awwal	One chevron
Private	Jundi	None

[1]In 1976 the highest rank—other than the amir and the commander in chief—was believed to be a colonel.

[2]All stars are five pointed.

[3]May be translated as brigadier or brigadier general and regarded as either depending on whether British or United States customs, respectively, are being followed.

Source: Based on information from U.S. Department of Commerce, Office of Technical Services, Joint Publications Research Service, "Law Concerning Military Pay and Allowances, Kuwait," 1972.

were from Western states—and those from other Arab countries. Bahraini dissidents have frequently protested the amount of influence and wealth those from the West have been able to accumulate, and these protests have been supported or sometimes instigated by non-Bahraini Arabs of nationalist and socialist orientations.

The elements of the Bahraini population that had grievances had in the past found sympathy from three radical Arab groups in particular. These were the Arab National Movement (ANM), the Popular Front for the Liberation of the Occupied Arab Gulf (PFLOAG), and the Iraqi branch of the Baath Party.

With the opening of the Assembly and the promulgation of the Constitution in 1973, many observers thought that the divisions within the country would be healed, but the experiment lasted less than two years. Not only were the divisions not healed, but the focus on the differences among the population in religion, wealth, and politics probably promoted a higher level of instability that led the amir to terminate the experiment of a legislative body.

The dissolution of the Assembly in August 1975 and the return of direct monarchical rule permitted the government to carry out its business but terminated, or at least suspended, the government's efforts to bring more of the citizens into the country's decisionmaking processes. Many dissidents were arrested after the closure of the Assembly, including leaders of the Bahrain National Liberation Front and members of the PFLOAG. The government claimed that both groups had extensive support from external sources, and in late 1976 some of those who had been arrested remained in jail. During the summer of 1976 the government decreed that everyone entering Bahrain would need a visa, but a seventy-two-hour visa could be obtained at the airport. Further the government began a campaign to discourage outside interference in its domestic affairs by warning that extreme penalties would be imposed on Bahrainis or others caught dealing with unfriendly foreign states or organizations.

* * *

Studies and statistics on Bahrain were difficult to obtain in late 1976. A brief survey of the economy was published in the *IMF Survey* of December 15, 1975. The *Financial Times* of London published several articles on Bahrain in the Februrary 10 and November 1, 1976, editions. Jared E. Hazleton's *Public Finance Prospects and Policies* for *Bahrain, 1975–1985* contains valuable data and insights, partly based on earlier World Bank studies. The government's annual *Statistical Abstract* contains a variety of statistics but lacks many that researchers would find useful, such as foreign trade and budget data. The government's annual publication *Foreign Trade* contains very detailed statistics for a single year, but individual annual volumes are required to develop a time series for even the simplest summary. A considerable amount of fragmentary information is published in newspapers and such periodicals as the *Middle East Economic Digest* and the *Arab Report and Record.* (For further information see Bibliography.)

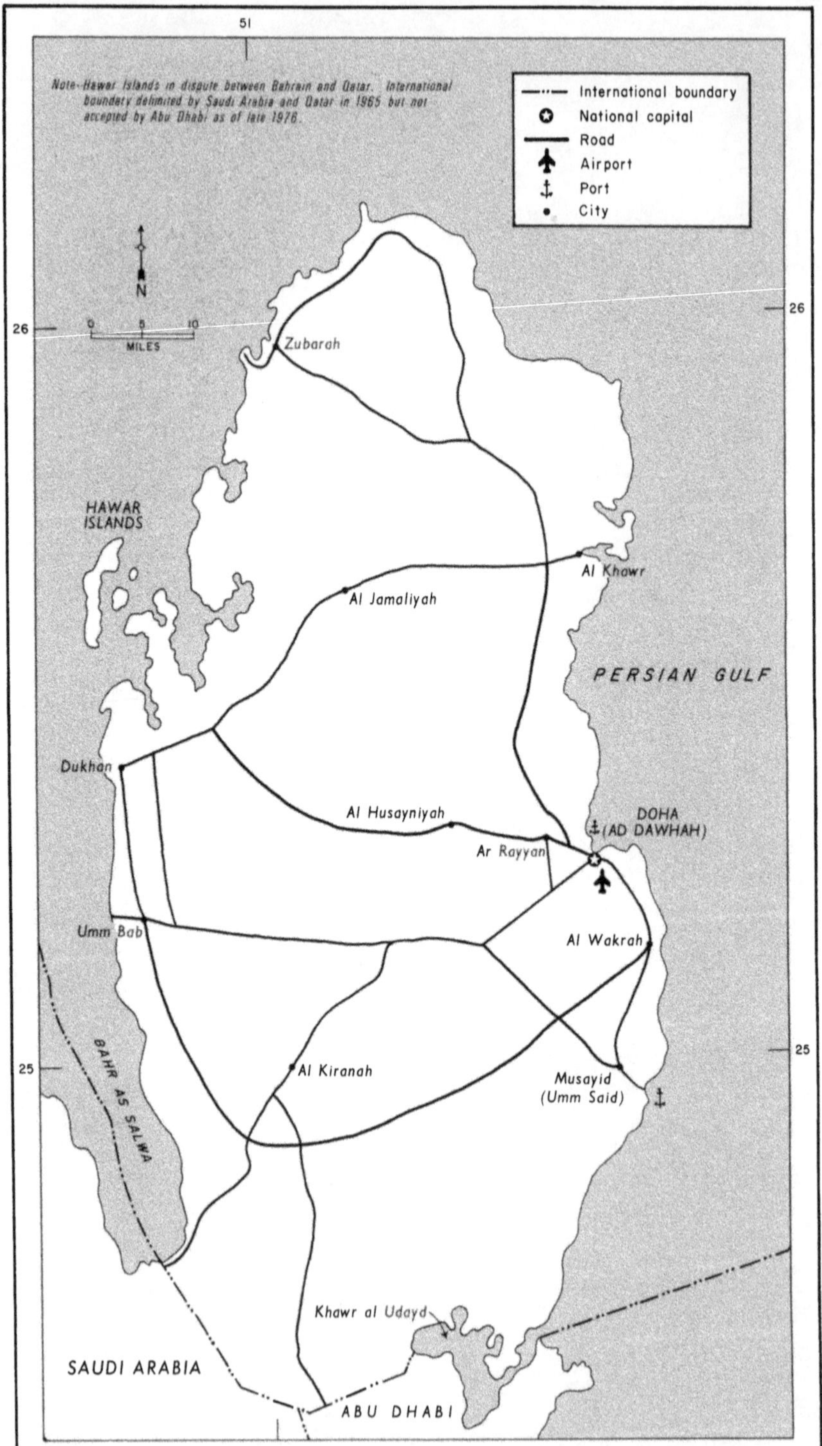

Figure 14. Qatar, 1976

CHAPTER 8

QATAR

The approximately 300 adult males of the Al Thani ruling family continued to dominate the Qatari society in 1976. The broader Al Thani tribe was estimated to number 20,000, perhaps more than half of the indigenous Qatari population. Qataris were beduin from Najd in the interior of the Arabian Peninsula who came to the Qatar Peninsula in search of forage for their animals (see The Founding of the Modern Gulf Polities: Gulf States in the Eighteenth and Nineteenth Centuries, ch. 2). Some continued their nomadic existence; others settled around wells to cultivate date palms and other crops, and some took up fishing. Doha (Ad Dawhah) became a small trading center but did not develop the entrepôt business and the importance of such gulf ports as Bahrain, Dubai, and Kuwait City.

The discovery of oil in 1940 and the beginning of its commercial exploitation in 1949 doomed the traditional pattern of existence. The oil industry, unhampered by local constraints, developed much faster than the rest of the economy. Wage scales were set in other countries, not locally. Higher pay in the oil fields and subsequently in trade, construction, and government employment attracted workers from lower paying traditional pursuits in Qatar and from other countries. The economy was rapidly restructured, and economic growth was paced primarily by the flow of oil revenues (see Oil Industry of Qatar, ch. 5).

By 1976 oil revenues made the country one of the world's richest in per capita terms, but income distribution was far from equal. There were extremes of poverty and wealth. The government's social welfare programs provided free schooling, free health care, and subsidized food, utilities, and housing that eased the lot of the poorest, although citizens fared better than expatriates.

In a 1916 treaty with Great Britain Qatar agreed not to enter into any relations with foreign governments without British consent (see The Twentieth Century, ch. 2). In return Great Britain agreed to protect Qatar from all attacks from the sea and to offer its good offices for negotiations in the event Qatar was attacked by land. Technically the 1916 treaty and a subsequent one signed between the two countries in 1934 pertained only to foreign relations. In fact Great Britain exercised considerable influence over Qatar's internal politics.

In 1968 Great Britain announced its intention of withdrawing from military commitments east of Suez, including those in force with Qatar, by 1971. Anticipating the country's complete independence, the ruler of Qatar issued a written provisional constitution in April 1970. The Constitution, which in late 1976 remained in force though still provisional, includes provisions for modern state administration in the form of a governmental ministry system, which reduced and consolidated the more than thirty departments into ten ministries. The Constitution also provides for a very limited extension of political participation and decisionmaking (see Government Process and Constitution, this ch.). In the realm of foreign relations the Constitution committed Qatar to joining Bahrain and the Trucial Coast states in forming the proposed Federation of the Arab Amirates. The original federation became a moot issue when, in large part because of mutual rivalries and suspicions, Bahrain and Qatar decided not to join. In August 1971 Bahrain proclaimed its independence; Qatar followed suit on September 1. The Trucial Coast states eventually united to form the seven-member United Arab Emirates (UAE).

GEOGRAPHIC AND DEMOGRAPHIC SETTING

Physical Environment

Qatar is located on a peninsula extending about 100 miles north from the eastern Saudi Arabian mainland and varying in breadth from thirty-five to fifty miles. The country contains over 4,000 square miles and separates the Persian Gulf from the Gulf of Bahrain and its lesser extension, Bahr as Salwa (see fig. 14). At various points the northwestern coast is less than twenty-five miles from the main island of Bahrain, and the Hawar Islands immediately off the peninsular coast remained the subject of a territorial dispute between Qatar and Bahrain. The southern shoreline of the Khawr al Udayd forms part of the land boundary that Qatar and Saudi Arabia defined and delimited in 1965, but as of 1976 the UAE had not recognized the delimitation.

The land is largely flat desert covered with loose sand and pebbles broken by occasional outcroppings of limestone. The western coast, where most of the oil fields are located, is marked by low cliffs and hills (see fig. on oil fields, ch. 5). In the south sand dunes and salt flats predominate. What little natural vegetation and cultivated land exists is confined to the north. Over 3,000 acres of cultivated land are located in this region (see Agriculture and Fishing, this ch.).

Doha is the capital of the country and the major administrative, commercial, and population center. It is linked to other towns and development sites by a system of more than 600 miles of roads—about one-third of which are hard surfaced—and contains the country's international airport. Doha has a deepwater port that is used for most shipping, although oil exports are usually piped to Musayid (Umm

Said), which is the major industrial center in the country. Limited port facilities are located at several other cities.

The climate is characterized by intense heat and humidity between June and September, when the temperature often reaches 120°F. The months of April, May, October, and November are more pleasant. During the winter the temperature may fall to 45°F, which is relatively cool for the latitude. Rainfall seldom reaches five inches a year; it is confined to the winter months and falls in short storms that often flood small ravines and the usually dry wadis.

The scarcity of rainfall and the limited underground water, most of which has such a high mineral content that it is unsuitable for drinking or irrigation, restricted the population size and the extent of agricultural and industrial development the country could support until desalination projects were started. Although water continued to be provided from underground sources, by the mid-1970s over half the water supply was obtained by desalination of seawater. Government plans called for increased investment in desalination facilities.

The desolate topography and desert climate restrict the variety and abundance of wildlife. The most common wild animals include such mammals as sand cats, jerboas, and bats. Once numerous, the gazelle has become increasingly rare as a result of indiscriminate hunting. The ruler keeps a small herd of Arabian oryx, no longer believed to be found wild; some scholars believe that a sketch of the profile of this animal may have been the source of the unicorn myth. About 100 different species of birds have been observed, but most are migratory. Only a dozen are believed to breed on the peninsula. Most of these are such water birds as the osprey, tern, heron, and plover; two species of lark are known to be desert breeders. Certain new species may be settling in new cultivation areas in the north. There are also numerous varieties of lizard.

Population and Working Force

As of late 1976 there had never been a national census, but the official estimate of the population in mid-1976 was 180,000. Other estimates report lower levels, ranging from 92,000 to 160,000, in part reflecting disagreement over the size of the expatriate labor force and how it should be reported. Improvements in health, welfare, and education were reflected in the annual rate of population growth, which had reportedly increased from about 2.3 percent in 1969 to about 3.0 percent in 1972. Continued growth at this rate would double the population by 1995. The increasing cost of providing social services and consumer goods for such an expanded population was of concern to the government, but no effort had been made to introduce family planning programs.

Specific information on the structure of the population was not available, but the population was believed to be relatively young, an estimated 45 percent of the total being under fifteen years of age. Life expectancy was about forty-seven years. About 71 percent of the population was classed as urban, and over half the population lived in the capital. Other population areas included six major towns, eighty small towns and villages, and over 100 small settlement areas. Almost all expatriate workers lived in the larger urban areas.

Before World War II there were very few foreign nationals living in the country. The discovery of oil in 1940 and the beginning of commercial production in 1949 provided oil revenues for investment in other industries (see The Economy, this ch.). The attractiveness of the labor market has resulted in a large influx of foreign labor. Although the number of expatriate workers was not known, reliable estimates indicated that the overwhelming majority of the economically active population was non-Qatari. Manpower needs were expected to increase steadily as the economy developed; for example, six new industrial projects in 1976 created a need for 2,000 more workers. About 70 percent of the labor force was employed by the industrial and service sector; the remainder was employed in commercial fishing and, to a lesser extent, grazing and small-scale cultivation.

Shifts in the structure of the economy since 1949 have had major effects on the Qataris. The percentage of Qatari nationals serving as manual laborers or working in the traditional industries of pearling and fishing has declined. Daily work habits have been altered by fixed hours and use of machinery. The dependency ratio among the Qataris is high. A legal basis for benefits for Qatari workers and their families has been provided, and a system of worker committees has been established to handle worker grievances.

The majority of the foreign work force were Iranian, Pakistani, and Indian. Arabs were a minority of the foreign workers. Most of them were from Egypt, Oman, and Yemen (Sana). Egyptians were employed mostly as teachers and clerks; Asians, Omanis, and Yemenis were more commonly employed as unskilled laborers and lived in compounds or communal residences. Few brought their families with them. The Iranians, with the exception of the Baluchis, were especially active in commerce, but the leading merchants were Qatari.

Living Conditions

During the early 1970s the Qataris began to experience a comfortable standard of living. Per capita gross national product (GNP) exceeded the equivalent of US$10,000 by 1974. Education and medical services were free. Urban centers and some towns possessed modern services—such public utilities as electricity and water were free—and modern architecture. A pension plan for employees in government and public institutions, including the military, was in operation, and there were subsidies for the impoverished. The standard of living was being eroded in the mid-1970s, however, by inflation, which was only partly offset by pay raises and government controls (see Money and Banking, this ch.).

An area of serious concern for the average family was the high cost of food. The dependence on imported foods exposed the economy to worldwide increases in the price of foodstuffs, and a good portion of the income of the average family was used for food. In 1974 the government imposed price controls on rice, sugar, edible fats, flour, bread, and animals imported for food. The country grows fruits and vegetables that are excellent sources of minerals and vitamins. Observers did not know to what extent the population consumed a balanced diet. Fish caught offshore was a favorite source of protein. Government price controls, however, resulted in a decline in the supply brought to local ports.

One of the serious concerns of the government was the provision of adequate housing for the population. The situation was continually exacerbated by the influx of expatriate labor. The government provided support for the construction of new housing. Dwellings intended for low-income families were available at nominal cost or were provided free of charge. Free land and long-term, interest-free financing were available, even for middle-income families interested in higher quality housing than they presently occupied. Construction often followed traditional patterns, but experimentation with new materials and techniques was common. An entirely new population center providing housing for about 10,000 was to be built on reclaimed tidal flats at West Bay, adjacent to Doha. In 1976 a new community for 30,000 inhabitants was under construction at Musayid.

Free medical services—including immunization, outpatient treatment, and hospitalization—were available for all residents of Qatar. In the mid-1970s the Ministry of Public Health was attempting to upgrade various facilities and increase the medical services locally available. Cardiac and neurosurgical cases continued to be sent abroad. By the early 1970s the infant mortality rate had been reduced to forty-two per 1,000 live births. Smallpox and cholera were controlled by vaccination and immunization programs. The major communicable diseases reported were measles, influenza, tuberculosis, and bacillary and amebic dysentery.

In the early 1970s there were about eighty doctors, 200 nurses, and 150 assistant nurses and nursing aides. By the mid-1970s health facilities had been expanded and included six public hospitals, several small health centers, and specialized clinics that provided such services as prenatal, dental, and psychiatric care. New hospital facilities were planned for Doha—with a 600-bed capacity—and Al Khawr.

The most significant problem in health care was the lack of an organized structure to supervise health education, industrial health and safety, sanitation, and food processing. In 1976 there were no public health laws. Specialized maternal and child health care services were inadequate. Rodent control had been initiated, but rodents continued to present a major threat to public health.

Education

Qatar has shown a strong interest in education and has one of the highest per pupil expenditures in the world. The first secular schools for boys were opened on a limited scale in 1952. Expansion and improvement of the educational system were undertaken by the government in 1955. Other developments in the mid-1950s included the establishment of separate schools for girls and the addition of secondary education. In the mid-1970s students continued to be segregated on the basis of sex and were instructed by teachers of their own sex. Education was free, and students received school supplies, clothing, meals, and transportation to and from school at no cost.

The educational system was composed of six years of primary school, three years of preparatory (intermediate) school, and three years of secondary school; a system of higher education was being developed. In addition to general academic courses, secondary education included technical, vocational, commercial, and religious training. Instruction throughout the system was in Arabic, but English was introduced in the last two years of primary school, and there were special language training programs for government personnel. Private facilities were available for kindergarten instruction.

With the exception of teacher training, higher education was provided by government-subsidized study abroad; in 1976 about 700 Qataris were taking advantage of this program. In 1973, however, the Ministry of Education and Youth Welfare initiated the upgrading of the teacher training institutes to college-level facilities. In 1976 they were designated the Faculty of Education of the new national university. Plans for a faculty of science and a faculty of civil aviation were under way. Most of the new campus was scheduled to be constructed by 1978; it was to accommodate 2,000 students. Segregation of the sexes would be maintained.

In the mid-1970s there were almost 28,000 students enrolled in the basic twelve-year system of education; there had been only about 1,000 in 1955. Of the total enrolled about 47 percent were female. About 72 percent were enrolled in primary school, 18 percent in preparatory school, and 10 percent in secondary school. Primary school enrollment included the majority of the primary school age-group. The eventual goal of the government was total enrollment of primary-school-age children and two-thirds enrollment of those of preparatory and secondary school ages by 1982. In addition to the existing ninety-seven schools, of which forty-five were for girls, twelve primary schools and fourteen secondary schools would have to be constructed. There were also 500 Qatari students enrolled in technical, vocational, and craft programs at the regional training center operated by the United Nations Development Program in Doha for students from various gulf states.

Estimates of the literacy rate ranged from 10 to over 20 percent in the mid-1970s. The government had established an adult literacy program and operated about twenty literacy training centers throughout the country. Religious restrictions on the intermingling of the sexes, however, had narrowed participation in the program to adult males. Efforts were under way in 1976 to provide special arrangements for the instruction of women.

THE ECONOMY

By 1976 the oil sector dominated the economy. The other commodity-producing sectors (farming, fishing, and manufacturing) were minor contributors to the total output of goods and services, although manufacturing became more significant with the completion of some modern plants in the 1970s. Apart from the production of crude oil, the economy was largely service oriented. The bulk of the labor force was engaged in trade, transportation, construction, and government services.

Expansion of the service sector followed naturally from the early stage of development of the economy and the dominance of the oil sector. Oil revenues, a very large part of gross national product (GNP), were in foreign exchange, which permitted imports per capita to be among the highest in the world. Transportation and distribution of the flow of imports engaged a substantial portion of the work force. In addition the government, the recipient of oil revenues, channeled the funds into correcting social and economic deficiencies. Schools and hospitals were built and staffed; the road network was expanded and the ports improved; and communications, electric power, housing, and sewage, and other facilities were constructed. It was relatively easy for unskilled Qatari farmers and fishermen to make the transition to unskilled jobs in construction, trade, and transportation, but many jobs requiring training and skills had to be performed by imported workers, although by 1976 education and vocational training were raising the skills of Qataris.

A lack of basic statistics hampered evaluation of the country's economy. Some observers estimated GNP at about US$2 billion in 1974 and per capita GNP at about US$10,000, but these were little more than guesses. The size of the labor force was unknown, although most observers agreed that expatriates substantially outnumbered Qataris. The government did not publish estimates of GNP or enough data for others to estimate it accurately.

Government officials decided on a major shift in economic development after the dramatic increase of oil revenues in 1974. The increased revenues were largely to be channeled into developing natural gas resources as a base for manufacturing industries. The shift had produced some temporary strains in the economy by late 1976 and required some foreign borrowing. The success of the decision, however, rested partly on the country's ability to attract and keep productive the large numbers of additional foreign workers that were needed to construct and operate the new facilities and partly on developments in foreign markets where the new commodities would be sold.

Agriculture and Fishing

Agriculture and fishing were the most important means of livelihood before the discovery of oil, but traditional techniques and equipment meant low incomes and a bare subsistence for the bulk of the population engaged in them. The discovery of oil and the subsequent expansion of construction, trade, and government services afforded employment opportunities in these sectors at higher pay and with urban amenities. Farmers and fishermen were drawn away from their traditional employment. By the 1960s farming and fishing were among the least important sectors of the economy. Nearly all food had to be imported.

Agriculture experienced a limited revival after the early 1960s, however, although statistics to measure the resurgence were few and of questionable reliability. In the mid-1970s there were over 450 farms having an average size of about 7.5 acres and employing about 1,400 workers. In 1974 the cultivated area amounted to between 3,000 and 3,500 acres, of which two-thirds to three-quarters was used to grow vegetables. In contrast only about 650 acres were cultivated in 1960, two-thirds of which was used to grow animal fodder.

The high incomes of urbanites and their willingness to pay for fresh vegetables contributed to the revival of agriculture. Vegetable growing had become the principal kind of farming, and production increased from less than 2,000 tons in 1960 to over 18,000 tons in 1974. Vegetable production may have exceeded 21,000 tons in 1975. By the mid-1970s the country was largely self-sufficient in vegetables and exported small quantities. Tomatoes, potatoes, beans, cabbage, and onions were among the vegetables grown. The other major crop was animal fodder (primarily alfalfa), production of which amounted to 3,400 tons in 1960 and 25,500 tons in 1974. Only about 1,000 tons of grain (largely barley) was produced annually in the mid-1970s, and imports of wheat and rice were needed to satisfy the bulk of consumption needs. About 165,000 palm trees supplied dates, but the trees were not nearly as prevalent as in other gulf states. An effort was under way to increase production of dates and other fruits.

Rainfall was extremely scarce, and there were no streams in the country. Cropping depended on irrigation from wells and springs, most of which were located in the more northerly portion of the peninsula. The combination of climate and irrigation often permitted two crops a year. By 1976 the water table was dropping quickly, though without the increased salinity experienced in Bahrain. The rapid expansion in the mid-1970s of desalination facilities to provide drinking water will ease the future drain on groundwater sources, but it will entail appreciable economic costs and some form of subsidy to agriculture. Purified water from a sewage treatment plant completed in 1975 near the capital was used on city gardens. Further development of treated water could supply water for farming close to the capital, but scarcity of water was a major constraint on agricultural development.

The government took an active part in trying to revive agriculture as part of a long-term goal of achieving self-sufficiency in food. Financial inducements included guaranteed prices for certain produce, grants for development of wells, and free plowing, seeds, and fertilizer. The government established agricultural research and training centers, undertook land reclamation and irrigation, and developed a 200-acre experimental farm about forty miles north of Doha. The experimental farm, which was started in 1963, tested and distributed seed varieties and provided advice to farmers on ways to increase yields. The farm also produced seedlings and supplied them to farmers for growing windbreaks to lessen the damage of the hot, searing winds.

A hydroagricultural survey was completed in 1973 by United Nations (UN) experts, largely from the Food and Agriculture Organization (FAO). The survey formed the basis for a series of projects to be implemented in the latter half of the 1970s. The major projects focused on expanding the country's meat supply.

The raising of livestock by nomads persisted even after oil was found. The animals subsisted largely on the coarse vegetation that managed to survive in the desert. Farmers also raised some livestock, which lived by foraging and on fodder grown by the farmers. In 1974 the estimated livestock population amounted to 5,600 cattle, 36,400 sheep, 42,300 goats, 8,100 camels, 300 horses, and 69,000 chickens.

A large, modern poultry farm was the first project undertaken by the government to expand the meat supply. The poultry farm, located about thirty miles north of Doha, began operations in 1975 with 15,000 imported chicks. When completed the automated poultry farm would use a combination of imported and local feed to produce 10 million eggs and 1 million broiler chickens a year. Production was expected to meet 80 percent of local demand. The second major project was the development of a modern, scientific sheep-rearing farm for an initial flock of 5,000; the first animals were to be imported in late 1976 or 1977. The site was near the Saudi border, where groundwater had been found. Cultivation of alfalfa had been started there in 1975 and was to be expanded as more irrigation became available. Local fodder was to be supplemented by imported concentrated feed. The farm was expected to provide 13,000 sheep annually for slaughter. The government intended to start a dairy farm in late 1976 or 1977 with 2,000 head of cattle. A privately owned dairy farm with 300 cows was already in operation in 1976. The government hoped to be self-sufficient in dairy products by the late 1970s.

Fish had long been a source of protein in the local diet and fishing a major occupation for Qataris. Fishing continued after oil was found but, except for shrimp fishing, without modernization. Using traditional techniques and equipment, the catch did not keep pace with population growth. Fish became scarce in local markets and had to be imported. In 1975 the government, pursuing its goal of self-sufficiency in food, joined with other gulf states in a three-year regional fisheries survey and development project to be carried out by FAO experts. The headquarters for the survey was in Doha. The survey was to identify the fish resources, potential annual catches, and the best kinds of equipment. When the survey was completed, the government would presumably take measures to modernize the local fishing industry. The government hoped that the industry would be able to supply local needs and make fishing a profitable occupation once again.

Shrimp fishing was modernized with the formation of the Qatar National Fishing Company in 1966. The government held 60-percent ownership and a foreign fishing concern the remainder. In 1976 the company had a fleet of six modern trawlers, which operated in Qatari waters, and shore facilities for processing, packaging, and freezing shrimp for export markets, primarily in the United States and Japan. Unlike many other gulf states Qatar had not overfished its shrimp beds, and the FAO commended the country's conservation measures. The company's shrimp fishing had been profitable. In 1973 shrimp exports amounted to 515 tons, valued at QR2.1 million (for value of the Qatari riyal—see Glossary). In 1975 the government enlisted the company's help in catching fish other than shrimp for domestic consumption. The company began operation of a modern trawler to catch fish, which the government marketed locally.

Industry

Modern industry began only after the discovery of oil. The government encouraged growth of the manufacturing sector in order to diversify the economy and reduce dependence on oil revenues. Development moved slowly, however, partly because roads, ports, utilities, and other parts of an industrial base had to be built. The labor force was uneducated and unskilled at first. Education and vocational training slowly upgraded the desired skills, but in the mid-1970s industrial expansion still required imported labor for both technical and menial work. Apart from crude oil and natural gas, the country's only known natural resource was limestone for the production of cement. Moreover the domestic market was small, requiring most modern manufacturing to be oriented toward foreign markets with established competitors. By the late 1960s the pace of industrialization began to accelerate, and by 1976 several plants had been completed, but manufacturing still was a small contributor to the total output of goods and services.

One of the fastest growing industrial activities was the production of electricity and fresh water. Population growth and industrial expansion greatly increased demand for both commodities. The government's installed electric generation capacity increased from seventy megawatts in 1970 to 160 megawatts in 1975. The main power generators were located near the capital. Smaller generators were associated with industrial plants located at the industrial park at Musayid and on the west coast. Transmission facilities and a grid system connecting most of the peninsula's populated areas were under construction in 1976, and completion of major portions was scheduled for 1977. Transmission lines were being extended to the industrial area at Musayid to provide power for projects under construction there until the industrial park's own large power station was completed.

Qatar lacked the groundwater sources of Bahrain and had built its first desalination plant in 1957. The plant was enlarged and modernized over the years. The distillation units were in tandem with electric generators that used associated gases from the oil fields. By 1975 water consumption amounted to about 11 million gallons a day, of which wells and springs supplied only about 3.2 million gallons. The rest was supplied by the electric power-desalination plant located near the capital. The government policy was to stop drawing on the groundwater sources as soon as possible and by 1979 at the latest.

Major expansion of electric power and desalination capacity was under way in 1976. The existing plant was being expanded modestly, which would raise production to about 10 million gallons a day by mid-1977. A major new plant was being built a few miles south of the capital. Its ultimate capacity would be 300 megawatts of electric power and 24 million gallons of water per day in 1979, but construction was phased so that some units would be operating by 1977. The government expected that total electric power capacity would be nearly 500 megawatts and desalination capacity 39 million gallons a day by the end of 1978. A large (over US$400 million) electric power and desalination plant (capacity of 450 megawatts and 8 million gallons of water a day) was under construction at the industrial park at Musayid. Water distribution facilities were also being considerably expanded in 1976.

The government welcomed foreign investors in joint ventures with Qatari interests, particularly where production increased exports. Legislation required 51-percent Qatari ownership. Only Qataris could own property. The government had participated in the larger industrial ventures. Exemptions from import duties and income taxes for a period of five years were granted to new investments. There were no exchange controls affecting settlement of commercial transactions or remittance of profits. The government was continuing to develop the infrastructure base for industry at the Musayid industrial park.

The largest manufacturing establishment in the country was the fertilizer plant at Musayid. It was owned by the Qatar Fertilizer Company (QAFCO), which in turn was owned by the government (63 percent) and foreign interests. A Norwegian concern, which had part ownership, managed plant operations and marketing. Construction started in 1969, commercial production began in late 1973, and the first exports were made in 1974. The plant used associated gases from the onshore oil field to produce urea and ammonia. Production figures were unavailable, and observers believed that start-up problems and the worldwide recession had kept production below the rated capacity of 1,000 tons of urea and 900 tons of ammonia a day through 1975. Even with these problems, by 1976 work had begun in order to double capacity by late 1978 at a cost of US$250 million. A mechanized jetty at the port of Musayid had sufficient capacity in the conveyor system to handle the plant's expanded capacity as well as the wheat imports for the adjacent flour mill.

The Qatar National Cement Company began operating a kiln on the west coast in 1969. The government initially held 60-percent ownership, which was down to 30 percent by 1976; private Qatari investors held the remainder. Annual capacity in 1976 was 266,000 tons of cement, and a third kiln was being added to raise capacity to about 366,000 tons per year by early 1977. A fourth kiln with a capacity of 900 tons per day was to be completed in 1978. The expansion also included additional electric power and desalination capacity because the west coast area around the oil field was not planned to be connected to the facilities near the capital in the immediate future. Cement production was consumed locally. Production figures were unavailable; however, because the building boom had created a tremendous demand for cement and cement had to be imported, production presumably was near capacity.

A flour mill was completed at Musayid in 1972 with a capacity of 100 tons of flour per day. The mill was owned by private Qatari investors, but the government had encouraged the establishment of the mill as part of the diversification process. In 1975 a small government-owned refinery, with a daily capacity of 6,200 barrels of crude oil, was completed to supply the local market with petroleum products. Its capacity was being tripled in 1976. A plant was also completed in 1975 to process and export products extracted from the natural gas associated with the production of crude oil from the onshore field (see Oil Industry of Qatar, ch. 5).

With the increase of oil revenues in 1974, the government increased budget allocations to diversify the economy. Diversification measures were the responsibility of the Technical Center for Industrial Development (TCID), which was responsible to the amir (ruler and head of state). The TCID had broad responsibilities, from identifying projects to meet the country's long-term needs to undertaking feasibility studies, supervising construction of projects, and coordinating the efforts of various ministries. The TCID, which became operational in 1975, had no overall development plan, only a series of projects for implementation.

The strategy of TCID planners was to develop heavy industry based on the country's natural gas resources. Part of the development required construction of a gas-gathering system from the offshore oil fields to supplement gas collected from the production of crude oil at the onshore field. The gas associated with crude oil production would last approximately forty years. A large gasfield unassociated with crude oil would provide a heat source and petrochemical feedstock for an additional period. The gas-gathering systems would lead to the Musayid industrial park, where manufacturing would be concentrated. During the late 1970s a new self-contained town for industrial workers was to be built at Musayid.

Other projects for Musayid included an export oil refinery that was under construction in 1976, an additional plant to process natural gas liquids, and a petrochemical plant to produce ethylene (see Oil Industry of Qatar, ch. 5). The petrochemical plant involved complicated arrangements with some French companies for management and marketing, for another petrochemical plant tc be built in France, and for the establishment of a shipping company. In 1976 work was under way under a contract with a Japanese company to build and manage a US$300 million integrated iron and steel complex using natural gas in a process for direct reduction of imported ores. The government's equity amounted to 70 percent; Japanese firms held the remainder. The steel mill will eventually produce about 400,000 tons of bars, shapes, and billets; about 75,000 tons of iron construction bars were initially scheduled for use in the domestic market, and at full production a large portion of the output probably would be exported. Initial production at the complex was scheduled for late 1977. The government had expressed an interest in an aluminum smelter capable of producing about 140,000 tons of ingots a year from imported alumina, but the project had not been definitely approved by late 1976. Construction of a pharmaceutical and cosmetics factory was under consideration.

A new variable was added to Qatar's industrialization plans by the formation in 1976 of the Arab Gulf Organization for Industrial Consultancy, with headquarters in Doha. The aims of the organization were to coordinate industrial planning, avoid duplication of projects, and reduce the dependency on foreign consultants in Qatar, Kuwait, Bahrain, Saudi Arabia, the UAE, Iraq, and Oman. An obvious need existed for the organization because all of the gulf states were attempting a number of similar projects, such as expanding refining capacity and collecting and processing natural gas. How effective the organization might become and whether changes might be made in Qatar's development plans could be determined only after the organization had been in operation for a while.

Foreign Trade and Balance of Payments

The country depended on imports for nearly all items of consumption and investment because of the low output of the economy except for crude petroleum. Oil revenues, however, permitted a very high level of imports, averaging the equivalent of US$2,250 per capita in 1975, one of the higher ratios in the world. Imports increased rapidly with the increase in oil revenues, more than doubling between 1973 and 1975 (see table 25). There were few restrictions on imports or foreign exchange transactions, and customs duties were a modest 2.5 percent on most imports. The country has never had the entrepôt trade of Bahrain or the UAE. Nearly all imports were for local use, and reexports, primarily to Saudi Arabia and the UAE, probably amounted to 10 percent or less of total imports. The bulk of imports arrived by ship, but some arrived by truck via Saudi Arabia when port congestion became severe throughout the gulf after 1973.

The jump in oil revenues and the expansion of development activities increased the proportion of imports of investment goods in the mid-1970s, but imports for personal consumption remained large. In 1975 imports of electrical equipment and appliances (primarily consumer appliances) amounted to QR302 million; machinery and spare parts, QR261 million; transportation equipment (largely passenger cars), QR254 million; steel products, QR123 million; textiles, clothing, and footwear, QR103 million; and food, QR150 million. The major industrialized countries supplied 58 percent of total imports in 1975: the United Kingdom QR342 million, Japan QR242 million, the United States QR202 million, and the Federal Republic of Germany (West Germany) QR151 million. Except in 1974, when imports from Japan were larger, the United Kingdom had always been the major supplier. Other industrialized countries supplied a large part of the remaining imports, although some imports were transshipped from Bahrain, Dubai, Kuwait, and Saudi Arabia.

Exports were primarily crude oil, which accounted for 98 percent of the value of exports in 1975. After completion of the chemical fertilizer plant in 1973, exports other than petroleum were primarily chemical fertilizer and liquid ammonia. These two commodities accounted for 76 percent of nonpetroleum exports in 1975. The remainder of nonpetroleum exports were probably reexports to nearby areas, particularly Saudi Arabia. The bulk of fertilizer exports went to Bangladesh, India, Vietnam, and the United States.

The dominance of the oil sector in the economy meant that a very

Table 25. Qatar, Summary of Foreign Trade, 1969–75
(in millions of Qatari riyals)[1]

Year	Imports (c.i.f.)[2]	Exports Total	Exports Crude Petroleum
1969	252	1,162	1,132
1970	306	1,204	1,160
1971	516	1,484	1,429
1972	607	1,740	1,674
1973	778	2,467	2,400
1974	1,069	7,956	7,814
1975	1,622	7,034	6,892

[1]For value of the Qatari riyal—see Glossary.
[2]Cost, insurance, and freight.

Source: Based on information from International Monetary Fund, *International Financial Statistics*, Washington, November 1976, p. 318.

large proportion of GNP was in foreign exchange. This had eliminated balance-of-payments problems for many years. The tremendous increase in oil revenues after 1973 created large surpluses in the balance of payments in spite of increased foreign aid extended by the government. Little information was available concerning the country's foreign aid, and the government did not publish balance-of-payments information. International Monetary Fund (IMF) economists estimated that the country's surplus in balance of payments exceeded US$1.2 billion in 1974 (see table 26). The surplus probably was smaller in 1975 because imports increased more than oil revenues in that year, but the country probably had a substantial addition to its foreign assets.

Table 26. Qatar, Balance of Payments, 1970–74[1]
(in millions of SDR)[2]

	1970	1971	1972	1973	1974[3]
Oil sector (net)	122	193	252	361	1.550
Non-oil exports	9	12	14	15	16
Imports	−64	−180	−128	−164	−226
Services, transfers, and capital movements (net)[4]	−56	−46	−54	−172	−290
Surplus[5]	11	−21	85	40	1,050

[1]Government oil receipts and surpluses were on a fiscal year basis (corresponding to the Muslim hijra year), which is ten to twelve days shorter than a Gregorian year (see table A).
[2]SDR—special drawing rights of the International Monetary Fund (IMF). The average yearly exchange rate was SDR1 equals US$1 in 1970 and 1971, US$1.09 in 1972, US$1.19 in 1973, and US$1.20 in 1974.
[3]IMF projections based on partial year data.
[4]Includes errors and omissions.
[5]Surplus is equivalent to the net increase in foreign assets.

Source: Based on information from "Qatar, with '74 Oil Income Approaching Past Decade's Total, Stresses Welfare," *IMF Survey*, Washington, August 19, 1974, p. 259.

Government Budget

In 1976 oil revenues continued to go to the amir, as they had since the first concession was granted (see Oil Industry of Qatar, ch. 5). This solidified the position of the ruler and freed him from any financial dependence on other segments of the society. As oil revenues rose, agencies of public administration were created to handle the revenues and extend government services, which in practice dispersed oil revenues into the economy. The government of Qatar had had a shorter experience in development than Kuwait and Bahrain, and by 1976 public administration remained highly centralized around the ruler. The amir reportedly signed all checks over QR100,000 (about US$25,000). Published budget data were sketchy and usually incomplete.

Oil revenues provided the bulk of government receipts, amounting to 89 percent in fiscal year (FY) 1970 and 98 percent in FY 1974 (see table 27). Import duties were kept low on most commodities because of the dependence on foreign goods, and the low taxes on international trade were unimportant sources of revenues. Businesses were taxed at progressive rates on profits earned within the country, but there was no tax on personal income in 1976. Income taxes produced minor amounts of revenues. Fees for such government services as water and electricity also produced minor revenues.

Budget expenditures were the primary means of injecting oil revenues into the economy. The government followed the pattern set by most other gulf oil states. Education, health, and housing programs were established to improve living conditions while infrastructure projects (roads, port facilities, and electric power and desalination plants) developed the economic base. Budget allocations for both current and capital expenditures jumped tremendously in FY 1975, the former doubling and the later tripling as a result of the increase of oil revenues the year before. Actual expenditures may have fallen considerably short of the budget allocations, however, because of complications in implementing the large investment program.

Capital expenditures were planned to more than double in the fiscal year starting on January 2, 1976, increasing from QR1.8 billion in FY 1975 to QR3.9 billion (nearly US$1 billion) in FY 1976. The primary cause for the increase was the substantial investment required to develop heavy industry. Budget allocations for industry (excluding electric power and desalination plants) amounted to QR1.5 billion in FY 1976 compared with QR305 million in FY 1975. An additional QR934 million (US$233 million) was allocated for electric power and desalination projects. Other projects in the FY 1976 capital budget included port expansion at Doha and Musayid; construction of housing, schools, and hospitals; road improvement; and completion of the space communication monitoring station. The capital budget for public administration, justice, and security increased

Table 27. Qatar, Budget Summary, Fiscal Years 1970–75[1]
(in millions of Qatari riyals)[2]

	Actual				Proposed	
	1970	1971	1972	1973	1974	1975
Revenues:						
Oil	515	837	1,104	1,616	5,378	n.a.
Other	64	108	126	104	118	n.a.
Total Revenues	579	945	1,230	1,720	5,496	7,135
Expenditures:						
Current	358	475	672	746	1,001	2,200
Capital	132	169	227	249	580	1,800
Aid to other countries	0	0	27	357	350	n.a.
Other	14	46	32	190	0	n.a.
Total Expenditures	504	690	958	1,542	1,931	5,302
Surplus	75	255	272	178	3,565	1,833

n.a.—not available.

[1]The government fiscal year (FY) is the Muslim hijra year, which is ten to twelve days shorter than a Gregorian calendar year (see table A). FY 1970 started March 9, 1970; FY 1971 started February 26, 1971; FY 1972 started February 15, 1972; FY 1973 started February 4, 1973; FY 1974 started January 23, 1974; FY 1975 started January 13, 1975; and FY 1976 started January 2, 1976.

[2]For value of the Qatari riyal—see Glossary.

Source: Based on information from International Monetary Fund, *International Financial Statistics*, Washington, November 1976, p. 318; and "Qatar, with '74 Oil Income Approaching Past Decade's Total, Stresses Welfare," *IMF Survey*, August 19, 1974, p. 258.

very little, from QR144 million in FY 1975 to QR173 million in FY 1976.

The government had acted prudently through FY 1974, keeping expenditures below revenues. The resulting surpluses went into state reserves. Little information was available on the size of accumulated state reserves; unofficial estimates by observers ranged up to US$2 billion by 1976. State reserves were managed by an investment board headed by the amir and his son, the minister of finance and petroleum (see Political Dynamics, this ch.). In 1975 an investment office was opened in London to prepare studies and offer advice on the placement of funds. The government also used experts for investment advice and management in several other major financial centers. Observers believed that Qatar's reserves were held in the major international currencies to minimize exchange risks but that most of the reserves were highly liquid, being placed on short call.

The decision in 1975 to channel the large increase in oil revenues into industrial projects meant a bunching of large capital expenditures over just a few years. In FY 1975 the budget surplus was still substantial although, because of greatly increased expenditures, it was only about half as large as in FY 1974. The substantial increase of capital expenditures in FY 1976—and they were expected to remain high through FY 1978—could cause budget deficits or require some external financing in the late 1970s. There was insufficient information in late 1976 to determine whether a budget deficit was likely, but there had been some borrowing during the year in European financial centers to secure partial financing of several of the country's development projects. It may turn out that Qatar has duplicated the experience of other oil-exporting countries by undertaking large development projects beyond the country's immediate ability to finance them. Foreign financing was available, however, because the country's credit standing was very good; the state reserves also could be drawn on to complete the projects. The temporary shortage of funds posed no problems as long as the investments generated the earnings anticipated in order to repay creditors.

Money and Banking

The Indian rupee was the principal currency until 1959, when it was replaced by a special gulf rupee to halt gold smuggling into India. In March 1966 Qatar and Dubai established a currency board to issue a Qatar-Dubai riyal. India devalued its rupee before the new riyal could be issued. As an interim measure the currency board borrowed 100 million Saudi Arabian riyals, which were circulated for a few months in Qatar and Dubai. The Qatar-Dubai riyal was introduced in September 1966 with a gold value of 0.1866 grams of fine gold, the same value as the gulf rupee before its devaluation. Until 1973 the Qatar-Dubai riyal was the principal currency circulating in Qatar; the riyal was also used by Dubai and, except for Abu Dhabi, the other amirates that had formed the UAE. In May 1973 the Qatari riyal was introduced, and the UAE also issued its own currency. The new Qatari riyal maintained the same gold value, 0.1866 grams of fine gold. The Qatari riyal had full foreign exchange backing and was a strong currency.

A central bank, the Qatar Monetary Agency (QMA), was established in May 1973 to issue the Qatari riyal. The central bank replaced a currency board that had been largely passive in monetary affairs. The QMA had full central bank responsibilities, including regulation of the commercial banks, but by 1976 its incomplete staffing limited the responsibilities it could effectively handle. In 1976 there were twelve commercial banks, the largest of which was the locally owned Qatar National Bank. Another Qatari-owned bank began operations in 1975. The rest of the commercial banks were foreign owned. Most banking personnel were expatriates, and few Qataris appeared interested in banking careers. Local interest rates were regulated by an agreement between banks, but booming economic conditions put a strain on the agreement in 1976.

Worldwide inflation sharply increased prices of Qatar's imports in 1973 and 1974 about the same time that oil revenues increased dramatically. As a result of increased oil revenues and development expenditures, most economic activity in the country picked up, particularly trade and construction. By 1976 there was real economic growth as well as upward pressure on prices. Unofficial estimates of inflation exceeded 25 percent in 1975. Rents, particularly for expatriate housing, were skyrocketing, and land speculation was extensive, financed in part by bank credit. Port congestion, which in October 1976 kept ships waiting 115 to 125 days to unload, added substantially to transport costs and disrupted construction schedules. The inflationary pressures that had at first been imported from abroad were being sustained by internal factors in late 1976.

How long the boom conditions would affect the course and cost of economic development was unpredictable. Some new port facilities would become available in 1977. More domestic cement would be available in 1977, and the iron and steel complex was scheduled to begin production of construction rods in 1977. Some of the investments that had contributed to the boom might eventually contribute to its reduction.

POLITICAL DYNAMICS

Under the terms of the 1916 and 1934 treaties between Qatar and Great Britain, British interests in Qatar were within the purview of the British political agent posted in Bahrain. In 1949, when Qatari oil production began, the British assigned a separate political agent to Doha. He was joined by other diplomatic and support personnel in the early 1950s and by a special British financial adviser to the Qatari amir, or ruler. Other British nationals served as commanders of the army and the police force. British extraterritorial legal jurisdiction was extended to cover not only British subjects in Qatar but also all non-Arab and non-Muslim foreigners.

British political and financial advisers provided assistance in organizing the state administration, as did an Egyptian, Hasan Kamil. When the post of British financial adviser was replaced with that of director of government in the late 1950s, Kamil filled the new position. He still held the post, retitled adviser to the ruler, in early 1976. By early 1970 more than thirty major governmental departments had been established. An effective vertical chain of command did not exist, however. Each of the departments was equal, and each department head reported directly to the ruler. The lack of hierarchical organization, absence of adequate communications between departments, and considerable duplication of functions led to significant bureaucratic inefficiency.

During the 1950s and 1960s the Qataris began developing state administrative organs, a development that was both necessitated and facilitated by increasing petroleum revenues, related industrial and commercial growth, and the desire to redistribute a major portion of the national income through various social welfare schemes. Modern state administrative machinery was also necessary because Qatar was assuming an increasingly important role in regional Arab political and economic affairs.

The Ruling Family and the Succession of Shaykh Khalifah bin Hamad Al Thani

In the mid-1970s the Al Thani ruling family comprised three main branches: the Bani Hamad, headed by Khalifah bin Hamad Al Thani (reigned 1972–); the Bani Ali, headed by Ahmad bin Ali Al Thani (reigned 1960–72); and the Bani Khalid, headed by Nasir bin Khalid Al Thani (the minister of commerce and economics in 1976) (see fig. 15).

Ahmad had succeeded his father, Ali bin Abd Allah Al Thani (reigned 1949–60), as Qatar's ruler, but neither had any particular interest in supervising daily government. Thus somewhat by default those duties had been assumed, beginning in the 1950s, by Ahmad's cousin Khalifah, the heir apparent and deputy ruler. By 1971 Khalifah not only had served as prime minister but also had headed the ministries or departments of foreign affairs, finance and petroleum, education and culture, and police and internal security.

On February 22, 1972, with the support of the Al Thani family, Khalifah assumed power as ruler of Qatar. Western sources frequently refer to the event as an overthrow, a takeover, even a bloodless coup d'etat. The Qataris, at least officially, regarded Khalifah's assumption of full power as a simple succession. This was because the Al Thani family notables had declared Khalifah the heir apparent on October 24, 1960, and it was their consensus that Ahmad should be replaced.

The reasons for the transfer of power are not entirely clear. Khalifah has reportedly stated that his assumption of power was intended "to remove the elements which tried to hinder [Qatar's] progress and modernization." Khalifah has consistently attempted to lead and control the process of modernization caused by the petroleum industry boom and the concomitant influx of foreigners and foreign ideas so that traditional mores and values based on Islam could be preserved. He and other influential members of the ruling family may well have grown impatient with Ahmad's indifference to government and his frequent vacation trips abroad. Khalifah and other family notables were also known to have been troubled by financial excesses on the part of many members of the Al Thani family. Ahmad was reported to have drawn one-fourth, and the entire Al Thani family together between one-third and one-half, of Qatar's oil revenues in 1971. The new ruler severely limited the family's financial privileges soon after taking power.

Family intrigue may also have played a part in the change of rulers. Factionalism, jealousies, and rivalries are not uncommon within ruling families, particularly those as large as the Al Thani. Western observers have reported rumors to the effect that Khalifah acted when he learned

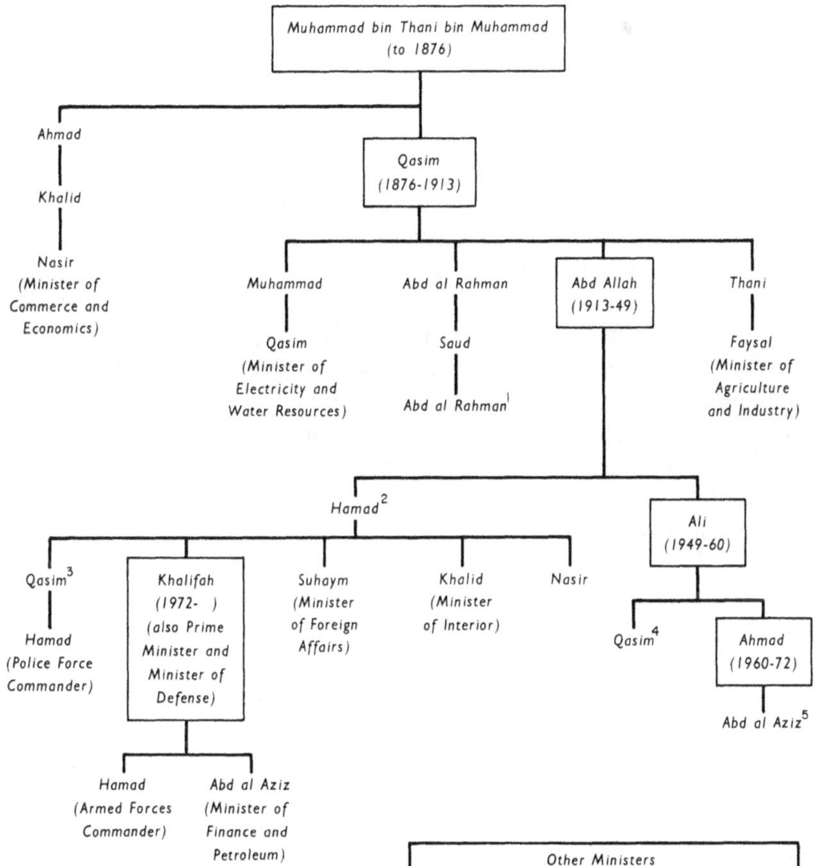

[1] Minister of justice under Khalifah through 1975.
[2] Designated heir apparent but died prematurely (1946).
[3] Minister of education and youth welfare until death (1976).
[4] Armed forces commander under Ahmad.
[5] Minister of public health under Ahmad.

Other Ministers

Isa Gharrim Al Kuwari, Information
Ali ibn Ahmad Al Ansari, Labor and Social Affairs
Muhammad ibn Jabr Al Thani, Municipal Affairs
Khalid Muhammad Al Mani, Public Health
Khalid ibn Abd Allah Al Atiya, Public Works
Abd Allah ibn Nasir Al Suwaydi, Transportation and Communications
Minister of Justice--vacant
Minister of Education and Youth Welfare--vacant

☐ Rulership
( ) Rulership period

Source: Based on information from John Duke Anthony, *Arab States of the Lower Gulf: People, Politics, Petroleum,* Washington, 1975, p. 78.

Figure 15. Qatar, Abridged Al Thani Genealogy and Government Ministers, June 1976

that Ahmad might be planning to substitute his son, Abd al Aziz, as heir apparent, a move that would have circumvented the declared consensus of the Al Thani family.

Government Process and Constitution

Qatar is governed by a regime that generally resembles a traditional monarchy. Because of the modernization process it probably is in some stage of transition, but to what is not yet clear. Qatar is a unitary state; it has no subnational political units possessing inherent authority. Municipal councils (in Doha, Al Wakrah, Al Khawr, Thakhira, Ar Rayyan, and Umm Salal), however, plan their own development programs while remaining directly responsible to the Ministry of Municipal Affairs. Sovereignty is, according to the Constitution, vested in the state, but to all intents and purposes it is vested in the head of state, or ruler. Although the ruler is supreme in relation to any other individual or institution, his rule is not in practice absolute.

Rule of Qatar is hereditary within the Al Thani family, but it is not automatically passed from father to son. Instead the ruler is designated by the consensus of family notables. Once a ruler loses that consensus, he will be replaced, as was illustrated in the transfer of power from Ahmad to Khalifah. A Qatari ruler is also guided, and to some degree constrained, by the ethics of Islam—particularly by the strictures of Wahhabi (see Glossary) Islam—which emphasize fairness, honesty, generosity, and mutual respect (see ch. 3). Islamic religious and ethical values are applicable both to the ruler's personal life and to his conduct of the amirship. Thus the ruler must retain the support of the religious community.

The state political organs include the ruler, the Council of Ministers, and the advisory council. The ruler makes all major executive decisions and legislates by decree. The 1970 Constitution institutionalized the executive-legislative process, in effect formalizing the supremacy of the ruler. The more important of the ruler's duties enumerated in the Constitution include convening the Council of Ministers, ratifying and promulgating laws and decrees, commanding and supervising the armed forces, appointing and dismissing senior civil servants and military officers (by decree), and reducing or waiving state penalties (also by decree). Finally the Constitution provides that the ruler possesses "any other powers with which he is vested under this provisional constitution or with which he may be vested under the law." This means, in short, that the ruler may extend or modify his own powers by personal decree.

The Constitution also provided for a deputy ruler, who was to assume the post of prime minister. The prime minister was to formulate government programs and exercise final supervisory control over the financial and administrative affairs of government. Although Khalifah was the heir apparent and the prime minister concurrently, the Constitution did not specify that the post of prime minister must be held by the heir apparent. As of December 1976 no heir apparent to Khalifah (who was only forty-one years of age) had been designated. Khalifah retained the post of prime minister when he became ruler.

The Council of Ministers, which resembles similar bodies in the West, forms the ruler's cabinet. In November 1976 the council included the heads of fifteen ministries. Seven of the ministers were Al Thani; three of these were members of the ruler's immediate family. They included his son, Abd al Aziz bin Khalifah Al Thani, the minister of finance and petroleum, and two brothers: Suhaym bin Hamad Al Thani and Khalid bin Hamad Al Thani, respectively the ministers of foreign affairs and interior.

The Council of Ministers was responsible collectively to the ruler, as was each minister individually. The ruler appointed and dismissed ministers (technically on the recommendation of the prime minister when that post was occupied by someone other than the ruler). Only natural-born Qataris could become ministers, and the Constitution prohibited the prime minister and other ministers from engaging in business or commercial activities while holding state office.

The duties of the Council of Ministers included proposing laws and decrees to be submitted to the ruler for ratification, issuing orders and regulations necessary for the implementation of ratified laws, and supervising state administrative and financial affairs. The discussions of the council were held in private; a decision required a majority vote and, if approved by the ruler, became binding on all members.

The advisory council debated laws proposed by the Council of Ministers before they were submitted to the ruler for ratification. If approved by the ruler, a law becomes effective on publication in the official gazette of Qatar. The advisory council also debated the draft budgets of major public projects and general policy on political, economic, social, and administrative affairs, as referred to it by a minister or by the prime minister. The advisory council could request from the Council of Ministers information pertaining to the policies it was debating, direct written questions to particular ministers, and summon ministers to answer questions on proposed legislation. Ministers had the right to attend and address advisory council meetings in which policy matters within their purview were being discussed. This constitutional guarantee seemed unnecessary because members of the Council of Ministers were also members of the advisory council.

The Constitution stipulates that for purposes of forming the advisory council Qatar would be divided into ten electoral districts. Each district would elect four candidates of whom the ruler would select two, making a total of twenty; these constituted the relatively representative portion of the council. The members would represent all the Qatari people, not their specific districts. The total advisory council also contained as many as ten additional members appointed by the ruler as well as the members of the Council of Ministers. Membership was limited to natural-born Qataris at least twenty-four years of age. All members served three-year terms. By December 1976 elections for the advisory council had not yet been held. Instead all members had been appointed by the ruler; their terms were to end in 1978.

The constitutional provisions for the advisory council indicated a recognition of the desirability of extending political participation, albeit in a very limited fashion. The Constitution also stipulates that the ruler may postpone council meetings for only a limited period and only once without the consent of the council during any given session. Although practically such provisions in no way limit the real power of the ruler, they do exhibit at least an embryonic recognition of the principle of institutional constraint of the individual leader, a novel principle, even if only in theory, in the Arabian Peninsula.

Before the implementation of the Constitution the ruler's legislative authority frequently overlapped or encompassed judicial functions as he personally adjudicated disputes and grievances brought before him. The Constitution apparently marked the beginning of an attempt to organize the judiciary. The organization had not been completed in 1976 but had resulted in a division of the Qatari judicial system into secular and religious elements.

The secular courts included a higher and a lower criminal court, a civil court, an appeals court, and a labor court. Civil and criminal codes, as well as a code of judicial procedure, were introduced in 1971. All civil and criminal law falls within the jurisdiction of these secular courts. The labor court was created in 1962, primarily because few of Qatar's existing judicial customs and codes were applicable to contemporary labor relations.

The sharia court was the oldest element in Qatar's judiciary. The court's law was based on the Hanbali school of Islam, wherein judges (qadis) adhere to a narrow and rigid interpretation of the Quran and the Sunna (see ch. 3). Originally the sharia court's jurisdiction covered all civil and criminal disputes between Qatari nationals and between all other Muslims. Beginning in the 1960s this jurisdiction was successively restricted by decree. In 1976 its responsibilities were primarily confined to family matters and religious morality. Non-Muslims were invariably tried in other courts.

The Constitution establishes the legal presumption of innocence and prohibits ex post facto laws. The Constitution also stipulates that "judges shall be independent in the exercise of their powers, and no party whatsoever may interfere in the administration of justice." The judiciary was apparently independent, not so much because of the constitutional guarantee as because its jurisdiction was most unlikely to confront the ruler's exercise of power. The secular courts adjudicated on the basis of the ruler's past decrees, and the religious courts were restricted to questions of family life and personal behavior. There was no provision for judicial review of the constitutionality of legislation or for landmark judicial decisions that might run afoul of the ruler's plans and wishes.

According to the ruler's preamble to the 1970 Constitution, the government of Qatar was undergoing a transitional or experimental stage of development. The Constitution was thus provisional and was to be replaced with a new constitution based on the results of the transitional period. Khalifah usually legitimates government changes he decrees by reference to the Constitution. By December 1976, however, there was no indication that the full implementation of the provisional Constitution was imminent (for example, the electoral aspects of selection to advisory council membership) or that the transitional period was ending and a new constitution might be forthcoming.

In addition to formally describing and delineating governmental authority, the provisional Constitution set forth such protections as equal Qatari citizenship regardless of race, sex, or religion; freedom of the press; sanctity of the home; and recognition of both private and collective ownership of property. Such guarantees, however, were limited by the public interest and had to be in accordance with the law—which was determined by the ruler.

The Constitution also includes a commitment to certain economic, social, and cultural principles, including state provision of health care, social security, and education. Housing, pension, medical, and educational programs were begun in the 1960s and expanded by Khalifah in the 1970s as increasingly substantial portions of the state's petroleum revenues were used to improve the living standards of Qatari citizens. No state taxes of any kind were imposed on individuals, and the state subsidized basic food prices to minimize inflationary rises. Although these programs appeared to reflect West European welfare statism, they were practical manifestations of the ruler's sense of duty based on obligations inherent in Islamic ethics.

Domestic Political Problems

Like other modernizing countries, Qatar faces two basic political problems as it attempts to develop. These are the lack of a skilled and fully organized administration and the possibility that impatience on the part of some quarter of the population with the slow rate of controlled change might result in political instability. The Al Thani family, the basis of any Qatari ruler's strength, also limits the ruler in that he must use appointments to government posts to retain family support and to balance diverse family factions. Although attempts had been made to improve government administration, clear lines of interorganizational and intraorganizational authority and responsibility had not been fully established. The resulting inefficiency was further underscored by the fact that the ruler continued to make all major decisions. He must undoubtedly slight the needs of some agencies while deciding the more pressing policies or issues of another.

The organized civil service is a relatively new phenomenon in Qatar, and it thus lacks a tradition of public service. Furthermore the lack of educated personnel plagues the entire system. The lack is partly offset by the employment of relatively skilled administrators from Egypt and other Arab countries and by the use of advisers, particularly from Great Britain. There has been some discontent over the employment of foreign nationals, and observers believe that pressure to staff the administrative and social services solely with Qataris will increase. Qatar's rapidly developing educational system will provide some of the personnel needed in lower level administrative work, but the lack of midlevel personnel will remain a problem for many years. Qatar sends students abroad for university education and, when they return, they are able to fill many of the higher administrative posts. Having been exposed to relatively liberal ideas and philosophies, returning university-educated students might become dysfunctional in the Qatari system because of impatience with the rigid, albeit paternalistic, rule of Khalifah and the Al Thani family notables.

A major demonstration, culminating in a general strike, occurred in April 1963. The incident began with an altercation between one of Ahmad's nephews and a crowd that was celebrating the proposed union of Egypt, Syria, and Iraq. A national unity front was formed and made several demands, which included restricting royal privilege, ending employment of foreigners by the state bureaucracy, establishing social welfare facilities, legalizing labor unions, and instituting municipal councils composed at least partly of elected members. The demonstrations were centered in Al Khawr, which some Al Thani family members reportedly wanted attacked by army artillery. Ahmad and Khalifah prevented such a reaction but also refused to accede to the demands. The incident involved discontent with, or resistance to, the primacy of the ruling family not by the masses but by non-Al Thani village elites (many of whom were employed by the oil companies). To what extent this element of the Qatari population constituted a threat to the Al Thani regime after several years of relatively enlightened rule by Khalifah was unknown in 1976.

Political parties and labor unions were prohibited in Qatar in 1976, although workers' committees, which attempted to settle grievances by means short of collective bargaining, were permitted. Most Qatari citizens were apparently satisfied with the social welfare programs of the ruler and had few grievances. Foreign nationals, however, who made up almost the entire unskilled work force and perhaps three-fourths of the total population, were not eligible for Qatari citizenship and received no benefits. Third-generation Qataris of foreign national origin were legally eligible for citizenship but at a status significantly lower than that of indigenous citizens. Foreign Arab elements of the population included Palestinians, Egyptians, Omanis, and Yemenis; the non-Arab elements were much larger, comprising primarily Iranians, Pakistanis, and Indians. The Arabs and, particularly, the Baluchis from Iran reportedly did the most arduous work for the least pay in Qatar.

At least two revolutionary groups, the National Liberation Front of Qatar and the Organization for the National Struggle of Qatar, were reported to be operating in Qatar in the early 1970s. Observers have also noted that the pan-Arab, nationalist, and socialist (but anticommunist) Baath Party had an organized branch in Qatar as late as 1975. If any of these or similar movements and groups did exist in Qatar in late 1976, their impact appeared to be insignificant. However, so long as there is a gap between the living standard of the minority of indigenous citizens and the majority of expatriate workers, and particularly if that gap continues to widen, the workers will probably provide the most fertile ground for internal radical ideological threats to the established regime.

FOREIGN RELATIONS

Qatar's primary foreign policy concern is the stability of the Persian Gulf area. It has consistently supported some form of unity among the Arab states of the gulf to foster regional stability. During the mid-1960s Qatar began advocating a unified monetary and customs policy and the creation of a common market. After the collapse in 1971 of plans for a political union comprising the Trucial Coast states, Bahrain, and Qatar, Khalifah began to reemphasize the original idea of regional economic unity, which in the mid-1970s remained one of his major policy goals.

Qatar apparently also believed that regional stability depended in large measure on a rough balance of power between Iran and Saudi Arabia. Qatar seeks friendly relations with both nations. Iran desires stable regimes in the area and consequently has supported Al Thani rule in Qatar. Foreign observers have speculated that Iran has used its influence to ensure that Iranian nationals in Qatar do not challenge the Al Thani regime. Iran was one of the first countries to recognize Khalifah's accession to rule.

Qatar and Saudi Arabia share the conservative, even puritanical, Wahhabi strain of Islam, which provides a positive link between the two countries. Qatar was one of the few Arab countries that observed the full forty-day mourning period after the assassination of Saudi Arabia's King Faisal in March 1975. Qatar has generally followed Saudi Arabia's lead in foreign affairs.

Like other states on the Arabian Peninsula, Qatar has opposed a naval buildup in the Persian Gulf and the adjacent Indian Ocean by either the United States or the Soviet Union. Relations with the United

States in 1976 were amicable but minimal; no major treaties were in force between the two countries. Relations have been somewhat strained because of American support for Israel. The Islamic religion, particularly Wahhabism, is antithetical to communist ideology; consequently Qatar did not recognize any communist countries, although it did engage in some relatively minor trade with East European nations and the People's Republic of China (PRC).

Qatar has emphasized close relations with Arab states and has championed Arab causes, particularly Palestinian claims. It has contributed unknown, but reportedly quite significant, amounts of financial assistance to the Palestinian Liberation Organization (PLO) and has demanded that the PLO be represented in any Middle East peace talks between the Arab states and Israel. Qatar has also shown increasing interest in the so-called third world developing states.

Bilateral Relations

In September 1976 Qatar maintained reciprocal diplomatic relations at the ambassadorial level with twenty-five countries: Algeria, Austria, Belgium, Egypt, France, India, Iran, Iraq, Japan, Jordan, Kuwait, Lebanon, Libya, Mauritania, Morocco, Oman, Pakistan, Saudi Arabia, Somalia, Sudan, Syria, Tunisia, the United Kingdom, the United States, and West Germany. Qatar had nonresidential diplomatic ties with sixteen additional countries: Afghanistan, Argentina, Brazil, Canada, Denmark, Ethiopia, Italy, Malaysia, the Netherlands, Norway, Senegal, Spain, Sweden, Turkey, Venezuela, and Yemen (Sana).

Numerous Arab countries have received development grants and loans from Qatar. Countries that occupy front-line positions against Israel, such as Jordan and particularly Egypt, have generally received the most aid. Agreements with Egypt have included provisions for the establishment of special training programs for Qataris at Egyptian universities. Relations with Abu Dhabi were strained for some time before 1969. In that year, however, the two countries agreed to an offshore boundary settlement that gave Qatar control of Halul Island and an adjacent oil field, both of which had been in dispute. Qatar also received half control of the large Al Bunduq oil field, which overlaps the offshore boundaries of both countries (see fig. on oil fields, ch. 5). Despite the satisfactory settlement of some of their boundary problems and generally friendly relations thereafter, Qatar and Abu Dhabi continued to compete for leadership of the lower gulf states in the mid-1970s.

Among the industrialized countries Great Britain continued to enjoy a special relationship with Qatar. The two countries maintained numerous economic and commercial ties, and Great Britain supplied training and technological assistance. When the 1916 treaty between the two countries was nullified by Qatar's proclamation of independence, it was immediately replaced by a new treaty of friendship. Qatari commercial and financial arrangements with France increased during 1975 and 1976. France was the only industrialized country that had, by 1976, received a loan (the equivalent of about US$150 million) from Qatar. Some observers have noted that the Arab states are favorably disposed toward France because France has been the most sympathetic of the major Western nations to Arab and Palestinian positions in confrontations with Israel. France has also played an intermediary role between petroleum exporters and importers and has consistently refused to join a united front of petroleum-importing nations, an idea periodically suggested as a means of resisting oil price increases by the Organization of Petroleum Exporting Countries (OPEC). Qatari-Japanese commercial ties, particularly in the realm of heavy industrial contracts, have also been increasing. Japan was a major importer of Middle East oil during the mid-1970s.

Chad was the first African country to receive financial assistance from Qatar. Senegal, Mali, Gabon, and Uganda were among other African states that had received Qatari loans by 1976. Qatar generally supported the African causes of anticolonialism and antiapartheid in return for African support against Israel's occupation of Arab territory after the June 1967 War. India and Pakistan were the major non-African third world nations that had received Qatari financial aid by 1976.

Multilateral Relations

Qatar is a member of the League of Arab States (Arab League) and has extensive multilateral relations with other Arab countries. It has made major contributions to the Arab Fund for Economic and Social Development (headquartered in Kuwait) and in May 1975 had agreed to finance, with Saudi Arabia and the UAE, the Arab Military Industries Organization (AMIO), an Arab arms industry to be based in Cairo. Qatar has fully supported the Arab commercial boycott of Israel. A member of the Organization of Arab Petroleum Exporting Countries (OAPEC), Qatar adopted the oil production cutback and embargo policies established by that group after the October 1973 War. Qatar has joined Saudi Arabia, Bahrain, Oman, Kuwait, the UAE, and Iraq in multistate commercial ventures, including an Arab maritime company. Other multilateral organizations to which Qatar belonged were the Arab Civil Aviation Organization, Arab Labor Organization, and Arab Union of Tourism.

As a member of OPEC Qatar has consistently recognized the interdependence of petroleum-exporting and -importing countries and thus the necessity of cooperation between them. At the same time Qatar has stressed the connection between the price of oil and the price of consumer goods. Its basic policy has been to favor rises in oil prices to counter inflationary price rises in the consumer goods and industrial equipment that it imports from industrialized countries.

Qatar became a member of the UN in September 1971, shortly after its proclaimed independence. It is a member of several UN specialized agencies, including the International Atomic Energy Agency; International Civil Aviation Organization; Food and Agriculture Organization (FAO); International Labor Organization; World Health Organization; Universal Postal Union; United Nations Educational, Scientific and Cultural Organization; International Monetary Fund (IMF); and International Bank for Reconstruction and Development (IBRD, also known as the World Bank).

Qatar frequently channels financial aid to third world countries through the World Bank. It has also contributed to the third world political groups, including the Conference of Nonaligned Nations held in August 1976 in Colombo, Sri Lanka.

Foreign Relations Problems

Qatar's most significant foreign relations problems in late 1976 were with two neighboring states: UAE member Dubai and Bahrain. Qatar and Dubai maintained very close relations until 1972, sharing a common currency and joining to challenge Abu Dhabi's influence in the area. The former Qatari ruler, Ahmad, was a son-in-law of Dubai's ruler, Rashid bin Said Al Maktum. Ahmad took up residence in Dubai after his loss of the Qatari rulership to Khalifah. Official relations between the two countries were cordial in 1976, if somewhat wary. The familial aspects of the relationship provided a continuing potential for strain, although in 1976 Ahmad appeared to have accepted his loss.

Problems between Qatar and Bahrain were more serious and were based on historical rivalry and jealousy between the ruling families of the two countries (see ch. 2). In 1937 Qatar took full control of the town of Zubarah, located on the northwestern Qatari coast. The ruling family of Bahrain, the Al Khalifah, had established a settlement at Zubarah and had begun their invasion and conquest of Bahrain from that point in the late nineteenth century. The Al Khalifah family, believing that it had retained sovereignty over the Zubarah area, has never accepted its loss to the Al Thani.

Bahrain also continued to claim the Hawar Islands, located off the west coast of Qatar. Third parties have periodically attempted to mediate the dispute but to no avail. In the interests of regional unity the Qatari ruler downplayed the Hawar problem in a January 1976 Bahraini magazine interview and stressed both the closeness of relations between the two countries and his personal regard for Bahrain's ruler. In March, however, the Qatari foreign minister, Suhaym, pointedly denied a claim reportedly made by his Bahraini counterpart to the effect that there was no longer a dispute over the islands. Suhaym stressed both Qatari sovereignty over the islands and confidence that fraternity between the two countries would ensure an eventual fair solution.

Although the Hawar Islands dispute has interrupted closer political relations between Qatar and Bahrain, it has not prevented increasingly close economic contact. In 1972 a joint economic committee was established to coordinate and develop the economies of the two countries. They joined in significant commercial ventures the following year. A Qatari plan to finance the building of a causeway to Bahrain was rejected by that country, however, because of the Hawar Islands dispute.

MASS COMMUNICATIONS

Qatar established government-owned and -operated radio and television services in 1968 and 1970 respectively. It is a member of the Arab States Broadcasting Union, a group that in 1975 included seventeen other Arab states, Palestine, two Islamic African states (Mauritania and Somalia), and associate members France, Pakistan, Spain, and Yugoslavia.

The Qatari radio station operated four amplitude modulation (AM) transmitters in 1976. These included a very powerful 750-kilowatt medium-frequency transmitter that used a six-mast directional antenna system to reach as far as Morocco to the west and Afghanistan to the east and a 100-kilowatt high-frequency transmitter. One thirty-five-kilowatt frequency modulation (FM) transmitter was also in operation. A 250-kilowatt high-frequency transmitter, planned to beam broadcasts to Europe and North Africa, was under construction by a British company. The Qatari radio system broadcast in Arabic an average of fifteen hours a day and in English approximately three hours a day. There were an estimated 35,000 radio receivers in Qatar in 1976.

The television station at Doha operated two transmitters with a total output power of 200 kilowatts. In mid-1976 a Norwegian company was constructing a high-power television complex on the west coast. In 1975 France entered into an agreement with Qatar to provide technical training for Qatari radio and television personnel. The television station averaged five and one-half hours daily of programming in Arabic, with news bulletins repeated in English. Approximately one-half of the programming was in color. The number of television receivers in Qatar was estimated at 29,000 in 1976, roughly one set for every five people.

Telephone and telecommunications systems were rapidly being developed in 1976, in large part because of the demands of the economic community in general and the petroleum industry in particular. Telephone communications were managed by a British company that was a minority shareholder in the Qatari telephone system. Doha's fully automated telephone exchange was being expanded in 1976 from 5,000 to 16,200 lines. Work was proceeding in mid-1976 on further line expansion in Doha and Madinat Khalifah, the construction of a 3,000-line-capacity telephone exchange in Musayid, and the extension of telephone service to other cities. The telephone system was planned to have a total capacity of 30,000 lines covering the entire Qatari Peninsula. The system had links to Europe and North America; its ratio of 12.4 telephones to 100 people in 1974 was the highest among Arab states. A computer-controlled automatic telex exchange, under construction by an American company, began service in August 1976 in Doha. It was planned to accommodate a projected annual traffic growth rate of 30 percent.

In 1976 work was completed by a Japanese company on a Qatari earth station that would beam transmissions to the internationally owned Indian Ocean communications satellite. The installation replaced Qatar's previous satellite link through Bahrain's ground facility. Initial operation of Qatar's station was planned to include thirty international telephone circuits plus additional radio and television traffic; seventy circuits were expected to be in continuous use by 1981.

Though high by regional standards, Qatar's 1976 literacy rate of approximately 20 percent meant that publishing was much less important than electronic media for domestic communications. The publishing that existed was centered in Doha. Periodicals had very limited circulation. The Ministry of Information's *Al Doha,* a monthly magazine in Arabic, had a circulation of 3,000. The weekly *Al Urouba* and the daily *Al Arab,* published in Arabic, had respective circulations of 12,000 and 7,000. Three other weeklies were *Al Ahd* and *Al Fajr,* both in Arabic, and the *Gulf News,* in English.

NATIONAL DEFENSE AND INTERNAL SECURITY

Qatari leaders had obviously determined that their country could not individually defend its territory in the event of a full-scale attack by a more powerful nation. They apparently felt that Qatar's security was connected with the broader stability of the surrounding region, stability that was best served by the maintenance of a balance of power between its relatively powerful neighbors Saudi Arabia and Iran. The recognition of Qatar's limited defense capability has prevented large armaments expenditures that would have had a detrimental effect on a national economy geared to economic development. Rather than a large military establishment, Qatar maintained small, mobile forces to preserve internal order and protect against small border incursions. As a member of the Arab League Qatar belonged to its two military components, the Arab Defense Council, formed in 1950, and the Unified Arab Command, formed in 1964. Qatar was also a signatory to international arms control treaties; in 1974, for example, it signed the Sea-Bed Treaty (1971), which prohibits the emplacement of nuclear weapons or other weapons of mass destruction on the ocean floor.

The ruler is the supreme commander of Qatar's armed forces. The Constitution provides for a defense council to advise him on the training, equipment, and placement of Qatari forces. The Constitution does not specify the composition of the council or its relation to the Ministry of Defense and the army command. The Qatari forces are reputed to be the best trained and most efficient in the lower gulf area. The security forces in 1976 included the internal security police force and the army, each of which numbered approximately 2,200 men. The army included an air force branch and a naval branch. On taking power in 1972 Khalifah immediately raised the pay of all military grades by 20 percent.

The police force was organized under the Ministry of Interior, which was headed in 1976 by one of the ruler's brothers, Khalid bin Hamad Al Thani. After becoming ruler Khalifah replaced the previous police commander, a British expatriate named Ronald Lock, with one of his nephews, Hamad bin Qasim Al Thani, a graduate of the Hendon Police College in Great Britain. Lock, however, was retained as an adviser. The police force included spotter helicopters and small patrol boats among its equipment.

Before Khalifah became ruler, the army had long been supervised by Ahmad's brother, Qasim bin Ali Al Thani, and commanded in the field by another British expatriate, Ronald Cochrane. Cochrane, a former Glasgow policeman, was converted to Islam and took the name Muhammad Mahdi. Khalifah ended Qasim's connection with the army and replaced Cochrane as its commander with one of his sons, Hamad bin Khalifah Al Thani, a graduate of Sandhurst, who held the rank of major general. Like Lock, Cochrane was retained as an adviser.

Ground troops of the army numbered approximately 1,600 in 1976; they were organized into a kind of brigade group, comprising one armored car regiment, one mobile regiment, and one infantry battalion. The army included a large number of Saudi, Yemeni, and Baluchi mercenaries who, until Khalifah became amir, had been paid directly from the ruler's personal funds; since that time they have been paid by the government. Most of the officers were British, and smaller numbers were Jordanians and Pakistanis. During the early 1970s the foreign elements in the army's ranks, particularly the Yemenis, were regarded with increasing suspicion by some Qataris. No large-scale replacements had been initiated by 1976, however.

Great Britain was Qatar's principal arms supplier in the early 1970s, although by 1976 Brazilian and French companies had also received Qatari contracts. Qatar became eligible to receive defense equipment from the United States in January 1973 under the Foreign Military Sales Act. By 1976, however, no major arms agreements had been concluded between the two countries.

All branches were steadily being resupplied with modern support equipment in the mid-1970s, especially in the realm of combat communications. The army's artillery was apparently not a high-priority item, being composed of old British 25-pounder fieldpieces (first designed in 1935) and 81-mm mortars. Other sections of the Qatari weapons inventory showed signs of modernization, particularly after the 1960s, when it became known that Great Britain planned to relinquish its military commitments in the area.

Between 1968 and 1971 Qatar received significant quantities of light armor from Great Britain. Included were nine Saracen six-wheeled, ten- to eleven-ton armored personnel carriers, ten Ferret five-ton scout cars, thirty Saladin eleven-ton armored cars, and ten Shorland Mark 2 light (three-ton) armored cars. The Saracen carried up to ten men plus crew and was armed with a 7.62-mm turret-mounted machine gun. The Ferret was armed with a pintle-mounted 7.62-mm machine gun. Armament for the Saladin included a 76-mm turret-mounted gun and two 7.62-mm machine guns. The Shorland Mark 2 was basically a long-wheelbase Land Rover with 8-mm armor plate; it carried a 7.62-mm turret-mounted machine gun and a 7.62-mm machine gun on a roof mount. Production of the Saracens, Ferrets, and Saladins began in the 1950s; the Shorland Mark 2 entered production in 1965. The newest armor in Qatar's inventory in 1976 was the EE–9 Cascavel armored reconnaissance vehicle. Twenty of the nine-ton, 434-mile-range vehicles were built by a Brazilian company during 1975 and 1976 for Qatar. Original armament was a 37-mm gun mounted in a rotating turret. Qatar's Cascavels were modified by a French company to mount a 90-mm gun, supported by an infrared night-firing guidance system.

Qatar's air force branch possessed both ground attack fighters and various kinds of helicopters. Thirteen British Hunter subsonic (Mach 0.92) jet aircraft provided a close air support capability (920-mile combat radius); no interceptor capability existed. The air branch also had one British Islander twin-piston engine transport. The rotary-wing aircraft included two British Westland Whirlwind Series 3 general-purpose helicopters; two Westland Commando Mark 2 assault helicopters; and two French Aerospatiale AS–341H Gazelle all-purpose lightweight helicopters. Troop carrier helicopters were on order. During 1970 and 1971 Qatar acquired fifteen Short Tigercat surface-to-air missile units from Great Britain. The Tigercat was designed for point defense of vital land targets; it could be employed in a mobile mode and had a cross-country capability. In mid-1975 Kuwait was reported to be negotiating with Qatar (and with Abu Dhabi, Saudi Arabia, and Bahrain) on a base dispersal plan, whereby Kuwait would station some of its United States-built McDonnell-Douglas A-4M Skyhawk attack bombers in those countries.

The army's naval branch was composed of coastal patrol vessels. During 1972 and 1973 four small patrol boats were secured from Great Britain. In 1975 and 1976 Qatar took delivery of additional craft from British companies, including six 103-foot, 120-ton patrol vessels, each of which carried a twenty-five-man crew and was armed with two 40-mm antiaircraft guns. During the same period two seventy-five-foot and three forty-five-foot patrol boats, all from British companies, were also procured.

* * *

In 1976 little information was available on the society and economy of Qatar. An authoritative but brief survey prepared by IMF economists was published in the *IMF Survey* of August 19, 1974. More recent information was contained in a series of articles in the June 23, 1975, edition of the *Times* [London]; the November 18, 1975, edition of *The Financial Times* [London]; and the September 16, 1976, edition of *Voice of the Arab World* [London]. The government published the *Yearly Bulletin of Imports and Exports,* but the latest information available in late 1976 was for 1973. The government occasionally placed large advertisements in such well-known publications as the *Times, The Financial Times,* and the *New York Times;* these presented some limited but up-to-date statistical and economic information. Fragmentary information is published in newspapers and such periodicals as the *Middle East Economic Digest* and the *Arab Report and Record.*

The scarcity of English source material on the politics of Qatar reflects the general lack of interest by Western scholars in the lower Persian Gulf area that existed until the region's petroleum exports became a topical issue. John Duke Anthony's *Arab States of the Lower Gulf: People, Politics, and Petroleum* and Muhammad T. Sadik and William P. Snavely's *Bahrain, Qatar, and the United Arab Emirates: Colonial Past, Present Problems, and Future Prospects* provide a good grounding in the politics of the region. Legal and constitutional information are provided respectively in Husain M. Al Baharna's "Qatar" in the *International Encyclopedia of Comparative Law* [The Hague], edited by Viktor Knapp, and Herbert J. Liebesny's *Qatar* in the Constitutions of the Countries of the World series, edited by Albert P. Blaustein and Gisbert H. Flanz. Information on the military situation is available in two Stockholm International Peace Research Institute publications, *World Armaments and Disarmament, SIPRI Yearbook 1975* and *Arms Trade Registers: The Arms Trade with the Third World;* the International Institute for Strategic Studies' *The Military Balance;* and the United States Congress, 94th, 1st Session, House of Representatives Committee on International Relations, Special Subcommittee on Investigations, *The Persian Gulf, 1975: The Continuing Debate on Arms Sales.* (For further information see Bibliography.)

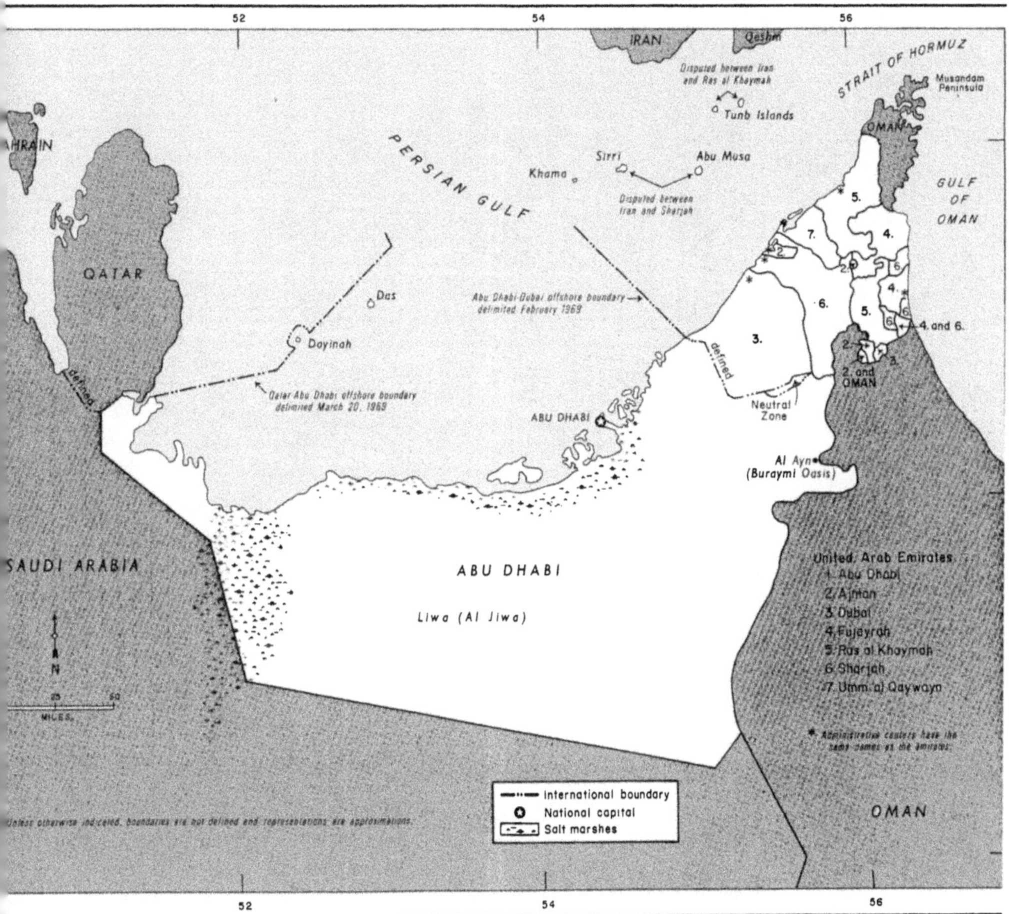

Figure 16. United Arab Emirates, 1976

CHAPTER 9

UNITED ARAB EMIRATES

Between the mid-seventeenth and mid-eighteenth centuries Portugal, which had been the dominant European commercial and naval power in the Persian Gulf area since about 1500, was displaced by Great Britain as part of the surge of expansionism in which the British established their empire in India (see The Entrance of the Europeans into the Gulf, ch. 2). In the gulf region British interests were primarily commercial and strategic—trade and naval stations for security of the sea route to India. The coast of the United Arab Emirates (UAE), for about 200 miles along the Persian Gulf from the town of Abu Dhabi northeast to the tip of the peninsula jutting into the Strait of Hormuz, was known as the Pirate Coast, harassed both by European and by seafaring Arab marauders. With the rise of the Islamic Wahhabi (see Glossary) movement in the Arabian Peninsula in the early nineteenth century, Arab seaborne depredations increased. In response Great Britain conducted punitive operations in 1806, in 1809 and, notably, in 1818. A treaty for suppression of piracy and the slave trade was concluded in 1820 between Great Britain and the Arab tribal shaykhs, and for a time a strong British naval squadron was based at Ras al Khaymah. Intertribal sea and land raiding again broke out, however, and Great Britain in 1835 negotiated a successful and lasting maritime truce with the shaykhs, who, in the further agreements of 1839 and 1847, undertook to prohibit slave traffic in their vessels and agreed to British enforcement of this prohibition.

This long series of treaties and agreements climaxed in May 1853 in the Treaty of Maritime Peace in Perpetuity between Great Britain and the Arab tribal rulers of what then became known as the Trucial Coast or the Trucial Oman. By consensus—the traditional, usually difficult, but most effective and necessary mode of joint action among the Arabs—the shaykhs, unable to trust one of themselves, entrusted an outsider, Great Britain, to supervise and enforce this maritime peace and to adjudicate alleged violations. Great Britain, in turn, undertook to perform this enforcement and to secure the Trucial Coast shaykhdoms against external attacks. Great Britain refrained from outright seizure or colonization of the inhospitable coast and from interference in the internal affairs and disputes of the shaykhs ashore. By the late nineteenth century, however, when France, Germany, and Russia began showing interest in the gulf area, the British imperial preeminence effectively established by the treaty of 1853 was strengthened further by identical, separate treaties between Great Britain and each of the Trucial Coast rulers. These prohibited the sale or disposal of any territories to any party except Great Britain and in effect gave the British control of the foreign relations of these states.

British control of the Persian Gulf in World War II was of major strategic importance to the Allied powers, although bases and stations in the Trucial Coast states had only a supporting role to the large Allied logistical installations at Abadan and other Iranian ports on the opposite side of the gulf. After the war Great Britain maintained a joint task force in the Trucial Coast area in continuation of its earlier obligations, which were unchanged. As part of the postwar adjustment period, however, and with a view to an eventual federation or union of the small shaykhdoms, Great Britain in 1951 set up the Trucial States Council of the seven rulers to meet at least twice annually under the chairmanship of the British political agent at Dubai.

As a result of changes in policy the British government in early 1968 announced that it would withdraw its force and terminate its special positions and obligations in the gulf by the end of 1971; in fact it did so. Representatives of Bahrain, Qatar, and the seven Trucial Coast states met in February 1968 and on March 30 announced the provisional formation of the Federation of the Arab Amirates. This arrangement did not last long, however, because of various boundary disputes, old dynastic quarrels, and inability to agree on details of precedence and organization. Bahrain and Qatar chose to remain separate and independent. Six of the shaykhdoms formed the UAE and adopted its provisional constitution on December 2, 1971; the seventh, Ras al Khaymah, acceded to the union in February 1972 (see fig. 16). (Although the name of the new country is usually given as United Arab Emirates, the member units are referred to here as amirates or as shaykhdoms.)

GOVERNMENT AND POLITICS

Constitutional Framework for the Federation

The first written constitution ever for any of the Trucial Coast states was signed on July 18, 1971, by Abu Dhabi, Dubai, Sharjah, Ajman, Umm al Qaywayn, and Fujayrah. It marked the inauguration of a constitutional experiment on December 2, 1971, when an independent federation, the United Arab Emirates (UAE), was proclaimed established under the Provisional Constitution of the UAE. Ras al Khaymah, the seventh member of the UAE, joined the federation on February 1, 1972.

The provisional charter, which was valid for five years, was to have been superseded by a permanent constitution in December 1976. On

November 28, 1976, however, the rulers of the seven shaykhdoms formally extended the Provisional Constitution for another five years, mainly because many of the existing constitutional provisions had yet to be put fully into effect. For these shaykhdoms constitutional government was still a revolutionary concept, the acceptance of which for practical application was believed possible only through a long period of transition.

The constitutional framework provides for the separation of powers into the executive, legislative, and judicial branches. Additionally it separates legislative and executive powers into federal and local jurisdictions. Certain powers are expressly reserved for the central government, residual powers being exercisable by the individual shaykhdoms.

The separation of powers remained nominal in 1976 in that the Supreme Council of Rulers—the Supreme Council for short—continued to function as the highest federal authority in executive and legislative capacities. Narrowly the executive branch consists of the Supreme Council, the Council of Ministers (the cabinet), and the Presidency (see fig. 17). The council is composed of the rulers of the seven amirates; the council elects from among its members a chairman and vice chairman, who serve for a term of five years. Its responsibilities include formulation of general policy; ratification of federal laws and decrees, including those relating to annual budget and fiscal matters; ratification of international treaties and agreements; and assent to the appointment of the prime minister and Supreme Court judges.

The rulers make decisions by a simple majority except on substantive issues; in that case a majority of five, including the votes of Abu Dhabi and Dubai, is mandatory. This requirement is in deference to the weighty role that the richest amirates were expected to play in the federation. Presumably any federal venture lacking the support of either Abu Dhabi or Dubai would be an exercise in futility. The Supreme Council carries out its work through a secretariat and may appoint an ad hoc committee.

The chairman of the Supreme Council is automatically the president of the UAE and the head of state. He is also the supreme commander of the federation in his capacity as the chairman of the Supreme Defense Council. The president convenes the Supreme Council and appoints the prime minister, deputy prime minister, cabinet ministers, and other senior civil and military officials. He is also empowered to proclaim martial law and to carry out a host of other functions usually associated with the chief executive of a modern nation-state.

The day-to-day management of federal affairs is the function of the Council of Ministers, whose original membership of fourteen was enlarged in 1973. The members must be citizens of the UAE and are individually and collectively answerable to the president and the Supreme Council. In addition to its executive duties, the Council of Ministers is responsible for drafting bills for formal enactment.

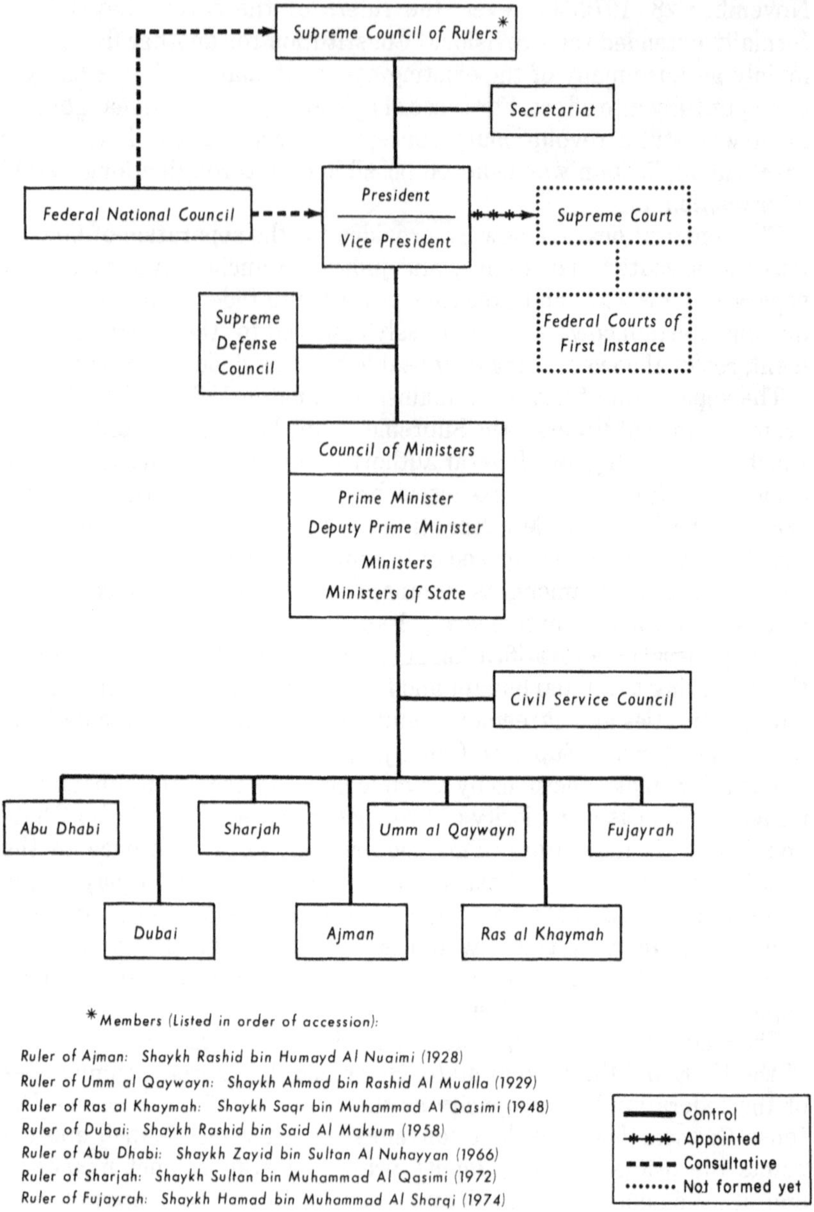

Figure 17. UAE, Government Organization, 1976

Under the constitution the Federal National Council is the principal legislative authority, but its actual role in the governmental process is limited to consultation. Its forty members are appointed for two years by the ruler of each amirate according to local conveniences and in accordance with a constitutionally fixed quota that reflects, albeit imperfectly, the wealth and size of Abu Dhabi and Dubai. The quota calls for eight members each from Abu Dhabi and Dubai; six each from Sharjah and Ras al Khaymah; and four each from Ajman, Umm al Qaywayn, and Fujayrah. The members must be citizens of the respective amirates, twenty-one years of age or older, and literate. They may not hold any other public offices. In an attempt to foster a spirit of federation, these members are urged, under the Constitution, to represent the people not only of their respective amirates but also of the federation as a whole.

The Federal National Council meets in a regular session for a minimum of six months beginning in the third week of November; a special session may be called by the president when necessary. The regular session is opened by the president with a speech on the state of the union. The council may express its "observations and wishes" concerning the speech in a reply submitted to the president for formal deliberation by the Supreme Council. The reply has no legal effect, nor do any of its decisions or recommendations that may be made known to the Council of Ministers, the president, and the Supreme Council on any legislative matters. The Federal National Council may discuss any government bills drafted by the Council of Ministers; it may agree with, amend, or reject such bills, but it may not veto them. These provisions include money bills. The council may be prohibited from discussing any federal matters when the Council of Ministers so decides in "the highest interests" of the federation. In any case, when the council's recommendations are not accepted by executive authorities, the Council of Ministers is required to provide its reasons to the council. In 1976 it appeared that the usefulness of the Federal National Council was essentially limited to its function as a training body for future lawmakers and as a sounding board for sentiments in individual amirates on the politics of federation making.

For purposes of administrative implementation the laws of the UAE are divided into two major categories: union law and decree. A bill drafted by the Council of Ministers for nonbinding deliberation by the Federal National Council and then submitted to the president for his assent and the Supreme Council for ratification becomes a union law when promulgated by the president. A decree is issued jointly by the president and the Council of Ministers between the sessions of the Supreme Council; it must be submitted to the Supreme Council within a week for confirmation. If not confirmed the decree becomes invalid, subject to certain reservations that may be expressed by the Supreme Council. The Federal National Council will be notified of any decree so issued, but only for "information."

The Provisional Constitution provides for the establishment of an independent judiciary under the Supreme Council. The highest court is vested with a power of judicial review and original jurisdiction over federal-amirate as well as interamirate disputes; the court is also empowered to try cases of official misconduct involving cabinet and other senior federal officials. Federal courts of first instance are also to be established to handle certain civil, commercial, criminal, and administrative cases, judgments from these courts being appealable to the Supreme Court. As of 1976, however, none of the constitutionally provided courts had been established. The administration of justice remained the primary responsibility of qadi (religious) courts in the amirates.

The executive and legislative subjects assigned under the Constitution to the federal government include foreign affairs, national defense, internal security, federal finances and taxes, communications, highways and aviation, nationality, education, public health, public information, census, currency, electricity, and weights and measures. In addition the federal government is responsible for legislation on labor relations, social security, banks, the penal code, civil and criminal proceedings in courts, printing and publications, copyright, the protection of cultural and industrial property, the delimitation of territorial waters, and navigation on the high seas.

The amirates retain responsibility for all subjects not reserved to the federal government. Among these residual subjects is the power of each amirate to conclude "limited agreements of a local and administrative nature with the neighboring state or regions." The Constitution states, however, that these agreements must be consistent with the interests of the UAE and that the circumstances relating to these accords must be provided in advance to the Supreme Council and other federal authorities.

The intent of the framers of the Provisional Constitution is clear concerning the primacy of federal authority over the amirates. Federal laws and international treaties and agreements are to be binding on all amirates. Any laws of the amirates inconsistent with the Provisional Constitution or federal laws and decrees are to be declared null and void through the judicial review of the Supreme Court.

Among other notable features of the constitution is a provision requiring each amirate to contribute a specified proportion of its annual revenues to the annual budget of the federal government. The Constitution also directs the federal government to establish a unified military force and a centralized internal security force.

The procedure for constitutional amendment is relatively flexible. The Supreme Council may first submit a draft amendment bill to the Federal National Council for its pro forma discussion. Regardless of the outcome of the debate, the bill is to be signed and promulgated by the president in the name of the Supreme Council. The constitution can be suspended in whole or in part only under martial law, which the president proclaims with the assent of the Supreme Council.

The Provisional Constitution designates Islam as the official religion of the UAE, Arabic as the official language, and Islamic sharia as a "principal source of legislation" throughout the federation. Until the permanent capital, to be named Al Karama, is built on a neutral site on the border between Abu Dhabi and Dubai, the provisional capital is the city of Abu Dhabi.

Among the basic social and economic principles of the UAE are those prescribing education as "compulsory in its primary stage" and "free at all stages." A comprehensive health care program is guaranteed "by the society for all its citizens." Private property is inviolable and may be expropriated only for due cause and on payment of fair compensation. The natural resources and wealth of each amirate are to be considered the public property of that amirate. The Constitution also lists fundamental rights and liberties that are familiar to most modern nations.

The State of the Federation in 1976

As 1976 drew to an end, the UAE was on a slow but steady course toward the goal of federalization. The amalgam envisioned in the Provisional Constitution was far from a reality. Admittedly much still remained to be done in removing obstacles in the way of a full-fledged union, but there were no immediate signs to bear out the gloomy prognostications at the time of the UAE's birth.

Challenges facing the UAE were formidable, however. The long tradition of rivalry and suspicion among the amirates was not expected to disappear overnight. The old tribal independence often slowed the efforts of federalists to organize an effective central authority. Although they acknowledged the importance of unity for common good, some of the rulers were disinclined to hasten the process of federalization, mindful that a strong central government in areas where they had retained autonomy would mean a loss of their power and prestige. Thus during the first several years of the UAE many rulers went through the motions of cooperation. There were some token achievements, but they fell far short of what UAE President Shaykh Zayid bin Sultan Al Nuhayyan had hoped to accomplish.

It came as no surprise that in August 1976 Shaykh Zayid, the amir of Abu Dhabi and the principal architect of the UAE federation, expressed his intention to step down from the presidency when his five-year term ended on December 2, 1976. He was believed to have been much disappointed at the lack of necessary support from his fellow amirs on several issues.

Among the thorny issues was federal funding. Most rulers were reluctant to contribute "a specified proportion of their annual revenues" to the federal budget as they had originally agreed to do when they signed the Provisional Constitution. In 1972 Abu Dhabi, by far the richest of the amirates, was to underwrite 85 percent of the UAE's initial budget, and the other six amirates were to share the remainder.

As it turned out, however, Abu Dhabi had to pay it all. The pattern was repeated in subsequent years, Abu Dhabi contributing close to 95 percent of the federal budget during 1976.

In addition there were other administrative and technical problems in the formulation of unified policies on several functional areas. These included immigration and naturalization, public information, centralized radio and television services, industrial projects, recruitment of government officials, and the federalization of police, intelligence operations, and defense forces. Disputes over interamirate boundaries also slowed the UAE's quest for full political integration.

By mid-1976 it had become obvious that the transition from tribal shaykhdoms to a federation could not be compressed into a period of only a few years. The constitutional order as envisaged in 1971 was not beyond the realm of possibility to the rulers of the Supreme Council. Realistically, however, they decided to opt for continuity rather than face the uncertainty of any drastic political innovation. Thus the rulers shelved the draft of a permanent constitution, which had been completed by mid-1976 after a year-long effort by a constitutional committee. If approved the draft would have gone into effect in December 1976, but the rulers apparently had many reservations. First and foremost they were reported to have strenuously objected to a provision that would have required individual amirates to contribute 50 percent of their respective revenues to federal funding. Moreover the draft document was believed to have substantially strengthened federal authority over other matters, Abu Dhabi being almost alone in supporting the adoption of the draft.

Nonetheless the rulers were apprehensive about the future of the fledgling federation, especially in light of Shaykh Zayid's threat to resign. His departure would have cast shadows over the politics of federation building because of Abu Dhabi's pivotal role in the evolution of the UAE. In early November 1976, therefore, the Supreme Council took several steps to avert a major constitutional crisis.

The first was to repeal the Provisional Constitution clause that permitted each amirate to have separate security forces, and an amendment was written into the constitution so that the ground, air, and naval forces would be established under the sole jurisdiction of the federal government (see National Defense and Internal Security, this ch.). This measure, which had been demanded by Shaykh Zayid, would likely hasten the process of integrating defense forces, the merger of which had been agreed to in May 1976 but had been delayed because of interamirate wranglings over the allocation of key command positions.

Among the steps taken in early November was one to empower the president to deal more effectively with immigration and alien residents and with public order and internal security throughout the UAE. Evidently this would mean that control of coastal areas, border posts, ports, and airports, which had previously been under the jurisdiction of individual amirates, would be relinquished to the federal government. In a similar vein the rulers of the Supreme Council also agreed to enact a federal law providing for the establishment of a single central security department reporting directly to the president. This new department would absorb all existing intelligence operations of the amirates.

In an attempt to foster a common political outlook and to focus popular loyalty on the UAE, the Supreme Council also decided to federalize the public information functions, including radio and television, of the amirates (see Mass Communication, this ch.). The federal Ministry of Information and Culture would have the power to scrutinize all political and news materials to ensure that they were consistent with the internal and foreign policies of the federation.

Perhaps most significant, the rulers pledged to share the burden of federal funding by allocating 20 percent of their annual revenues. This decision was made official on November 30, 1976, when Shaykh Zayid agreed to stay on as president for a second five-year term. The continuity of leadership was formalized on the same day; the Supreme Council unanimously reelected him and his deputy, Vice President Shaykh Rashid bin Said Al Maktum, the amir of Dubai. Undoubtedly the compromise worked out during November by the Supreme Council represented a major breakthrough in federation building. Certainly the spirit of cooperation among the amirates was more evident than ever before; it was manifest in late November when Dubai and Sharjah agreed in private to settle their long-standing border disputes.

The show of unity in November justifies some new optimism for the future of the UAE. It not only laid a groundwork for improved understanding among the amirates but also assured a leading role for Abu Dhabi. The federation would have faltered without Abu Dhabi's unflagging financial support inasmuch as this amirate had been instrumental in bringing modern facilities of every kind to many parts of the UAE, which until the mid-1960s were in effect little more than coastal towns and oasis villages.

Abu Dhabi's money, derived from its enormous oil revenues, was generously parceled out to less affluent amirates. Schools, roads, hospitals, low-cost housing, factories, and public utilities were being constructed everywhere. According to an American reporter who visited the UAE in 1976, the transformation of the UAE, as measured by sheer physical growth, was "incredible."

The dazzling progress was achieved also partly through parallel efforts of individual amirates, which for their own prestige invested considerably in the development of socioeconomic infrastructures. These efforts were often undertaken, however, without coordination or consultation with the federal government. Given the rulers' disposition to run their respective domains with as few central restraints as possible, there was an inevitable duplication of projects. In 1976, for example, there were four new cement factories and four modern international airports (and two more soon to be built) in an area the size of Maine (see The Economy, this ch.). It was unclear in late 1976 whether the Supreme Council would seek to avoid wasteful duplication through more effective central control. Such need was occasionally voiced publicly in Abu Dhabi, but from all indications the ideal of economic interdependence among the amirates was still an expression of faith in the future.

Although Shaykh Zayid was unquestionably first among equals, he was by no means an autocratic or authoritarian chief executive. As far as could be determined, the decisionmaking within the Supreme Council was through consultation and consensus. Islamic concepts of brotherhood, equality, and self-reliance as tempered by beduin tribal customs continued to influence the way decisions were made and then translated into policies and actions. In the case of serious disagreements over important policy matters, delay or compromise rather than confrontation was sought. The pattern of consensus politics was evidently occasioned by mutual concern that adversary relationships would undermine the fragile union.

The largely collectivist and moderate vein in leadership at the federal level rubbed off on the UAE's strategy for modernization. The hereditary rulers of the amirates refrained from advocating any revolutionary or doctrinaire solutions. They were content with a pragmatic and gradual approach, eschewing radicalism and all forms of virulent nationalism. Their primary concern in 1976 was to improve the social and economic life of their respective shaykhdoms through an elitist political structure, which remained under the highly paternalistic and patrimonial control of tribal shaykhs and their personal staffs. Political development through mass participation, electoral competition, and organized group activities remained on the lowest rung of federal and local priorities.

The decisions of the Supreme Council were executed by the Council of Ministers. Chaired by the prime minister, the Council of Ministers had as of January 3, 1977, a deputy prime minister, fifteen regular ministers, and five ministers of state. Generally the number of cabinet posts assigned to an amirate was in proportion to its wealth and economic resources.

The performance of the federal ministries varied considerably depending on national priorities as reflected in funding. In view of the urgent need for infrastructural development for which federal funding usually was necessary, the ministries in charge of education, public health, road building, electricity, water, low-cost housing, and so forth were more openly embraced by individual amirates than were some of the other ministries. For example, the ministries responsible for petroleum, defense, police, and finance often found themselves unable to extend their influence beyond the city limits of Abu Dhabi. The decisions of the Supreme Council in November 1976 to lower administrative barriers among the amirates were therefore encouraging and should in time bear some fruit.

As of mid-1975 the federal ministries had a total of 16,600 officials in various categories. The Ministry of Health and Education had by far the largest single bloc, about one-half of the total number. A majority of government officials were foreign nationals on temporary contract. In general UAE citizens with advanced schooling and an indeterminate number of qualified Egyptians, Palestinians, Lebanese, and Jordanians held senior-and middle-level positions in government ministries. The presence of foreign expatriates was especially noticeable in the ministries of defense, education, and information and culture, and in various positions requiring highly specialized training and experience. Government officials were subject to the control of the Civil Service Council, which reported directly to the office of the prime minister.

The indigenization of civil service appeared likely to be a slow process. With a gradual increase in the number of qualified UAE citizens, the proportion of foreigners would undoubtedly decrease. The continued presence of foreign expatriates in key federal positions, however, might prove beneficial in the short run inasmuch as these officials would be relatively unaffected by parochial loyalties to individual amirates. The politics of federation building, which in effect meant the removal of barriers across the boundaries of the amirates, would require a neutral federal bureaucracy. In any event the number of UAE citizens enrolled in foreign universities—many in the United States—was about 2,000 in the mid-1970s. The University of the United Arab Emirates, the first institution of higher learning in the federation, was originally scheduled to open in October 1976 but had to delay its opening to late 1977 at the earliest.

Political Process

In the Islamic political culture as influenced by beduin tribal customs —of which the amirates of the UAE remain an integral part—politics and government were inseparable and continue to be so in the 1970s without any appreciable modification. These functions were the exclusive preserve of the tribal ruling families. The power to regulate the lives of pastoral nomads in the desert was perceived principally in terms of the paramount shaykh and the web of personal relations he maintained within his kinship group. Loyalty and obedience were focused on the patriarch of the ruling household less as the head of a formal administrative structure with fixed territorial boundaries than as the paternalistic guardian, protector, and adjudicator. The ruler was the epitome of authority, personifying the unity of religious and political authority. This Islamic, beduin cultural heritage inhibited the rise of any notion that worldly affairs could be separate from religious faith, the latter being the sole source and repository of all authority. Politics and government, viewed as the ruler's prerogatives, could be neither differentiated nor delegated. The concept of a territorially bounded state or of a formal administrative structure was still embryonic through the early 1970s.

In 1976 the hold of tradition was still pronounced in the society of the amirates. It resulted at least in part from the long isolation of the Trucial Coast states from the societies and peoples of different cultures. Even during the more than a century of British hegemony in the area, the traditional way of life went on unruffled. Under special treaty relationships with the British, tribal shaykhs ruled their fiefdoms according to their personal predilections, a function of the British policy of noninterference in the internal affairs of the shaykhdoms. For their part the paternalistic tribal chiefs pledged not to deal with any foreign power other than the political agents of the United Kingdom. These special ties enabled the British to enjoy dominant influence in the Trucial Coast to the exclusion of other foreign powers, namely, Russian, French, and German. The British protection ensured the continuance of political stability in the area, shielding the shaykhdoms not only from foreign encroachments but from the turbulence of changes and revolutionary fermentations that affected other parts of the region in the post-World War II era. The sociopolitical institutions of the Trucial Coast states remained all but frozen in their feudal ways through much of the 1960s.

After December 1971 the Trucial Coast states assumed a new collective identity, the UAE. They agreed on constitutional principles and guidelines partly to adapt themselves to the rapidly changing, uncertain regional environment and partly to join forces for the complex task of modernization. In so doing, however, the shaykhs of the amirates apparently did not contemplate any substantial changes from their age-old pattern of personalized rule. The process of adaptation would be effected if at all by gradual and educational means. The shape of transition from old to new would be fashioned according to the benign mentorship of the tribal shaykhs and their personal advisers. There was no haste if only because there was little grass-roots pressure for change.

In early 1976 President Shaykh Zayid stated that the UAE would need a transitional period before any parliamentary election was allowed. He did not specify how long this transition would last, nor did he elaborate on the mechanics of election. In any case the process of defining issues and channeling popular demands independently of the shaykhs was still alien to the society. The constitutional guarantee for political freedoms was academic. As 1976 drew to an end there was as yet no political party, either under official sponsorship or in opposition, anywhere in the amirates.

The only potential source of demand for change from below was the growing number of intellectuals and professionals who were educated abroad and politically sophisticated. But it appeared unlikely that they would become activist for any antiestablishment cause, individually or collectively. For one thing almost all of these individuals were gainfully employed as advisers and technocrats at the federal and amirate levels. For another many were foreign nationals who were recruited for temporary service and whose contracts were extended on the basis of technical merit and performance. Their activities were monitored closely by the UAE authorities. Moreover, given the lack of any tradition of organized popular politics in the states, it would be a formidable challenge for anyone to arouse the traditionally apolitical merchants, farmers, fishermen, and desert nomads to political activism.

The power structures were preempted by the hereditary ruling families—the central political and economic institutions—in all amirates. The members of these extended families or clans held nearly all important levers of power, and only rarely were members of nonruling tribes allowed access to power—and then only to the periphery of the power structure. At the center of power hierarchy was the paramount shaykh of the ruling household, whose preeminence was buttressed by the legitimacy of lineage and reinforced by such leadership qualities as courage, prudence, wisdom, wealth, benevolence, and fairness. Family unity was always important in preserving power but equally critical was the effectiveness of rulership, because in the tribal political culture the shaykh was accountable for his mandate to results as measured by peace within a given shaykhdom and material well-being among the influential members of the shaykh's immediate clan.

The transfer of power would generally take place within the ruling family. In order to forestall a palace coup or internecine violence, the ruling shaykh would appoint a son as heir apparent after consulting his family elders. The heir apparent might double as deputy ruler if he was over eighteen years of age. Otherwise a deputy ruler would usually be named from among the ruler's brothers.

In 1976 administrative development varied considerably from one amirate to another. Abu Dhabi, Dubai, and Sharjah, were far ahead of the others; Ajman, Umm al Qaywayn, and Fujayrah were the least developed. Administrative specialization depended on a combination of such factors as the disposition of a ruling shaykh and the ruling family's wealth, specifically, on the extent of revenues derived from oil, commerce, and agriculture.

The development of Abu Dhabi was portended by the discovery of oil in 1960, but it was not until after Shaykh Zayid's accession to rulership in August 1966 that the amirate witnessed a rapid increase in the number of government departments. At the time the UAE was formed, Abu Dhabi already had ministries responsible for such functions as finance, defense, education, public health, public works, police, communications, petroleum, electricity and water, and justice. These formal bodies scarcely affected Shaykh Zayid's personal authority inasmuch as they were placed mostly under the charge of members of his dynasty. Besides, the shaykh had his own personal advisory cabinet, diwan, whose small membership was answerable to him. These advisory members were chosen by the ruler from among the elders of the ruling family, influential tribal leaders, and prominent commoner families.

In July 1971, under a major governmental reorganization, the sixteen-member Council of Ministers was formed, and several months later a national advisory council was inaugurated as a consultative legislature. Key portfolios in the Council of Ministers were assigned to the influential members of Shaykh Zayid's family, and lesser figures were appointed to the legislature for a term of two years. Abu Dhabi's experiment with these bodies ended in December 1973, when most of the ministries were merged with those of the federal government and the advisory council was abolished. This step was taken to strengthen central authority by eliminating duplication in government staffing; until that time the officials of Abu Dhabi acted in dual capacities as officials of both the amirate and the federal government. The 1973 merger did not affect the municipal authorities of Abu Dhabi and Al Ayn, or some of the ministries that Shaykh Zayid decided to retain as autonomous units of his amirate.

Political power resided in the Al Nuhayyan family clan, a subtribe of the Bani Yas tribal confederation, which traced its origins to Liwa and became dominant settlers in the oasis region of Liwa, in the town of Abu Dhabi, and in the Al Ayn section of the Buraymi Oasis region. The richest of all tribal families, the Al Nuhayyan had some thirty shaykhs, the most prominent and powerful being Shaykh Zayid, who before his assumption of rulership in 1966 had been governor of Eastern Province.

Among the influential figures assisting Shaykh Zayid were his sons, Khalifah and Sultan, the former being heir apparent, deputy ruler, and the federal deputy prime minister; and several nephews, sons of the late Shaykh Shakhbut (see fig. 18). In 1976 the balance of power or harmony within the Nuhayyan remained essential to the political stability of the amirate. According to John Duke Anthony, if the traditional pattern of succession politics held, a descendant of the Khalifah line might someday advance "his own right or that of his son to the rulership" of Abu Dhabi.

Political dominance was maintained partly by cultivating the support and allegiance of other leading, albeit nonruling, tribes. Among these tribes were the more than ten other subtribes of the Bani Yas; the Manasir tribe, traditionally influential in the interior of the western region of the amirate; the Dhawahir, in the Al Ayn area of Eastern Province; and the Awamir, in an area west of the Buraymi Oasis and south of Al Dafrah (Dafir). In addition the Suwaydis, Utaybahs, Habru-

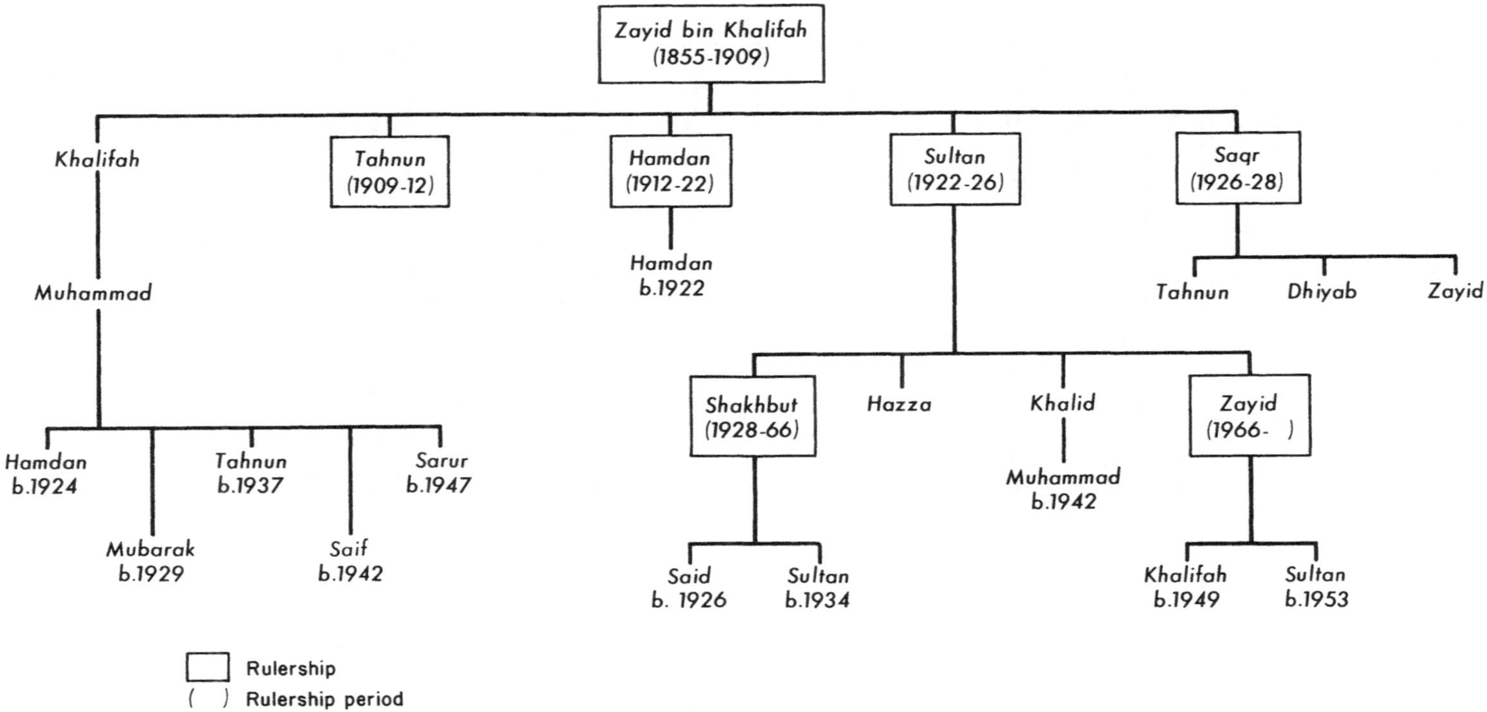

Source: Based on information from John Duke Anthony, *Arab States of the Lower Gulf: People, Politics, Petroleum,* Washington, 1975, p. 129; and Donald Hawley, *The Trucial States,* New York, 1970, pp. 352–353.

Figure 18. UAE, Abridged Genealogy of the Al Nuhayyan Family of Abu Dhabi

shis, and Al Kindis were among the distinguished commoner families whose importance was considerable because of their commercial acumen, professional and managerial talents, and bureaucratic experience.

The administration of Dubai was less elaborate in comparison with that of Abu Dhabi but was long noted for its efficiency and laissez-faire commercial policies. The discovery of oil in 1967 in Dubai and the ensuing socioeconomic development led to a modest administrative expansion. In 1976 Shaykh Rashid ruled, as before, in an informal yet effective manner with as few administrative or regulatory strings as possible to ensure Dubai's thriving free enterprise system.

Shaykh Rashid presided over a power structure in which his immediate household—the Al Maktum—was the dominant force, controlling key political and commercial interests in several government departments, the Dubai Municipal Council, and such successful enterprises as the Dubai Electricity Company and the Dubai State Telephone System. The ruler was ably assisted by a number of expatriate advisers and personally maintained close touch with key non-Arab and British members of business and banking communities.

Among Shaykh Rashid's more influential supporters were his three sons: Shaykh Maktum bin Rashid Al Maktum, heir apparent and the federal prime minister; Shaykh Hamdan bin Rashid Al Maktum, the federal minister of finance and industry; and Shaykh Muhammad bin Rashid Al Maktum, the federal minister of defense. In addition the shaykhs of the Rashid and Suhayl branches of the Al Bu Falasah tribe appeared to exert considerable influence (see fig. 19).

Sharjah, which had three enclaves on the eastern coast of the Gulf of Oman and claimed sovereignty over two islands near the Strait of Hormuz, had a fairly elaborate administrative system combining traditional and modern features. Shaykh Sultan bin Muhammad Al Qasimi, the supreme ruler, attended to affairs of the shaykhdom personally through his diwan and also by holding a daily majlis, an informal audience. In addition Shaykh Sultan presided over a number of departments whose operations were centrally coordinated by the municipality of Sharjah, which had a commendable managerial record dating back to the 1920s. As in the other amirates, key administrative positions were entrusted to members of the ruling dynasty.

In 1976 Sharjah was the seat of the Union Defense Force Headquarters; before 1971 British army and air bases and the headquarters of the Trucial Oman Scouts were located there. The presence of British residents in Sharjah influenced the development of a modern court system. Apart from the traditional sharia court, there was a separate court for civil and criminal matters involving non-Muslim foreigners. The amirate was also noted for its excellence in the field of education and agriculture; Sharjah had the UAE's highest proportion of university graduates. Sharjah's socioeconomic advance was considerably

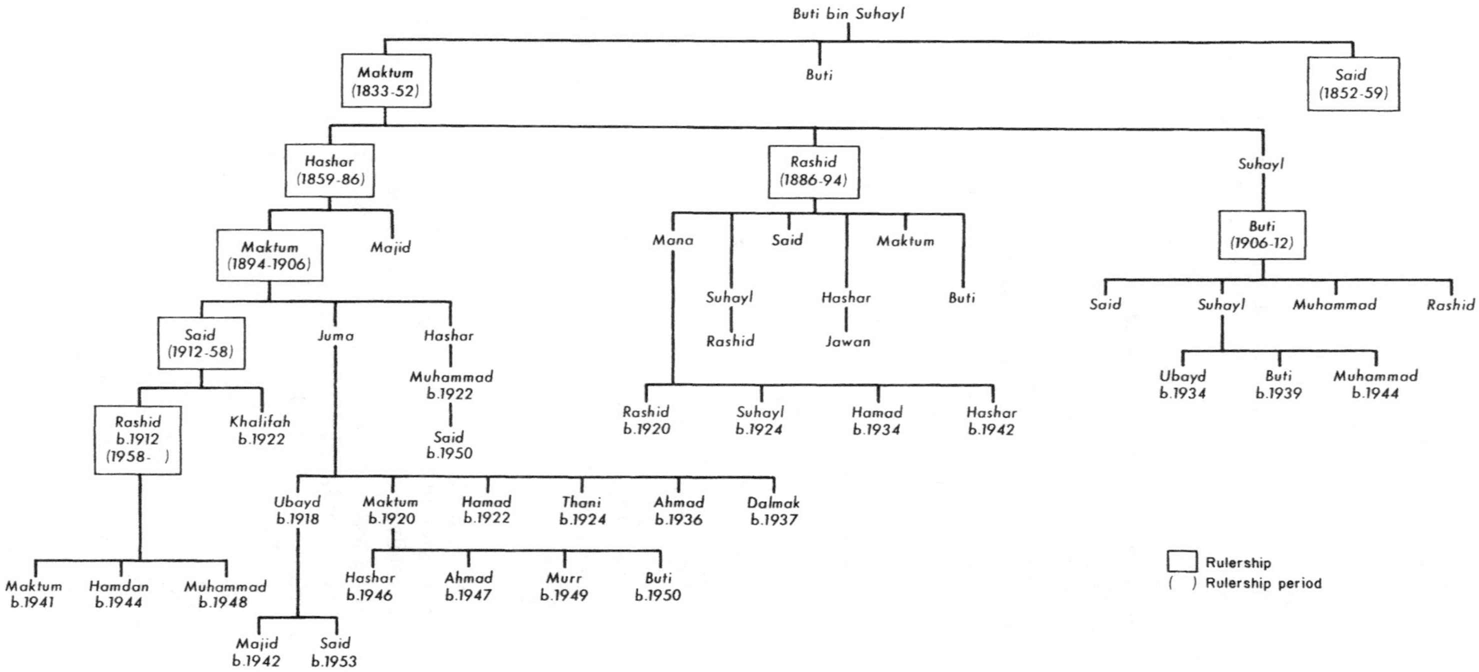

Source: Based on information from John Duke Anthony, *Arab States of the Lower Gulf: People, Politics, Petroleum*, Washington, 1975, p. 157; and Donald Hawley, *The Trucial States*, New York, 1970, pp. 354–355.

Figure 19. UAE, Abridged Genealogy of the Al Maktum Family of Dubai

boosted in 1974 when oil began to be produced for export (see Oil Industry of the United Arab Emirates, ch. 5.).

In 1976 the power group of Sharjah was the Al Qasimi tribe, whose three collateral branches accounted for most of the important figures (see fig. 20). Among the influential supporters of Shaykh Sultan were Shaykh Saqr (deputy ruler) and Shaykh Abd al Aziz of the Muhammad line; Shaykh Khalid, Shaykh Muhammad, Shaykh Ahmad, and Shaykh Saud of the Sultan line; and Shaykh Hamad of the Majid line. These shaykhs depended on the Al Madfa and the Taryam, two of the most prominent commoner families, for much of the administrative and managerial expertise as well as on a sizable number of expatriates including British, Egyptians, and Americans.

In the years after his accession to rulership in 1972 Shaykh Sultan, who at that time was the federal minister of education and was personally close to Shaykh Zayid of Abu Dhabi, actively supported the cause of the UAE federation. In 1975 he was the first amirate ruler to adopt the UAE flag in place of Sharjah's, and he handed over his militia, police, and courts to the federal jurisdiction. Relations with adjoining Dubai had been more competitive than cooperative, but they were expected to improve substantially.

Ras al Khaymah, despite its reputation as the breadbasket of the UAE, was hampered in its socioeconomic development because of its limited internal revenues. The discovery of oil was reported in 1976, but its commercial potential was as yet unclear.

The two principal governing institutions were the Al Qasimi ruling family, which belonged to the same Qasimi tribe as ruled Sharjah, and the municipality of Ras al Khaymah. The ruling dynasty had two competing collateral branches, the Muhammad line, headed by Shaykh Saqr bin Muhammad Al Qasimi, who was the ruler of the amirate, and the Sultan line, whose Shaykh Sultan had reigned from 1921 to 1948 (see fig. 21).

Shaykh Saqr's heir apparent and deputy ruler was his son, Shaykh Khalid, who was educated in the United States. Some of the prominent Ras al Khaymah family members active at one time or another in federal government service were Shaykh Abd Allah bin Humayd Al Qasimi; Shaykh Abd al Malik Al Qasimi; Said Ahmad Ghubash; Said bin Abd Allah bin Salman; and Sayf bin Ghubash.

The Al Qasimi shakyhs of Ras al Khaymah, like some of their cousins in Sharjah, take pride in their former imperial dominance in the Trucial Coast, which often was fittingly called the Qasimi coast. The Qasimi's fortune sank precipitously after the British invasion of Ras al Khaymah in 1819. In the years after 1971 the Al Qasimi of Ras al Khaymah were known to wish for a bigger share of power at the federal level. Their feeling was sustained in part by their claim that Ras al Khaymah had more indigenous Arabs than any other amirate and hence was entitled to a rightful place commensurate with tradition and contempo-

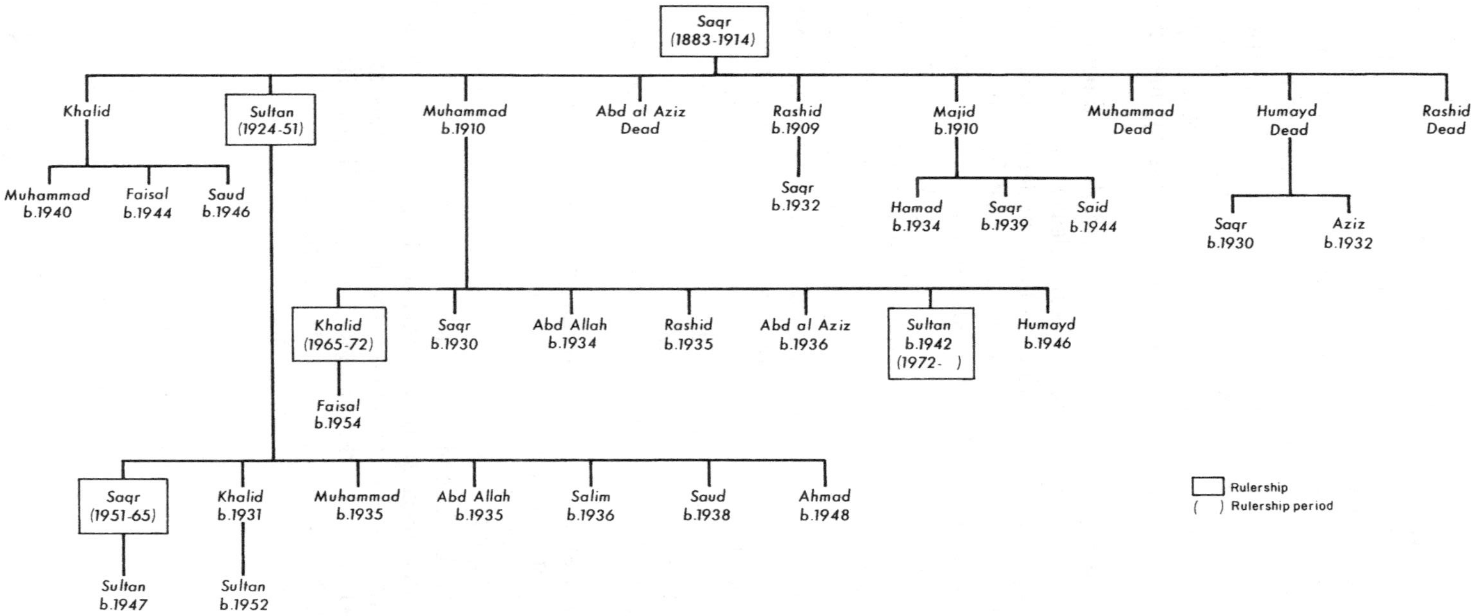

Note--Saqr deposed Shaykh Salim bin Sultan (1868-83), whose son Sultan won recognition as independent ruler of Ras al Khaymah in 1921 (see Abridged Genealogy of the Al Qasimi Family of Ras al Khaymah). Shaykh Saqr was succeeded in 1914 by his nephew Shaykh Khalid bin Ahmad (not shown on this chart), who in turn was succeeded by Sultan bin Saqr in 1924.

Source: Based on information from John Duke Anthony, *Arab States of the Lower Gulf: People, Politics, Petroleum*, Washington, 1975, p. 175; and Donald Hawley, *The Trucial States*, New York, 1970, pp. 356–357.

Figure 20. UAE, Abridged Genealogy of the Al Qasimi Family of Sharjah

rary numerical status. Predictably Arab-centered sentiments tended to run deeper in Ras al Khaymah than in other parts of the federation and yet were so diffuse that they failed to manifest themselves in any form or direction. The notable exception was Iran, which won the enmity of many Ras al Khaymans by seizing control of the two Tunb Islands in the Strait of Hormuz in 1971. The islands, historically claimed by Ras al Khaymah, remained a source of strain between the UAE and Iran. Ras al Khaymah's refusal to join the UAE in 1971 stemmed in part from its unhappiness over the reluctance of the British and other amirates to support unequivocally Ras al Khaymah's case against the Iranian occupation.

Ajman, Umm al Qaywayn, and Fujayrah—the smallest and poorest shaykhdoms—depended heavily on federal subsidies. They were sometimes likened to village states or described as being under one-man rule in that their administrative functions were exercised by the rulers' kinsmen and a few expatriate advisers. Reform was still in a nascent stage if undertaken at all; where it was initiated, it was prompted mainly by the need for coordination with the federal authorities responsible for development projects. Invariably the municipal officials in charge of capital towns dispensed limited social and economic services.

Administration in Ajman and Fujayrah was complicated somewhat because these amirates had enclaves scattered inside the territories of other amirates. According to Anthony, these enclaves contained "a village here, a date palm oasis there, or a well or a grazing area," to which their rulers claimed "the right to collect the zakat, or Islamic tax, from various nomadic tribes" migrating between two or more amirates.

Information was scarce about the internal structures of the ruling families in the three amirates. Ajman was ruled by the Nuaimi family, which Shaykh Rashid bin Humayd Al Nuaimi headed from 1928 onward, the longest rulership in the UAE. The dominant tribe in Umm al Qaywayn was the Al Mualla under the chieftainship of Shaykh Ahmad bin Rashid Al Mualla; because of his advanced age, Crown Prince Humayd was taking an active role in both federal and amirate affairs. Fujayrah was ruled by the youthful and well-educated Shaykh Hamad bin Muhammad Al Sharqi who, at twenty-six years of age in 1976, presided over the dominant Al Sharqi tribe.

Mass Communication

The potential utility of mass media as an agent of political socialization was early recognized by the founding fathers of the UAE; but as with other federal issues, the business of bringing mass communication to bear on the process of political integration was hampered by the ambivalence of some of the amirs. These tribal chieftains grew apprehensive that central control of press and broadcasting would further

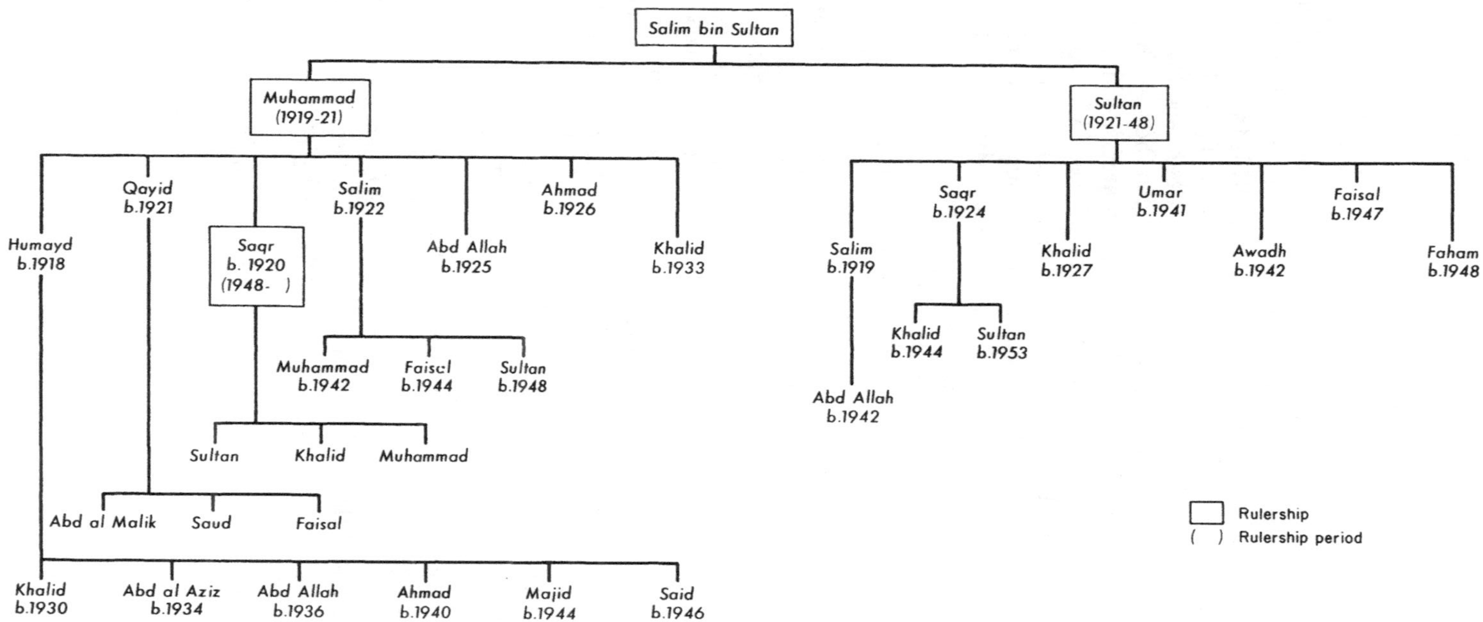

Source: Based on information from John Duke Anthony, *Arab States of the Lower Gulf: People, Politics, Petroleum,* Washington, 1975, p. 194; and Donald Hawley, *The Trucial States,* New York, 1970, p. 358.

Figure 21. UAE, Abridged Genealogy of the Al Qasimi Family of Ras al Khaymah

bolster what they believed was the already formidable position of Abu Dhabi and Dubai in UAE affairs.

The media of public communication are essentially a post-1971 development that was propelled by the realization that elitist as well as popular political orientations should be redirected toward a common political focus—the UAE. Under the Provisional Constitution, public information and communication media were to have been placed under federal jurisdiction and, in fact, through the mid-1970s these functions were consistently among the several top priorities in federal efforts to establish an administrative superstructure. Abu Dhabi and Dubai and, to a lesser degree, Sharjah and Ras al Khaymah led the way. By 1976, aided by a sharp increase in the number of transistor radios in the UAE, they had done much to broaden the horizons of the UAE population.

In 1972 the federal Ministry of Information and Culture began publishing *Al Ittihad* (Unity) as the official daily newspaper in Arabic—the first daily ever in any of the trucial shaykhdoms. Also in 1972 radiobroadcasting stations went into service under governmental control in Abu Dhabi, Dubai, and Ras al Khaymah and commercially in Sharjah, broadcasting daily in Arabic. The Abu Dhabi station also broadcast in English. Television broadcasting in color was started in Abu Dhabi and Dubai in the same year. The number of radio and television receivers was estimated, as of the mid-1970s, at 50,000 and 16,000, respectively.

In 1973 a bill was enacted placing the press under fairly rigid government control, as was common in most other Arab capitals of the region. From the beginning official policy was to disseminate information or political messages that were in accord with the UAE's domestic and foreign policies. Evidently critical communications, though widely accepted by Western democracies as necessary to the functioning of their political systems, would not be encouraged in the UAE because of their potentially destabilizing effects on traditional values and institutions. The federal government's restricted view of the mass media was consistent with local intolerance toward any attempt, actual or imagined, either to propagate religions other than Islam or to discredit the traditional Islamic way of life.

The anticipated development of common UAE-centered political orientations made little headway through 1974. Therefore in mid-1975 the Ministry of Information and Culture initiated a plan to establish a centrally directed national news service, the UAE News Agency, with its head office in Abu Dhabi and branches in Dubai, Sharjah, Al Ayn, and Ajman. In addition in December 1975 the federal government started publishing the *Emirate News*, the first English-language daily newspaper in the trucial shaykhdoms, replacing the five-year old English weekly, the *UAE News*. The new daily was intended not only for foreign consumption but also for a number of non-Arab foreigners in the UAE among whom English was the lingua franca.

Moreover for the development of independent newspapers and publications, the federal authorities provided assistance including interest-free loans, prefabricated buildings to house newspaper offices, housing for newspaper workers, and yearly subsidies. Much to their dismay, however, the subsidies were often diverted to quick-yield commercial ventures, and newspaper owners more often relied on low-paid and incompetent nonjournalists. The result was admittedly disappointing even by the modest standard set by the government. A notable exception was *Al Wahdah,* an Arabic daily newspaper founded in 1973 and published in Abu Dhabi under the expert direction of an Egyptian editor on loan from the Cairo newspaper *Al Ahram.*

The difficulty of independent publications also stemmed, ironically, from the government patronage of *Al Ittihad* and the *Emirate News.* These newspapers accepted advertisements and were distributed gratis to all government offices; nongovernmental newspapers therefore found it difficult to compete for readership. Thus in mid-1976 the Ministry of Information and Culture decided to detach the two dailies from its direct control and place them under the autonomous, but state-owned, Al Ittihad Press and Publishing Corporation. The board of directors of this body was to be chaired ex officio by the minister of information and culture.

At the end of 1976 it was as yet unclear whether the government's action had any beneficial effect on independent newspapers. It was clear, however, that there had been major strides toward the federalization of all mass media. The rulers of the seven amirates agreed in November 1976 that the Ministry of Information and Culture should be vested with the power to scrutinize all newscasts, programs, political commentaries, and discussions dealing with the foreign and domestic policies of the UAE before they were broadcast. Additionally under their accord all radio and television stations were henceforth to identity themselves as the Voice of the United Arab Emirates.

Patterns of Foreign Relations

For a newcomer to the perilous arena of international politics, the UAE quickly established itself as a significant force, especially in the regional affairs of the Middle East. This feat was accomplished almost entirely with a dexterous application of the federation's oil-generated wealth to the conduct of diplomacy.

The UAE, calling itself a nonaligned and developing nation, joined the League of Arab States (Arab League) on December 6, 1971, as its eighteenth member and was admitted to the United Nations (UN) three days later as its one hundred thirty-second member nation. It has since joined a number of international organizations, including the International Bank for Reconstruction and Development (IBRD, also known as the World Bank), the International Monetary Fund (IMF), the Food and Agriculture Organization, the International Labor Organization, the World Health Organization, and the United Nations Educational, Scientific and Cultural Organization. The UAE holds membership also in the Organization of Petroleum Exporting Countries (OPEC) and the Organization of Arab Petroleum Exporting Countries (OAPEC).

Diplomatic relations were established with some fifty countries, including all Middle Eastern countries and selected states in Western Europe, the Western Hemisphere, Asia, and Africa. In 1976 the UAE had no relations with Israel or with any communist-controlled nations. It had, however, limited trade ties with the German Democratic Republic (East Germany), Hungary, the People's Republic of China (PRC), and Romania.

In 1976 the UAE had friendly relations with neighbors and distant countries alike. The decade-old dispute with Saudi Arabia over the Buraymi Oasis was settled in 1974 with the agreement on a new border by the UAE, Oman, Saudi Arabia, and Qatar. The compromise settlement gave the Saudis a corridor to the gulf east of Qatar, but as of 1976 the work of delineating the new border had yet to be started. The 1974 agreement heralded a new era of very friendly relations between the two countries. Relations with Iran, strained in the early 1970s because of the Iranian occupation of the Tunb Islands, appeared to have improved. The two countries reportedly reached an unspecified working arrangement over the disputed islands. Furthermore their war of semantics—Iran's insistence on Persian Gulf and the UAE's on Arabian Gulf—was apparently ending with both sides prepared to call it simply the Gulf.

The conduct of foreign relations has been geared mainly to the Arab world, of which the UAE regards itself as an integral part because of shared heritage in religion, language, culture, and history. The federation has unflaggingly aided the Arab cause in all its manifestations, as evidenced in part in its generous assistance to the so-called confrontation states (Egypt, Jordan, and Syria) and the Palestinian Liberation Organization (PLO) against Israel. In July 1976 the UAE proposed to extend technical and financial aid to the PLO under a formal cooperation agreement, the first of its kind between an Arab state and the PLO. In November 1976 the UAE sent a seventy-man unit as part of the Arab peacekeeping force to civil-war-ravaged Lebanon.

The UAE's principal instrument of foreign policy is international aid, which in 1975 accounted for no less than 25 percent of its actual budget revenues—the highest ratio in the world. The aid is administered through the Abu Dhabi Fund for Arab Economic Development (ADFAED), which was founded in 1971 originally to help Arab countries in need and which soon won credit for the UAE as one of the most generous donors of low-interest loans and grants to the world's poor countries. Continuing its substantial aid to Syria and Egypt and other member states of the Arab League, the UAE has since 1974 gradually expanded its aid operations to Asian and African countries, including India, Pakistan, Bangladesh, Malaysia, Tanzania, Burundi, and Rwanda. It has also made sizable contributions to the World Bank. On the question of oil pricing the UAE has argued for moderation. At the OPEC conference meeting in Qatar in December 1976, it joined Saudi Arabia in adopting a price increase of only 5 percent.

Relations with the United States have been close and friendly. The United States was among the first nations to recognize the UAE and sent a resident ambassador in 1974 to Abu Dhabi. One of the top three trading partners of the UAE, the United States in 1975 signed an economic cooperation agreement with the UAE; among the areas of cooperation considered under that accord were joint ventures in gas, fertilizer, and steel projects with possible American technical assistance for electricity and desalination projects. As of early 1976 as many as 2,000 Americans were residing in Dubai alone.

GEOGRAPHIC AND DEMOGRAPHIC SETTING

The federation, including its numerous islands in the Persian Gulf, encompasses approximately 32,000 square miles. Because of disputed claims to some of the islands, the lack of precise information as to the size of many of the islands, and many undefined land boundaries, the exact size of the country is unknown. Although by 1976 all territorial disputes between and among the seven amirates had theoretically been resolved, the several enclaves, some of which were under so-called shared or joint administration, retained considerable potential for interamirate disputes (see fig. 16).

Despite the nearly 400 miles of coastline, the country had only one large natural harbor, Dubai. Other ports were being built or planned in the mid-1970s, both on the Persian Gulf and on the Gulf of Oman (see The Economy, this ch.) There are numerous small islands, reefs, and shifting sandbars that menace navigation, and strong tides and occasional windstorms further complicate ship movements near the shore.

South and west of Abu Dhabi the vast rolling sand dunes merge into the Rub al Khali (Empty Quarter) of Saudi Arabia. Large sections of the Abu Dhabi coast are salt marshes *(sabkhat)* that extend for miles inland. Inland from the Gulf of Oman coast the terrain is sharply different. The western Hajar Mountains—rising in places to 7,000 or 8,000 feet—run down close to the shore in many places. Ras al Khaymah, Fujayrah, and the eastern part of Sharjah are hilly or mountainous regions topographically distinct from Abu Dhabi and Dubai, which together account for more than 87 percent of the territory. Only in a few scattered areas is agriculture possible (see Agriculture and Fishing, this ch.).

The climate is generally hot and dry. Between June and September the daytime temperature frequently reaches 120°F on the coastal plain. In the mountains temperatures are considerably cooler, a result both of increased altitude and of vegetation. During the late summer months a humid wind known as the *sharqi* makes the coastal region especially unpleasant. The average annual rainfall in the coastal area is less than five inches, but in the mountains annual rainfall often reaches fifteen inches. Rain in the coastal region falls in short torrential bursts during the summer months, sometimes resulting in floods in ordinarily dry wadi beds. Supplies of groundwater are very limited.

Population

The preliminary report on the first nationwide census gave a total population of almost 656,000 at the end of 1975 (see table 28). In December 1976 the UAE government estimated the population to have grown to about 690,000. These data reveal a massive increase in population since 1968, when the British conducted a census that indicated a population of about 180,000. In 1976 over 90 percent of the population was classified as urban, and over 60 percent of the population resided in Abu Dhabi and Dubai.

By 1976 illegal immigration had become a major problem. During one brief period in late 1976 over 600 people attempting to enter the country illegally were arrested. The government estimated that an additional 800 aliens evaded government officials and entered illegally during the same period. The government was sending a delegation to certain Asian states in order to obtain their cooperation in controlling the smuggling of foreign nationals into the country. By 1976 the working force was believed to be about 280,000; about 42 percent were employed by the construction industry and the government. Expatriates from neighboring states and countries in Asia played a major role in the construction industry and in the commercial sector, but a specific breakdown by nationality was not available (see The Economy, this ch.).

Table 28. UAE, Area and Population by Amirate, 1968 and 1975

Amirate	Area (in square miles)	Population 1968 Census	Population 1975 Census
Abu Dhabi	26,000 to 30,000[1]	46,500	236,000
Dubai	1,500	59,000	207,000
Sharjah	1,000	31,500	88,000
Ras al Khaymah	650	24,500	57,300
Fujayrah	450	9,700	26,500
Ajman	150	4,200	21,500
Umm al Qaywayn	300	3,800	17,000
TOTAL	32,000[1]	179,200	653,300[2]

[1]Estimates of total size range from 30,000 to 33,000 square miles; the size of Abu Dhabi is variously given by the UAE government and other sources in a range from 26,000 to 30,000 square miles.

[2]Presumably 2,300 residents, most of them students, were abroad, accounting for discrepancy between rounded total of 653,000 and overall figure of 656,000.

Education and Welfare Services

With the exception of a few religious schools attached to mosques, in 1952 there were no schools in the seven shaykhdoms. In 1953 a school was opened in Sharjah that provided free education to about 450 boys. By 1973 there were an estimated 140 schools, twelve of which provided boarding facilities. In 1976 there were over 61,000 students (about one-third of them girls) and over 5,000 teachers, and the Ministry of Education announced that construction was to begin on 117 new schools and twenty kindergartens. Over 2,000 students were studying abroad, and a national university was scheduled to open in 1977.

A 1971 law made primary education compulsory for children above the age of six. All educational costs, including study abroad, were borne by the federal government. Expenditures for education have regularly been the largest item in the federal development budget, and the expenditures have grown dramatically. In fiscal year (FY) 1973 UD89 million (for value of the dirham—see Glossary) was spent on education, in 1974 UD321 million, and in 1975 UD475 million.

In addition to the more than 61,000 students enrolled in public primary, intermediate, secondary, and technical schools in 1976, about 16,000 adults were reportedly enrolled in evening literacy classes, and over 17,000 students were enrolled in private schools. Most of the private schools were operated for the children of Indian and Pakistani medical personnel, but some were for aliens from Western Europe and North America.

The projected National University for the UAE will reportedly have three faculties for men—science, the arts, and commercial management—and one for women. In early 1976 another facility for higher education was established when the National College of Choueifat in Beirut was relocated in the UAE. When the college was in Beirut, most of its students had come from outside Lebanon, and because of the civil war the number of students had fallen by about 75 percent. Fifty teachers arrived from Beirut and began classes in temporary quarters for 600 boarding students and 1,800 day students. Many of the boarding students were from neighboring Arab countries. Although specific information was not available, foreign observers assumed that the federal government was providing substantial assistance to this new educational effort.

The government would like to staff its faculties with UAE citizens and encourages its young people to go abroad for teacher training. For the foreseeable future, however, most teachers will be foreigners. Egyptians, Jordanians, Lebanese, and Palestinians, usually serving on four-year contracts, dominated the teaching profession. The government wished to increase and improve its technical education facilities, but achievement of this goal was also expected to take several years.

The pace of change during the mid-1970s was so great that information quickly became outdated. For example, the August 1976 issue of the *UAE News*, published monthly by the UAE embassy in Washington, noted that construction of a 638-bed hospital in Dubai at a cost of US$90 million would begin in 1977 and was scheduled to be completed in 1980; that US$7.5 million had been allocated for housing construction; that contracts had been signed for about US$14 million for two mosques, two intermediate schools, houses for four doctors, a radio transmission station, a post office, coast guard and other government offices, playgrounds, and roads; that work had begun on a US$750-million harbor near Dubai; that a new US$11 million earth station for satellite communication was scheduled for completion in late 1977; and that 300 UAE students had arrived in the United States for education paid for entirely by the UAE government. The embassy newsletter also reported that Ramada International, an American firm, would build a US$90 million hotel "shaped like an inverted pyramid" in Abu Dhabi. Other issues of the embassy newsletter contained comparable reports on the expenditures of large sums of money on a wide range of projects.

Data on the existing facilities and their adequacy were either unavailable or incomplete in late 1976. The government announced that the completion of the 638-bed hospital in Dubai would meet that city's medical needs for the 1980s, suggesting that the facilities available in the mid-1970s were relatively good. In early 1976 there were hospitals in the cities of Abu Dhabi, Dubai, Al Ayn, Ras al Khaymah, Umm al Qaywayn, and Dibbah. In addition to the second hospital in Dubai, two hospitals in Abu Dhabi, two in Al Ayn, and a tuberculosis hospital near Al Ayn were scheduled for completion by 1979. According to the government all the new hospitals are to be comparable in technical facilities and equipment to hospitals in Western Europe and the United States.

In addition to the hospitals most of the smaller towns have clinics, and clinics and health centers are located throughout the major cities. Doctors and dentists are reported to be readily available; most medical personnel, including nurses and technicians, are Indians or Pakistanis.

Health care is free for all residents and apparently for visitors as well. For citizens of the UAE this free health and medical care includes all costs of sending a patient abroad for treatment.

According to the UAE government malaria is the only endemic disease; it is generally found only in the foothills of the western Hajar Mountains near the Gulf of Oman coast. Visitors to the UAE are advised to have the typhoid inoculation, however, and are required to possess proof of current cholera inoculation and smallpox vaccination.

In 1976 hotel accommodations for business visitors and tourists were often inadequate, and housing for residents was critically inadequate.

In 1976, for example, free and suitable housing was demanded by striking expatriate workers. The government was encouraging private contractors to construct housing, and in 1975 public housing projects with a total of 1,700 dwellings were completed. This was far below demand, however, and observers believed that the housing shortage would continue well into the 1980s.

THE ECONOMY

Before the discovery of oil economic life was essentially the same in the separate amirates that make up the UAE. People clustered around the few places where water was available. Where water was relatively plentiful, as in the Buraymi and Liwa (Al Jiwa) oases and the plains of Ras al Khaymah, settled agriculture developed. Cultivation of date palms and fodder predominated. Nomadic herdsmen also used the wells and oases for their base camps, moving their animals to other forage areas during the course of a year. Settlement occurred on the coast where the water supply permitted; the inhabitants usually went to sea for their livelihood. Fishing and trade became the sources of income. Pearling was a major occupation until the worldwide depression and Japanese cultured pearls disrupted the market in the 1930s.

The people clustered around the available water sources formed relatively isolated and autonomous communities. Production was geared to the small markets and was mainly for subsistence. Trade between communities and with the outside world was difficult. Only Dubai and Sharjah, because of the shelter afforded shipping by inlets, depended on trade and merchant activities. Except for a few wealthy merchants, the bulk of the population until the 1960s was extremely poor by international standards and lacked such basic amenities as electricity and modern transportation.

The discovery of oil set in motion economic changes that were still accelerating in late 1976. Oil was first found in Abu Dhabi in 1960, and oil revenues began to amount in 1963 (see Oil Industry of the United Arab Emirates, ch. 5). Abu Dhabi's reserves and oil revenues greatly exceeded those of the other amirates. Oil revenues amounted to about US$3.8 billion in 1975 for a population in the amirate of about 236,000. Oil revenues became significant in Dubai in 1970 and in Sharjah in 1975. Oil was found in Ras al Khaymah in 1976, but its commercial viability had not been determined by the year's end. The disparity in resource endowment led to uneven economic development in the individual amirates; the development that had been accomplished by late 1976 had essentially occurred within the previous ten years—even within the previous five years for the amirates without oil.

Oil revenues made the UAE one of the world's wealthiest countries in 1976. Relatively few statistics were available for the UAE, and estimates of gross national product (GNP) were educated guesses at best. An estimate by a United States government agency placed GNP at US$5.8 billion in 1975, yielding a per capita GNP of US$8,800. The estimate appeared reasonable but certainly lacked precision; per capita GNP could only be said to be in that neighborhood. National income was not evenly distributed, however. Some of the people still following traditional methods of farming probably had incomes equivalent to less than US$100 a year, and many urban dwellers in the poorer amirates had incomes only slightly higher. Nevertheless the opportunity for higher incomes existed because citizens had preference in employment, and the government stood ready to employ any citizen in need of a job.

Transformation of economic life started in Abu Dhabi. It spread to Dubai at first because of the trading activities of Dubai merchants and, after 1969, because of Dubai's oil revenues. Economic development of the other amirates largely started in 1972 after formation of the UAE. The ruler of Abu Dhabi favored a federation and almost completely funded the federal budget through 1976 as well as financing individual projects in the amirates. The expanding federal expenditures for economic development financed much of the improvement in living conditions in the amirates lacking oil. By 1977 modern roads linked all of the amirates and joined roads in Saudi Arabia that connected with systems farther north. Schools and hospitals had been built. Electricity and piped, pure water was available to an increasing proportion of the population. Diet had considerably improved through imported foods sold at subsidized prices. Subsidized housing and utilities were available to nationals. The larger towns were rebuilt and expanded, and a start had been made in industrialization and in modernizing agriculture.

Economic development was not without its problems. High wages in the oil fields first attracted workers from farming and fishing. Mounting oil revenues provided funds for imports, new construction, schools, and other activities formerly precluded by poverty. The structure of the economy changed, and people had to work at new jobs under vastly different conditions. The economy shifted from production of commodities to orientation toward services. Trade, transportation, construction, and government services became the main employers and the largest contributors to GNP after oil. The transition was not easy for many of the uneducated and unskilled. The government offered employment to nationals in need of a job and became probably the largest employer and, after oil, the largest contributor to GNP. Development required a variety of skills and trained personnel, however, and these were extremely scarce in a society where schools had been unavailable until economic development began. Moreover the indigenous population was small for the number of jobs that a rapidly expanding economy needed filled.

Development was achieved by importing foreign labor. The population of the UAE grew at an average rate of over 20 percent per year between 1968 and 1975, primarily because of immigration. The 1975 census showed a population of about 656,000, two-thirds of whom were in Abu Dhabi and Dubai. Expatriates far outnumbered nationals, particularly in the two most developed amirates. Figures for the labor force were not available, but they would be even more skewed. By 1976 Abu Dhabi and Dubai were undertaking large development projects that would require more imported workers in the future. Many local and foreign observers were concerned about how to handle the problems associated with expatriates in order to preserve local values and institutions.

Rapid economic development was accompanied by inflation. Inflationary pressures were at first imported, caused by a rapid increase of international commodity prices after 1972. After UAE oil revenues jumped—more than fourfold between 1973 and 1974, government spending and business activity accelerated, creating domestic sources of inflation. Imports increased tremendously, but bottlenecks caused shortages and higher prices. In late 1976 all ports were overtaxed, and 200 ships were waiting to unload at Dubai, the largest port. Warehousing was inadequate, and spoilage contributed to the shortages. Figures were unavailable to measure the rate of inflation, but some observers thought it was about 30 percent each year. Inflation and the influx of workers and businesses had a pronounced impact on property values and rents; they were very high, causing particular difficulties for low-paid foreign workers.

In the opinion of most observers the boom conditions were temporary. Eventually the inflationary pressures would ease and more normal economic growth set in, but no one was certain how soon. Until then the main towns in the UAE could best be visualized as massive construction sites.

Budgets and Government Policy

The lack of unity within the UAE in 1976 was reflected in the separate budgets and fiscal autonomy of the individual amirates. For thousands of years the rulers of groups of people, such as each of the amirates, had had financial autonomy. The ruler had income, whether it came from property, trade, raids on neighbors, or taxes on traders passing through his territory on land or sea, that he disposed of as he saw fit. Taxing British ships passing through the gulf led to the British presence in what became known as the Trucial Coast (see The Founding of the Modern Gulf Polities: Gulf States in the Eighteenth and Nineteenth Centuries, ch. 2). In more recent times the rulers received income from additional sources, such as operation of the British base in Sharjah, issue of unusual stamps purchased by philatelists in foreign countries, and rental payments from oil companies for exploration concessions. The British, the Kuwaitis, the Saudis, and others also financed some development projects that individual rulers in the amirates wanted. The tradition in the area was one of financial autonomy for the ruler, and the tradition was not immediately dispelled in 1971 when the amirates joined in a federation.

Tradition began to weaken with the discovery of oil. Abu Dhabi, which had the earliest and largest oil revenues, had to expand government to provide the services and facilities needed by the people. As government grew, so did the need for financial controls. By 1976 Abu Dhabi had developed the most modern public administration and budgetary process in the UAE, even though the ruler still retained considerable independence; nevertheless it was not possible to determine precisely the fiscal position of Abu Dhabi from the information available. In Dubai public funds and the ruler's purse were indistinguishable. A budgetary process had started, and some financial information was available, but Dubai's fiscal position was not clear. The ruler sometimes gave his personal guarantee for government commitments, and little was known about the revenues of the ruler and of the state. Scant information was available about finances in the other amirates.

In 1976 the data available were insufficient to determine how the UAE was using its money or whether it was spending more than it earned. There were indications that development expenditures since 1973 had at times exceeded revenues in some of the amirates, requiring substantial borrowing from abroad. The UAE prospects and credit were good. State reserves had been set aside in Abu Dhabi and Dubai, and funds were easily available in overseas money markets. Repayment of the credits would depend in part on the financial success of the projects for which the loans were made.

In 1976 economic planning or even simple project coordination did not exist between the amirates. The individual amirates had a large degree of autonomy in selecting the direction of development even under the federation. Unfortunately for the best use of scarce resources, some competition and duplication of effort had occurred. The example most frequently cited concerned international airports, the amirate status symbol in 1976. Three, those in Dubai, Sharjah, and Ras al Khaymah, were within a half-hour drive of each other, and the large airport at Abu Dhabi was less than two hours' drive away from the most distant. Dubai was enlarging its present airport and considering a second, and Abu Dhabi was constructing a new and larger international airport. Other areas of possible excess investments included port facilities, hotels, and cement plants. Observers also questioned Dubai's investment in an extremely large and costly dry dock that will compete with a more modest one under construction in Bahrain financed by several Arab countries, including the UAE (essentially Abu Dhabi).

Senior UAE officials were not unaware of the adverse effects of the individualistic approach to development. One reflection of a broader approach was a survey by federation officials of planned expenditures during 1976 by all of the amirates, the first such survey. Another reflection was the establishment in mid-1976 of a federal investment authority to handle foreign and domestic investments and loans. Journalists reported that the investment authority was set up after officials learned that individual amirates were borrowing from abroad at much higher interest rates than was earned on federal funds; thus development costs could be reduced by the amirates' borrowing from the central investment authority, which would also gain a measure of control over investments. In mid-1976 the federal government took the further step of establishing a planning council patterned after that of Kuwait (see The Economy, ch. 6). The prime minister was chairman, and several ministers involved with economic affairs were members. A number of private citizens were to be named to the council. It would take time for the planning council to organize and to acquire a staff, but the real key to successful planning would be the degree of autonomy the rulers of the amirates agreed to give up.

Part of the staff for the federal planning council would probably be drawn from planning officials in Abu Dhabi. Abu Dhabi started planning relatively early, and its first national plan covered the years 1968 to 1973. Budget constraints kept the level of investment expenditure substantially below that planned. An industrial survey was conducted after the first plan to determine ways the amirate could reduce its dependency on oil. A three-year plan for 1977 through 1979 was in preparation in late 1976, but its contents had not been released.

Development in Dubai initially focused on building up the commercial base that had long been an important source of income and employment in the amirate. A large modern port was constructed that served as the main port for UAE imports. Other projects were selected after a thorough feasibility study; only projects with sound commercial prospects were approved. Development in the smaller amirates was partly financed by external aid, largely from Arabian oil states. Kuwait built schools and medical facilities, for example, and Saudi Arabia financed the Ras al Khaymah airport. The ruler of Abu Dhabi privately financed some development projects. Since 1971 the federal budget has provided most of the development in the northern amirates, particularly roads, schools, hospitals, water and electric facilities, and communications.

Federal Budget

In 1976 the federal budget had limited significance because of the financial autonomy retained by the amirates. A budgetary process had been developed, however, which could be useful when and if unity becomes more of a reality. A major deficiency lay in the sources of revenues. According to the UAE constitution each amirate was to contribute to the federal budget, but through 1976 almost all revenue had been provided by Abu Dhabi. The federal government was also empowered to introduce taxes and fees. By 1976 no federal taxes were levied, and only small amounts were collected through fees from the postal service and passport and immigration departments. In December 1976 UAE members took up the problem of contributions from each amirate and a strengthened federal budget. Initial indications were that each amirate would contribute 20 percent of its revenues to future federal budgets—an important step in the unification process.

As Abu Dhabi's oil revenues increased, particularly after 1973, its support of the federal budget increased. Federal activities and expenditures expanded accordingly. Federal expenditures went from UD164 million (for value of the dirham—see Glossary) in 1972 to UD795 million in 1974 (see table 29). Planned expenditures increased to UD2.8 billion in 1975 and UD4.2 billion in 1976. Current expenditures increased rapidly as personnel and maintenance costs rose because of the opening of new roads, schools, hospitals, and other facilities, but development expenditures rose even more rapidly. By 1976 planned development expenditures amounted to UD2.0 billion, almost equal to current expenditures of UD2.1 billion. The large increase in federal expenditures undoubtedly helped foster the idea of federation, particularly in the northern amirates, which would otherwise have been short of funds to provide services to their population.

Federal budget expenditures were incomplete because a large part of the personnel costs for federal ministries were paid directly from the Abu Dhabi budget. The importance of this budget support was shown in Abu Dhabi's 1976 budget, in which costs of Abu Dhabi departments working for the federal government amounted to nearly UD4.3 billion, a sum larger than total federal budget expenditures.

Federal expenditures had generally been less than planned in the early 1970s, and similar shortfalls probably continued through 1976. Development expenditures were difficult to estimate with any precision. Abu Dhabi reportedly turned over funds to the federal budget as they were needed to pay for expenditures. Thus the amount of state reserves was uncertain.

A complete and consistent breakdown of federal expenditures was not available, but a general pattern was discernible. Defense and security consistently received large allocations, but education frequently had the largest planned expenditure. Health, housing, and public works (including electricity and water) also had high priority. In 1975 planned expenditures by ministry included UD457 million for education, UD344 million for security and defense, UD267 million for health, UD256 million for electricity and water, and UD189 million for public works.

The federal budget was an effective means of distributing Abu Dhabi's oil revenues to the poorer amirates while promoting unity at the same time. Government employment at a decent salary was open to any citizen, establishing a minimum standard of living. Government

Table 29. UAE, Summary of Federal Budget, 1972–74
(in millions of dirhams)[1]

	1972	1973	1974
Revenues:			
From Abu Dhabi	196	403	799
Other	5	17	16
Total	201	420	815
Expenditures:			
Current	149	311	638
Development	15	73	157
Total	164	384	795
Surplus	37[2]	36	20

[1]For value of the dirham—see Glossary.
[2]Includes UD18 million for the country's International Monetary Fund quota.

Source: Based on information from "United Arab Emirates, Aided by Oil Income, Continues Effort Toward Central Institutions," *IMF Survey,* Washington, August 11, 1975, p. 238.

salaries were frequently raised to compensate for inflation and affected wage scales throughout the country. Subsidies through the budget provided free education and health care and reduced costs for food, utilities, and housing for all citizens. Fee schedules, such as those for electricity, made it more costly for the amirates that attempted to construct or retain independent facilities. A national road system ensured that main areas were linked, not just the areas desired by an individual ruler. The budget helped in the rapid transformation of living conditions in the northern amirates.

Abu Dhabi Budget

In 1974 about 98 percent of Abu Dhabi's revenues were from oil. Oil revenues were paid to the ruler, who kept a portion and turned the rest over to the government. The ruler's share was unknown but was reported to be declining in the 1970s. The ruler frequently used his own funds to finance projects in other amirates and in foreign countries. These transactions were not necessarily included in the amirate's budget. Other revenue sources for the government were small amounts from the modest import tariffs and fees collected for various services. Abu Dhabi approved income taxes as far back as 1965, but the taxes were imposed only on oil companies. There were essentially no direct taxes on businesses other than oil companies or on individuals. Most revenues other than oil came from unusual transactions, such as oil company help in building a refinery and some of the ruler's payments toward projects. Earnings on investments were another source of revenue, but they were probably small before 1975.

Oil revenues increased more than fourfold in 1974, relieving the fiscal pinch caused by growing expenditures (see table 30). Details of Abu Dhabi's current and development expenditures were unavailable, but the pattern was similar to the federal budget. Defense, education, roads, ports, water and electricity, housing, and health were the kinds of activities and facilities allocated the most money. The very rapid expansion of the amirate's population, at an average rate of 26 percent a year between 1968 and 1975 (caused largely by the influx of foreign labor), created part of the need for more activities and services. Government employment also increased rapidly, partly in response to the greater activities and partly because of a policy of assuring citizens of employment.

Abu Dhabi was exceptionally generous among the countries of the world, excessively so according to some critics, in providing foreign assistance to poorer developing countries. Foreign aid accounted for about 38 percent of actual budget revenues in 1973, 15 percent in 1974, and more than 25 percent in 1975. Moreover the bulk of the aid, 87 percent in 1974, was in the form of grants, freeing the beneficiary from the burden of repayment. Most aid was provided to Arab countries, particularly Egypt and Syria, although aid to other Asian and African countries was increasing by 1976. Abu Dhabi's aid improved the image of oil-rich shaykhdoms and increased the UAE's influence among developing countries. Most observers expected that Abu Dhabi's aid would diminish during the late 1970s as domestic investments increased and that more of it would be in the form of loans instead of grants.

The government established the Abu Dhabi Fund for Arab Economic Development (ADFAED) in 1971 as a mechanism for providing developing countries with loans. The ADFAED was patterned after and was assisted in establishing itself by the successful Kuwait Fund for Arab Economic Development. The ADFAED's original capitalization of UD500 million was increased to UD2 billion in 1974, and its lending operations were extended beyond the Arab world to other countries in Asia and Africa. ADFAED's staff evaluated requests for financing in terms of a project's commercial prospects. Financial terms were concessionary; interest rates generally ranged between 3 and 6 percent, and repayment might extend over more than twenty years with generous grace periods. The ADFAED had played a minor role in Abu Dhabi's aid program, however, because most aid by the mid-1970s had been in the form of grants.

By 1976 the ADFAED had committed about US$160 million to about twenty projects in some twelve countries, but actual disbursement of funds reportedly amounted to only about US$15 million because project evaluation was slow and project implementation in recipient countries even slower. Disbursements were expected to accelerate after 1976 because of commitments already made and improvements in the project evaluation process. The ADFAED was participating in joint financing with other Arab and international lending institutions, such as

Table 30. UAE, Summary of Abu Dhabi Budget, 1971–76
(in millions of dirhams)[1]

	Actual				Proposed[2]	
	1971	1972	1973	1974	1975	1976
Revenues						
Oil[3]	1,599	2,075	3,042	13,927	n.a.	n.a.
Other	52	106	179	361	n.a.	n.a.
Total Revenues	1,651	2,181	3,221	14,288	n.a.	18,401
Expenditures						
Current	569	725	972	2,010	3,278	4,296
Development	369	371	525	928	3,500	5,000
Contribution to federal budget	0	196	403	799	2,000	4,000
Foreign aid extended	143	332	1,212	2,137	3,500	4,500
Other[4]	23	112	279	1,004	806	409
Total Expenditures	1,104	1,736	3,391	6,878	13,084	18,205
Surplus or Deficit (−)	547	445	−170	7,410	n.a.	196

n.a.—not available.
[1]For value of the dirham—see Glossary.
[2]Proposed data for 1975 and 1976 derived from various news reports.
[3]Excludes ruler's share.
[4]Largely funding of domestic firms and institutions.

Kuwait's fund and the International Bank for Reconstruction and Development (IBRD, also known as the World Bank). Some observers anticipated that Abu Dhabi would channel a larger proportion of its aid through the ADFAED in the late 1970s.

Abu Dhabi had generally followed a prudent fiscal policy. In 1969, when development expenditures began to exceed revenues, development spending was reduced; a more cautious expansion followed. In 1973, however, Abu Dhabi incurred a budget deficit of about UD170 million because of a sharp increase of grant aid, largely to Arab countries, and an unexpected, temporary slowdown in oil exports. In fact Abu Dhabi borrowed heavily in European money markets late in the year to obtain the funds to aid Egypt and Syria after their October 1973 War with Israel.

In 1974 the huge jump in oil revenues created a large budget surplus of over US$1.8 billion. In 1975 budget revenues were reportedly about UD2.0 billion higher than estimated and expenditures about UD1.7 billion less than estimated, even though foreign aid outlays amounted to about UD4.0 billion instead of the UD3.5 billion budgeted. Unofficial data for 1975 suggested that a budget surplus of nearly UD4.0 billion (US$1.0 billion) was achieved. In 1976 a small surplus was budgeted, but a combination of higher oil revenues than planned and a shortfall in planned investments appeared likely from data available during the year, suggesting a surplus in 1976 somewhat higher than budgeted.

The imprecision of available data led observers to differing conclusions about the size of Abu Dhabi's foreign assets. Unofficial estimates ranged between US$2 billion and US$6 billion at the beginning of 1976, but the extremes appeared unlikely on the basis of budget figures. Observers, basing their estimates on apparent budget surpluses, believed that state reserves held abroad amounted to about US$3 to US$4 billion in early 1976. These reserves were held partly in foreign exchange and short-term securities and partly in long-term investments. The long-term investments were largely in bonds and stocks; investments in property were small. The Abu Dhabi Investment Authority was established in March 1976 to handle the amirate's investments, formerly handled by the Ministry of Finance and an investment office based in London.

Agriculture and Fishing

The bulk of the land is either mountainous or desert, and all of it is arid. Ras al Khaymah, the area of most rainfall, may receive eight inches in an exceptional year and much less in most years. All farming depended on irrigation. Most irrigation water originally fell as rain on the Hajar Mountains (see Geographic and Demographic Setting, this ch.). Some cultivation used the direct runoff of the rainfall for irrigation, and by the mid-1970s dams had been built to impound more of the runoff for irrigation. Most irrigation, however, tapped underground reservoirs created by impervious rock that trapped the runoff of rainfall as it sank through porous strata above. The underground reservoirs were dispersed, accounting for the isolated patches of cultivation. Ras al Khaymah and Abu Dhabi had the more important agricultural areas, although agricultural production was more important to the economies of the northern amirates—those without oil.

The underground reservoirs were frequently in areas unsuitable for cultivation because of topography or soil conditions. Centuries ago the inhabitants had developed a primitive though ingenious system of tunnels to lead the water to fertile soils. The tunnels sloped gently from the reservoir to the fields. Some were twenty miles long and more than fifty feet underground. Vertical shafts, which provided air for the diggers, means of removing dirt from the excavation, and openings to obtain water and to clear silt after the tunnel was in operation, were placed about ten yards apart. One tunnel, under construction in the 1940s, required eighteen years to complete slightly less than a mile. Al Ayn had about sixteen such tunnels (although six were no longer functioning) that provided its water, estimated at about 11 million gallons a day in 1976.

The area cultivated in 1973, according to the latest data available, amounted to 21,000 acres. Cultivation reportedly was largely by private farmers on small holdings. A breakdown of the cultivated area by amirate was unavailable, but Ras Al Khaymah was the principal agricultural region with about 15 percent of its area cultivated. In 1973 about 59 percent of the cultivated area was devoted to fruit trees, primarily date palms, although some citrus and other fruit was grown. Production data for dates were not available, presumably because most of the dates were consumed where they were grown and not formally marketed. Vegetables were grown on 28 percent of the cultivated land, and production reportedly amounted to 22,000 tons in 1973. Alfalfa was grown on 6 percent of the cultivated land; about 15,000 tons was produced. About 8,000 tons of tobacco was raised on a little more than 1,000 acres in 1973. A number of minor crops were grown on the remaining acreage. In addition the Abu Dhabi government had irrigated and planted trees on 1,700 acres along the highway to Al Ayn to lessen soil erosion and provide some protection from dust storms.

Before the discovery of oil the raising of livestock had been integrated with settled farming. The nomadic herdsmen, primarily raising camels and goats, had depended on settled communities for some food and equipment even though ranging widely to find forage for their animals. During the hot months the nomads settled around perennial wells and springs, where there usually was some settled farming.

The growth of oil revenues introduced some changes in agriculture, but statistics were lacking to determine the extent of change. Higher paying jobs attracted farmers and nomads, particularly the young, from traditional pursuits. The government tried with various degrees of success to settle the nomads. The government also encouraged commercial farming, particularly of vegetables and fruits (other than dates) instead of the less remunerative dates and fodder that had formerly linked the nomads and settled farmers. Commercial poultry farms had been established away from traditional agricultural regions and close to towns.

Many observers believed that the nomadic population and the numbers of animals had dropped sharply by the mid-1970s. There was less certainty about the number of settled farmers. Government investments in irrigation had expanded the area cultivated, but statistics were lacking to determine whether the number of farmers had increased or decreased. It seemed likely, however, that hired, expatriate labor had replaced local farmers in many of the larger cultivation centers.

It was readily apparent to the rulers after oil revenues began to affect the economies of the various amirates that the agricultural sector needed help to raise productivity. Agricultural production had to be increased to meet the growing urban demand for food and to raise the incomes and standard of living of those engaged in farming. Before they left in 1971, the British provided aid to renovate some of the irrigation tunnels and helped in other ways. The individual amirates also undertook various programs. Abu Dhabi, partly because of growing oil revenues and partly because Al Ayn, in the Buraymi Oasis, was the home and first administrative post of Shaykh Zayid bin Sultan Al

Nuhayyan, provided considerable help in developing agriculture. Al Ayn's underground water tunnels were improved and additional ones built, an agricultural center was established, and roads were improved to move produce to Abu Dhabi town.

The most innovative of Abu Dhabi's measures was the establishment of the Arid Lands Research Center on an island adjacent to Abu Dhabi town in 1969. It followed some experiments in Mexico undertaken in part by the University of Arizona and financed by the Rockefeller Foundation. The University of Arizona helped with the project in Abu Dhabi, although it was financed by the local government. The original idea was that solar energy, usually abundant in arid lands, could be used to produce water to make the desert bloom. In Abu Dhabi the first plastic greenhouses did not have enough fresh water, and the diesel engines that produced electricity for the center and a nearby fishing village were modified to produce desalinated water also.

By 1976 the center had fifty-two plastic greenhouses covering about five acres. Most were air supported. Incoming air was filtered and moistened by straw to avoid the sometimes dry and salty air common on the gulf coast. The controlled cool air permitted cultivation throughout the year. Trickle irrigation was used on most plants with precautions to avoid overwatering or waste of water. Nutrients were added to the irrigation water. The center primarily grew vegetables, and yields were extremely high, sometimes two, three, or even five times more than yields of similar field-grown crops in the United States. By 1976 vegetable production at the center probably was about 400 tons a year, a small part of which was exported. Cost figures were not available, but there was a strong presumption that the center required subsidies. By 1976, however, a commercial vegetable-growing venture of fifteen acres of plastic greenhouses using some of the center's techniques was being developed in the desert near Al Ayn.

The Agricultural Trials Station in Ras al Khaymah was established in 1955 to promote modernization of farming. From humble beginnings it progressed to a large complex encompassing an agricultural school and intensive, experimental cultivation on nearly 400 acres. The station introduced new crops and new techniques based on its research and experimentation. Information was disseminated through the students of the agricultural school, classes for local farmers, and extension services. The station also operated experimental units in other amirates.

Several other amirates had ongoing agricultural projects, and both federal and local authorities encouraged agricultural production by providing farmers with services and financial support. Wells had been drilled, underground tunnels repaired, and dams constructed to store rainfall and reduce damage from flash floods. Farmers received seeds, pumps, and mechanical equipment at subsidized prices and on concessionary repayment terms. Water, housing, and land were sometimes provided free to farmers. Herdsmen were provided payments based on the number of their animals in 1976. The Abu Dhabi government was the most liberal, providing most inputs free of charge, guaranteeing prices to farmers, and marketing crops.

The federal and local aid to agriculture stemmed from no illusion that the desert could be made to bloom or that the country could become self-sufficient in food. Water and labor were too limited for either to happen. Grains and other kinds of food would probably always have to be imported; self-sufficiency could only be achieved in a limited range of foods. The encouragement aimed instead at making the best use of local resources, essentially water, and at the same time improving the living conditions of those engaged in agriculture. Similar efforts in other developing countries showed that the transformation of traditional agriculture to commercial farming takes time. Modern transport and marketing organizations are a prerequisite, and they were at an early stage of development in the UAE. An adequate truck route between the east and west coasts was completed only in 1976, and some agricultural areas still lacked modern transportation. Better use of available water and higher incomes for farmers were in the future and would come about with a reduction in the number of date palms, because vegetables, other fruits, and perhaps even fodder produced higher incomes without appreciably affecting the consumption of water.

Fishing was the main source of employment and income before the discovery of oil. Since then fishermen, like farmers, have left their traditional pursuits for higher paying jobs in other sectors. Surveys during the 1960s indicated that the annual incomes of fishermen were substantially higher (although still low) than those of farmers, causing some observers to believe that fishing remained a more important source of employment and income than agriculture in the mid-1970s. Ras al Khaymah was the exception. Its relatively large cultivated area provided more employment than fishing. Fishing was much more important to the economies of the northern amirates than to Abu Dhabi and Dubai. It was a major sector of the economy of Fujayrah, on the east coast.

The fish catch was believed to have expanded rapidly in the 1960s as some boats were motorized, equipment was improved, and better techniques were adopted. By the mid-1970s the catch probably was in the neighborhood of 50,000 tons. Only a small part of the catch, probably less than 10,000 tons, was used for human consumption, however. Many fishermen sold their catch in the local market, and there were few facilities for preserving fish or moving it to markets away from where it was landed. A large part of the catch was sun dried and used for fertilizer and animal feed, particularly the very small fish caught close to shore.

Fish were plentiful in the waters around the UAE, and fishing could be considerably expanded. The UAE participated in a regional fishing survey by the UN Food and Agriculture Organization to determine the kinds of fish available, the optimal size of catches to avoid overfishing, and the equipment best suited to the conditions. The results of the survey, completed in the mid-1970s, would probably guide government efforts to develop the fishing potential in the late 1970s. The federal government had begun in the early 1970s to provide fishing equipment at subsidized prices and on concessionary repayment terms. Perhaps of more immediate importance was development of the road system and such shore facilities as ice factories and refrigeration services through which fish could be marketed over a wider area. The completion of a modern road between the east and west coasts in 1976 made possible the shipment of fresh fish from Fujayrah to the population centers on the west coast. Development of marketing was as crucial as improved equipment if fishing was to reach its potential.

Industry and Development

Apart from oil and gas the amirates have few known natural resources. Limestone is abundant and was being exploited for cement production. Ajman has a high grade of marble that was being quarried; most of it was used locally. Ras al Khaymah possesses good quality construction rock that was used in building the port at Abu Dhabi city and was also exported to Saudi Arabia for use in construction. Surveys had discovered chrome, manganese, copper, bauxite, magnesium, gold, and asbestos in small quantities, but the possibilities of commercial exploitation had not been determined. In late 1976 further surveys were scheduled in various amirates to determine the feasibility of commercial production of some of the more promising deposits.

The lack of electric power and fresh water imposed constraints on industrial development at the initial stages, but expanding the supply of basic utilities received early priority. At the beginning these essential utilities were under the control of the individual amirates, which have tended to retain control even though the federal government funded much of the investment and development in the northern amirates after 1971. Statistics were not available to measure growth in the production and consumption of electricity and fresh water, but it must have been extremely rapid, because of the rapid population increase through immigration and the extension of service to more areas.

Development of electric power and fresh water systems started in Abu Dhabi and Dubai and moved northward. Only in the mid-1970s did Fujayrah become able to provide electricity and fresh water to a significant number of consumers. Development of these services was piecemeal in all of the amirates, to meet immediate problems. Small, inefficient diesel generators were added to supply electricity to Al Ayn and other villages in the Buraymi Oasis, for example, but a pipeline was built to supply water from Al Ayn to Abu Dhabi city (although the planners envisioned a reverse flow of desalinated seawater sometime in the future). In other places a generator was added or a well bored to meet immediate problems.

A national system for power generation and water supply was not planned in 1976 although transmission lines were being extended from local power plants. It was symptomatic of the lack of a system that in mid-1976 the cement mill at Ras al Khaymah suffered from power problems because it needed more than half of the local supply and could not install needed loading equipment; the cement mill at Al Ayn had been built with auxiliary generators to meet potential power deficiencies. The power and water systems were being tremendously expanded in late 1976 with large electric power and desalination units, but development of an overall system did not appear imminent.

By 1976 the existing manufacturing sector was very small. Most plants were small scale, producing soft drinks, dairy products, baked goods, furniture, wooden boats, and building materials. Among the larger concerns were cement plants in Abu Dhabi (capacity 200,000 tons per year) and Ras al Khaymah (capacity 250,000 tons per year), a small plant producing iron bars for construction (Abu Dhabi), and a fabrication shop for some oil field equipment (Dubai). The two largest manufacturing concerns were in Abu Dhabi and were associated with the oil sector. A small refinery to supply part of the amirates' need for refined products was completed in early 1976, and a large-scale natural gas liquefaction plant on Das Island was completed in late 1976 (see Oil Industry of the United Arab Emirates, ch. 5). The latter project cost over US$500 million, employed at full capacity at least 2,000 workers, and produced for export. The governments of the individual amirates had at least part ownership of the larger concerns.

The manufacturing sector was small for many reasons. The population was small and conditions were primitive when oil revenues began to mount. Initial investments went into improving social and physical conditions. This meant building schools, hospitals, roads, ports, electric power stations, and water and sewerage systems.

As the basic infrastructure expanded, larger expenditures were allocated to economic development, particularly as oil revenues increased after 1973. By late 1976 the list of development projects was extensive, but there was no certainty that all of them would be built. Abu Dhabi intended to build an industrial complex near the oil port at Az Zannah based on the collection and processing of gas from the onshore oil fields. A fertilizer plant, an export refinery, an iron and steel mill, and petrochemical plants were being considered for the complex. A new town would be required with its own large electric power and seawater desalination plant. The port at Abu Dhabi was being expanded, a new airport was to be constructed, power and water facilities were to be increased, and the cement plant at Al Ayn was to be enlarged.

In 1976 Dubai was enlarging its port and constructing a new one about seventeen miles south along the coast at Al Jabal al Ali, which was to be a complete free zone with several large industrial plants and a residential city. A liquid petroleum gas plant was to be constructed to process the amirate's offshore gas. Most of the gas would be consumed by a US$500 million aluminum smelter under construction in 1976. Production at the smelter was to start in 1979 and full production of about 135,000 tons to be achieved in 1981. An extrusion plant was to be built (capacity of 3,000 tons a year) to process a small part of the smelter's output. A cement plant was being constructed (daily capacity of 1,500 tons) and was scheduled for completion in late 1977. A steel mill using sponge iron from India was being considered; the proposed capacity was 300,000 tons of mild steel billets. In 1976 Dubai was also constructing what was claimed to be the world's largest dry dock facility. It will be able to accommodate tankers up to 1 million deadweight tons when completed in 1979. This facility would be competing with the dry dock being constructed in Bahrain.

In 1976 development projects in the other amirates were more modest. Ras al Khaymah was expanding its cement plant, and Sharjah planned to build a cement plant with a capacity of 250,000 tons a year. Sharjah was developing a port for container ships and planned to construct a second container port in its enclave on the east coast (the Gulf of Oman). Other amirates were expanding their ports, increasing power and fresh water supplies, and constructing hotels and tourist accommodations. The UAE was encouraging tourist development, and Fujayrah on the east coast was considered to have the best potential because of its exceptional scenery.

The federal and local authorities welcomed foreign businesses that built up the economy. A local partner or sponsor was required in Abu Dhabi and Dubai but not in the other amirates. Foreigners could not own property in Abu Dhabi. Incentives were offered to new industries meeting specific conditions. The incentives included a five-year tax holiday, freedom from import duties, and nominal or free rent of sites. In 1976, however, most import duties were quite low and, although Abu Dhabi and Dubai had income taxes, they were levied only on oil companies. Generally laws regulating businesses were few.

Foreign Trade and Balance of Payments

Before the discovery of oil Dubai and Sharjah were the commercial centers in the amirates. By the 1960s Dubai dominated foreign trade, a dominance that continued in the mid-1970s. Completion of Dubai's modern port in 1972 and an active merchant community suggested that Dubai would remain the major trading center for UAE through the 1970s. Other amirates—particularly Abu Dhabi—were expanding their port facilities, however, and the proportion of UAE imports entering via Dubai would probably decrease. Moreover completion of the road linking the UAE to Qatar and Saudi Arabia opened the possibility of truck transport from the Mediterranean coast and Europe.

Other than oil the UAE had little to export. Most non-oil exports were reexports, largely by Dubai merchants, continuing a centuries-old tradition of entrepôt trade. The gold trade had once been a significant part of entrepôt activities, but it had essentially disappeared by the mid-1970s. Dubai merchants along with other merchants in the gulf had imported gold from Europe and then smuggled it into India, where the price was higher than in Europe. Transactions were facilitated for Dubai and other merchants that lived in areas using the Indian rupee. India and Great Britain introduced a special gulf rupee in 1959 in an unsuccessful effort to halt the trade. In 1970 Dubai merchants reexported 258 tons of gold, most or all of it destined for illegal sale in India. When gold prices reached exceedingly high levels in the early 1970s, however, it became too expensive for Indians as a way of holding their wealth, and the trade virtually ceased.

Petroleum products accounted for nearly 94 percent of UAE exports in 1975 (see table 31). This overstated the importance of petroleum exports, however, because the government valuation used was the posted or market price (see ch. 5). The foreign exchange the UAE received for oil exports was substantially less than the value of petroleum exports at posted prices. Non-oil exports originating in the UAE were mainly dried fish, dates, hides, and scrap metal, and their value was minor. About 87 percent of the non-oil exports in 1975 were reexports of commodities previously imported. Non-oil exports went primarily to the neighboring countries of Oman, Saudi Arabia, Bahrain, Kuwait, and Iran.

Imports rose rapidly with the rise in oil revenues, increasing more than eight times between 1970 and 1975. The high-income economy and nearly unrestricted imports meant that consumer goods predominated even though they declined in relative importance. There was a limit in practice to the amount of food and other consumption imports the economy could absorb. As a result capital goods and intermediate goods (partly cement, steel rods, and other materials used in construction) increased in importance. Imports of live animals and fresh fruits and vegetables came mainly from nearby countries. The remainder of imports came from industrialized countries. In 1974 Japan was the most important source of imports, supplying 18 percent, followed by the United Kingdom with 15 percent, the United States with 13 percent, and the Federal Republic of Germany (West Germany) with 7 percent. Other countries of Western Europe supplied much of the rest. In 1974 Dubai accounted for two-thirds of total UAE imports and Abu Dhabi one-third. Imports by the other amirates amounted to less than 1 percent; they received imports via Dubai and Abu Dhabi.

By 1976 the UAE had not published, and probably had not been able

Table 31. UAE, Summary of Exports and Imports, Selected Years, 1970–75
(in millions of dirhams)[1]

Exports	1971	1973	1974	1975
Oil exports[2]	3,411	7,459	27,525	26,838
Non-oil exports	579	1,149	1,696	1,799
TOTAL	3,990	8,608	29,221	28,637
Imports	**1970**	**1974**	**1975**	
Foods	199	934	1,173	
Other consumer goods	546	2,576	3,904	
Intermediate goods	124	896	1,333	
Capital goods	330	2,187	4,388	
Oil field materials	75	412	0	
Other	17	48	112	
TOTAL	1,291	7,053	10,910	

[1]For value of the dirham—see Glossary.
[2]Valued at posted or market price; receipts of foreign exchange were substantially less than this valuation.

Source: Based on information from *Economist,* London, December 4, 1976, p. 84.

to compile, balance-of-payments information. Only Abu Dhabi compiled and published relatively complete information on trade and financial transactions. International Monetary Fund (IMF) economists, using data available from UAE sources, compiled a rough estimate of international transactions (see table 32). This first approximation of the UAE's balance of payments suffered from many deficiencies. Major gaps included the amount of Dubai's reexport trade, the size of remittances by expatriate workers, and capital transactions by the amirates. The latter could be significant. It was known, for example, that Dubai borrowed abroad for some of its large projects, such as the aluminum smelter, and that other amirates had borrowed for their projects. Abu Dhabi had borrowed in Europe to finance its US$200 million aid grants to Syria and Egypt in late 1973. UAE borrowing in Eurocurrency markets in the first quarter of 1976 amounted to US$150 million. Even with deficiencies the estimate of the UAE's balance of payments showed that the dramatic increase of oil revenues in 1974 had largely removed any balance-of-payments constraints for a few years if foreign aid commitments did not become excessive.

Money and Banking

One area in which immediate unity was achieved by the amirates was in a common currency. Various notes and coins had long circulated in the region, but the Indian rupee had been the official medium of exchange in most of the twentieth century. When the rupee was devalued in 1966, Abu Dhabi adopted the Bahraini dinar as legal tender. Dubai and the northern amirates opted for the Qatar-Dubai riyal, which was issued in 1966, but for a few months the Saudi Arabian riyal was substituted until the new riyal could be issued. At independence and federation in 1971 the Bahraini dinar and the Qatar-Dubai riyal were the principal currencies in circulation.

The amirates agreed that a common currency was needed. In May 1973 the federal Currency Board was established, which issued the UAE dirham (UD). The dirham was defined as equal to 0.1866 gram of fine gold, corresponding to 0.21 special drawing rights (SDR—see Glossary) and US$0.2533 at the time of issuance (for later values of the dirham—see Glossary). The currency conversion was officially completed by November 1973, although the settlement of claims with Qatar was delayed until 1974. The UAE had very few exchange restrictions, and the dirham was freely convertible into international currencies by businesses and individuals. Since the bulk of GNP was in the form of foreign exchange for petroleum exports, the dirham was a strong, stable currency.

Banks had been active in the amirates for many years, particularly in Dubai and Sharjah. As Abu Dhabi's oil revenues mounted in the 1960s, some banks established themselves there. The amirates had little formal control over banks, however. The bankers in Dubai and Abu Dhabi had separate agreements among themselves concerning interest rates, bank charges, and similar matters. Establishment of the federal Currency Board in 1973 was the start of a regulated monetary system.

The Currency Board had many of the functions of a central bank but

Table 32. UAE, Summary of Balance of Payments, 1971–74[1]
(in millions of United States dollars)

	1971	1972	1973	1974[2]
Oil sector receipts (net)	399	573	923	4,285
Non-oil exports	138	194	280	439
Imports (c.i.f.)[3]	–275	–424	–746	–1,500
Services and private transfers (net)	–29	–47	–76	–115
Current Account Surplus[4]	232	296	381	3,108
Official transfers and nonmonetary capital (net)	–24	–71	–85	–1,075
Other transactions (net)[5]	–99	–40	–130	–152
Overall Surplus	109	185	167	1,881

[1]Estimated by International Monetary Fund economists from incomplete data.

[2]Estimated from partial year data.

[3]Cost, insurance, and freight.

[4]Because of data problems, official transfers were included in capital movements.

[5]Includes errors and omissions, private nonmonetary capital, and some allowances for military and other transactions not included in imports.

Source: Based on information from "United Arab Emirates, Aided by Oil Income, Continues Effort Toward Central Institutions," *IMF Survey*, Washington, August 11, 1975, p. 239.

lacked others. The board was responsible for note issue and had some control of commercial banks, but it lacked full control of commercial banking and credit. The federal government and the individual amirates maintained part of their deposits with banks instead of the Currency Board (at least well into 1975), enabling the banks to expand credit beyond the direct control of the board. Efforts were under way to give the board full central bank functions, but this had not been accomplished by late 1976.

The number of banks expanded rapidly after the Currency Board was created, from twenty in early 1973 to fifty with 217 branches in mid-1976. The rapid expansion of the number of banks, both foreign and domestic, caused the board to impose a moratorium on licensing new banks in early 1975. The moratorium was briefly lifted to license a few banks desired for particular reasons, but a further two-year moratorium was established in early 1976. The authorities feared unsound competition if there were too many banks for the small population. The board had established regulations and requirements for banking operations that provided considerable control. Part of this control was designed to ensure sound banking and part to eliminate differences between amirates and to encourage interbank transactions. Since 1973 the board has required standardized statistical reporting from banks, which has vastly improved banking information.

In early 1976 the Currency Board approved the establishment of tax-free offshore banking units. These units would receive a license restricting them to foreign currency operations and excluding them from domestic transactions. The number of units was initially limited to about twelve. Some observers attributed the small number of offshore licenses to the board's concern not to jeopardize Bahrain's efforts to develop offshore banking (see The Economy, ch. 7). These same observers thought that offshore units would prefer to settle in the UAE for several reasons, the most important of which was that UAE customers would be major users of them. UAE communication facilities did not yet equal those in Bahrain, however, and good communications were essential for foreign exchange dealings. It would take time to determine the significance of offshore banking units in the UAE, and a decision by Saudi Arabian monetary officials to use the facilities in Bahrain or in the UAE would particularly affect developments.

Bank deposits and credit expanded rapidly as oil revenues increased in the 1970s. Dubai was the main banking center in the amirates because bank credit was used extensively to finance foreign and domestic trade. Abu Dhabi was the next most important center, where bank credit primarily financed construction. Sharjah encouraged banking activities, but in 1976 it lagged considerably behind the other two oil amirates as a banking center. The booming economic conditions in all the amirates since 1973 had fostered land speculation and development that bank credit helped to finance. The substantial profits possible from property and commodity transactions under the boom conditions had retarded the outward flow via the banks of private funds into foreign investment.

UAE banking focused largely on short-term financing, and the economic boom reinforced the tendency. Consequently private agricultural and industrial projects had limited access to credit and almost none to medium- and long-term financing. The United Arab Emirates Development Bank was formed in 1974 to meet this need. The bank was empowered to join in equity financing or to provide long-term loans at subsidized, concessionary rates to promote development of industry, agriculture, and fishing by the private sector or the individual amirates.

Apart from commercial banks, development of financial activities was at an early stage. There was a need, for example, for investment houses, markets for securities and other financial paper, and a stock exchange. Dubai opened a stock exchange, and others were planned for Abu Dhabi and Sharjah, but Dubai's exchange had had little trading by late 1976 because owners seldom sold shares. Sharjah hoped to make its exchange more active when it opened by encouraging new companies to sell stock to the public, thus increasing the shares available. Growth of financial activities also required laws on bankruptcy, standards of accounting, and credit investigation agencies among other things to protect investors. These activities were encouraged but developed slowly, partly because of the low risk and high profit in land and commodity transactions since 1974 and partly because they required a change of attitude by investors.

NATIONAL DEFENSE AND INTERNAL SECURITY

Background of Military Development

The numerous treaties that Great Britain concluded with the several amirates in the nineteenth century provided, among other things, that the British were responsible for defense and foreign relations. From then until the early 1950s the principal military presence in the Trucial Coast states consisted of British forces that were literally Arab security police and personal bodyguard forces of the rulers. This colorful bodyguard was still maintained in the UAE in 1976, although the long-barreled beduin musket and saber of earlier years were apt to have been replaced by a chromium-plated automatic assault rifle.

In 1966 the British stepped up development of the military base at Sharjah to supplement and eventually replace the one at Aden; by 1968 Sharjah had become the principal base in the gulf. The British joint task force, headquartered in Bahrain, had a total strength of about 9,000 in the late 1960s. The army component numbered about 4,700, consisting of a reinforced brigade group, one battalion of which was stationed at Sharjah. The Royal Navy maintained a squadron of three destroyer escorts and six coastal minesweepers. The Royal Air Force (RAF) component consisted of two jet fighter squadrons and supporting facilities, one at Bahrain and the other at Sharjah—where the RAF had maintained a base since 1940.

In the early 1950s the British formed the Trucial Oman Scouts (sometimes called the Trucial Oman Levies), replacing an earlier security force. By independence in 1971 the Trucial Oman Scouts was a mobile force of about 1,600 men trained and led by about thirty British officers. Approximately 40 percent of the men were Arabs recruited from the Trucial Coast states, about 30 percent were Omanis, and the rest were of Iranian, Pakistani, and Indian origin. The British personnel were on assignment (seconded) from the Royal Marines, and Great Britain paid for the force at an annual rate equivalent to about US$2.4 million. Organized along armored cavalry lines, the Trucial Oman Scouts used British armored cars, vehicles, and infantry weapons and was responsible to the British political agent as chairman of the Trucial Council in maintaining order among the tribes and shaykhdoms. The force had been generally successful and was well respected. British observers pointed out that the strength and advantage of the scouts lay in their independence; that is, they were not under the command of any one of the seven shaykhs.

As independence approached, plans for defense and security forces to replace the British necessarily received serious attention. Major General Sir John Willoughby, designated the first defense adviser to the UAE, proposed in 1969 that UAE forces should be built around the Trucial Oman Scouts as their nucleus. Under the Willoughby plan top control was to be vested in the UAE Supreme Council and was to be exercised in practice by a commander, or chief of staff, designated by the council. Conservative British opinion held, even so, that the only way the force could be accepted by all the rulers—and therefore be successful—would be if military control was in non-Arab hands. Although considerable verbal support for unified UAE forces was expressed by the rulers, sufficient resources to support the project fully were not forthcoming; the states of the union—especially Abu Dhabi—made it clear that they intended to maintain the individual forces that some of them had begun building in the mid-1960s.

Oil had been discovered in Abu Dhabi in 1958 and by 1962 was providing previously undreamed-of revenues that continued to increase. By 1970 Abu Dhabi had a separate force of about 4,500, mainly officered by British and Jordanians on contract, along with some Indians, Pakistanis, and Palestinians. This force was composed of two infantry battalions, an armored car battalion with British Ferrets and Saladins, and an artillery battery with British 25-pounders. An air wing became operational in late 1968 with two transports and two helicopters, and procurement of air defense missiles and fighter and transport aircraft was initiated in 1969. Training and maintenance services were provided by a British contract firm, and in 1970 twelve rebuilt Hawker

Hunter fighter-bombers were secured, to be flown by former RAF pilots on contract and by seconded Pakistani Air Force pilots. A sea defense wing (navy) was formed in 1968, when four fast patrol boats were bought from Great Britain; several British naval officers were engaged on contract.

At independence the defense structure of the UAE appeared a compromise between the various preceding plans. The Trucial Oman Scouts was renamed the Union Defense Force (UDF), keeping its headquarters at Sharjah, and was made responsible to the federal minister of defense, the Supreme Defense Council, and ultimately to the president of the UAE, Shaykh Zayid bin Sultan Al Nuhayyan of Abu Dhabi. Although British officers and noncommissioned officers continued to provide operational leadership and training and management of logistics and maintenance, their gradual withdrawal was set as a goal. Separate shaykhdom forces were also authorized to allay both initial apprehensions about the power of the central government and fears that British officers in the UDF might be suddenly withdrawn. The largest separate force was that of Abu Dhabi; Dubai had a much smaller force and Ras al Khaymah and Sharjah even smaller ones. The role of the UDF—with additional forces on call from the separate shaykhdom forces—continued to be essentially the same as that of the Trucial Oman Scouts: police and paramilitary duties in patrolling villages and desert regions to maintain order and prevent local disputes from escalating into intertribal or intershaykhdom conflicts.

In 1971 the UDF and shaykhdom forces stopped accepting enlistments of Omanis from the Dhofar Province of Oman and of men from Yemen (Aden), on grounds that they might include radicals intent on spreading dangerous revolutionary doctrines in the UAE. Other security actions soon after independence included UDF assistance in suppressing a coup attempt in Sharjah in January 1972 and, in the following month, settling a classic tribal fight over well-rights in which about twenty Sharjan and Fujayran tribesmen were killed in one week. The UDF was assisted in that campaign by the Abu Dhabi Defense Force (ADDF), and a UDF battalion was stationed permanently in the area of the disturbance, the most serious to occur in the UAE up to late 1976.

Abu Dhabi—the largest, the most populated, and by far the richest of the seven shakyhdowms—has had to provide most of the funds and resources for the government of the UAE and its military establishment (see table 33). It was inevitable, therefore, that Abu Dhabi should be the predominant though not exclusive political power center of the new state and that Shaykh Zayid should seek to enhance the central authority as the government of an integrated and united state rather than simply a federated state. Progress in that direction has taken place gradually since 1971. The separate Abu Dhabi government structure was done away with in December 1973 and was accompanied by a shift of some ministers into the central UAE cabinet, reorganizing

Table 33. UAE, Structure and Composition of Military Forces, 1975

Military Force	Strength	Comments
Union Defense Force	3,500	Organized into six mobile squadrons (or battalions) with light tanks, light armored cars, and 81-mm mortars; also an air detachment with seven helicopters
Abu Dhabi Defense Force	10,000 (Army: 8,600; Navy: 200; Air Force: 1,200	Main equipment items of the army included heavy and light armored cars, 25-pounder artillery, and Vigilant antitank missiles; of the navy, nine patrol craft; and of the air force, fourteen Mirage V and twelve Hawker Hunter jet fighter-bombers and a transport element
Dubai defense force	1,500	Organized into three infantry battalions, one armored car battalion, and one support battalion employing light tanks, armored cars, and 81-mm mortars; also an air wing with ten light aircraft and helicopters
Ras al Khaymah defense force .	300	Company formations with light armored cars
Sharjah defense force	250	Company formations with light armored cars

Source: Based on information from International Institute for Strategic Studies, *The Military Balance: 1976–1977*, London, 1976.

that body and increasing central authority. In May 1975 the seven amirs, sitting as the UAE Supreme Council, approved in principle further steps toward full unification. In November 1975 Sharjah merged its small military force with the UDF and turned over its communications, police, and courts to the UAE ministries of communications, interior, and justice, respectively. Sharjah, Fujayrah, and Abu Dhabi also gave up their separate flags and adopted the UAE flag with its vertical red stripe along the staff and, from the top, equal horizontal stripes of green, white, and black—the traditional Arab colors.

The next major step in unification occurred in early May 1976, when the formal merger of the UDF with the separate forces of Abu Dhabi, Dubai, and Ras al Khaymah was announced. Not surprisingly Shaykh Zayid remained supreme commander; Shaykh Rashid bin Said Al Maktum of Dubai, the vice president of the UAE, would act for Zayid in his absence. Shaykh Khalifah bin Zayid Al Nuhayyan, the deputy prime minister, was named deputy commander of forces, and Shaykh Muhammad bin Rashid Al Maktum continued as minister of defense. A Jordanian, Major General Walid Muhammad al Khalidi, who had been engaged on contract, was affirmed as chief of staff (see Defense Missions and Constitutional Provisions; Command and Organizational Relationships, this ch.).

External and Internal Threats

The UAE became independent in December 1971 in a climate of general international acceptance. Admission to the United Nations (UN) and the League of Arab States (Arab League) and recognition by the major Western powers were quickly accorded. The seven shaykhdoms had been established as political entities under the British, and at independence there were no claims against the UAE. In 1976 there were no serious external threats to the existence of the UAE, although certain land and offshore boundary problems with neighbors remained. Outright external aggression to seize the country would so disturb the political balance of power in the gulf region that it was unlikely to occur. Foreign observers note that, although the UAE enjoys cordial relations with all of its neighbors, its considerable oil reserves and its small population make it a tempting target.

In early 1970, as part of the preparation for independence, border disputes were settled by Qatar and Abu Dhabi and by Abu Dhabi and Dubai. At about the same time—in April 1970—Iran, the Muslim but non-Arab giant across the gulf, accepted a UN survey showing that predominantly Arab Bahrain wanted independence. Iran relinquished its centuries-old claim to that island and demarcated offshore boundaries with Bahrain and Qatar. On November 30, 1971, however, Iran asserted its long-known claim to Abu Musa and the two Tunb Islands (Greater and Lesser)—three small islands in the Persian Gulf—by militarily occupying them; the islands were recognized as belonging to Sharjah and Ras al Khaymah, respectively. By previous negotiated arrangement the ruler of Sharjah did not oppose the Iranian occupation of Abu Musa; the ruler of Ras al Khaymah had made no such agreement, however, and his small police detachment on the Tunb Islands resisted, causing some casualties on both sides.

Some observers believed that the shah of Iran had seized the three islands as a one-time demonstration of imperial might to offset his earlier relinquishment—reportedly not well received in Iran—of the claim to Bahrain. In any case the seizure of the Tunbs had several effects. Ras al Khaymah initially declined to join the UAE but did so in February 1972 when it was clear that specific assistance to recover the tiny islands would not be forthcoming from Great Britain or the Arab countries. The Iranian action, however, generated a furious verbal reaction in the Arab world, especially among so-called radical states (notably Libya); in the UAE it caused resentment of the sizable Iranian minority of the population. That subsided, but in the mid-1970s an undercurrent of resentment of Iran remained, especially in Ras al Khaymah. However, no further incidents have occurred, and political and commercial relations across the gulf have improved. The UAE, like the other gulf states, necessarily recognizes that Iran is by far the strongest power in the region and that accommodation is mutually beneficial. Although some apprehension exists about increasing Iranian influence, and although the offshore delimitations between Iran and the UAE remained in doubt in 1976, there was no known serious and responsibly held fear of invasion. Some tentative discussions of a collective security pact among the Arab gulf states have taken place, but there had not been any substantial results by late 1976.

Another serious problem confronted by the UAE at its inception was the old question of the Buraymi Oasis—with its water and oil deposits—in the Abu Dhabi-Oman border region (see Patterns of Foreign Relations, this ch.) (see ch. 3). Saudi Arabia claimed it because of the presence of tribesmen and villages of the Wahhabi (see Glossary) sect of Islam who have historic religious and tribal ties with the Saudis. In 1955, after three years of dispute, the Trucial Oman Scouts drove a Saudi police detachment out of that part of the oasis area claimed by Abu Dhabi and recognized as such by the British; in 1966 the shaykh of Abu Dhabi and the sultan of Oman formally confirmed the de facto division of the oasis between them. The Saudi claim was not abandoned, however, and in 1970 King Faisal of Saudi Arabia called for a new determination. Four years of intermittent but highly complex negotiations followed, concluded by an agreement on July 29, 1974, in which uncontested Abu Dhabi sovereignty was recognized in its part of the oasis, the oil revenues from that area were divided between Abu Dhabi and Saudi Arabia, and Saudi Arabia was granted access across Abu Dhabi to a gulf outlet at Khawr al Udayd. Saudi Arabia then extended diplomatic recognition to the UAE.

Some reports have claimed that the agreement of 1974 is marred by serious technical defects—allegedly discovered after 1974—in the supporting maps and documents and that the defects had not been corrected as of late 1976 and might cause a revival of the dispute. The participants in the 1974 action, however, clearly intended to achieve a peaceful settlement, and the 1974 agreement was generally regarded as having resolved the Buraymi question; relations between the UAE and Saudi Arabia have subsequently improved.

The principal internal threat to the UAE in 1971 was not radical subversive insurgency but the prospect that old dynastic rivalries between the shaykhdoms (and their patterns of often-bloody power seizures by ruling family rivals within individual shaykhdoms) would reemerge, tearing the new state apart and leaving a dangerous power vacuum. The pressures for rapid modernization of a traditional Arab society thrust onto the world stage were also apparent, and some observers described the UAE as a "political laboratory." However, the receptive international attitude and absence of major external threats, the size and wealth of Abu Dhabi as compared with the other member states, and the experience and vigor of Shaykh Zayid all argued for stability. With the possible exception of Dubai, none of the six other shaykhdoms could exist alone, and their choices were to join Abu Dhabi in the UAE or—an unpalatable alternative—to align themselves with some other power. In late 1976 internal threats to stability were not entirely absent but appeared to be more those of illegal entry, the rising economic expectations of the poor, and the localized village or tribal altercations of the traditional sort (see Internal Security, this ch.).

Defense Missions and Constitutional Provisions

The missions of the armed forces are the classic ones of land, air, and sea defense. Except for Iran, Saudi Arabia, and Iraq, other national forces in the gulf are not vastly superior in numbers to those of the UAE, and in some cases they are smaller. The country's vast wealth enables it to buy, in the competitive arms markets of the world, as much of the most sophisticated modern conventional weaponry as it can handle. Thus the size and composition of UAE forces and their financial resources give them the opportunity to develop the training and equipment needed to conduct successfully brief localized operations or to impose severe initial losses on an attacker until diplomatic and outside military help can be mobilized.

Like the Trucial Oman Scouts and the UDF, the UAE armed forces have the familiar mission, for which their capabilities are more than adequate, of maintaining internal security in rural areas and reinforcing the police in towns. The naval element has a coast guard function.

The Provisional Constitution of 1971 contains a number of articles relating to defense and security. Shaykh Zayid is known to favor the replacement of the document with a permanent constitution, as projected in the preamble, but this step had not been accomplished by late 1976. Observers believed, however, that the substantive defense provisions of the 1971 constitution were likely to be retained in a permanent document, but in early November 1976 the Supreme Council agreed to delete a constitutional provision that had allowed each amirate to maintain separate armed forces.

The 1971 constitution states that the goals of the union include preserving independence, sovereignty, security, and stability and repelling aggression. It declares that "the Defense of the Union is a sacred duty of every citizen. Military service is an honor for every citizen and shall be regulated by law." The constitution also establishes the ministries of defense and interior as part of the federal Council of Ministers (see

Constitutional Framework for the Federation, this ch.). Aggression against any member of the union is defined as aggression against all, and all federal and local forces must cooperate to repel it. Land, sea, and air forces are formed by the union under one command and with a "unified standard of training"; the commander in chief and chief of staff are appointed by federal decree. Federal security forces (police) are authorized separately. The Council of Ministers is responsible to the president and the Supreme Council for all defense and security matters, and regulations and rules of discipline for the services are established by law.

Offensive war is prohibited; defensive war is declared by presidential decree after approval by the Supreme Council. Martial law may be declared by presidential decree if approved by the Supreme Council, must be reported to the Federal National Council at its next meeting, and is lifted by decree approved by the Supreme Council.

The Supreme Defense Council provides advice on all matters relating to defense and security, equipping and training the armed forces, and stationing the forces. Its membership includes the president as chairman and, as members, the vice president, prime minister, ministers of foreign affairs, defense, finance, and interior, the commander in chief, and the chief of staff. Nothing in the constitution precludes the employment of noncitizens as members of federal or local forces.

Command and Organizational Relationships

The scope and clarity of the constitutional provisions for the armed forces envisaged a truly national military establishment, with forces much larger and more diversified than those of the Trucial Oman Scouts or the UDF. The expansion of the ADDF and the increases in military expenditures in 1974 and 1975 were followed by an agreement on May 6, 1976, to merge the UDF and amirate forces under a single central command called the General Command of the Armed Forces. The designation by the Supreme Defense Council of President Shaykh Zayid as commander in chief was a logical and constitutional step simplifying the line of control to the chief of staff as functional military commander and military adviser to the president. The minister of defense is part of the chain of command; as a member of the Council of Ministers he has direct responsibility to the president and the Supreme Council for armed forces matters; the chief of staff is responsible to him. The role of the ministry in practice, however, appeared to be more one of administration, budgetary management, procurement, and allocation of resources than of close participation in operations and training.

In late 1976 the chief of staff was Major General Khalidi—a Jordanian of impeccable beduin origins and a graduate of the British military academy at Sandhurst, the United States Army Command and

General Staff College at Fort Leavenworth, and other British and American schools. He was an experienced veteran of field and combat service. As a young officer in the early 1950s Khalidi served under the legendary Englishman, Glubb Pasha, in the Jordanian Arab Legion. A generation later Khalidi had a similar role in the UAE. He has, however, at least two great advantages that Glubb Pasha lacked—he is an Arab and a Muslim, and the country he serves is rich.

After the 1976 merger agreement the headquarters of the General Command of the Armed Forces was established in a new and well-equipped facility in Abu Dhabi. The largest installation and alternate headquarters, however, is at Sharjah, where the base and airfield installations first developed by the British and since improved are located. The army is by far the predominant service, and the army headquarters staff is in effect the top military level for all services. The air and naval elements have a limited degree of autonomy and their own operational bases and staffs but are responsible to armed forces headquarters and dependent on it for command guidance and support.

UAE forces were organized under three commands: Western, Central, and Northern. The designaiton, establishment, and coordination of these commands has been a major part of General Khalidi's task in developing a centralized military authority and in implanting among them a sense of common identity and unity. The largest force in the UAE at the time of merger was that of Abu Dhabi, a fact that facilitated Khalidi's task. The integration of these elements, however, requires time, training, and acclimatization to the new patterns and organizations that are gradually being introduced.

Manpower, Expenditures, and Procurement

According to the provisional figures of the 1975 census, the population of the country was 655,937. The two most populated shaykhdoms were Abu Dhabi (235,662) and Dubai (206,861). Like most of the developing countries of the world, they have predominantly young populations. Contracting and enlisting foreigners from minorities in the population or from outside the country has been a customary practice. The society is characterized by traditional Arab values and attitudes, and military service is regarded as an honorable profession. Constitutional authority for conscription is clearly provided, but as a result of voluntary service and the contracting of foreigners the draft system had not been required up to late 1976.

The chief problems in establishing a force have occurred because of the low levels of general education and the lack of technical training and military leadership experience; presumably these can be overcome by effective training programs. The total strength of the UAE forces in late 1976 was given by the International Institute for Strategic Studies as 21,400—apparently reflecting a marked expansion from the total of about 16,000 believed to be in service in early 1975. Comparing the 21,400 with the 1975 population figures shows a ratio of thirty-three men in service per 1,000 people. Figures of the United States Arms Control and Disarmament Agency (ACDA) show an even higher military proportion. By comparison, the ACDA figures for Turkey show only fourteen per 1,000; for Great Britain, six; for the Soviet Union, about two; for the United States, ten; for Iran and Iraq, about ten; for Egypt, twelve; for Jordan, twenty-seven; and for Israel, forty-eight. These comparisons suggest strongly that the UAE may have pressed its own sources of military manpower as much as it can, short of an increased use of foreigners. Therefore increases in total UAE military strength appear unlikely in the late 1970s. However, slower and gradual increases—possibly proportionately more in the air and naval elements than in the ground forces—and a program of improvement in administration, training, equipment, and overall capability of the newly unified UAE military establishment are likely.

Initially foreign officers and noncommissioned officers in shaykhdom and UAE forces were mostly British or Pakistanis on contract or secondment. By 1976 British representation in the naval element and British and Pakistani representation in the air force remained strong but had declined in the army, where British presence was increasingly limited to logistical and technical fields. Jordan had become the main single source of foreign military personnel, especially in the combat arms. A compatible Arab monarchy also having a background of British tutelage, Jordan has experience in modern conventional warfare and a record of successful resistance to radical insurgency. For Shaykh Zayid and the other rulers Jordan was therefore a logical and politically astute choice to request assistance from. The number of Jordanians in the UAE forces in 1976 was not accurately known but had been estimated at about thirty officers and as many noncommissioned officers.

Military expenditures up to 1976 had been mainly those of Abu Dhabi and Dubai; figures for the merged military forces as part of the national UAE budget for 1976 and 1977 were not reliably known. The ACDA has estimated that UAE military operating expenses were the equivalent of about US$19 million and US$22 million in 1972 and 1973, respectively, increasing sharply to US$51 million in 1974 and about the same in 1975. The Stockholm International Peace Research Institute (SIPRI) estimates that the value of arms deliveries to Abu Dhabi as a result of previous and current orders reached US$63 million in 1974 compared with an annual average of about US$10 million in the preceding five years.

In December 1974 Abu Dhabi placed an order of about US$80.5 million with the British Aircraft Corporation (BAC) for a Rapier air defense missile system using the Blindfire radar, and three air defense battalions were formed in 1975. Also a US$16-million order was placed with France for SS–11 and Harpon antitank missiles and an order (value unknown) for six Mirage V fighters, adding to the fourteen on hand. In 1974 Dubai ordered five counterinsurgency strike aircraft from Italy. These data, though incomplete, support the conclusion that expansion occurred in the 1974–75 period, probably in anticipation of the merger of May 1976.

The UAE was not a recipient of grant military aid, being well able to pay for the equipment and training services received from abroad. Some equipment was transferred from Great Britain to the UAE at the time of independence. The principal supplier on purchase to the shaykhdoms since 1968 and to the UAE since 1971 has been Great Britain, although a good share of the market has also gone to France. Between 1965 and 1975 Great Britain accounted for about 44 percent of arms imports, France for about 31 percent, and other suppliers—particularly Italy and West Germany—for the rest. In an exchange of notes on June 15 and June 21, 1975, the United States formally agreed to sell defense articles to the UAE subject to the provisions of the United States Foreign Military Sales Act. Actual United States sales totaled about US$437,000 in 1972 and only US$56,000 in 1973; in the succeeding years up to late 1976 the only purchase had been some minor equipment for the Union Defense Force. The UAE had made no military purchases from the Soviet Union or other communist states and had no military-training exchange or diplomatic relations with them.

Organization and Administration

The Army

Total army strength in 1976 was estimated at about 18,800. Principal formations included one Royal Guards Brigade of three battalions (mobile); three mixed tank and armored car battalions; seven separate infantry battalions; three field artillery battalions; and three air defense battalions. Equipment (British unless otherwise indicated) included twenty-seven Scorpion light tanks, about fifty Saladin and Shorland armored cars, thirty-six Ferret light armored cars, some French Pouhard armored cars, at least sixteen 25-pounders, sixteen to twenty United States 105-mm howitzers, sixteen French AMX self-propelled 155-mm howitzers, an unknown number of Rapier surface-to-air missile launchers, and unknown numbers of Vigilant and French SS-11 and Harpon antitank missiles.

The Air Force

Air force strength was estimated at 1,800; most of the pilots were Pakistanis and British under contract. The aircraft inventory in late 1976 included twenty French Mirage V jet fighters, twelve British Hawker Hunter jet ground-attack fighters, eight Italian MB–326 jet counterinsurgency strike aircraft, two United States C–130 heavy transports, nine light transports, and about twenty-five assorted French and Italian helicopters. Organization was not clearly known but appeared to consist of one interceptor fighter squadron, one fighter ground-attack squadron, one transport squadron, and one helicopter squadron. Some of the helicopters may be suitable for gunships and for medical evacuation.

The Navy

The navy had an estimated strength in 1976 of 800, up markedly from the 200 estimated in 1974 and 1975. Senior officers were British. Equipment included six new Vosper-Thornycraft large (110-foot) long-range patrol craft mounting two 40-mm guns each; three Keith Nelson fifty-seven-foot patrol craft mounting two 20-mm guns; five Fairey Marine Spear thirty-foot fast patrol craft mounting two machine guns; and six older forty-one-foot Keith Nelson patrol boats, also mounting machine guns. All of these vessels are of British origin. The size and equipment of this force provided capability for patrol and reconnaissance in coastal waters and for intercepting smugglers and illegal entrants, although not for conventional naval combat at sea. Naval tactical exercises using live ammunition were conducted at sea on November 1, 1976. These exercises, attended by the colonel commanding the Western Command, were reportedly the first performed by naval elements.

Uniforms, Ranks, and Insignia

Uniforms in all services closely resemble the summer uniforms of the corresponding British services, a consequence of the long period of British influence. Nevertheless, distinctively Arab items of dress and accouterment are seen in some units, particularly among bodyguards and ceremonial detachments.

The system of officer, noncommissioned officer, and other ranks is the British system, modified and generally standardized among the Arab countries of the Middle East and North Africa (see table 34). The Arabic words for the various ranks are the same for all services, including the police. Specific distinction may be made for the particular service by adding the appropriate service modifer: for example, *naqib al shurta* is a captain *(naqib)* of police *(shurta).* Insignia of rank also follow the basic British system in modified form, employing combinations of a star (corresponding to the British pip), a crown, and crossed sabers.

Pay Rates

New codifications of personnel pay and allowances were expected in support of the new centralized command and nationalization of forces of 1976. Base pay levels, in-grade step increases, and special allowances for the UDF and ADDF, though not the highest, were among the more generous in the Arab world and were a major factor contributing to morale and loyalty. It was unlikely that new pay scales for the merged national forces would be less ample. The following examples were approximate monthly base pay rates in United States dollars for selected grades in the ADDF in 1973: colonel, US$670; major, US$450; first lieutenant, US$315; warrant officer, US$290; sergeant, US$175; corporal, US$165; and private, US$140.

Internal Security

The Police

Federal police and security guard forces are authorized under the 1971 Constitution and were established by a decree law on April 24, 1974. They are subordinate to and part of the Ministry of Interior. Total police and security guard strength in 1976 was not accurately known but was estimated at about 5,000; the police were located in the cities and towns, and the security guard forces were stationed in rural villages.

Under the 1974 decree UAE citizenship is a requirement for police

Table 34. UAE, Armed Services Ranks and Insignia, 1976

Rank	Arabic	Insignia[1]
Major General[2]	Liwa	Crossed sabers and one star
Brigadier General	Amid[3]	Crown and three stars
Colonel	Aqid	Crown and two stars
Lieutenant Colonel	Muqaddam	Crown and one star
Major	Raid	Crown
Captain	Naqib	Three stars
First Lieutenant	Mulazzim Awwal	Two stars
Second Lieutenant	Mulazzim Thani	One star
Warrant Officer	Wakil	One gold bar
First Sergeant	Raqib Awwal	Crown and three chevrons
Sergeant	Raqib	Three chevrons
Corporal	Arif	Two chevrons
Private First Class	Jundi Awwal[4]	One chevron
Private	Jundi[4]	None
Recruit	Jundi Mustajid[4]	Do.

[1]All stars are five pointed.

[2]Grades above major general not in regular use in 1976.

[3]May be translated as brigadier or brigadier general and regarded as either depending on whether British or United States customs, respectively, are being followed.

[4]In the police, replace the word *jundi* (soldier) with the word *shurti* (policeman).

Source: Based on information from U.S. Department of Commerce, Office of Technical Services, Joint Publications Research Service, "Federal Bill Defines Police, Security Forces Jurisdiction," 1974; and U.S. Department of Commerce, Office of Technical Services, Joint Publications Research Service, "Interview with Shaykh Sa'd al Abdullah, Minister of Defense and Interior," 1974.

service, and a minimum age of twenty-eight years is required. Appointment to grades of lieutenant colonel or above must be made by federal decree and below that grade may be made by ministerial directive. Police grades and insignia were the same as those of the armed forces with minor variations. Applicants for commissioning as officers must have at least a high school education and must be graduates of police-training courses approved by the Ministry of Interior unless they are university graduates. No educational requirement was specified for noncommissioned officers and other ranks.

The national police are charged by law with protecting the state from within, maintaining public security and order, and combating crime. The UAE police, especially in the special investigative branch, employed a number of British and Jordanian advisers.

Crime and Punishment

The UAE criminal law is predominantly that of sharia law with some British influences. The 1971 constitution established the office of attorney general for the union; his staff and his functions as public prosecutor are set forth in specific laws. Until 1976 police and prosecution matters were chiefly the concern of the individual amirates. Because of the trend toward centralization that was under way in that year, a wider role for the UAE attorney general could be expected.

Traditionally intertribal conflict and raiding and localized village disputes over water, cattle, or land were the main causes of public disorder, but by the mid-1970s these were largely absent. The most serious problems were those of coastal smuggling and—in the late 1960s and to late 1976—the illegal entry of indigent Pakistanis transported by dhows whose masters unload them surreptitiously at points on the Musandam Peninsula. From there, if they are not intercepted, the illegal entrants make their way on foot to Abu Dhabi. Few succeed. Harsh measures to repress this exploitative traffic have been taken, but it remained a matter of concern in late 1976.

Subversion and Political Security

Because of the still largely traditional characteristics of its society and because of close police security, the UAE has shown little vulnerability to radicalization and intrusion by such organizations as the Popular Front for the Liberation of the Occupied Arab Gulf (PFLOAG) and the Popular Front for the Liberation of Oman (PFLO). As of 1976 the immigrant communities, despite recurring predictions to the contrary, had not generated dissidence or disaffection in a degree approaching that of the large and restive Palestinian minority in Kuwait. The immigrant merchants, in particular, tend to identify with the existing regime.

The possibility of a sudden coup or assassination attempt that would seriously disturb the political evolution of the UAE and its emergent society cannot be categorically ruled out. Nevertheless, a well-informed observer and student of the lower gulf area, John Duke Anthony, summed up the security situation in the UAE as follows in 1975:

> The governments of the shaykhdoms have been encouraged by their fortunate internal security situation.... Problems thus far have been minimal. This is due largely to the low level of political consciousness in most of the shaykhdoms and the embryonic state of the educational and socialization processes, on the one hand, and the contribution of British and Jordanian expatriates as police and special branch personnel, on the other. These efficient officials have maintained dossiers on nearly all the revolutionaries and radical nationalists over the past two decades. In addition, they continue to keep a watchful eye on suspected new dissidents who occasionally enter the shaykhdoms from abroad.

* * *

Statistical data and studies of the society and the economy were difficult to obtain in late 1976, partly because statistical collection was at an early stage of development. A brief but authoritative survey was contained in the *IMF Survey* of August 11, 1975. The *Financial Times* of London published a number of economic articles in the May 22, 1975, May 10, 1976, and December 10, 1976 issues. Michael Tomkinson's *The United Arab Emirates* contained considerable economic information along with descriptions of many places in the UAE. Fragmentary information is published in newspapers and in such periodicals as the *Middle East Economic Digest* and the *Arab Report and Record.* The UAE section in *The Gulf Handbook, 1976–77* (and subsequent editions) might also be consulted.

The preparation of the section on government and politics was substantially aided by John Duke Anthony's seminal work, *Arab States of the Lower Gulf: People, Politics, Petroleum,* which focuses on the political setting, structure, and processes not only of the federal level but also of the seven constituent amirates of the UAE. For a general introduction to the socioeconomic setting of the UAE, K.G. Fenelon's *The United Arab Emirates: An Economic and Social Survey* was useful.

Among the more reliable sources on armed forces, equipment, and procurement patterns are such standard reference works as the annual issues of *The Military Balance* by the International Institute for Strategic Studies and the appropriate annual publications by Jane's Yearbooks, especially *Jane's Fighting Ships, Jane's Weapon Systems,* and *Jane's All the World's Aircraft.* The publications of the Stockholm

International Peace Research Institute are also particularly valuable, especially *World Armaments and Disarmament: SIPRI Yearbook 1975,* and subsequent editions. Useful data and insights on defense and security matters may also be found in John Duke Anthony's *Arab States of the Lower Gulf.* (For further information see Bibliography.)

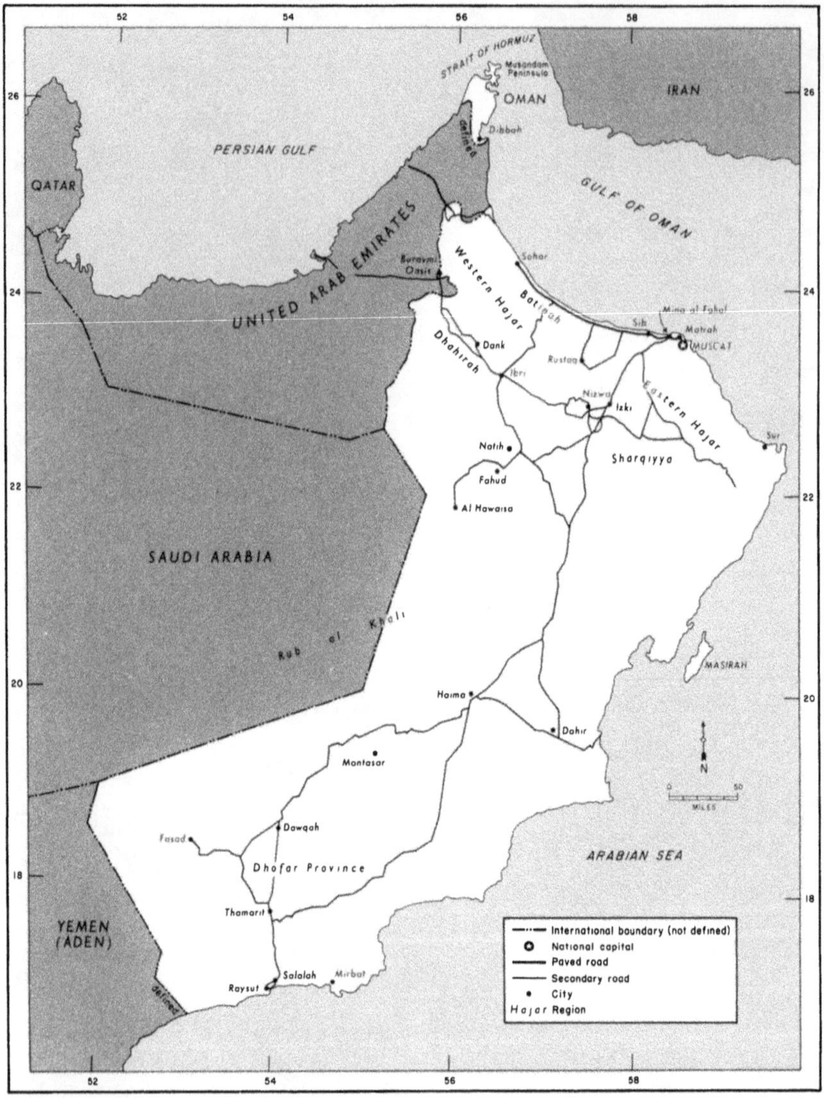

Figure 22. Oman, Regions, Roads, and Major Cities, 1976

CHAPTER 10

OMAN

In late 1976 the Sultanate of Oman continued to function as an absolute monarchy. Its ruler was Sultan Qaboos bin Said bin Taimur Al Bu Said, the fourteenth ruler of the Al Bu Said dynasty, which had established its power in the country in the mid-eighteenth century. November 18, 1976, marked the sixth National Day of Oman. National Day commemorates the anniversary of the country under its new name, the Sultanate of Oman, and the accession to rule of Sultan Qaboos. It is indicative of the focus of the new government that National Day does not fall on the anniversary of the coup d'etat that occurred on July 23, 1970, in which Qaboos deposed his father and seized power, but rather on his birthday, thus underlining Qaboos' central and absolute authority and the mirrored association of him with the state.

Oman has no constitution, legislature, political parties, or elected assemblies on the federal level; nor is there suffrage. The sultan serves as his own prime minister, minister of defense, and minister of foreign affairs. He is also commander in chief of the armed forces and the Royal Oman Police, is the only source of new legislation, and embodies in his person the highest judicial power.

HISTORICAL BACKGROUND

In addition to the central subject of independence, defense and security concerns in Oman have historically involved the closely related matters of Muslim religious doctrine, intertribal warfare, personal power struggles, foreign influence and intervention and, in the 1960s and 1970s, protracted attempts by Marxist-oriented, radical Arab nationalist factions to overthrow the traditionalist monarchy through an externally assisted guerrilla war. The historical interplay of these factors has governed the political life of the country, its socioeconomic development, and the evolution of its military and security forces, supported since 1964 by modest oil revenues.

The Arab tribes of the Omani region, having been converted to Islam by the mid-seventh century A.D., fell under the rule of the Umayyad caliphal dynasty in Damascus. At about the same time the schismatic Kharadjite movement (or Kharadjite heresy) appeared in Iraq, North

Africa, and elsewhere as an outgrowth of the dynastic warfare and theological contentions of early Islam (see ch. 3). This belief held as its main tenet that the office of caliph, or earthly successor to the Prophet Muhammad as ruler of the Muslims, should be filled by consensual choice and should not pass in hereditary line or be limited to one family or tribe, or even necessarily to Arabs. The Kharadjite movement soon became fragmented into numerous subsects, one of which was headed by the Iraqi theologian Abd Allah ibn Ibad al Murri al Tamimi and whose members became known as the Ibadis. Fleeing from suppression by the Umayyads, the Ibadis from Iraq made common cause with Omani tribes who were also resisting the Damascus caliphate. The Ibadi brand of Islam thereafter became entrenched in Oman, becoming in effect the Omani ideology.

When the Abbasid faction of Sunni Islam overthrew the Umayyads and set up their capital at Baghdad in the mid-eighth century, the Omani tribes successfuly withstood Abbasid attempts to establish permanent sovereignty throughout Oman. In consonance with Ibadi practice the tribes elected an imam (spiritual and temporal leader—see Glossary), in 749, who is often regarded as the first Omani ruler after the coming of Islam. Although subject temporarily to invasions by several external powers, including Iran, Oman generally maintained independence and attained substantial maritime power until 1507, when Portugal seized control of the coastline. In 1650 an Omani faction defeated the Portuguese forces in the region. Independence is generally reckoned from that date, making Oman the longest continuously independent Arab state in modern times.

In the Omani renaissance and resurgence after 1650 the imams extended their maritime conquests to the East African coast, incorporating Mogadiscio, Mombasa, and Zanzibar into the empire. During the first half of the eighteenth century, however, civil war between rival contenders for the office of imam shook the country. Although one of the unsuccessful contenders was supported by an Iranian military force, the victor was Ahmad bin Said, who was duly elected imam in 1749 and became the founder of the Al Bu Said dynasty, which has ruled continuously since then in part or all of Oman. At about this time the head of the Al Bu Said family began to be known as the Sultan, although he frequently was also the imam. Another period of prosperity followed, and in 1786 the capital was moved from Nizwa and Rustaq to the principal Omani port city of Muscat. This move led to later political differentiations between the interior areas and the coast, and these differentiations became further identified with religious and tribal contentions and renewed contests for power.

In historical perspective possibly the most important long-term development contributing to the stability and continuance in power of the Al Bu Said sultanate began in the nineteenth century, when Oman and Great Britain commenced a special relationship that continued thereafter and remained strong in 1976. Treaties of friendship between Oman and Great Britain were negotiated in 1798 and 1800, and a full exchange of consular relations followed in 1839. Although treaty relations were established with other states outside the Middle East (for example, with the United States in 1833), the most consistent and influential connection has been with Great Britain. It can not be neatly described in conventional foreign relations terms, but it illustrates the British capability of long-term, workable improvisation according to time and place.

In contrast to the Persian Gulf shaykhdoms, Egypt, Jordan, and other areas in the Middle East, Oman has not been in a colonial, protectorate, or mandated arrangement with Great Britain. Formal relations have rested on the treaties of friendship, commerce, and navigation of 1891, 1939, and 1951, and British military presence and assistance have never been based on treaty obligations. Nevertheless Great Britain from time to time has had regular British units, naval bases, and air stations in Oman; the last such base was scheduled to close in March 1977.

The British presence as political and military advisers and as active members of the Omani military forces has been distinctive. The arrangements have been ad hoc agreements implemented by contracted or attached (seconded) individual British military personnel, provision of training and equipment, and the stationing of small forces in Oman. Consequently the principal characteristics of the regular Omani military and security forces in uniforms, organization, equipment, and training became British (or adaptations of British usages) and continued to be so in 1976, when some informed observers maintained that the British connection was still the "key to the Sultan's survival."

During the reign of Said bin Sultan Al Bu Said (1804–56) the Omani domains attained their greatest extent—Dhofar Province being added in 1826 (see fig. 22). Said had a strong navy of fifteen Western-style warships, including a seventy-four-gun ship of the line. In Oman Said could call on perhaps 5,000 men from tribal levies, and he had a mercenary regular force of about 1,500. The practice of using regular forces of mercenaries thus has long-standing customary sanction in Oman. At Said's death the empire was divided between two of his sons, one becoming sultan of Zanzibar and the other of Oman. This major historical development and the sharp reduction—brought about through British influence—of the slave and arms trade through Muscat reduced Omani revenues and the economy in the second half of the nineteenth century and caused unrest. Two major insurrections occurred and many minor disturbances. Owing to British assistance and the traditional rivalry between the Ghafiri and Hinawi tribal confederations of the interior, the Al Bu Saids maintained a precarious authority (see The Twentieth Century, ch. 2). The Hinawi of the southeast were strict, conservative Ibadis. The Ghafiri of the northwest were also Ibadis but slightly more receptive to outside influences. When these groups temporarily united, however, the sultanate probably could not have survived without British help.

Over the years the conservative tribesmen of the interior had grown progressively disaffected from the Al Bu Saids, who had established hereditary rule contrary to Ibadi traditions and had moved the capital. To protest against the growing secularization of what had once been a largely religious office, the tribesmen resorted in the late eighteenth century to electing another imam. This led to a situation that lasted for roughly 150 years in which there was an imam of Oman and a sultan of Muscat and Oman. Sometimes, as for example between 1868 and 1873, the sultan was also acknowledged as the imam; sometimes there was no imam at all. Throughout this period and later it was the sultan in Muscat who was recognized by the major maritime powers: Great Britain, France, and the United States.

In 1913 Sultan Faisal bin Turki Al Bu Said died and was succeeded by his son, Taimur bin Faisal Al Bu Said. A different imam was elected in the interior, however, and there was a general rising of tribes from both federations against Sultan Taimur. The tribesmen regarded the sultan's cooperation with Great Britain to restrict the arms traffic in Oman as subservience to Great Britain. Encouraged by propaganda emanating from Berlin and Istanbul, the tribesmen besieged Muscat in 1915 only to be routed by a British force. Eventually in 1920 the sultan and the tribes reached accommodation through the Treaty of Sib. In this agreement the sultan recognized the elected imam's spiritual authority and also allowed him temporal authority over those tribesmen who accepted his jurisdiction. The sultan did not relinquish his claim to full sovereignty.

Although the Treaty of Sib had left many fundamental questions unanswered, it endured unbroken for thirty-four years. Imam Muhammad Abd Allah al Khalili was an archconservative, Ibadi to the core, firmly committed to Omani particularism, and was as hostile to foreign incursion as were the Al Bu Said sultans. When Saudi Arabia seized the Buraymi Oasis in 1952, the imam supported efforts to oust the Saudis and provided Omani tribesmen to the sultan's army.

Imam Muhammad died in May 1954. In the election that followed the winner was Ghalib bin Ali of the Hina tribe of the Hinawi federation. His success was possible with the support of Sulyman bin Himyar, chief of the powerful Bani Riyam tribe and strongest figure in the Ghafiri federation, and Sahib bin Issa of the southern district of Sharqiyya. These three leaders commanded enough support across the tribal spectrum to establish a formidable grip over the interior. Encouraged by foreign assistance, Ghalib, through his brother, Talib bin Ali, applied to the League of Arab States (Arab League) for full membership for Oman. The new imam's application for membership in the Arab League in 1955 reopened the whole question of the Treaty of Sib, particularly in regard to the autonomy or independence of the imam in the conduct of foreign relations.

By 1955 Sultan Said bin Taimur Al Bu Said, who had become sultan in 1938, became convinced that his regime was threatened by a Saudi-backed coalition and that Imam Ghalib had, by his representations to the Arab League, violated the Treaty of Sib. In 1954 the sultan's forces occupied the town of Ibri, isolating the imam at his base in the interior from the Saudi-controlled Buraymi Oasis. In the meantime Said's army had been reorganized into four task forces, each of about battalion size, under British command: the Dhofar Force at Salalah; the Muscat and Oman Field Force at the oil camp of Petroleum Development (Oman) (PDO) near Fahud; the Batinah Force at the coastal town of Sohar; and the Muscat Levies, garrisoned near the capital.

From the advance base at Ibri the Muscat and Oman Field Force moved rapidly against Nizwa and occupied it on December 15, 1955, in an almost bloodless coup de main. Two days later the Batinah Force stormed Rustaq, which fell after determined resistance by the imam's brother, Talib. The civil war appeared over, and the sultan embarked on an unprecedented 600-mile truck trip from Salalah to Nizwa. At Nizwa the sultan allowed Ghalib to retire to his village and accepted a pledge of fealty from Sulyman. The annexation of the interior was justified on the grounds that the imam had violated the Treaty of Sib and had collaborated with Saudi Arabia.

The sultan's triumphant entry into Nizwa in 1955 ended only the first phase of the civil war. The imam's brother, Talib, who had escaped to Saudi Arabia and thence to Cairo, where he had operated a publicity office for the imamate, returned to Oman in June 1957. Aided by an arms flow from abroad, Talib raised the tribes again, asserting the imam's authority in the interior. Sultan Said immediately requested British aid. Nizwa and most of inner Oman were reoccupied in September 1957. The imam, Talib, and Sulyman retreated to Sulyman's mountain stronghold at the summit of the Jabal Akhdar (Green Mountain); desultory fighting continued until early 1959, when the sultan's troops and 300 British regulars attacked the stronghold, bringing the war to its conclusion. Although Ghalib, Talib, and Sulyman escaped to Saudi Arabia, the sultan's authority was recognized throughout Oman. He promptly decreed the Treaty of Sib terminated and the office of the interior imam abolished (see National Defense and Internal Security, this ch.).

Sporadic acts of terrorism continued for several years, but the sultan's security forces gradually extirpated the imam's political apparatus. In the international arena, however, the imam's cause continued to find some support for various, often seemingly incongruous reasons. In October 1960 ten Arab countries succeeded in placing the "Question of Oman" on the agenda of the General Assembly of the United Nations, but a draft resolution calling for the imamate's independence did not secure the necessary majority in December 1961. A United Nations

Commission of Inquiry visited Oman in the spring of 1963. The commission's report refuted the imam's charges of oppressive government and strong public feeling against the sultan. The case was reopened at intervals until October 1971, when Oman was finally admitted to full membership in the United Nations (UN).

Oman had a new ruler when it was granted full membership in the UN. Despite the cajoling of the British, Sultan Said had not diminished his isolationist stance, nor had he mitigated his savage and tyrannous practices. In the second week of June 1970 there was an outbreak of guerrilla activity in the interior of Oman, and Shell Oil pressed the British government for some action to ameliorate the rapidly deteriorating situation. On June 20 there was a change of government in Great Britain, and in July the coup d'etat in Oman occurred.

The events of the month before the coup have not been made available to the public. The British government noted that the sultan's son Qaboos, with the aid of loyal Dhofaris, had peacefully deposed the sultan on July 23 and then had him flown into exile in England. Qaboos, Said's only son, had been under house arrest since 1966, when he had returned from military training in Great Britain. Great Britain recognized Qaboos as sultan on July 29, and the British Foreign Office denied any knowledge of the coup or complicity in it. Observers noted that it would be highly improbable that such an event could have occurred without British knowledge and at the very least their tacit approval, because British officers held all the key positions in the armed forces.

GOVERNMENT AND POLITICS

Sultan Qaboos, who celebrated his thirty-seventh birthday on November 18, 1976, was born in Salalah, the capital of Dhofar Province, where his father had isolated himself. His mother is a Qara, a tribe of non-Arab ethnic origins, and Qaboos was his parents' only issue. He was sent to Sandhurst for training and served for a time in the British Army of the Rhine.

Qaboos did not marry until March 1976. He had rejected the bride chosen for him by his father and married Kamilla, the daughter of his only surviving uncle, Tariq bin Taimur Al Bu Said. Foreign observers speculate that he may have postponed marriage well beyond the average age in order to devote himself fully to the consolidation of his political position. He may of course only have been waiting for Kamilla to complete her education abroad.

Qaboos is thought of by many foreign observers as an enigmatic ruler. Although he governs as an absolute monarch—his rule being more absolute than any other of the traditional rulers of the Arabian Peninsula—he is at the same time an extremely progressive social developer. Foreign observers believe that Qaboos is aware that he must develop the country as quickly as possible. In 1976 it was estimated that Oman could expect only ten more years of relatively small oil revenues (see The Economy, this ch.). To this end Qaboos relies on the advice of international specialists and on former Omani expatriates who were educated abroad in places ranging from Western Europe to the Soviet Union. The sultan's policy of staffing the civil service and the armed forces with Omanis (a policy known as Omanization) was being implemented in the mid-1970s without waiting for the young to complete higher education and without waiting for the slow social evolution that might in several generations place women in the working force.

The sultan's political goals appeared to be to detribalize the society, break down regionalism, allow no threats to his political autonomy, and pacify Dhofar Province completely. Qaboos was aware of the desire among the progressives in his country to have a more modern political form but responded by saying, "A parliament whose members we will choose can be created; we can create a phony parliament to give the impression of a semblance of democracy in our country. All this is possible, but does it correspond to the aim for which a parliament is supposed to exist? We need more time to reach this stage."

In 1976 Oman did not have a majlis, the traditional peninsular institution whereby citizens can directly petition a ruler. The majlis was discontinued in 1958 when Said bin Taimur ensconced himself at the palace at Salalah; Sultan Qaboos has not revived it, perhaps because of the danger of assassination attempts. The palace diwan, or royal court, is the main forum for petitioners, who are usually from the lower or upper classes. An average of fifty people daily petition the court. The minister of the royal court is assisted by a deputy and by an administrative director, the director for employee affairs (civil service), the head of the protocol department, and a press adviser.

Government Structure at the National Level

Although Sultan Qaboos remained the head of state and source of all authority, by late 1976 sixteen ministers had been appointed by him to be responsible for various governmental functions (see fig. 23). In addition to the ministers the sultan had a council of advisers. The two most influential advisers were the sultan's uncle, Tariq, and Sayyid Thuwayyni bin Shihab, the sultan's cousin. Sayyid Thuwayyni was also wali (governor) of the capital area and was acting sultan during Qaboos' absences from the country.

Qaboos' most pressing concern after his assumption of power was to gain support for his coup and to bring peace to Dhofar Province. The establishment of the ministerial cabinet was therefore a lengthy and careful process. One of Qaboos' first political acts was to call Tariq out of exile and appoint him prime minister on August 9, 1970. The first cabinet was established by Tariq's decree on August 16, 1970. The cabinet consisted of ministers of interior, education, health and justice,

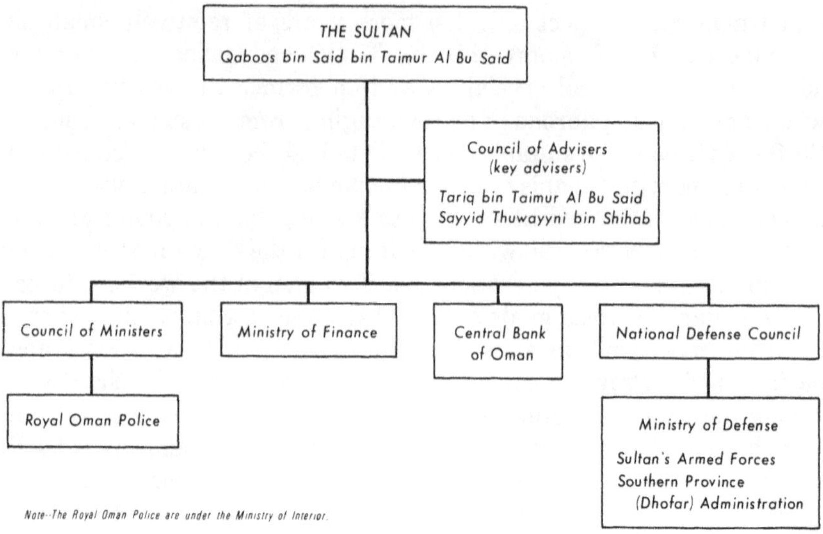

Figure 23. Oman, Government Structure, 1976

and foreign affairs. Ministers of information, social affairs, and labor were appointed on December 15, 1970, and a minister of economy on March 1, 1971.

Tariq enjoyed great popularity among the people because he was the only member of the royal family to have fought against the imamic insurgency in the 1950s (see Historical Background, this ch.). He was known and respected more than the young sultan who, because of the house arrest imposed upon him by his father, had seldom been seen by the people. However, soon after his return Tariq became a controversial figure. He wished the country to have a formal written constitution, and he hoped that in time Oman would become a republic. He believed the institution of the sultanate to be anachronistic and intrinsically ill designed to serve Oman's entry into the modern world. On January 1, 1972, Tariq resigned his post, and his duties as prime minister were assumed by the sultan. Qaboos reorganized the cabinet, which then included interior and justice, foreign affairs, health, education and culture, economy, defense and communications, labor, and agriculture.

Further reorganizations and expansions occurred, the latest on April 12, 1976. By that time the cabinet was composed of sixteen ministries (see table 35). The two newest ministries were the Ministry of Youth Affairs (incorporated with the duties of the deputy minister of defense) and the Ministry of National Heritage. Combining the responsibility for the Ministry of Youth Affairs with the duties of deputy minister of defense was natural considering the Dhofar insurgency (see National Defense and Internal Security, this ch.). Since its inception the new government has placed great emphasis on the youth of the country.

The government wants to train young people as quickly as possible to take responsible positions and to inculcate in them a sense of identification with the orientation and goals of the sultan's government. Many programs were aimed at detribalizing the young, that is, giving them an allegiance greater than the one to their tribe and fostering in them a sense of individual responsibility toward the state.

The Ministry of National Heritage has been designed to protect natural features of the country and its historical landmarks. In the rush of development many fine buildings were destroyed. The ministry will seek to prevent unnecessary destruction that might damage the tourist industry the government hopes to develop. Traditional crafts will also be protected. Artisans, such as the *khanjar* (traditional dagger) makers, are paid monthly subsidies by the government to encourage them to continue their work.

The ministry also seeks to inculcate in the Omani citizen a sense of Oman's unique historical legacy. The idea of a nation with fixed boundaries is a relatively late concept in the Arabian Peninsula, occurring only in the middle of the twentieth century or, in the case of Oman, only since 1970. It is vital for Oman's internal security that Omanis learn to identify themselves with the state as opposed to locality or tribe.

*Table 35. Oman, Council of Ministers, December 1976**

Minister	Ministry
Hamad bin Hamad Al Bu Said	Palace Affairs
Muhammad bin Ahmad Al Bu Said	Interior
Fahar bin Taimur Al Bu Said	Youth Affairs; Deputy Minister of Defense
Faisal bin Ali Al Bu Said	National Heritage
Fahad Mahmud Al Bu Said	Information and Culture
Asim al Jamali	Land Affairs and Municipality
Ahmad Abd Allah Al Ghazali	Education
Abd al Hafiz Salim Rajab	Communications
Mubarak Khaduri	Health
Muhammad al Zubayr	Industry and Commerce
Hilal bin Hamad Al Sammar	Justice
Khalfan bin Nasir al Wahaybi	Labor and Social Affairs
Karim Ahmad al Harami	Public Works
Walid bin Zahir al Hinai	Awqaf and Islamic Affairs
Said Ahmad al Shanfari	Agriculture, Fisheries, Petroleum, and Minerals
Qais Abd al Munim Zawawi	Minister of State for Foreign Affairs
Burayk bin Hamud al Ghaffari	Minister without Portfolio; Governor of Dhofar Province

*Sultan Qaboos bin Said bin Taimur Al Bu Said serves as prime minister, defense minister, and foreign minister.

Regional and Local Government

In late 1976 Oman was organized into thirty-seven divisions called *wilayats,* one province, and the Municipality of the Capital (see table 36). The capital area—Muscat, Matrah, Ruwi, Bashur, and Sib—was administered by a powerful wali who was also a personal adviser to the sultan. Dhofar Province was administered by a governor who was also a minister without portfolio. The municipal area and Dhofar Province were the two most important political areas of the sultanate and consequently received the most attention. The *wilayat* is a system borrowed

Table 36. Oman, Regional Administration by Geographic Location, 1976

Wilayat*	Location	Wilayat*	Location
Barka		Bahla	
Al Masanaa		Nizwa	
Suwaiq		Birkat al Mawz	The Interior
Al Khabura	Batinah coast	Manah	
Saham		Adam	
Sohar			
Liwa		Al Mudaibi	
Shinas		Ibra	
		Wadi Bani Khalid	Sharqiyya
Khasab		Al Mudhayrib	
Dibbah	Musandam Peninsula	Biddiya	
Bukha			
		Bilad Bani Bu Hasan	
Buraymi		Kamil and Wafi	
Mahadhah	Jau and Buraymi	Wadi Dimma	Jalan and Sur
		Jalan and Bilad Bani Bu Ali	
Dank		Sur	
Ibri	Dhahirah		
Nakhl and Wadi		Quriyat	
Maawal		Bid Bid	Eastern Hajar
Rustaq	Western Hajar	Sumail	
Awabi		Izki	

*A *wilayat* is a regional and administrative division.

from the Ottoman Turks that has prevailed in Oman for many centuries. The sultan appoints the wali (governor of a *wilayat*), whose responsibilities include representing the national government, implementing sultanic and ministerial directives, settling disputes, and administering justice. Each wali is assisted by an Islamic judge, a qadi (see Justice and Legislation, this ch.).

Dhofar Province has a municipal council of eight elected members who meet at the provincial capital of Salalah. Nizwa, Sur, Sohar, Sumail, Rustaq, Ibri, Buraymi, and Masirah have acquired municipality status, although they still come under the *wilayat* system.

The wali of an area works closely with the *tamimahs,* the paramount shaykhs of his area. The *tamimahs* are the walis' chief avenue of communication with the tribes, many of which were widely scattered over inhospitable terrain. Many tribes continued to function as autonomous units when a matter purely of internal tribal interest occurred. The tribes, however, were slowly being drawn into the network of the state. The walis were sensitive to tribal concerns and interfered in tribal affairs only when necessary. In turn the tribes recognized the wali as their political superior. A wali's suzerainty over the tribes was reinforced chiefly because of his sole ability (acting with a qadi) to settle intratribal disputes and because he lobbied for the needs of his people with the central government. Because the tribesmen were finally experiencing beneficial effects from the state, the old animosities toward Muscat appeared to be softening.

Justice and Legislation

Oman's legal system is based on the Ibadi interpretation of sharia, Islamic religious law, which contains very precise dicta for virtually every aspect of a Muslim's personal and private life. Sharia varies from school to school, but the Ibadi interpretation is similar to that of the four orthodox schools. Ibadite sharia primarily stresses the Quran and Sunna as the sources of law (see ch. 3). The Sunna is the collection of the traditions of the early caliphs' and the Prophet Muhammad's rulings. Jurisdiction is administered locally by the wali in conjunction with the qadi, a judge who has attained that position either by graduating from an Islamic law college or by advanced study with local religious legal experts. The majority of all cases are decided locally. Although primarily guided by sharia, jurisdiction at the local level aims at arriving fairly at a decision or compromise that is acceptable to all parties. Invariably tribal law has become mixed with religious law at that level. Because of the nearly universal illiteracy, witnesses are given predominance over written evidence in both fact-finding and judgment.

In cases of such major criminal offenses as murder, which is extremely rare in Oman, or of appeal by one party, the verdict is given in an appeals court in Muscat presided over by four qadis of equal rank. In extraordinary cases a direct appeal may be made to the sultan. The minister of justice concerns himself with the administration of the total system.

Legislation other than sharia is in the form of sultanic decrees. Occasionally, though not necessarily, the proclamation may be preceded by consultations with the Council of Ministers or specific ministers whose areas of jurisdiction are involved. These decrees are perceived as emanating entirely from Islamic law but accommodate modernizing trends in areas not covered by Islamic law. The decrees, many of which are rules and regulations as well as general directives, cover many areas—matters of public interest as well as private moral concerns. In 1974 and 1975 decrees were issued giving precise guidance for persons, commercial firms, and government officials. Included were a foreign trade and investment law, a commercial code, and a decree establishing an autonomous currency board.

Special judges, although bound to the fundamentals of Islamic jurisdiction, will preside at new courts to be established in 1977. Such matters as traffic laws, rules covering the exploitation of natural resources, banking laws, and labor regulations will then receive attention.

Succession, Royal Family, and Tribes

The sultan has no brothers and as of late 1976 no successor had been designated. The succession issue was a matter of considerable discussion in the sultanate until March 1976, when Sultan Qaboos married his first cousin, Kamilla. Kamilla was twenty years old at the time of her marriage and had been educated in Geneva. The marriage is a politically astute one, primarily because Tariq, Kamilla's father, is second only to the sultan in the respect he commands among the people and because, in the absence of a son by Qaboos, Tariq would be the most likely candidate for the throne. If Kamilla and Qaboos produce a son, any reservations some of the population may have about the absolute authority possessed by the sultan may diminish. Since Kamilla is the daughter of the sultan's father's brother, she is Qaboos' *bint amm* (daughter of father's brother). Such a marriage is the most highly regarded contract in the Middle East and therefore has great appeal to the traditionalists. Because Kamilla is Western educated, the progressives are also pleased.

Qaboos' father had arranged a marriage for him in April 1970, three months before the coup d'etat. The prospective bride was the uneducated daughter of Shaykh Ahmad bin Muhammad, *tamimah* of the Al Hirth, one of the most powerful tribes in the country. Qaboos' father hoped to buttress his political position by forming an alliance with a tribe that could call up large numbers of military reserves. Although tribal support remained necessary in 1976 because virtually the entire population continued to be tribally oriented, the sophisticated military machine developed by the sultan made such marriage alliances less necessary. Further, the sultan does not appear to wish to recognize one tribe over another or in fact even to suggest that the tribes have some de facto power that must be recognized.

This principle operates within the royal family, the clan of the Al Bu Said. The sultan's relatives do not have the political clout enjoyed by many of the royalty of the Arabian Peninsula. The sultan's two chief advisers were relatives; in addition to them there were five other relatives in cabinet positions. Members of the royal family are frequently given in the press with the title Sayyid before their names. This is meant as a token of respect and does not have the usual connotation of the title—a descendant of the family of the Prophet Muhammad. In 1976 there were approximately ten walis who were related, some quite distantly, to the sultan.

The royal relatives who were advisers or in the cabinet were described by foreign observers as men of proven ability, and those who were walis or who were employed in lesser positions in the civil service held these positions not as a mark of the sultan's acquiescence in the nepotism typical of traditional peninsular societies but as a mark of the sultan's patronage. It was understood that they were only his representatives. One Omani official observed, "For us there is only one member of the royal family and that is the Sultan."

The tribe was the basic social and political unit of traditional Omani society. Tribal power has always been localized. When the tribes united for action in a common interest, however, they could at the very least exclude a sultan from their areas of sovereignty. In the mid-1970s the tribes still possessed the potential to do that if they were to act in concert; lacking unity, they could be highly disruptive, as evidenced by the rebellion in Dhofar (see The Military and Foreign Policy Aspects of the Dhofar Insurgency, this ch.). However, with the advent of an armed force equipped with sophisticated weapons, an enormous number of tribes would have to coalesce to make any significant impact.

The power of the tribes to function as pressure groups within their community remained strong in 1976, although that function was increasingly diluted by the inevitable extension of the political machinery of the central government. Every year large numbers of tribesmen leave to seek the greater opportunities to be found in the municipality of Muscat or in other large towns; this also decreases tribal authority. Generally the more remote the location, the more tribal leaders retain their traditional authority.

John Duke Anthony sought to identify the major tribes by location and political importance in 1975. Clearly the Al Bu Said, the sultan's tribe, was the most important, although at least a dozen tribes were greater numerically. In the interior the Bani Ghafir, Bani Amr, and Hawasin were the prominent tribes. Their numbers were not great, but their influence stemmed from their proven reputation as militarily powerful tribes and from their staunch loyalty to the sultans. They were frequently used as *askars* (security guards) for the ruling family and for the sultan's representatives. The Dura was the preeminent tribe of the Dhahirah, the region where most of the country's oil wells were located. They were long-term supporters of the sultan in Muscat and were very cooperative with oil industry developers.

The relationship between the central government and the Al Hirth, the major tribe of the huge Sharqiyya territory, has been an unhappy one. The Al Hirth shaykh had traditionally enjoyed virtual political autonomy in Sharqiyya with the blessing of the sultan in exchange for military assistance when necessary.

The Al Hirth supported the followers of the imam in the 1950s and 1960s, however, a move that undoubtedly prompted Sultan Said to propose an alliance by marriage to prevent such allegiances in the future. The Al Hirth were reportedly displeased that their shaykh's daughter was not accepted by Sultan Qaboos, a matter that amounts to a violation of contract. The geographic proximity of this tribe to the Muscat municipality and its lack of sympathy with the office of sultan made it clear that its power had to be broken.

Other tribal groups that have less than harmonious relations with the central government are the Shihuh and the Habus of the Musandam Peninsula (Ras Musandam). These tribes have close ties with the amirates of Ras al Khaymah and Fujayrah of the United Arab Emirates (UAE). In 1975 some of the Shihuh and Habus accepted an offer of citizenship from Ras al Khaymah. Despite the sultan's efforts to develop the long-neglected Musandam area, he can not match the kinds of goods and services available to the citizens of the UAE. The acceptance of citizenship was particularly unnerving to the central government because supplies and recruits for the Dhofar insurgency were known to have come from the Musandam area.

The three most powerful tribes of Dhofar Province were those involved in the rebellion. They present special difficulties to the central government. The Mahrah and the Al Kathir share the cultural and social orientation of the tribes of the Hadramaut in Yemen (Aden). Until the rebellion they had very little interaction or involvement with the rest of Omani society. The Qara, who have distinguished themselves by their indefatigable warriors, are not of Arab origin. Their language differs from that spoken in Oman and, despite the fact that the sultan's mother is from that tribe, it will undoubtedly be a long and arduous process to integrate them into Omani society.

The Merchant Families

Before the reign of Said the flow of foreign merchants into the country from surrounding areas was considerable. Originally the merchants were the local representatives of trading houses in their own countries. Because of the economically advantageous position of the Gulf of Oman, many of the merchants settled in the Muscat area and began new mercantile ventures. During Said's reign these merchant families acquired considerable social and political power. They were Said's link with the outside world, ensuring the supply of items necessary to the marginally subsistent economy. There was an unspoken agreement between Said and the merchants: the merchants would not import items for the general market that smacked of modernization or westernization—such as eyeglasses, books, medicines, or radio transmitters—and in return the merchants had carte blanche to expand their own commercial houses and to engage in any trade that would not have undesirable repercussions for Omani society. They were also permitted to practice their religions unmolested, and the Indians and Pakistanis were even permitted to have a private school for the education of their daughters. None of the prominent merchant families were Ibadi Muslims in 1970.

Since the coup small numbers of Ibadi Muslims who were educated abroad have established modest import companies and in the mid-1970s were becoming involved in internal commercial activity. However, the commercial houses that had consolidated their power during Said's reign continued to amass great fortunes, largely through monopolistic or quasi-monopolistic franchises. None were directly involved in the oil business, but together they were the major suppliers of goods to the government, local contractors, expatriate firms, local consumers, and the oil industry. Most of the valuable distributorships for consumer and capital services and goods were under their aegis. Close cooperation between the merchants and Said has evolved into a mutually protective relationship with civil servants in Qaboos' government. Because their trading houses had established political ties, they were able to use their interest with the government, particularly in development schemes. And, because they had developed accepted spheres of influence among themselves, it was difficult for outsiders to establish competitive markets. In late 1976 the most important trading houses were the Khimji Ramdases, who are Hindu; the Bahwans and the Zubayrs, who are Sunni Muslims; and the Ali Sultans, the Zawawis, the Gurdanases, the Abd al Rahmans, the Al Asfoors, the Darwishes, and the Abd al Rahims, who are all Shia Muslims of either Arab or Persian ancestry.

Women and Youth in the Politics of Development

In late 1976 Oman appeared to have the greatest success among the states of the Arabian Peninsula in initiating women into the developmental processes of an emerging nation. Nevertheless the full participation of women in political, economic, and social affairs will undoubtedly be a lengthy process. Widespread illiteracy and the traditional perception held by both men and women of the proper sphere of activity for the woman, which was necessarily a limited one, hampered governmental attempts to integrate women into the full life of the society. However, the will of the sultan was clear, and efforts by various governmental ministries had yielded significant and positive results. Women are a potentially important labor resource and, because Oman has only a small income from oil revenues and few natural resources, Qaboos deems it necessary that women enter the labor market as quickly as possible. More important, it is vital to demonstrate that a traditional absolute monarchy need not have the anachronistic social and economic trappings generally associated with it.

Women were particularly prominent in communications; for a time all radio and television announcers were women by command of the sultan, and in late 1976 they still dominated the field. Women were also prominent in banking, foreign service, and internal civil service. In 1976 the director general of education in the Ministry of Education was a woman. There were women in the Omani police force and a few commissioned women officers in the armed forces. Two women from the Ministry of Foreign Affairs and one from the Ministry of Education were Omani delegates at the World Conference on International Women's Year held in Mexico City in 1975.

Like many Omani civil servants in responsible positions and in the foreign ministry, most of the women conspicuous in public service were former expatriates, raised and educated outside the country. They served as important role models for the girls and women of the country and demonstrated to the male population that women in the work force need not behave in ways antithetical to Islamic mores.

Before Ahmad Abd Allah Al Ghazali was appointed minister of education, he initiated a series of informal programs in Sur—many of which have spread to other areas—that have directly or indirectly advanced the position of women. In the villages he encouraged the people to form committees to raise funds and collect equipment for evening classes for women. Women learn basic literacy skills, first aid, nutrition, and hygiene. Many husbands initially opposed these centers but, as the rise in the standard of living in the families where the women attended classes became obvious, more and more women were encouraged to attend.

One of Ahmad Abd Allah Al Ghazali's more startling innovations was the organization of public meetings in mosques to discuss development and modernization in the light of Islamic teachings. Women did not attend the meetings, but men were encouraged to ask questions from the (usually) four religious leaders who presided over the discussions. Questions and answers were published weekly by the education department. Two British filmmakers were present at one session when the qadis were asked to define the attitude of Islam toward the education of women. Many Quranic injunctions and hadith were brought forth stressing the Muslim's religious obligation to educate his daughters and citing the punishments that would accrue if the parents did not fulfill God's will in this matter. Such meetings were believed to be more productive in changing attitudes than any civil injunction emanating from the central government. Omanis are a deeply religious people. Most Omanis, however, naturally confuse ingrained cultural mores with Islamic law. When a governmental program is reinforced by religious authority, any reservations about the propriety of a program diminish considerably.

Omani youths—who in the 1970s were the first generation to have the opportunity of a secular education—were perceived by the government as the country's most important resource. In 1974 Omanization, as the government calls it, was begun, whereby Omani citizens were given special consideration for appointment to government positions. Moreover the age range of candidates for various jobs was lowered to accommodate students and school dropouts. The Ministry of Information and Culture noted that "a country in a hurry cannot afford to wait too long for practical skills to be polished or academic excellence to remain theoretical." In addition to a wide range of employment opportunities, Omani youths were encouraged to identify with the goals of the state through a program of community affairs and social activities that stressed shared responsibility and decisionmaking.

The Department of Youth Affairs within the Ministry of Labor and Social Affairs has established a network of clubs and social institutions open to both girls and boys. In 1976 forty such clubs existed in Oman: ten in the capital area, nine along the Batinah coast, eight in Sharqiyya, six in the interior, four in Salalah, and three in the Musandam Peninsula. Each club was formed around a team sport, and soccer was the most popular. The government provided books and sporting equipment and paid for the club's operating expenses, including a coach. Organization and administration of the club were left to its members.

Organized sports like many other things in Oman were new, but by the mid-1970s teams had already participated in international tournaments. Hockey and soccer teams were sent to the Arab Games in Kuwait in 1974, and Oman applied for inclusion in the Asian Games to be held in Pakistan in 1978. Soccer, hockey, and volleyball teams were popular among the boys, while the chief sporting activities for girls were table tennis, volleyball, and basketball; a broader training program was planned for both.

Oman joined the scouting movement in 1972, and by the end of 1974 there were 3,000 boy scouts, 200 cubs, 128 girl guides, and 504 brownies. In 1974 the government allocated OR6,000 (for value of the Omani rial—see Glossary) for scouting activities, and the government sponsored attendance by boy scouts at jamborees in other Arab countries. Scouting activities in Oman are similar to those in the United States with perhaps a greater emphasis on community affairs. Omani scouts have frequent one-day outings at which they repair and paint schools, clean streets, and engage in other community action projects.

The goal of the government was to generate a sense of national unity and of loyalty to the sultan as the personification of the state. Although the number of youths involved in the various programs was only a small fraction of all Omani youngsters, the government viewed the programs as an important part of the nation-building and detribalizing efforts.

Mass Communications

In 1976 the Oman Broadcasting Station, managed by the Ministry of Information and Culture, had two mediumwave transmitters, each of

100 kilowatts. Radio Oman, broadcast from Muscat, had seven hours daily of programs in Arabic; Radio Salalah had programs in both Arabic and Dhofari dialects. The British Broadcasting Corporation (BBC) had built a mediumwave relay station on Masirah Island that broadcast in Arabic, Farsi, and Urdu. The BBC also had two low-powered twenty-kilowatt transmitters; one provided programs in English for the British servicemen stationed on Masirah until March 1977 and the other, in Muscat, for the English-speaking expatriate population. The English-language channels operated twelve hours daily, and there were plans to increase the service to a 100-kilowatt transmitter that could carry programs to other parts of the Persian Gulf, the west coast of India, North Africa, and southern Iran.

There were two color television stations, one opened in Muscat in 1974 and one in Salalah in 1975. In view of the almost universal illiteracy of the adult Omani population, radio and television programs performed a critical service in educating and generally informing the people of governmental objectives.

Radios and color television sets have been installed in many community centers in Dhofar Province. Government policy statements and addresses by the sultan were prefaced by entertaining programs, frequently featuring animal and circus acts and animated cartoons that were great favorites.

There were two weekly Arabic newspapers published by the Ministry of Education—*Al Watan* and *Oman.* Two English-language weeklies, *Gulf Mirror* and *Times of Oman,* were published in Muscat. The content and the advertisements suggest that they were aimed as much for the indigenous population as for the foreigner. There were seven periodicals, all lavishly illustrated, that with one exception appeared to be primarily didactic. The one exception, *Tijarat Oman* (Commerce of Oman), was a quarterly trade magazine. The others were *Al Aquida,* published weekly; *Al Muzari,* published monthly by the Ministry of Agriculture, Fisheries, Petroleum, and Minerals; *Nahwa Tarbiya Afdal,* a monthly; *Jund Oman,* published monthly by the Ministry of Defense; *Al Usra,* a weekly; and *Al Nahda,* published fortnightly.

GEOGRAPHIC AND DEMOGRAPHIC SETTING

Geography and Population

Because the country's boundaries remain unfixed, its exact size was not known but was estimated to be between 82,000 and 100,000 square miles, roughly the size of the state of Kansas. Geography has made the sultanate a virtual island, bordered mostly by the sea and the Rub al Khali (Empty Quarter) (see fig. 22). Historically the country's contacts with the rest of the world have been largely by way of the sea, which not only provided access to foreign countries but also linked the coastal towns of Oman. The Rub al Khali, difficult to cross until the era of modern desert transport, has served as a barrier between the sultanate and the Arabian interior. Insularity has been reinforced by the formidable Hajar Mountains, which form a belt between the coast and the desert from the Musandam Peninsula (Ras Musandam) to the city of Sur. The desert and the mountains have combined to keep Oman remarkably free of foreign encroachments—either military or cultural—from the interior of the peninsula. In 1976 the international boundaries remained undefined, but Oman had no serious boundary disputes.

The country is divided by natural features into several distinct areas: the Musandam Peninsula; the Batinah plain; the Muscat-Matrah coastal area; the Oman interior, comprising the Jabal Akhdar (Green Mountain), its western foothills, and the desert fringes; Dhofar Province in the south; and the offshore island of Masirah.

The northernmost area, Ras al Jabal, extends from the Musandam Peninsula to Dibbah. It touches the Strait of Hormuz, which links the Persian Gulf with the Gulf of Oman, and is separated from the rest of the sultanate by a belt of territory belonging to several states of the UAE. This area consists entirely of low mountains forming the northernmost extremity of the Western Hajar. Two inlets, Elphinstone and Malcolm, cleave the coastline about one-third of the distance from the Strait of Hormuz and at one point are separated by only a few hundred yards of land. The coastline is extremely rugged, and the Elphinstone inlet, ten miles long and surrounded by cliffs 3,000 to 4,000 feet high, has frequently been compared with fjords in Norway.

The UAE territory separating Ras al Jabal from the rest of Oman extends almost as far south as the coastal town of Shinas. From the point at which the sultanate is entered again to the town of Sib, 150 miles to the southeast, runs a narrow well-populated coastal plain known as Batinah. Across it a number of wadis descend from the Western Hajar to the south, which are heavily populated in their upper courses. A ribbon of oases, watered by wells and underground channels, extends the length of the plain about one mile inland from the coast. Dates, limes, mangoes, and other fruits are grown in irrigated coastal gardens, and along their edges small quantities of cereal grains are produced. On the landward side of the Batinah plain there are acacia trees, beyond which a barren, pebbly plain slopes gradually to the foothills of the Western Hajar about ten miles away. The gray partridge is common in coastal gardens; in the colder season flocks of sand grouse appear in adjoining plains, and ducks are seen in pools and creeks where the wadis run to the sea.

South of Sib the coast changes character. For about 110 miles from Sib to Ras al Hadd, it is barren and bounded almost its entire length by cliffs; there is no cultivation and very little habitation. Although the deep water off this coast renders navigation relatively easy, there are few natural harbors or safe anchorages. The two best are at Muscat and Matrah, where natural harbors gave rise to the growth of cities centuries ago.

West of the coastal areas lies the tableland of central Oman. The Hajar Mountains form two ranges: the Western Hajar (Hajar al Gharbi) and the Eastern Hajar (Hajar al Sharqi). They are divided by the Wadi Samail, a valley that forms the traditional route between Muscat and the interior. The general elevation is about 4,000 feet, but the peaks of a high ridge known as the Jabal Akhdar—which is considered a separate area but is actually part of the Western Hajar—rise to 10,000 feet above sea level in some places. The Jabal Akhdar is the only home for a unique species of wild goat, the Arabian tahr. In hopes of saving this rare animal, Sultan Qaboos has declared part of the Jabal Akhdar a national park.

Behind the Western Hajar are two inland regions, Dhahirah and inner Oman, separated by the lateral range of Jabal al Khawr. Both regions become stony desert before meeting the Rub al Khali. Inland of the Eastern Hajar are the sandy regions of Sharqiyya and Jalan, which also border the desert.

The desolate coastal tract from Jalan to Ras Naw has no name. Low hills and wastelands meet the sea for miles on end. Halfway along this coast and about ten miles offshore is the barren island of Masirah. It is forty miles long, virtually uninhabited, and distinguished only by a small air base (see National Defense and Internal Security, this ch.).

Dhofar Province extends from Ras Sharbithat on the coast to the border of Yemen (Aden). Its exact northern limit has never been defined, but the territory claimed by the sultan includes the Wadi Muqshin, about 150 miles inland. The southwestern portion of Dhofar's coastal plain is regarded as one of the most beautiful in Arabia, and its capital, Salalah, was the permanent residence of the former sultan and the birthplace of Sultan Qaboos. The southwestern coastal plain contains splendid vegetation and birdlife, but only ten miles inland the rugged foothills of the Jabal al Qara begin. The highest peaks are between 3,000 and 4,000 feet. They gradually slope down to a narrow pebbly desert adjoining the Rub al Khali to the north.

With the exception of Dhofar Province, which has a monsoon climate and receives cool winds from the Indian Ocean, Oman's climate is extremely hot and dry for most of the year. Summer begins in mid-April and lasts until October. The highest temperatures are registered in the interior, where readings of 130°F in the shade are common. On the Batinah plain summer temperatures seldom exceed 115°F, but because of the low elevation the plain experiences up to 90-percent humidity. The mean summer temperature in Muscat is 92°F, but the *gharbi* (literally, western), a strong wind that blows from the Rub al Khali, can raise temperatures for the towns of the Gulf of Oman by 20°F. Winter is mild and pleasant with mean temperatures ranging between 60°F and

73°F depending on location. Muscat also experiences oppressive heat on summer nights, when the black mountains that surround the town radiate the heat they have absorbed during the day.

Precipitation on the coasts and on the interior plains ranges from one to five inches a year and falls during the middle and late winter. Rainfall in the mountains, particularly over the Jabal Akhdar, is much higher and may reach fifteen inches. Because the plateau of the Jabal Akhdar is porous limestone, rainfall quickly seeps through it, and the vegetation, which might be expected to be more lush, is extremely meager. However, a huge reservoir under the plateau provides springs for low-lying areas. Additionally an enormous wadi channels water to these valleys, making the area agriculturally productive in years of good rainfall. Dhofar, benefiting from a southwest monsoon between June and September, receives between twenty-five and thirty inches of rainfall annually and has constantly running streams, which make the province Oman's most fertile area.

As of late 1976 no census had ever been taken in Oman. Estimates of the 1976 population varied between 750,000 and 1.5 million. For planning purposes the government used the 1.5-million figure. The annual rate of growth was estimated at 3.1 percent by the International Bank for Reconstruction and Development (IBRD, also known as the World Bank). Density of population was presumed to be approximately eight per square mile, but in the interior it was probably less than one per square mile. About one-third of the population lived in the Batinah and in coastal cities, the Batinah alone having 150,000 people. The remote Musandam Peninsula had approximately 10,000 people. The interior was the residence for about one-half the total population, one-sixth of whom were presumed to be beduin. The population was predominantly Arab, but in the towns designated by the government as the capital area—Muscat, Matrah, Sib, and Bashur—there were many Indians, Pakistanis, Baluchis, and East Africans. Muscat and Matrah had a combined population of approximately 25,000 and Dhofar Province, 50,000.

Living Conditions

Food

The foods consumed are typical in both kind and preparation of those common to the Arabian Peninsula. The basic dichotomy between the food intake of beduin and of settled agriculturists and urban dwellers also obtains: the beduin have a higher protein intake but a much lower total caloric intake. Rice, vegetables, fruit, and bread form the bulk of the diet for city dwellers. Meat is a luxury, consumed regularly only by the rich. For the majority of the population meat is reserved for festive occasions, such as honoring a guest or celebrating a religious festival, a wedding, or a son's circumcision. Typically food is eaten with the right hand from shared dishes, a practice that underlines the communal aspect of the traditional life.

A typical main dish is an enormous tray of rice cooked with raisins, seasoned with a hot sauce, and sprinkled with chopped tomatoes, onions, and lime juice. Flat bread is liberally consumed. If it is a special occasion, mutton, roast goat, chicken, or fish—in order of descending cost—will be under the rice and will be ripped or hacked into serving pieces by the host. By the mid-1970s canned foods were available and slowly finding favor with those who could afford them. The main meal is eaten shortly after noon. Breakfast consists of bread, cheese or yogurt, and sweetened tea. Supper, late in the evening, is a light meal. Thick, bitter coffee and sweetened tea are drunk throughout the day and are always offered even to the most casual guest. Nutritional surveys were not available in 1976, but it appeared that the most balanced diet was enjoyed by coastal dwellers, who could buy a variety of fish cheaply.

Housing

Housing is basically of three kinds, two of them traditional. In Muscat and other large towns houses are thick walled and built of stone and mud. The houses, which may be three stories high, have deep windows, frequently with bars but without glass. The deep-set windows provide air circulation without admitting the sun, and the mud in the walls permits the rooms to "breathe," so that homes are usually fairly comfortable despite the ferocious heat. The flat roofs are sometimes used as sleeping places in the summer. Less elaborate establishments have the thick stone and mud walls but not the masonry roof. These one-storied houses have roofs of plaited leaves supported by wooden beams that are carved with geometrical patterns or calligraphy and painted in bright colors.

The most common dwellings are the *barastis,* which are particularly common in Dhofar and on the Batinah coast. A *barasti* is built like a stockade. Corner posts are wooden, and walls are woven palm fronds. Inside the walls is the central courtyard, where most family activities take place. It may have a few simple pieces of furniture and always contains the kitchen equipment. A few rooms used mainly for sleeping and also built with palm fronds form the center.

Modern concrete buildings, sometimes making up a whole housing project, were increasingly common in the mid-1970s. The disadvantages of concrete in a climate like Oman's has quickly become apparent, and the government has suggested that new housing designs be modeled along traditional lines and that local structural materials be used whenever possible.

Clothing

The traditional dress for Omani men and boys is an ankle-length white gown, the *dishdasha,* similar to but not identical with a caftan. Some men wear a long, front-pleated skirt with an overshirt. A small embroidered skullcap is worn by boys. Men wear either this cap or a loosely wrapped turban. At maturity each Omani boy receives a *khanjar,* a large curved dagger in an ornamental sheath that is worn in the belt regardless of occupation. Men frequently carry an old and usually unworkable rifle or a camel stick. Blue jeans, slacks, and Western shirts are seen but much less frequently than traditional clothing.

Women wear loose-fitting, ankle-length cotton dresses with long sleeves and a long head veil, which is usually loosely draped over the head and shoulders. The head veil can hang down the back or be drawn around the body. Both dress and head veil are of brightly colored printed cloth, often imported from Indonesia or India. In towns along the Gulf of Oman the *abayah*—a black, voluminous, sheetlike cloak that envelops the head and body—is worn by women in public. Little girls wear sized-down versions of their mothers' clothes, usually without the head veil. Schoolgirls wear light cotton slacks with tunics. When girls reach adolescence, they may wear head scarfs. Some women wear semitransparent, masklike face veils, but it is not a universal practice. Married women may pull their head veils over their faces in the presence of strange men; girls almost never follow this practice, nor do they wear face veils. Schoolgirls and girl scouts appear in public ceremonies with neither face veil nor head covering, in conformance with a government policy of fostering an atmosphere more conducive to the full participation of women in the social and economic life of the country.

Health and Medical Facilities

In 1976 there were neither morbidity nor mortality statistics available for Oman, but experts presumed that the average life span was between forty and forty-five years and that the infant mortality rate was high. The disease characteristics of the country were incompletely understood. Typical diseases of the Arabian Peninsula, such as trachoma, enteric diseases, and helminthiasis, were prevalent, particularly in Dhofar Province, which has a monsoon climate. In addition poliomyelitis, tuberculosis, protein-calorie malnutrition, and diseases caused by vitamin deficiencies were common. There was also a small incidence of leprosy, cholera, and smallpox. In 1976 free medical facilities were provided for the population as quickly as the government could afford, staff, and build them, but the government's explanation of the high incidence of disease before 1970 remained applicable in 1976 for many areas, particularly in the hinterland: "diseases [are] caused mainly by primitive living conditions, non-existent sanitation and insufficient knowledge of personal hygiene."

Many of the most life-threatening or debilitating of these diseases are preventable by immunization. Others, such as certain of the enteric diseases and trachoma, could be eliminated by better sanitation facilities and improved knowledge of disease prevention. Much of the population is nomadic, however, and their widely scattered locations are difficult to reach because of the inhospitable terrain and the lack of roads into the interior. The nearly universal illiteracy, the almost total ignorance of the germ causation of diseases, and the application of folk remedies that may exacerbate health problems complicate the government's efforts to bring minimal health care to citizens.

In 1970 the country had had only three small clinics, but by 1976 medical facilities included five hospitals in the capital area, two of which were for referral cases; eleven health centers; and forty health dispensaries. Together they provided approximately 930 beds. Development plans included expansion of the two major referral hospitals, Al Khoulla and Al Nahdah, both located in the capital area, and the replacement of the eighty-bed district hospital at Salalah with a modern 250-bed building. When all renovations are completed, a total of 1,280 beds will be available. Al Khoulla, formerly the private hospital of Shell Oil, treats serious obstetrical and surgical cases. Al Nahdah, which has the best laboratory equipment in the country, handles complex medical and cardiac problems. There have been preliminary studies for a 600-bed hospital bearing the sultan's name in the capital area. This hospital would contain the country's only psychiatric unit and would house specialties not existing in the sultanate in 1976.

Although such services are seriously needed, the government appeared to be somewhat hesitant about committing OR12 million (for value of the Omani rial—see Glossary) to one institution when public health care, health maintenance, and preventive medicine clearly need to be developed further. Small units, such as dispensaries and health centers, appeared to be the best vehicles for treating the nonurban population; there were proposals to expand the existing number during the 1976–81 period. Fifteen new health centers were scheduled to be built, each containing twenty-four beds and staffed by two physicians and ten nurses. According to government plans the centers will have X-ray machines and laboratory facilities and will be equipped to handle obstetrical confinements and minor surgery. Forty new dispensaries, each administered by a health technician with a knowledge of pharmacy, will be scattered throughout the more remote areas. Mobile sea and land dispensaries will also be increased and will be a critical medical resource for the nomadic beduin.

The availability of competent medical staff remained a serious problem in a country with only a six-year history of primary schools. Of the 200 physicians working in Oman in 1976 only twenty were Omani, and they had received their entire education outside the country. Most physicians were Pakistani or Indian, who on the average earned twice what they would in their homelands. Approximately 125 other Indians and Pakistanis have set up private practices in Oman, but most of them were government sponsored. Most nurses and health technicians were also from India or Pakistan. Oman had an embryonic nursing school, but it was expected to be many years before it had a medical school.

The Public Health Department of the Ministry of Health, formerly perceived as less important than the primary-care facilities, received increasing attention after 1974. Because a majority of the country's fatalities, particularly those seen at hospitals, are caused by diseases that are to a great extent preventable, development of public health services will probably receive paramount attention during the late 1970s and early 1980s. In 1971 public health authorities were eventually able to control a cholera epidemic that had spread throughout the country. Since 1971 they have attempted to carry out mass immunizations for smallpox and other infectious diseases. Their area of responsibility was enormous and included all inoculations and vaccinations, disease control, maternal and child care, health education, school health programs, sanitation, food hygiene control, antimalaria campaigns, and nutrition. In 1976 eight public health compounds were scheduled for early operation, and the department planned to increase its numbers of midwives and midwives' assistants, then at sixty-four and twenty-four respectively.

Midwives were primarily from other Arab countries, but their assistants were recruited locally. Besides providing a much-needed service, they advised against certain harmful local postnatal practices and gave some indication of what hygienic delivery conditions entail. This is particularly important because it is highly unlikely that a woman will have professional assistance for every birth. It is hoped that a woman who has received such help will also be better able to advise her neighbors and, having been positively reinforced in the ways of modern medicine, more likely to seek and accept guidance for other health problems. The idea of an informed community sharing the responsibility with professionals for its standard of living is the keystone of many Omani development programs.

Education

By late 1976 free educational facilities had greatly increased in both quantity and quality since the impoverished period before the accession of Sultan Qaboos in 1970. In early 1970 there were only three primary schools in the sultanate—in Muscat, Matrah, and Salalah. These were reserved for approximately 900 boys handpicked from among many applicants by the sultan. Additionally, in Muscat there were a religious institute with an enrollment of fifty boys, three private schools for

Hyderabadis (Indians), and one American missionary school serving fifty foreign girls.

In mid-1976 there were 207 schools: 181 primary (four years), twenty-three intermediate (four years), and three secondary schools (four years). A total of 55,752 students were enrolled in the system. Of this number 54,457 (39,640 boys and 14,817 girls) were in primary school. At the intermediate level there were 925 boys and 170 girls, and at the secondary level there were 143 boys and 57 girls. Because of the virtual absence of education before 1970, enrollment in the secondary schools was proportionately the smallest category, having only 1 percent of the potential age-group enrolled. Enrollment is expected to expand significantly by 1979, when students who began primary school in 1971 reach secondary level. Matriculants will by no means, however, match the number enrolled in primary and intermediate grades.

In the mid-1970s an annual academic screening test was used because the number of teachers and facilities was inadequate to serve all Omani children. At the end of the 1975–76 academic year the screening process resulted in a 7-percent dropout rate in primary schools and a 25-percent dropout rate in secondary schools. In 1976 approximately 55 percent of Omani children appeared to be receiving education at some level.

In the 1975–76 academic year sixty-five schools were scheduled to be built, but this number was projected to taper off to about twenty-five a year. Most of the new schools were to be built in remote areas of the interior, which in 1976 had no educational facilities. Some of these new schools will be combined with health facilities. The minister of education announced that future development plans would concentrate on the quality of education rather than the number of new buildings.

By 1976 Oman had demonstrated a remarkable flexibility in tailoring programs to fit the needs of its young citizens. United Nations Children's Fund (UNICEF) and United Nations Educational, Scientific and Cultural Organization (UNESCO) experts suggested in 1972 and 1973 that the curriculum content be updated to provide more basic instruction in literacy and numerical skills and to downplay history and classical Islamic subjects. The suggestions were immediately implemented, and teaching methods were also changed. According to the government, instead of the rote method that prevails in most of the Arabian Peninsula, Omani students are encouraged to do research, write essays, and deliver ten-minute lectures.

Between 1971 and 1975 only 7.4 percent of the total government development expenditure was on health and education. Because the country's development needs were as diverse as they were enormous, in 1976 it seemed unlikely that there would be a significant increase in the allocation of funds for education. The government was therefore husbanding its funds and relying on the advice of international experts as to the most efficient and judicious use of its spending.

In 1976 about 85 percent of the 2,000 teachers were from Arab countries; Egyptians, Jordanians, and Sudanese filled most of the positions. To encourage foreign Arab teachers, the government paid them approximately OR150 per month, which was four times as much as the average Egyptian teacher would earn at home. Recruitment for the more remote areas was difficult; teachers willing to go to the almost completely undeveloped districts received increments of as much as OR50 per month. In the larger towns teachers had to work double shifts because of the insufficient number of school buildings.

In cooperation with international agencies, Omani officials planned to open the first of three teacher training institutes by the beginning of the 1976–77 academic year, and Oman hoped to have three such institutes fully operational by 1980. One would provide training for 200 men annually in Salalah, in troubled Dhofar Province; one for 200 women in Muscat; and one for 300 men, also in Muscat.

Projects for a national university had been delayed because of the urgent need to provide vocational training institutes. The only government vocational school functioning in 1976 was the Omani Vocational Training Center in Matrah. Formerly a trade school administered by Shell Oil, it became the responsibility of the government in 1970. Approximately forty young men, who were live-in students, studied mechanics, metalworking, welding, motor engineering, electrical installation, and carpentry in addition to Arabic, English, trade theory, and Islamic subjects. The Institute of Public Administration, which opened in 1973, offered a one-year program to train students as clerks and secretaries. English, Arabic, bookkeeping, and typing formed the curriculum. In 1976 five vocational schools were under construction or in the planning stages: a commercial school for girls in the Muscat area; a technical school in Muscat; and agricultural schools in Salalah, Sohar, and Nizwa.

Adult education programs were also in the planning stages in 1976. With assistance from the World Bank and United Nations Development Program (UNDP), two adult education centers at Muscat and Sohar offering vocational and academic subjects will be the starting point for adult education throughout the country. No facilities existed for teaching the handicapped, but the government requested that parents or guardians of blind or deaf children register with the Ministry of Education so that such children are eligible for enrollment and scholarships at the school for the blind in Bahrain and similar institutions in Saudi Arabia.

Omani students who wished to pursue higher education did so outside the country. In late 1976 statistics on the number of students studying abroad were not available, but the Ministry of Education noted that a large number of young Omani men were studying in the United States, Europe, and Arab countries, particularly Jordan, Syria, Egypt, and Iraq. Most of the students in Europe and the United States were reported to be studying engineering or medicine. Qatar may finance some higher education facilities in Oman. The government of Qatar has indicated a desire to build an agricultural college that would function as the Agriculture Faculty of Qatar's own planned university.

THE ECONOMY

Before 1967 the stagnant economy was based almost entirely on subsistence agriculture and fisheries. Per capita income was probably the equivalent of less than US$100. There were only a few literate people in a population that may have totaled 600,000.

Oil exports began in 1967, resulting in a manyfold increase in foreign exchange earnings and government revenues (see ch. 5). But Sultan Said bin Taimur Al Bu Said opposed change, and little of the new wealth was spent on development. When Sultan Qaboos bin Said bin Taimur Al Bu Said seized power from his father in 1970, he immediately pressed forward an economic development and modernization program. Priority was on building up the country's almost nonexistent infrastructure. Despite a high level of military expenditures, largely because of rebellion in Dhofar Province, investments in development projects in the 1971–75 period totaled OR383 million (for value of the Omani rial—see Glossary). Nearly 20 percent was spent for roads, 16 percent for utilities, 11 percent for communications, 9 percent for harbors and airports, 7 percent for buildings, and 13 percent for development projects in Dhofar. In addition 5 percent was spent on health projects, 2 percent on agriculture, and 2 percent on education; miscellaneous projects absorbed the remaining 15 percent.

Much of the development effort has been concentrated in the capital area, consisting of Muscat, Matrah, and the Ruwi Valley (see fig. 22). The population of this area grew from 25,000 in 1971 to 80,000 in 1976. The province of Dhofar, long the most backward and isolated region of a backward and isolated country, has also received special attention for development (see ch. 4).

By 1976 the country had 285 miles of asphalt road, compared with only six miles in 1970. Another 3,700 miles had been graded. A deepwater port, Mina Qaboos, had been completed at Matrah, a harbor for small trading and fishing vessels at Raysut in Dhofar, and an international airport at Sib. Numerous government office buildings had been erected. The electric-generating capacity had increased manyfold. A desalination and electric-generating complex was nearing completion near Muscat. Other major projects under way included improvement and expansion of the water and sewerage systems in the capital area, the installation of a modern dial telephone system, the construction of a new town of 1,000 housing units near Muscat, and the building of several modern hotels. From three schools with 900 students in 1970, the school system has expanded to 207 schools with 55,752 students.

In 1974 the Development Council was established by the government and given the responsibility for approving a national development plan for the 1976–80 period, defining overall strategy, and setting priorities. In late 1976 the Development Council was composed of the governor of Dhofar and the ministers of foreign affairs; communications; interior; health; industry and commerce; agriculture, fisheries, petroleum, and minerals; labor and social affairs; and land affairs and municipality. Technical assistance to help establish a planning system was provided by the International Bank for Reconstruction and Development (IBRD, also known as the World Bank).

In contrast to the emphasis on infrastructure during the first half of the decade, priority for the 1976–80 period was given to income-generating projects. By mid-1976 the five-year plans of some of the ministries had been approved. Each ministry was responsible for development projects within its sphere.

Oman has sought technical guidance from numerous foreign sources. Emphasis has been placed on resource surveys and industry feasibility studies. Most have been carried out by well-known international corporations.

For technical aid Oman has turned mainly to the United Nations Development Program (UNDP). There were thirty-nine UNDP experts in the country in mid-1976 working on such diverse assignments as agriculture, education, telecommunications, public administration, and postal services. Although the Omanis seemed well pleased with the program, there were no plans for any material expansion.

Although Oman's economy is in many respects market oriented because of the income it receives from oil, the government accounts for about two-thirds of national income and spending. The primary momentum for economic growth has therefore been provided by government development programs, and government ownership and control predominates in most large-scale enterprises, such as oil, utilities, and ports. But the government generally prefers joint ventures to wholly government-owned firms, and its policy is to leave the establishment and operation of small-scale businesses to private enterprise. Private merchants have begun to broaden their activities beyond traditional trading to real estate, repair facilities, and light assembly plants.

The government welcomes foreign investments, and participation by Omani citizens in all kinds of business is required by law. Omani participation in capital and profits may not be less than two-thirds for press and information companies; 51 percent for public utility, real estate, shipping, and airline companies; and 35 percent for other business endeavors. The sultan has the power to decree exemptions, and in the mid-1970s foreign firms operating in Oman under contract with the government were exempt.

The country's foreign exchange and trading systems favored the growth of private enterprise. Neither import nor export licenses were required, and there were no controls on the exchange of currencies. The gold content of the Omani rial had not been changed since it was instituted as the national currency in 1970. In mid-1976 its dollar equivalent was US$2.89524.

Although it made its first contribution in 1967, oil accounted for an estimated 68 percent of the nation's gross domestic product (GDP) in 1975; however, fewer than 4,000 were employed in this sector. Apart from oil, construction—at 10 percent—made the largest contribution to GDP in 1975. Trade accounted for 5 percent and transportation and communications 2 percent.

The role of government has expanded significantly. Its contribution jumped from 2 percent of the small 1967 GDP to 8 percent of the greatly multiplied 1975 GDP. Government employment shot up twentyfold in that decade.

Because of the acute shortage of trained Omanis, the country's economic progress has largely depended on expatriate managers and workers; in 1976 there were 36,000 expatriates in a nonrural labor force estimated at 56,000. But the government was pressing for gradual replacement of expatriates by Omani citizens, and by 1976 some trained Omanis had returned from abroad to take up the new employment opportunities that had developed (see ch. 4).

Because of long isolation from the rest of the world, Omanis felt no need to collect statistics. Therefore the quantitative estimates of the various sectors of the Omani economy must be used with caution; many are merely guesses. Although progress has been made since 1970, a really reliable system of national statistics had yet to be developed.

Effective January 1, 1976, Oman officially adopted the metric system. Before that date the government had purchased a large order of kilogram and gram weights from India and sold them to merchants and others at a fraction of their cost.

Agriculture

Farming is the main source of livelihood of an estimated three-fourths of the population. Before the opening of the oil fields in 1967, agriculture accounted for over half of GDP; farm products and fish were almost the only earners of foreign exchange. Although farm production changed very little, by 1976 agriculture's contribution to the oil-swollen GDP amounted to only 3 percent, and shipments of farm products accounted for about 0.2 percent of total exports.

Subsistence farming dominates Omani agriculture; practically no system exists for marketing surplus produce. Farming units are almost universally small; and most family holdings consist of two or more separate plots.

Farmland ownership and tenancy rights are based largely on tradition. Three main forms of land tenure prevail: short-term lease of land for the growing of annual crops, long-term lease for the planting of tree crops, and full ownership. In addition nomadic herdsmen have a vague but generally recognized right to move their herds over certain areas. Traditional tribal grazing ranges often cross the ill-defined national frontiers.

Only about 90,000 acres—little more than 0.1 percent of the land surface—was used for crops. Rainfall is scanty, and water for irrigation is essential for crop production. Therefore new sources of water must be developed before the farming area can be extended.

More than one-third of the land is under date cultivation, and the harvest—estimated at 55,000 tons in 1975—accounted for about one-fourth of the value of all crops harvested. Dates are an important item in the Omani diet and are traditionally a leading export item.

The second crop in area, alfalfa, which is cut and used as forage for livestock, accounts for nearly half the value of all crop production. Alfalfa production in 1975 was estimated at 140,000 tons. The crop is extremely important to the mixed crop-livestock systems that prevail in most areas.

About 10 percent of the cultivated area was devoted to limes, the principal non-oil export commodity. About the same acreage was in onions. Between 3 and 4 percent of the total cultivated area was in wheat—annual production was about 5,000 tons. Other crops included bananas, coconuts, grapes, mangoes, oranges, papayas, sorghums, millets, tobacco, okra, eggplant, and tomatoes.

Although estimates of Oman's livestock population vary, it appears that in 1976 there were approximately 150,000 goats, 60,000 cattle, 50,000 donkeys, 30,000 sheep, 5,000 camels, and 100,000 chickens. There were also a few horses.

Goats are raised throughout the country for milk and meat and sheep for meat and wool. The hides of both are valuable by-products. Donkeys are used primarily for transportation; camels are used for both transportation and plowing. In the latter function, however, they have been largely replaced by tractors provided by the government for a nominal rent.

Cattle are kept primarily for milk production. Dhofar Province is the cattle-raising area of Oman; elsewhere very few are seen.

There are four distinct agricultural regions. These are the interior oases, the valleys, and the high plateau of the eastern region; the Batinah coast, extending about 150 miles along the northern region on the gulf from Shiraz in the north near the border with the UAE southeast to Sib; Dhofar Province in the south, along the narrow coastal strip from near the border with Yemen (Aden) to Ras Nus and the mountains to the north; and the detached Musandam Peninsula.

Interior farming areas account for nearly three-fifths of the country's cultivated land. Rainfall, though greater than along the coast, is insufficient for the production of crops. Precipitation records are few, but most of the region probably receives about ten inches a year, with exceptional precipitation of fifteen to twenty inches on the higher elevations. Air streams from the Mediterranean that reach Oman in winter by way of the Persian Gulf, where some rejuvenation occurs, provide nearly all the rainfall.

Most of the water for irrigation is obtained by use of the millennia-old *falaj* system. This system consists of a vertical shaft dug from the surface to reach water in porous rock. From the bottom of this shaft a gently sloping tunnel is dug to tap the water and allow it to flow to a point on the surface at a lower level or into a cistern or underground pool from which it can be lifted by bucket or pump.

A *falaj* may be many miles in length, requiring numerous additional vertical shafts in order to provide fresh air to the tunneler and permit the removal of the excavated rock and soil. A *falaj* requires a tremendous expenditure of human labor for maintenance as well as for construction, and observers doubt that new ones will be built. The future tapping of underground water in the interior of Oman will probably use modern well-drilling machinery and motor pumps. Nevertheless existing *falaj* will undoubtedly continue to play an important role.

In the mid-1970s subsistence farming continued in the interior. At most locations the leading crop was dates. Alfalfa for forage was extremely important because practically every family kept a few goats for milk and meat for family consumption. Donkeys and sheep were also common; and wheat, millets, and sorghums were grown.

The cooler climate on the high plateau of the Jabal Akhdar allows production in the more favored spots of apricots, grapes, peaches, and walnuts. These products were transported by donkey down steep trails to the markets of the small towns at the foot of the mountain. In the mountain region and to the west a few beduin—estimated at about 25,000—wander with their herds of camels, goats, and sheep (see ch. 4).

The Batinah plain, accounting for nearly two-fifths of the land under cultivation, is the most concentrated farming area. The annual rainfall along the coast is only about three to four inches, but moisture falling on the mountains percolates through permeable strata to the coastal strip and provides a source of underground water only six to eight feet below the surface. Small diesel motors are used to pump water from these shallow wells for irrigation.

In the mid-1970s the water table along the coast was reportedly dropping and the salt content of the water in the wells increasing. The cause was the cultivation of land too close to the sea combined with the pumping of more well water than was being resupplied by nature, thus permitting seawater to seep in. In an effort to remedy the situation the government was trying to persuade villagers to move their fields and wells back from the coast.

Dates were the principal crop grown in numerous villages. Other crops included alfalfa, limes, lemons, oranges, bananas, papayas, mangoes, onions, melons, tobacco, eggplant, tomatoes, and potatoes.

The section of Dhofar Province inland from Salalah has more rainfall than any other region. Good natural grazing is made possible by the monsoon rains that move inland from the Arabian Sea from June to September. As a consequence in the mid-1970s some 50,000 head of cattle were maintained in the area. Although vegetation dries up quickly after the seasonal rains, cattle can get along by grazing dried grass until the next rainy season brings renewed growth.

Insufficient drinking water for stock during the long dry season has limited the growth of the cattle industry. In its attempt to overcome this handicap the government, as part of its development program, has a project for drilling wells at key centers to provide water for animals.

Little more than 2 percent of the cultivated area of the country is in Dhofar. Coconuts were the most important crop. Alfalfa, sorghums, millets, and bananas made up the bulk of the remaining farm production, but such vegetables as okra and tomatoes were also grown. For millennia Dhofar has been the world's leading source of frankincense. It is obtained by cutting the bark of the trees, which are low and squat, and collecting and drying the gum that oozes from the gash.

The Musandam Peninsula has only about 1,000 acres of irrigated farmland. Dates and vegetables were the main crops.

The rudiments of agricultural experimental and extension programs were launched in Oman in the first half of the 1970s. In 1976 experimental farms were in operation at Rumais and Sohar on the Batinah coast, at Nizwa and Wadi Quriyat in the interior, and at Rabat and Salalah in Dhofar. There was also an animal breeding center at Salalah. There were twenty agricultural extension service centers, each with one farm adviser and one trainee. Small quantities of chemical fertilizers, at government-subsidized prices, were distributed to farmers by these centers. With help from the World Bank plans were under way in 1976 to establish an agricultural school at Nizwa that will accommodate 180 boarding students.

Since cattle have been kept mainly for milk, the practice in Dhofar has been to slaughter bull calves shortly after birth. In its development program for that province the government has established a cattle feeding center at Salalah. Bull calves are purchased from the herdsmen and raised for meat. It is believed that Dhofar can, by this means, produce enough beef to satisfy the nation's requirements and that Oman can even be an exporter of beef to nearby countries.

In mid-1976 a five-year agricultural development plan for the 1976–80 period called for funding at the equivalent of US$200 million. The main goals of the plan are to increase the area under cultivation from 90,000 to 114,000 acres, improve water supplies, and double total crop production. Production is to be aimed toward a gradual shift from the traditional diet, which relies heavily on dates, fish, and rice, to a diet incorporating more dairy products, meat, and vegetables. The primitive —almost nonexistent—agricultural marketing system is to be improved. It is planned that food exports will counterbalance food imports. Dates, dried limes, and coconuts are envisioned as the main agricultural exports. Another objective of the plan is to halt the movement of young people away from farms by making rural life more profitable and satisfying.

The country is to be divided into seven agricultural districts, and thirty government centers will give advice on machinery and fertilizer and other inputs. To carry out the plan the government foresaw the need to import 300 tractors, 1,500 water pumps, and 12,000 sprayers over the five-year period.

Fisheries

Fishing has traditionally been second only to farming as an economic activity in Oman. Both the Gulf of Oman and the Arabian Sea abound in a variety of fish; sardines predominate, but bluefish, mackerel, shark, and tuna are also plentiful. Abalone, lobsters, and oysters reportedly are abundant.

Fish are important in the diet of the peoples of the coast and of the interior. Although fresh fish are available only to those living near the sea, dried and salted fish are sold throughout the country. Dried sardine meal has long been used for fertilizer and for cattlefeed.

There are about 15,000 fishermen in Oman, who fish the waters near the coast using a traditional, small seagoing canoe to which an outboard motor has been added. The annual catch was about 40,000 tons in the mid-1970s. This was below the level of the mid-1960s, because some fishermen had turned to more remunerative employment provided by the oil boom. While the catch declined, local demand expanded. In consequence exports of salted and dried fish, once substantial, became negligible.

In its long-term economic planning the government stresses modernizing and expanding the fishing industry and developing it into an important export industry that will provide expanded employment opportunities. A foreign consulting firm, Mardela International, was employed to investigate the fishing potential and, after a four-year survey, confirmed that there were substantial resources in the seas bordering Oman and suggested the possibility of a tenfold increase in the annual catch.

The first major fishing project was launched in April 1976 with the award of a three-year concession to a consortium of Japanese firms, Taiyo Fishery Company and Nissho-Iwai Company. Apparently firms from India, Iran, and the Republic of Korea (South Korea) provided stiff competition to the Japanese in bidding for the concession. Nevertheless no additional concessions to foreign firms were expected, because the Oman government has stated that it plans to keep the fishing industry in national hands.

The concession granted the Japanese firms allowed them to fish a deepwater area extending from Masirah Island to the Kuria Muria Islands. This is off the country's desert region and thus in an area not usually fished by Omanis. It was expected that this project alone would increase the annual catch by 15 to 20 percent. The agreement provides that the government of Oman will receive 40 percent of the catch; most of this will be sold back to the Japanese. It therefore seems likely that nearly all of the catch will be exported to Japan.

Beginning with an allocation of OR500,000 in 1976, the government, under its five-year (1976–80) plan for fisheries, established a loan fund to enable Omani fishermen to purchase modern equipment. Fishermen were required to repay only 75 percent of the amount they borrowed, and interest rates were low. The amount of money allocated to this loan fund was expected to be increased substantially during the plan period and in subsequent years.

Mineral Resources

In the mid-1970s government officials and advisers in Oman talked optimistically of the country's becoming a major minerals producer and exporter in the 1980s. There were some solid bases for encouragement, but not enough was known of Oman's mineral resources for unguarded optimism.

Passages in the Bible, Roman documents, and travelers' accounts of long ago seem to confirm the claim that in ancient times substantial quantities of copper, silver, and coal were mined in Oman. Nevertheless, except for clay for pottery and tile and marble for building, no minerals other than oil were being exploited in 1976. But limestone was to be used in a cement plant scheduled for construction near Muscat; the plant, Oman Portland Cement Company, was jointly owned by the government (51 percent) and foreign investors.

Surveys have indicated numerous minerals—asbestos, chrome, coal, copper, iron ore, lead, manganese, nickel, silver, and zinc—and there may be extensive deposits of several. Only copper, however, had been confirmed to exist in commercially exploitable deposits.

Surveys by Prospection of Canada have reportedly resulted in proven copper reserves of 12 million tons. Some minerals experts think much more will be discovered. Oman Mining Company, owned jointly by the government (51 percent) with Prospection and with Marshall Company of the United States, expected to have a copper-mining venture under way in the mountains near Sohar by 1979. From rock having an average copper content of 2.1 percent, it was expected that the initial output per month would be about 1,800 tons of copper plus small amounts of silver. The necessary facilities—roads, crushers, port facilities at Sohar, and related infrastructure—were lacking in 1976 and would need to be built before Oman could export copper.

Asbestos of good fibrous quality, reportedly in commercially exploitable deposits, has been discovered in the northwest mountains. A joint mining and asbestos-cement plant venture seemed to be a possibility. Although there was talk of other mining ventures, none appeared to be based on concrete facts or plans. More minerals surveys would be required before projects could be launched.

Industry

At the beginning of the 1970s there were no modern industries in Oman with the exception of the oil industry (see Oil Industry of Oman, ch. 5). By 1976, however, a few light industries had been established. In operation were date-packaging plants at Nizwa and Rustaq with daily capacities of fifty tons each. New factories manufacturing furniture, aluminum products, soft drinks, milk products, and industrial gases had also gone into operation. Scheduled to become operational by 1977 at Matrah was a flour mill with a daily capacity of 150 tons and adjacent grain silos with 30,000 tons storage capacity. Also under construction in mid-1976 were fish-processing facilities at Sohar, Matrah, and Salalah and a milk reconstitution plant at Qurum.

Still operating by primitive methods were a few small-scale traditional industries, such as the production of ghee (clarified butter) and the drying of fish, dates, and limes. Some handicraft industries remained, although their importance has steadily declined. Silversmiths continued to ply their trade in almost every town. Bahla was an important center for the production of household pottery, goldsmiths were found in and around the capital area, and a few areas turned out low-quality handmade cloth from locally produced wool. Boats were still being built in the coastal towns.

The government has retrenched from plans of the late 1960s to establish a fairly substantial amount of industry. Budgetary restrictions and an acute shortage of skilled labor were the main reasons for abandoning the more ambitious industrial schemes. It seemed unlikely that Oman could be competitive with neighboring countries in petrochemicals.

Projects that had apparently been completely abandoned included a fertilizer plant and a gas liquefaction plant. Feasibility studies have also been made for a cement plant, a sugar refinery, a brickworks, asbestos products factories, a sponge iron plant and steel rerolling mill, and a dry dock and floating dock. Preliminary feasibility studies included proposals for plants to manufacture cigarettes, nails and screws, plasticware, and glass products.

But in mid-1976 the only industrial proposal that seemed certain to be completed was the cement plant, which was projected to cost the equivalent of US$50 million. The plant will have an annual capacity of 350,000 tons, and it was expected to start operating in 1979. The Omani market is likely to absorb the entire production. (In fact, even when the plant is in full production, imports will probably still be necessary.) Cement imports in 1975 exceeded 310,000 tons, and the construction industry was expected to continue to expand.

Future industrial projects may center on industries based on Oman's mineral wealth—if it proves to be as extensive as the more optimistic reports indicated. Some possibilities that were under consideration in the mid-1970s included a copper smelter plant with ancillary plants manufacturing copper products, asbestos products plants, and manufacturing projects using Oman's high-quality marble. Even if financing becomes available, a shortage of skilled labor will continue to be a major constraint on industrial expansion.

Tourism

Oman has the potential for developing a tourist industry, but as of 1976 there had been little support for tourism in the government. Perhaps this was because the society was closed for so long and because, for security reasons, entry into the country was still strictly controlled. In the event that oil and mineral resources prove to be less than hoped for, a tourist industry could be a source of foreign currency earnings.

The country's winter climate is nearly ideal. There are vast stretches of clean, sandy beaches, and fishing is superb. There are colorful towns, historical sites, a variety of natural scenery, and a growing number of good hotels.

Until 1975 Al Falaj in Ruwi, with seventy-six rooms, was the only international-class hotel in Oman. But in late 1975 and 1976 about 500 rooms were added in four hotels in the capital area—three new hotels and an extension to the Al Falaj. Another 300-room beach hotel was under construction at Qurum. And in 1977 Salalah, which has had only a thirty-room hotel, will have a new 140-room hotel with twenty adjacent villas.

Water and Power

The country's water resources are a key to the economic future of Oman, and development will require a lot more water than has been available in the past. Rainfall is so scant that crop production is not possible without irrigation. Livestock raising is restricted to areas having a dependable supply of drinking water for animals. Any substantial expansion of agricultural production will therefore necessitate the development of new sources of water. Industrial expansion, the development of tourism, and an increasing standard of living will combine to boost the requirements for water.

Several preliminary water-resource surveys have been carried out, but in 1976 there were still few data on the water resources of Oman. Such survey data as were available led to the tentative conclusion that the country does have some valuable underground water deposits, especially in the interior, but the discoveries to date have not been in the quantities hoped for.

The government has actively pushed projects to make more water available as rapidly as possible. Wells for village water supplies have been drilled throughout the country, and supplies in the capital area and in Salalah have been expanded and improved.

The most ambitious rural water supply program has been in Dhofar Province, where in isolated areas hundreds of wells have been drilled to provide drinking water for cattle as well as for the human inhabitants. In 1975 alone some seventy-five wells were drilled in twenty-seven locations.

In 1976 work was under way to improve and expand the water distribution and sewerage systems in the capital area and in Salalah and surrounding areas. When the desalination plant at Al Ghubra reaches capacity operation, it will supply the capital area with six to seven times the 358 million gallons it received in 1975 from sixteen deep wells. In 1976 there was also a desalination plant under construction on Masirah Island.

More water for crop production is another matter. A significant, but relatively costly, source of more water would be to trap by means of ponds and dams the water that now runs into the sea. As the farmers of Oman are said to be wasteful in the use of irrigation water, improved irrigation techniques should provide a limited quantity of additional water for expanding the area devoted to crops.

The Food and Agriculture Organization (FAO), in cooperation with several foreign consulting firms, was scheduled to complete a survey of Oman's water resources by late 1976. The survey involved setting up 700 observation points in the country to measure water flow. After a study of the results of the survey the Oman government may establish a national organization for water resources.

Between 1971 and mid-1976 more than OR60 million was spent installing electric-power generating facilities, including OR25 million on the desalination and electric-generating complex at Al Ghubra. Between 1971 and 1974 the generating capacity of the government-owned power plants, which serve the capital area and Salalah, expanded elevenfold, from four megawatts to forty-four megawatts—thirty-seven in the capital area and seven in Salalah. In 1974 there were also thirty-one megawatts of generating capacity under the control of Oman's oil company, Petroleum Development (Oman) (PDO)—twenty-three at Fahud and eight at Mina al Fahal (see Oil Industry of Oman, ch. 5). Privately owned generators supplied the limited additional electric power available in the country.

In addition to producing fresh water, the power and desalination complex at Al Ghubra—officially opened in November 1975—was designed to have a generating capacity of thirty-eight megawatts at full capacity. But since its official opening the plant has been bedeviled with technical problems, and reportedly further major alterations will be required before it can achieve anywhere near its full capacity. Once the Al Ghubra power station has been brought to its planned capacity, it will probably be expanded to a capacity of seventy-six megawatts.

In 1976 two foreign firms were installing small diesel generating stations in about thirty towns and rural areas in the north of the country. In addition the capacity of the Salalah power station was scheduled to be increased.

Public Finance

Before the accession of Sultan Qaboos virtually no distinction was made between the finances of the sultan himself and those of the sultanate. Oman's first government budget, therefore, dates back only to 1971.

Between 1971 and 1975 budgeted revenues increased more than ninefold, from OR50 million to OR468 million; expenditures advanced even more rapidly, from OR46 million to OR521 million, an elevenfold gain

Table 37. Oman, Government Finances, 1971–75
(in millions of Omani rials)[1]

	1971	1972	1973	1974	1975[2]
Receipts					
Oil revenues	48	50	61	292	387[3]
Other revenues	2	3	4	12	11
Grants	—	—	4	28	70
Total Receipts	50	53	69	332	468
Expenditures					
Current					
Defense	16	28	43	138	238
Civil	10	14	20	63	84
Oil company operations	...	...	...	6	7
Total	26	42	63	207	329
Development	20	30	30	179	192
Equity payment to PDO[4]	—	—	—	(36)	—
Total Expenditures	46	72	93	386	521
Surplus or Deficit	4	−19	−24	−54	−53

. . . means not applicable; — means none.

[1]For value of the Omani rial—see Glossary.

[2]Preliminary.

[3]Does not agree with figure reported elsewhere for government oil revenues, probably because of different method of accounting and exchange rate.

[4]Petroleum Development (Oman).

Source: Based on information from "Oman Seeks Modernization of Economy, Emphasizes Income-Generating Projects," *IMF Survey*, Washington, February 2, 1976.

(see table 37). Deficits increased rapidly after 1972. Provisional figures issued in June 1976, however, indicated that, because of increased income from oil and larger foreign loans, government receipts would be about 20 percent higher in 1976. Increased receipts plus stringent control on expenditures led to the forecast of a small budgetary surplus, the first in five years.

Petroleum revenues moved upward at a relatively moderate pace from the beginning of oil sales in 1967 to a level of OR61 million in 1973. In 1974 revenues leaped to OR292 million because of the sharp rise in international oil prices and increased revenues accruing to the government after its acquisition that year of a 60-percent share in PDO (see Oil Industry of Oman, ch. 5). Oil revenues totaled an estimated OR387 million in 1975.

In 1975 revenues from non-oil sources accounted for less than 3 percent of the total. They consisted mainly of custom duties levied on imports, an income tax on corporate bodies, charges for public utilities, and investment income on government assets placed abroad. Although a relatively small part of the total, revenues from these sources seem destined to increase rapidly, as a result of anticipated growth in imports and rapid expansion of the commercial sector. Nevertheless, in anticipation of the time when oil revenues will decline, analysts suggested that it would appear prudent for Oman to widen its tax base by introducing personal income and property taxes.

Despite the sharp climb in revenues, the government experienced a record deficit and a cash flow crisis in 1974 because of a lack of proper fiscal controls. A temporary freeze was placed on new nondefense contracts. This halted the growth of the deficit, but earlier commitments were of such a magnitude that few new development projects could be initiated before 1977.

The high level of expenditures for defense, boosted by the rebellion in Dhofar, placed a heavy burden on the government's financial resources in the first half of the 1970s (see National Defense and Internal Security, this ch.). In 1976, however, military measures combined with development activities appeared to have halted the insurrection. Unless guerrilla activity springs up again in Dhofar, a smaller share of future budgets may be required for defense expenditures, permitting a larger share for development.

By 1976 a real effort was being made by the government to improve budget accounting procedures and to regulate and coordinate contract signing by its various agencies. Progress was being made, and the growth in deficits had been halted. As it struggled to get its financial house in order, the government in 1976 was seeking more expert guidance from the International Monetary Fund (IMF), had hired a financial management specialist, and intended to retain an international firm of accountants to assist with an improved system of fiscal reporting and coordination.

Money, Credit, and Prices

In late 1976 the Omani rial, which replaced the gulf Indian rupee in May 1970, continued to be a freely convertible currency. The gold content of the rial has remained unchanged since its introduction. The rial was in effect pegged to the United States dollar, whereas the exchange rate for other currencies was based on the London market.

Under a banking law issued in November 1974 the government on April 1, 1975, established the Central Bank of Oman. The new institution replaced the Oman Currency Board, which since 1972 had been responsible for issuing currency, managing government accounts, and carrying out banking transactions with commercial banks and international institutions. The Central Bank, capitalized at OR2 million, was placed under the management of a board of governors appointed by the sultan. The board possessed full authority to discharge all the functions required for the operation of the Central Bank and the supervision of commercial banking.

The Central Bank's responsibilities include the management of the government's foreign assets. It may make advances to the government to cover temporary deficiencies in recurrent revenues (up to maximums of ninety days and an estimated 10 percent of recurrent revenues). It may purchase government treasury bills and securities with a maximum maturity of ten years. It may make advances to commercial banks and is empowered to buy, sell, discount, and rediscount commercial paper.

The number of commercial banks operating in Oman increased from three in 1972 to six in 1974 and to fourteen in 1976. Ten were subsidiaries of foreign-owned banks. The fourteen commercial banks in mid-1976 had forty-five branches, thirty of which were in the capital area. Under the 1974 banking law, in addition to the initial capital requirement of OR500,000 and reserve requirement of 5 percent of deposits, each bank must maintain an interest-bearing account with the Central Bank of 0.1 percent of its resources, including, for foreign institutions, the total assets of the parent bank. The minimum deposit is OR50,000 and the maximum OR500,000.

Commercial bank credit, which had grown rapidly since 1970, spurted sharply in 1974 and 1975. Credits to the private sector increased from OR4 million in 1970 to OR19 million in 1973, OR65 million in 1974, and an estimated OR85 million in 1975 (see table 38). Loans to the government advanced even more rapidly between 1973 and 1975, from OR7 million to an estimated OR65 million. Deposits also rose sharply; foreign liabilities increased even more rapidly. About three-fourths of the credit to the private sector has been channeled to trade, 17 percent to construction, 3 percent to transportation, and 5 percent to industry and services. A number of loans, particularly in construc-

Table 38. Oman, Monetary Developments, 1970–74[1]
(in millions of Omani rials)[2]

	1970	1971	1972	1973	1974
Money supply	8	13	19	25	51
Quasi-money	28	32	26	22	37
Net foreign assets	82	95	79	51	53
Foreign assets, Central Bank of Oman	5	10	15	17	46
Foreign assets, commercial banks	29	37	22	16	31
Foreign liabilities, commercial banks	1	2	5	3	54
Credit to private sector	4	4	7	19	65
Credit to public sector	n.a.	n.a.	7	7	49

n.a.—not available.
[1]Figures are for December of each year.
[2]For value of the Omani rial—see Glossary.

tion, were highly speculative, and the banks have been forced to write off several. As a consequence by 1976 more conservative banking practices were being followed.

Commercial banks were the only financial institutions operating in Oman in 1976. There was, however, a detailed study being carried out for a planned government development bank to extend credit to agriculture, fisheries, and industry.

Oman experienced a high rate of inflation in the 1970–76 period. By the first quarter of 1972 the cost of living was estimated (there being no index of prices) to be 50 percent above the level of two years earlier. Prices went up by an estimated 80 percent in 1973. The inflation rate slowed to an approximately 20-percent annual increase in 1974 and was probably a little lower in 1975 and 1976.

A major cause of inflation was the very rapid increase in the money supply stemming from government spending and the sharp rise in private sector credit. The money supply grew at an average annual rate of about 40 percent in the 1971–73 period and then more than doubled in 1974.

Inflation was fed by a shortage of housing and by the need to import most consumer goods. As skilled workers moved in from abroad, rents for the few rooms and apartments available in the capital area doubled and redoubled. Very large markups on imported goods were common. By 1975 shortages were not so acute, and the inflationary pressures on some basic foodstuffs were indirectly controlled by government subsidies.

Foreign Trade and Balance of Payments

Oil shipments have accounted for well over 99 percent of the value of Oman's export trade. Development goods and defense items dominated imports; but increasing personal incomes, especially of the growing number of expatriates, resulted in a rapid boost in the demand for and import of consumer goods.

All of Oman's petroleum is exported as crude oil since there are no refineries in the country; gasoline, diesel fuel, and lubricants are imported for local consumption from nearby Persian Gulf countries (see ch. 5). Oil exports for 1976 were estimated at 139 million barrels, up from 125 million in 1975. Payments to the Oman government for the oil extracted totaled OR370 million in 1975 and OR292 million in 1974.

Japan was the main destination for Oman's crude oil; exports to that country accounted for 38 percent of the total in 1975, about 35 percent in 1974, and 36 percent in 1973. In 1975 the Netherlands accounted for 21 percent of exports, displacing France as the second most important destination. Other countries to which oil was shipped in 1975 included Trinidad and Tobago, France, the United Kingdom, Singapore, the United States, and Canada.

Exports of commodities other than oil totaled slightly over OR1 million in 1975, a substantial increase over previous years (see table 39). Dried limes accounted for more than half of the total value of non-oil exports, and dried dates were in second place. Fresh tomatoes, figs, a little tobacco, and a negligible quantity of dried and salted fish made up the remainder of non-oil exports. In the mid-1960s fish rivaled dates and limes as an export, but the catch was down and domestic demand up, leaving virtually none available for export.

In 1975 shipments to Dubai and Abu Dhabi in the UAE accounted for 93 percent of total non-oil exports. India and Iraq were the only other export destinations. Small shipments were recorded to Bahrain, Iran, Kuwait, Qatar, and Yemen (Aden) in some earlier years.

Oman's recorded imports of OR231 million in 1975 probably understate actual imports by at least 50 percent. Imports of defense goods were not reported, and a substantial amount of other imports went unrecorded.

Although a few light industries have recently been established, Oman must still import most industrial goods and a growing proportion of its food requirements. Since 1973 the category of machinery and transport equipment has been the leading one, but there has been a phenomenal increase in imports in all categories.

The United Kingdom, traditionally Oman's leading source of imports, moved back into first place in 1975 after being displaced for some years by the UAE. Practically all imports come from Asia, Australia, Western Europe, and the United States—there is virtually no trade with Africa or Latin America.

Cement, at 310,000 tons, was the leading import by weight in 1975. Other leading imports included an estimated 93,000 tons of steel, 40,000 tons of timber, 110,000 tons of other building and construction materials, 8,500 motor vehicles, and a wide range of food and other consumer goods.

Table 39. Oman, Foreign Trade, 1970–75
(in thousands of Omani rials)[1]

	1970	1971	1972	1973	1974	1975
Exports (excluding petroleum):						
Fish and fish products	9	3	3	19	6	4
Dried dates	49	28	105	99	54	233
Dried limes	275	331	270	269	338	594
Other fruits and vegetables	11	45	6	179	13	211
Goat skins and hides	14	n.a.	n.a.	n.a.	4	n.a.
Tobacco	16	21	11	35	12	27
Other	15	2	--	8	15	9
Total Exports	389	430	395	609	442	1,078
Imports;[2]						
Food and live animals	3,000	4,998	5,645	9,748	15,033	26,765
Animal and vegetable fats and oils	--[3]	--[3]	--[3]	179	704	869
Beverages and tobacco	141	420	666	814	1,584	3,087
Crude materials	76	263	293	782	2,763	5,572
Mineral fuels and lubricants	340	955	910	1,759	5,073	10,799
Chemicals	183	608	785	1,911	5,016	8,999
Manufactured goods	1,813	2,233	3,630	8,016	29,567	48,519
Machinery and transport equipment	1,284	3,451	5,223	12,721	53,710	95,774
Miscellaneous manufactured goods	450	666	946	2,645	11,060	18,043
Other	305	190	615	2,100	11,067	12,591
Total Imports	7,592	13,784	18,713	40,675	135,577	231,018

n.a.—not available.
[1]For value of the Omani rial—see Glossary.
[2]Excludes government imports of defense goods; imports by private contractors not included until the second half of 1973.
[3]Before 1973 included under "Food and live animals."

Source: Based on information from "Unveiling Oman's Economic Realities," *Middle East Economic Digest*, London, February 20, 1976, p. 7.

Food imports in 1975 included an estimated 33,000 tons of rice, mainly from Pakistan; 40,000 tons of wheat flour, mainly from Australia; 5,000 tons of beef and mutton from Australia; 1,000 tons of frozen poultry from Western Europe; 4,000 tons of sugar from India; potatoes, onions, and other fresh vegetables from India and Lebanon; US$1 million of processed and packaged food items from the United States, including US$200,000 of soft drink ingredients for the new bottling plants; 9,000 tons of milk and confectionery products; and 2,000 tons of coffee. Virtually all the country's cloth requirements are met by imports from the People's Republic of China (PRC) and India.

In spite of a rapid rise in oil revenues between 1972 and 1975, the balance of payments was adverse, necessitating substantial foreign assistance. Several factors contributed to the unfavorable international payment situation. The government's acquisition of equity in the oil company operating in Oman required diversion of part, though a small part, of oil revenues to pay for the government's equity shares and the company's operating expenses. The value of imports increased sharply, partly because worldwide inflation raised import prices steeply and partly because of a larger volume of imports. Imports of consumption goods increased as incomes rose, particularly incomes of the large number of expatriates working in the country; imports of investment goods increased as the government undertook development projects; and imports of military equipment increased as the government sought to suppress the rebellion in Dhofar Province. Moreover the increasing number of expatriates in Oman caused the outflow of remittances to swell as they sent part of their earnings home.

Oman has made substantial contributions to several international organizations. Since 1972 it has purchased US$38 million of World Bank bonds, contributed US$13.4 million to the IMF oil facility, and made a US$1 million contribution to the Arab Bank for Economic Development in Africa. In 1972 it extended a US$2 million grant to Jordan.

Because of its budgetary and balance-of-payments problems, Oman has been the recipient of large amounts of foreign grants and loans. In 1973 and 1974 it received grants totaling US$93 million from Saudi Arabia and Abu Dhabi. In 1974 the government borrowed US$100 million from a foreign bank to finance its budget deficit. Grants from Saudi Arabia and Abu Dhabi totaled about US$300 million in 1975. The combination of this aid and drawing on the country's foreign exchange reserves covered the deficit on current account between 1972 and 1975.

Future Development

Certain elements in the situation seem favorable for relatively rapid economic development. Oil revenues are likely to continue high for the next few years at least. By 1976 a lot of valuable infrastructure had been put in place, fishery resources seemed large and relatively easily exploitable, the development of tourism offered considerable potential, the climate for business was favorable, and foreign participation in business ventures was encouraged.

NATIONAL DEFENSE AND INTERNAL SECURITY

The Military and Foreign Policy Aspects of the Dhofar Insurgency

Context of the Insurgency

Although the UN Commission of Inquiry of 1963 cleared Sultan Said bin Taimur Al Bu Said of the charge of outright oppression, the general situation in Oman in the early 1960s suggested a likelihood of further disturbances (see Historical Background, this ch.). Sultan Said, operating through his small group of British advisers and supported by his British-led armed forces, was ultraconservative and isolationist. He sought in effect to exclude the twentieth century from his domain, but contemporary pressures mounted against him.

In July 1952 Gamal Abdul Nasser, Anwar al Sadat, and their companions had overthrown the monarchy of King Faruk in Egypt and aroused a radical Arab nationalism that spread in varying forms and with varying success to all parts of the Middle East and North Africa. The two principal forms of this nationalism were Nasser's Arab socialism and that of the Baath (Resurrection) Party of Syria and its variant in Iraq. All were characterized by intense hostility to "Western colonial imperialism" and to monarchical institutions; they advocated social justice and often had distorted Marxist overtones. Some were compatible with the egalitarian teachings of Islam and were not necessarily in open conflict with religion, but the movements were primarily secular and political rather than religious.

In time the ideas of radical Arab nationalism penetrated even into Oman and helped kindle an antiregime insurgency—sometimes called the Dhofar Rebellion—that continued at varying levels of intensity from 1963 until it was virtually snuffed out in December 1975. In this contest both the established sultanate and the insurgents attempting to overthrow it through a "war of national liberation" had certain definite strengths and weaknesses.

Advantages and Disadvantages of the Contestants in the 1960s

One of the main advantages held by the insurgents was the remote location and rugged topography of their base in the Jabal al Qara, the mountain area lying irregularly north of Salalah and along the eastern border of Yemen (Aden), which is under a radical, Marxist-oriented regime. This regime actively supported the Dhofar insurgents, providing them with a cross-border sanctuary, a radio propaganda base at Aden, and a channel for external aid from the PRC and the Soviet Union. Consequently, under the doctrine of wars of national liberation as propounded by Mao Tse-tung, Vo Nguyen Giap, Ernesto (Ché) Guevara, and similar revolutionaries, the base area of the Dhofar insurgency was virtually ideal. Other advantages included external aid in money, arms, and training from a number of radical Arab states and communist states, the main source always being the Soviet Union. Further, the ubiquity of the guerrilla threat forced the small Omani regular forces to be divided and dispersed throughout the country, initially preventing a heavy concentration of government forces in Dhofar.

The insurgents, however, had few experienced leaders, lacked a broad base of popular support, had continual factional and organizational problems, and were unable to generate and sustain a high level of action outside Dhofar Province. At first the insurgency was much like that of a traditional Arab local tribal demonstration against the central authority; the movement soon adopted a radical ideology, but the task of animating large numbers of the traditionalist, illiterate Omani population in these terms proved too difficult.

The central government had never exercised strong control over the people of the Jabal al Qara region, who had not viewed themselves as subjects of the sultan and whose pastoral-agrarian society was characterized by age-old family and tribal warfare among themselves and with the people of the Salalah coastal plain. The insurgents were able to exploit some of the mountain feuds to advantage, but their textbook tactics of Viet Cong-style collectivization and taxation tended to alienate the traditionally independent mountain groups. Among those who subscribed to or were coerced into supporting the insurgent cause, the tasks of education and training were quite difficult; for the mountain people are illiterate, and their ancient dialect is local and obscure (see ch. 4). Moreover the two political causes that were generally well understood throughout Oman were the old ones of the sultan on the one hand and of the Ibadi imam on the other (see Historical Background, this ch.). The Dhofar insurgents' cause, being neither of these, was strange and foreign to most Omanis.

The sultan had the advantage of control of the established government, the armed forces, and the much smaller but effective police and security service. He had two further major advantages from the outset. The first lay in the agreement of July 25, 1958, with Great Britain. Under this agreement Sultan Said allowed the British Royal Air Force continued use of the fields and facilities at Salalah and Masirah Island; and Great Britain undertook to assist in strengthening the sultan's army, to second regular officers from the British army to serve as an integral part of the sultan's forces, to provide training facilities and advice, to assist in establishing an air force and make personnel available to it, to continue the development of civil aviation, and to assist in certain civil development programs.

The sultan's second major advantage arose from the discovery of oil in commercial quantities in 1964 and the receipt of sizable revenues from it by 1967 (see Oil Industry of Oman, ch. 5). Thus almost at the outset of their effort the insurgents received a heavy blow, for the possibility that they could impose a destructive financial strain on the sultan was eliminated.

It has been estimated that Sultan Said allocated about 60 percent of the annual budget for military and security purposes before substantial oil revenues began to flow in and about 40 percent thereafter. By the late 1960s the British-led forces of the sultan numbered about 2,500 men and eighty British officers and were organized as a brigade group but not concentrated as such. Headquarters were at Bait al Falaj, about five miles west of Muscat, and detachments at Sohar on the Batinah plain, in the Buraymi Oasis area, at Ibri, and at Nizwa in the heart of the old imamate region. Larger units were deployed in Dhofar against the insurgents and on guard duty at the oil installations of north-central Oman between Al Hawaisa and Natih. Among the troops, in addition to Omanis, were many first-and second-generation Baluchis and a number of Pakistanis. The new Omani air force had eleven British pilots and was equipped with six armed jet trainers, five armed piston aircraft, and four light transports. A naval element had been started, with one coastal patrol craft. The national police force consisted of about 300 men, including about fifty uniformed urban police in the adjoining cities of Muscat and Matrah.

Perhaps the greatest handicap to the cause of the sultanate was Sultan Said himself. He was and had long been estranged from Saudi Arabia, and he had not cultivated strong links with other monarchies in the central Middle East: Jordan, Kuwait, and Iran. As summarized by the United States Department of State in 1975, "Oman had only limited contacts with the outside world, including neighboring Arab states." Sultan Said's reactionary isolationist resistance to changes in government, public education, and beneficial public services impeded full use even of the carefully measured British assistance that was available and became an embarrassment to Great Britain and the sultan's conservative British advisers.

The effect was severely to limit external support for the cause of the sultanate against the Dhofar insurgency, although none of the neighboring monarchies would have viewed with favor the establishment in strategically located Oman of a revolutionary radical government, particularly as they—and the Western powers—noted the growth of the Soviet Union's influence and access in Yemen (Aden) and Somalia by 1970. The handicap of Said's ruling style was suddenly removed in July 1970 when the government of Oman was changed, not by the insurgents but by a palace coup at Salalah in which Said's son, Qaboos bin Said bin Taimur Al Bu Said, ousted his father and assumed power as the ruling sultan. Said withdrew to London, where he died in 1972.

Phases of the Insurgency

Viewing the Dhofar Rebellion from the standpoint of both the insurgents' operations and the government's response, analyst D.L. Price in 1975 summarized events in four phases.

First Phase: 1963–67.

The Dhofar Charitable Association (DCA) was formed in 1962 as a cover for the Dhofar Liberation Front (DLF), an offshoot of the clandestine Dhofar branch of the Arab National Movement (ANM). The motives of the DLF, as Price relates, were "mixed—nationalist, Marxist, Nasserite—but primarily it was opposed to the Sultan's rule and to the British connection." The revolt was at first directed with great intensity against Sultan Said. Overt manifestations of the dissident movement first appeared in early 1963. Oil company vehicles were ambushed, and antiregime leaflets were distributed. A leader of the movement from the Al Kathir tribe of north-central Dhofar was captured but escaped and made his way to Saudi Arabia. There, in an action showing the ideological fuzziness of the movement, he made contact with the exiled Imam Ghalib bin Ali (see Historical Background, this ch.). Eventually he went to Iraq, where he received further assistance and gained some adherents. By late 1964 the DLF had grown to include this core group and the DCA, the Dhofar ANM branch, and another clandestine body called the Dhofar Soldiers' Organization (DSO). The total strength of the insurgents then has been estimated at about 250.

The sultan's security service identified and captured about forty of the DLF in Dhofari towns during the spring of 1965. This serious blow forced the insurgents out of the towns and back into the central interior of Dhofar. No general uprising had taken place, and the movement might then have been said to be in early phase one—survival, organization, and limited guerrilla action—of Mao's classic three phases of insurgency. In the interior the insurgents formed an executive committee and issued their first formal statement in June 1965. This manifesto had three short parts: the "poor classes" were to form "the backbone of the organization"; the "imperialist presence" was to be destroyed; and the "hireling regime" of Said was to be destroyed.

The guerrillas divided Dhofar Province into eastern, central, and western sectors and attempted to recruit on an individual basis. Operations were confined to attempted sabotage, sniping, and harassment of the main road from Salalah to Thamarit (sometimes also called Midway) and police posts on the plain. Near Salalah, on April 26, 1966, Dhofari irregular soldiers recruited by the DLF and the DSO attempted to assassinate Sultan Said. The attempt failed, the sultan became even more of a recluse, and government forces tightened security measures in the Dhofar towns. During most of 1967, then, the guerrillas were dispersed in the hills, not tightly organized, on the defensive, and short of money, arms, and ammunition.

The general strategy of the Sultan's Armed Forces (SAF) against the guerrillas was to isolate them from supplies and from the population. This strategy was at first carried out through a policy of defensive attrition that included imposing losses when possible, keeping the guerrillas out of the towns, and later developing an administration for the mountain area that could preempt the insurgents. During the first phase the SAF held Salalah and the coastal plain but conducted no sustained offensive operations into the hills. The guerrillas were generally isolated, but this was a result as much of their own weakness as of government pressure. If the SAF was not particularly effective, neither were the insurgents; had the status quo of mid-1967 continued indefinitely, it is likely that the insurgency would gradually have died out as a result of the sultan's policy of simply holding the line.

Second Phase: 1967–70.

The status quo, however, did not endure; late 1967 saw two developments that changed it markedly. One of these favored the government—the receipt of the first substantial oil revenues. Military expenditures were increased, but Sultan Said failed to capitalize on his new advantage by intensive and sustained offensive operations promptly directed against the insurgents. The other major development favored the insurgents. When Yemen (Aden) became independent in December 1967, the radical government there provided the Dhofar insurgency with a secure, contiguous sanctuary and training area and channel for the renewed supply that came from the PRC and the Soviet Union. The nature of the Dhofar uprising then changed dramatically, not only in the new support resources acquired by the insurgents but in its motivating rationale and leadership.

Probably in 1965 the DLF leadership began to be infiltrated by a Marxist-Leninist organization from Bahrain called the Popular Front for the Liberation of the Occupied Arab Gulf (PFLOAG). At a congress in interior Dhofar in September 1968 the DLF took this name, and its earlier leadership was replaced by a new general command of twenty-five members, headed by Muhammad Ahmad al Ghassani. The doctrines of organized revolutionary violence and scientific socialism were adopted, and the congress, as Price described it, "marked the start of an attempt to escalate what was a localized, tribal revolt into an ideological movement with mass popular support throughout the Gulf."

In the Dhofar mountain area the PFLOAG quickly set about extending control over the population by convincing them that the PFLOAG was the wave of the future. This task was aided greatly by the weakness or complete absence of government opposition. Agricultural committees and land collectives were formed under PFLOAG commissars. Many recalcitrant tribesmen were shot, or their cattle were seized. Recruitment and impressment for the growing guerrilla forces were stepped up. Arms and supplies poured in, and young men were sent to the PRC, the Soviet Union, and the Democratic People's Republic of Korea (North Korea) for political-military insurgency training. Hundreds of Dhofari children were taken away to a "Lenin school" at Huaf, on the Yemen (Aden) coast just over the border. A radio station called the Voice of Oman and a PFLOAG headquarters were openly set up in Aden—although Yemen (Aden) and Oman were not then or later in a formal state of war. By 1970 most of the Dhofar hill population was actively in support of the insurgents or passively subject to them. During this phase the government response had not kept pace and, as the mortar and light rocket (recoilless rifle) bombardment of the Salalah plain became increasingly heavy, the military and political initiative appeared to have passed to the insurgents.

Third Phase: 1970–73.

In early 1970 the progress of the insurgency seemed to have reached the stage of equilibrium in which insurgent units have become numerous and strong enough to fight conventional battles with government troops and to move on the cities. Until this time the Dhofar uprising had been largely confined to Dhofar, although the bulk of the population and important cities lay in the northern part of the country. In June 1970, however, in northern Oman a mixed bag of small clandestine organizations whose members varied from young, ideological Omanis educated abroad to old, uneducated Omanis who simply hated the sultan merged to form an organization with a name at least as cumbersome in English as in Arabic: the National Democratic Front for the Liberation of Oman and the Arab Gulf (NDFLOAG). This group, identifying itself as "part of the world national liberation movement," was backed by Iraq. It had two units of about thirty men each (one group was to operate northwest of Muscat and the other to the southeast) and a smaller terrorist squad at Sur.

The NDFLOAG launched its campaign to open a northern front with a number of raids on interior towns, including Izki and Nizwa, in June 1970. The security forces in these areas, however, were alert, and all attacks failed. Interrogation of prisoners yielded valuable intelligence, and almost all the NDFLOAG leaders were captured at Matrah on July 19. Within a week the insurgency received, indirectly, what was probably its fatal blow when Qaboos deposed his father.

At once a new vigor and determination appeared in the SAF and security forces, as the new sultan opened his well-publicized dual campaign to defeat the insurgency and get national development under way. In October 1970 tribesmen in the eastern Dhofar mountains began to resist the forced collectivism and harsh rule of the PFLOAG, and it was at that time—another significant turning point—that defections to the government side began. The insurgents' position, however, though badly shattered in the north, remained strong in Dhofar. After a third party congress in June 1971, the insurgents issued a lengthy twenty-nine-point program. In December 1971 the PFLOAG absorbed what was left of the NDFLOAG and changed its name by dropping the word *Occupied* and substituting the word *Oman* to form the Popular Front for the Liberation of Oman and the Arab Gulf (PFLOAG).

Possibly stimulated by these accomplishments and doubtless annoyed by the deteriorating political situation in their mountain base area, the PFLOAG then went for the classic goal symbolizing insurgency strength—the military seizure of a government town. After careful preparation a company-size task force on July 19, 1972, attacked the town of Mirbat on the coastal plain about fifty miles east of Salalah. This operation proved to be what Price has called the "watershed in the military campaign." For the PFLOAG it was a disaster. Their attack force was annihilated, about 20 percent were killed, and the tide of the insurgency was pushed back into the mountains, where more aggressive government air and ground action began compressing it toward the Yemen (Aden) border. The war was not over, but the government and its forces under Sultan Qaboos' leadership were well on the way to recovering the initiative and before the year was out handed the insurgents another stunning defeat. The PFLOAG, trying to snap back from the disaster at Mirbat, attempted to reactivate the urban sabotage and terrorist infrastructure in the north. Security forces detected the network, however, and in late December 1972 in a five-day roundup called Operation Jason seized ninety people and several arms caches of Soviet and PRC automatic weapons, light mortars, and mines.

Fourth Phase: 1973–75.

During 1973 the military initiative passed to the government, and foreign aid came in from several sources; during that and the next year the planned upgrading of the SAF in numbers and equipment became a reality. Virtually wiped out in the north, the PFLOAG was forced on the defensive in Dhofar and gradually lost strength during 1973. Reports at the fourth PFLOAG congress in January 1974 showed a high rate of defection to the government side, supply shortages, and breakdown of communications. Internal dissent and disagreement on policy at this congress enervated the leadership from January to August. On August 9 the Voice of Oman radio announced that the movement would thereafter be called the Popular Front for the Liberation of Oman (PFLO). This meant that the PFLOAG, which had suffered a number of defeats elsewhere in the gulf region as well as in Oman, had become limited to the Dhofar insurgency.

During 1973 and 1974 economic problems in Yemen (Aden) required that country to devote more attention to its internal affairs. The facilities and oil provided to the PFLO were by no means cut off, but the degree of collaboration diminished. During 1973, possibly because of the lack of results, aid to the PFLO from the PRC decreased, and by mid-1974 it became insignificant. The gap, however, was filled by the Soviet Union, which became the principal supplier and the main source of external propaganda support.

Despite growing adversity the PFLO in mid-1974 still held much of the Dhofar mountain area and had an estimated strength of 800 hard-core fighters and about 1,000 part-time militia organized in local groups. The hard-core troops continued to be organized in 100-man companies, each with a commander, a political commissar, a second in command, a heavy weapons officer, three platoon leaders (in a fully manned company), and signal, supply, and medical noncommissioned officers. As they had from the start, the insurgents divided Dhofar into three territorial sectors from west to east, each allocated to a regiment. These regiments (in fact no more than battalions at best and always fluctuating in strength) were called the 1 April, 9 June, and 14 October regiments, after noteworthy anniversaries in PFLO history. These names were the fourth set that had been used since 1963, the various changes having apparently been made to confuse the enemy. Armament included 82-mm, 81-mm, and 60-mm mortars; RPG–7 recoilless rifles; the famous Soviet AK-47 (Kalashnikov) automatic assault rifle and older model rifles; antipersonnel and antitank mines; and by 1974 the Soviet Katyvsha 122-mm rocket, which the PFLO had begun, ineffectively, to use.

On the government side the strength of the armed forces more than doubled between 1972 and 1974; military expenditures jumped, and new equipment (mostly British) was delivered in response to orders placed by Sultan Qaboos (see Expenditures and Procurement; Manpower and Personnel Policies, this ch.). By mid-1974 the total SAF regular strength (not all deployed in Dhofar) was about 10,000, and there was an auxiliary force of about 1,500 Dhofari irregulars. Airfield facilities at Salalah, Muscat, and elsewhere had been expanded and improved, and the sultan's air force had a squadron of BAC–167 Strikemaster ground attack jets operating effectively against the insurgents, as well as sufficient transports and helicopters to make a substantial contribution to logistic support. Orders for the British Jaguar jet fighter and Rapier air defense system were placed. Air force pilots continued to be contract personnel, mostly British.

Until 1973 government operations in the Dhofar mountains had fluctuated according to the seasons. During the monsoon months of July to September the SAF reduced its presence to a few static outposts, allowing the PFLO almost a free hand until the return of the dry season again permitted sustained air strikes and logistic support of regular units in the mountains. In May 1972 a fortified strongpoint was established at Sarfeet (or Simba) on the Oman-Yemen (Aden) border and, starting in the summer of 1973 after support capabilities had been improved, units of the SAF stayed on the mountain at Sarfeet and elsewhere during the monsoon and kept pressure on the enemy.

Down on the plain British military engineers in late 1973 began supervising the construction of a barbed-wire barrier starting on the coast about fifty miles west of Salalah. By the end of June 1974 this barrier, called the Hornbeam line, had been pushed thirty-five miles inland to the northwest. The wire entanglement was mined, equipped with unattended ground sensors, and defended by platoon strongpoints at intervals along its length. Patrols covered the intervals between strongpoints, operated to the north of the line, and set out ambushes from bases on the line. The Hornbeam barrier cut the PFLO's camel train supply routes from Yemen (Aden) into Dhofar and forced supplies to be moved by man pack over longer and still vulnerable routes around the northern end of the wire. By the end of 1974 the barrier and the operations based on it had virtually isolated the PFLO of the central and eastern Dhofar from their sources of supply and command to the west.

As SAF regular units became increasingly capable of defending urban areas and conducting conventional operations and as the Hornbeam line was being pushed northward, several other developments turned the insurgents' own tactics against them. Between October 1970 and May 1974 the number of PFLO fighters who defected to the government was about 800 and by the end of the year it had increased to 1,000. (The number of PFLO fighters killed during the same period was between 400 and 433.) Rewards were offered for PFLO leaders dead or alive, and monetary inducements stimulated defection to the government, the amount being increased if the defector brought in weapons or could provide accurate information of arms caches. At the end of 1974 the rate of defection to the SAF was believed to be exceeding the rate of PFLO recruitment. Interrogation of defectors showed that their reasons for joining the PFLO had little to do with political ideology; far more often PFLO recruitment was based on simple intimidation or the influence of tribal loyalties.

During 1974 a detachment—reportedly about 100 men—from the British army's Special Air Services began training Omani irregular counterinsurgency units. Each of these units was called a *firqat,* and they have usually been conveniently (if wrongly) referred to in the plural as *firqats.* Individual members are sometimes called *firquers.* At the year's end there were eighteen of these units, organized on tribal lines and varying in size from about fifty to 150 men, of whom about 80 percent were defectors from the PFLO. Armed with rifles and small arms and regularly paid, the *firqats* provided excellent service as tribal police and defense forces for the mountain people and their cattle in secured and marginal areas where insurgents still sought to operate. Their knowledge of the terrain and the enemy also made them valuable as scouts for the regular forces and in tracking down infiltrators through the Hornbeam line.

In addition to the *firqats* civil action teams were developed to dig wells and establish government centers around well and water sites. Some population resettlement was required. Each center was to include such facilities as a market, school, and clinic, the whole center lying within a defense perimeter held by *firqat* forces. This pacification program, continuously developed and expanded in 1975 and 1976, demonstrated government presence and viability and simultaneously promoted political, military, and economic objectives. Food supply in the shrinking guerrilla-held areas became more and more difficult; guerrilla targets were vulnerable to intensifying air and artillery bombardment; and the PFLO's harsh measures of control in its areas turned the people against it. In short, as the insurgents increasingly lost favor with the mountain people, the sultan's government gained favor.

Critics of the *firqat* system maintained that, though effective in the short run, it had dangerous long-term implications because it tended to revive and perpetuate the old tribal structures with all their ancient disputes and blood feuds. Other spokesmen maintained that the war against the PFLO had to be won and that in any case tribalism was so deeply ingrained that decades would be required to reduce it under any system. The maintenance of government authority, security, and economic well-being in the hills was seen as the key to long-term stability and social progress.

During the 1973–75 period more changes in external relations occurred that affected the fortunes of both the PFLO and the sultan's government. In 1973 the PFLO gained an external ally that it might better have done without when Tudeh, the outlawed communist party of Iran, declared its support. This declaration was convenient for Sultan Qaboos, who requested troop support from Iran—which promptly sent a battle group of about 1,200 men to Dhofar in November 1973. On the night of December 19, 1973, this force, called the Imperial Iranian Task Force (IITF), occupied positions along the important road from Salalah northward some seventy-five miles to the key road and trail junction at Thamarit against only light resistance. This operation effectively opened and secured the road, previously hazardous and only intermittently secure, and drove another wedge into the PFLO in central Dhofar. Later SAF units took over positions along the road, and the IITF (for political rather than military reasons) was withdrawn from Oman in early October 1974.

In early 1975, however, the Iranian force returned to Dhofar and was expanded to about 2,500 men; in late 1976 it was still there. By January 1975 the Salalah-Thamarit road was secured by *firqats* and gendarmerie and, aside from its administrative and logistical units in Salalah, the IITF was deployed principally with the SAF in western Dhofar against the hard-core PFLO regulars and militia still in that desolate region.

Sultan Qaboos' diplomatic campaign for regional support produced substantial results, so much so that the situation in the 1973–75 period was effectively opposite to that of the pre-1970 isolation. In addition to the long-standing support of Great Britain and the more recent matériel and manpower aid from Iran, Oman was receiving annual financial aid the equivalent of about US$200 million from Abu Dhabi of the UAE to assist with military and civil development budgets and about US$2.5 million from Saudi Arabia, with whom relations had been much improved. Training spaces in military schools were provided for

SAF personnel by Great Britain, Jordan, Saudi Arabia, Egypt, and Pakistan. The UAE and Jordan from time to time provided troop units (fewer than Iran) for guard duty in the north, thus releasing SAF units for service in Dhofar. Some Jordanian officers have reportedly also been seconded to the SAF.

As the Qaboos government gained strength in regional external relations, the PFLO suffered reverses. In early 1974 a PFLO political delegation lobbied several Arab governments, Arab remnant communist parties in several states, the Palestinian Liberation Organization (PLO), and the League of Arab States (Arab League) in Cairo. Their attitude in contracts with the Arab League was reportedly one of doctrinaire arrogance, but they did invite Arab League members to visit their "liberated areas." To do this, however, visitors had to move through Yemen (Aden). The Arab League formed an eleven-member delegation drawn from six Arab states and headed by the league's secretary general to visit Yemen (Aden) and Oman. The radical government of Yemen (Aden), however, refused to permit this body to set foot in Aden. The delegation was well received in Oman in May 1974; the delegation's report to the Arab League supported Oman's position.

Nevertheless arms support to the PFLO from the Soviet Union continued, and in October 1975 the guerrillas were reported for the first time to have the Soviet SA–7 man-portable, shoulder-fired air defense missile. This upgrading in technology was too little and too late, however, to deflect the coordinated government offensive that followed. On December 11, 1975, Sultan Qaboos announced complete military victory and declared that the twelve-year insurrection was ended.

The Security Situation in 1976

Although the government was clearly victorious in the overall military campaign in Dhofar by the end of 1975, a few guerrilla fighters were still at large, and the PFLO political offices and facilities in Yemen (Aden) still existed. The IITF—then at a strength of about 3,000—remained in Dhofar, and PFLO broadcasts from Aden continued to report alleged guerrilla activity. During January 1976 Omani refugees from Yemen (Aden) began to return, and the government set up a reception center near the border. In an interview with a Kuwait newspaper the shah of Iran stated that fifty Iranian soldiers had been killed in Dhofar and one jet fighter lost. He stated that he would be glad to see the IITF in Oman replaced by Arab troops and added that the neighboring Arab states were "happy about what we are doing there, for it is not in the interest of the Arabs that Oman should fall."

Early in February 1976 Iraq—a former supporter of the PFLO—established diplomatic relations with Oman. This development, foreshadowed by the March 1975 accord between Iraq and Iran, had been under preparation since 1975. President Sadat of Egypt visited Sultan

Qaboos on February 26 and reportedly indicated that he wanted Egypt to mediate between Oman and Yemen (Aden). The PFLO leader Ghassani insisted to the Lebanese press that the war was continuing unabated; the PFLO lacked, however, the large cache of mortars, RPG–7 rocket launchers, and SA–7 missiles seized by government forces in February. In general it seemed that, once the sultan's forces and their allies had produced a credible military victory, the surrounding states of the Middle East found it desirable and possible to initiate political and economic action to try to separate Yemen (Aden) from the PFLO. The Soviet Union, though apparently willing to continue supply and supporting propaganda for the insurgents, had no role to play in the internal convolutions of regional inter-Arab diplomacy.

On March 10, 1976, Saudi Arabia established diplomatic relations with Yemen (Aden). On the same day Sultan Qaboos extended an amnesty "to all Omani citizens who have been misled by the false propaganda, to allow them to return to Oman without punishment to pursue a normal life as good citizens." This amnesty was to be effective from March 11 to May 11, 1976. The PFLO radio, broadcasting from Aden, rejected the amnesty, but 332 guerrillas had surrendered by May 11, 1976. Also during March Oman negotiated loans from Saudi Arabia to finance new development projects in Dhofar. By late March the reported guerrilla strength was about 400 in Yemen (Aden) and sixty to 100 in Dhofar.

During April 1976 the PFLO radio in Aden maintained that the revolution in Oman was passing through a sensitive and critical phase but claimed that its forces were continuing the struggle. Late in May the insurgent radio called for a new national front, and a new factor was added to the situation as contacts were reported between the Cuban embassy in Aden and the PFLO. The Soviet-Cuban operation in Angola had recently been successfully concluded, and observers speculated that Cuban advisers or even troops might become available to revive the insurgency. The PFLO celebrated its anniversary in Aden on June 9, but later that month Iraq ordered all PFLO offices and activities in that country closed and reportedly issued a government circular saying that the Dhofar revolt was over.

During a private visit to Great Britain in early July 1976, Sultan Qaboos met with senior British defense officials in preparation for the announcement that followed on July 19 to the effect that by March 31, 1977, the British Royal Air Force units on Masirah Island and at Salalah would be withdrawn by mutual agreement between the two countries. The numbers involved were reportedly about 350 men on Masirah and 150 at Salalah. It was stated, however, that this move would not affect the "close and friendly relationship" between Great Britain and Oman and did not apply to those individual British military personnel seconded to the sultan's forces. Neither were those engaged on individual contracts affected. The total number of British military personnel in Oman in late 1976 was variously reported. As a rough estimate, observers believed the withdrawal of the 500 from Masirah and Salalah would leave about 1,000 on secondment or contract.

The PFLO and Yemen (Aden) denounced the announcement of British unit withdrawal, alleging that British presence and influence would not actually be reduced. During September and October Oman announced the surrender of several groups of armed guerrillas sent back into Oman from Yemen (Aden) and, although the PFLO reported engagements with helicopter-borne SAF and Iranian troops in northwestern and eastern Dhofar, the Oman government clearly had the province under control.

In a speech to the General Assembly of the UN on October 6 the foreign minister of Yemen (Aden) reaffirmed his country's support of the PFLO and condemned foreign military interference in Oman. The Omani foreign minister in reply told the General Assembly that this statement was evidence of interference in the internal affairs of Oman by Yemen (Aden), which was continuing to export subversion, and he decried the necessity for independent, developing nations like his own to spend valuable resources to "defend their existence against infiltrators and saboteurs."

Toward the end of 1976 Sultan Qaboos reviewed the security situation and defense policy in two interviews published October 15 and November 19 in British and French journals respectively. Qaboos, referring to Yemen (Aden) as Aden, stated that "military action has ended inside Dhofar and along our borders with Aden. But there is still a Marxist military presence in Yemen (Aden). The Soviet, Chinese, and East German military advisors have been replaced by hundreds of Cuban counterparts. I don't want a repetition of the Angola tragedy in my country." Qaboos maintained that efforts by Saudi Arabia and Egypt to promote stability in the area were being spoiled by Libya, that an axis had been developed between Libya, Aden, and Somalia, and that the Soviet Union was using Libya as its "agent for Communist strategy in the Arab world." The sultan asserted that in 1967 Soviet strategists chose Oman "as their route to the Gulf's oil wells, and the starting base for future control of the region." The insurgent leaders in Oman then tried "to replace the existing leadership with a communist regime unacceptable in the region and completely divorced from our people's ideology and understanding of the Islamic tradition." Iran, the sultan added, was defending itself in Oman as well as defending Oman; he would not ask the Iranian force to withdraw until Yemen (Aden) expelled the Cuban advisers, ceased its hostile acts, and recognized Oman "as a friendly neighboring state." On the question of a Persian Gulf multilateral security pact, he replied that the subject needed much study and agreement on fundamentals by the heads of state before details could be worked out.

In the interview of November 19 Qaboos said he did not intend to reduce the level of military spending, because he was "determined to have forces capable of defending the country without resorting to foreign aid, which we now receive from friendly countries, to whom we are grateful." In response to a question about Omanization of the civil and military arms of his government, he replied that this had largely been achieved on the civil side but that "in the armed forces the process will take longer, especially in the air force. But there are already several Omani officers of the rank of lieutenant colonel sharing responsibilities with their British comrades." As for the air base on Masirah Island, he said that after the British withdrawal it would be "essentially a training base for our own forces" but that friendly countries might be permitted to use its facilities.

The sultan had visited the United States in January 1975 and in October 1976 and, again in response to a question, said that, if an official request were received from the United States for use rights on Masirah, the request would be considered. Qaboos added, "Although the Aden regime has now undertaken, at the instigation of Saudi Arabia, to stop the Dhofari terrorists from crossing the border and reportedly intends to normalize relations with the sultanate . . . I remain on my guard."

At the end of 1976 the principal external security threat to Oman was Yemen (Aden) and its support of the PFLO, which still had a few score activists in Dhofar and was endeavoring to rebuild its movement in the Yemen (Aden) sanctuary with new and old sources of support, including the Soviet Union, Cuba, and possibly Libya and the PLO. The gravest danger to internal security lay in the possibility that the PFLO might be able to develop an effective campaign of urban terrorism and sabotage, although their past efforts along this line were never particularly successful, or to conduct a fast and intensive wave of assassinations aimed at the sultan and key officials and thereby to shatter the government. The possibility of subverting a preponderance of the well-paid and -equipped Omani armed forces appeared slight.

Expenditures and Procurement

Military expenditures began to increase with the assumption of power by Sultan Qaboos in 1970. Before that time annual outlays were less than the equivalent of US$50 million a year. Expenditures increased in 1971 and 1972 and more than doubled between 1973 and 1974, reflecting the expansion of the SAF and the procurement of new equipment for the intensified campaign against the Dhofar insurgency. In 1975 and again in 1976 defense expenditures more than doubled, but they were believed to be leveling off (see table 40).

According to estimates by the United States Arms Control and Disarmament Agency (ACDA), Oman by 1974 was one of only fourteen countries in the world whose military expenditures exceeded 10 percent

Table 40. Oman, Military Expenditures and Forces Strengths, 1973–76[1]

	1973	1974	1975	1976
Expenditures:				
(in millions of				
United States dollars)	77.5	169.4	361.9	767.2
Regular Forces Strength:				
Army	9,000	9,000	12,900	13,200
Navy	200	200	200	400
Air force	400[2]	500[2]	1,000[3]	1,000[3]
Total	9,600[4]	9,700[4]	14,100[5]	14,600[5]
Other Forces Strength:				
Gendarmerie	1,000	1,000	1,000	1,000
Irregulars *(firqats)*	1,000	1,000	1,500	2,000
Total	2,000	2,000	2,500	3,000

[1]Figures do not include units of foreign allies or the Royal Oman Police.
[2]Includes about 150 expatriates.
[3]Includes about 600 expatriates.
[4]Includes 500 to 650 expatriates in all three services.
[5]Includes about 1,000 expatriates in all three services.

of the gross national product (GNP). These expenditures have been about 30 percent of annual national budget appropriations, including the substantial but indeterminate portion caused by the Dhofar operations (see table 41). At least through 1976 Omani oil revenues and supplementary foreign aid allowed both these levels of expenditure and the national development projects undertaken by Qaboos.

From 1965 to 1970 the average annual value of Oman's arms imports was only about US$1.4 million; between 1970 and 1974 it rose to about US$9 million. It jumped to about US$67 million in 1974 and roughly doubled in 1975 and 1976. By the end of 1976 the Dhofar insurgency, though not completely extinguished, was essentially defeated. The principal tasks of the armed forces then became those of sustaining their deployment and alert status against a renewal of the insurgency, absorbing and maintaining the new equipment received, and perfecting their training. Analysts believed that future expansion of the forces and the procurement of new weapons systems were likely to be more gradual than the surge of the 1974–76 years. Plans for an increase of total strength to 23,000 by 1985 have been reported, however.

Weapons and equipment are predominantly of British origin and have been procured in three principal ways: directly from the United Kingdom, by transfer from British forces and stocks in Oman, and by transfer from other countries, such as Jordan and Abu Dhabi. Some items of arms and equipment of non-British manufacture have also been procured; supplier states included Canada, Italy, the Netherlands, Iran (transfers), Switzerland, the United States, and several others; France was beginning to enter the market in 1975.

Table 41. Oman, Defense Budgetary Authorizations, 1974[1]
(in millions of United States dollars)

Expenditure Category	Amount
Capital Expenditure	48.6
Operations and Maintenance	
Salaries and allowances	31.9
Other operations and maintenance	88.8
Total	120.7
Total Defense	169.2[2]
Capital Expenditure as Percent of Defense Total	28.7
National Budget Total	581.9
Defense Expenditure as Percent of Budget Total	29.1

[1]Initial budget figures.
[2]Figures do not add to total because of rounding.

Source: Based on information from "Royal Decree No. 2774 Concerning 1974 Draft Budget," *Al-Jaridah Al-Rasmiyah*, Oman, April 15, 1974, pp. 4–9.

The larger arms procurement transactions in 1974 and later included orders placed with Great Britain for twelve British Aircraft Corporation (BAC) Jaguar supersonic jet fighters at about US$83 million; a BAC Rapier air defense missile system of twenty-eight fire units (one battalion of three batteries) at US$108.1 million; a BAC Blindfire radar and communications system to handle Rapier and Jaguar at about US$60 million; four BAC-167 counterinsurgency jets; eighteen assorted transport aircraft; four fast patrol boats at about US$14 million; forty Saladin twelve-ton armored cars; and thirty-six of the new 105-mm light guns capable of firing United States 105-mm howitzer ammunition as well as the ammunition designed for them. From Jordan thirty-one Hawker Hunter fighter-bombers were secured on a transfer sale, and Abu Dhabi provided twenty-eight Saladin armored cars as a gift. Orders were placed with Italy for fifteen Aresta-Bell helicopters, with France for an unknown quantity of Matra R-550 Magic air-to-air missiles, and with the Netherlands for two large patrol boats at US$1.5 million.

All of these were in addition to the inventories of land, air, and sea armaments already on hand in early 1974. Except for the Jaguar fighters, Rapier surface-to-air missiles, Blindfire radars, fast patrol boats, Matra missiles, and 105-mm light guns, deliveries were believed to have been completed by the end of 1976; the rest were due in 1977, except for the Jaguar-Rapier-Blindfire system, which probably would not become operational until 1978. Some commentators viewed this high-technology system as beyond Oman's capabilities and needs; others pointed out, however, that it was intended to fill a gap in the general defense posture and not necessarily keyed to insurgency operations.

In 1976 Oman continued to be eligible to buy from the United States.

Arms purchases from the United States, however, have played only a minor role in Oman's procurement; the only major purchase was of TOW antitank missiles in 1974 at a value of about US$560,000. Earlier purchases in 1972 and 1973 totaled only about US$239,000. Since 1974 no training programs have been in effect, and orders have been limited to minor items.

In addition to capital expenditures for armaments and the personnel and maintenance costs of a military establishment about three times as large in 1976 as in 1970, heavy outlays have been required for new military construction, including a defense headquarters at Qurum, an armed forces headquarters at Raysut, a naval base, an airfield, and port facilities in the Muscat and Salalah areas and elsewhere, and such facilities as military barracks, shops, warehouses, clinics, and schools. Some defense expenditures have no doubt been occasioned by the service in Oman of Iranian and Jordanian units, but the details and costs of these arrangements were not publicly available.

Manpower and Personnel Policies

Military service is voluntary, and the mid-1976 population of about 800,000 was predominantly young and growing at an annual rate of 3.1 percent. The number of military-age men was ample to provide volunteers for the armed forces at their total regular strength of about 15,000. Further, in traditional Arab society military service was regarded as an honorable pursuit, and relations between the civil population and the military were generally good. Three related factors, however, show why the armed forces in 1976 were still commanded by expatriates (mostly British) and included a substantial number, if not a majority, of minority elements and foreigners.

First, the use of foreign mercenaries in the SAF has a long history in Oman. These forces have been the sultan's forces, and their loyalty has been to him. He was ideally the embodiment and protector of the monarchical-Islamic society. This attitude continued to exist in 1976; the military establishment was called the Sultan's Armed Forces rather than the Omani Armed Forces. There is no constitution and hence no constitutional provision—nor is there a statutory provision or strong tradition—that military service to the state is an obligation and may be required. The ideas of modern nationalism, as differentiated from the traditionalist, monarch-oriented Islamic way, and service in the armed forces as an obligation of citizenship were not deeply ingrained, although they were beginning to take root as the opening up of society progressed.

Second, the strong expatriate and minority representation in the armed forces, especially in the command structure but also among the troops, has served the sultans of Oman well by decreasing the chances that the armed forces might be subverted and turned against the regime. After World War II, and continuing in 1976, apolitical command became institutionalized with British officers seconded from their own services or employed under contract. The case for the expatriate system, as related by Price, holds that "Omanization of the Sultan's Armed Forces will, in the long term, render it politically vulnerable to intrigue and factionalism, especially as the Omani officer cadre works its way through the system. The presence of expatriate advisers insulates SAF against political influence, so, in the long term, because the stability of Oman and the region has strategic significance—both locally and for the Western alliance—this presence is desirable."

Third, the low level of literacy in Oman effectively prevented the development of Omani officers, even had the old sultan been inclined to appoint them. There were few suitable to be appointed and even fewer with adequate backgrounds to move effectively into the higher ranks. Sultan Said, the captive of his own paranoid medievalism, not only wanted no Arab colonels but wanted in general as few people as possible in his kingdom who could read and write. A body of Omani noncommissioned officers and soldiers capable of handling modern equipment was therefore precluded in his time and extremely difficult to develop later. The country's legacy from Sultan Said of enforced illiteracy severely penalized the armed forces and the society as a whole. Sultan Qaboos made the building of a school system one of his first priorities in 1970, but at least a generation would be required to improve the situation substantially.

The national blight of illiteracy required the continuing use in 1976 of expatriates for many command, administrative, and technical positions and largely accounted for the fact that the air force was manned by British and Pakistani pilots and was likely to be for some time to come. Advertising in the British press in late 1975, for example, Oman offered British pilots three-year contracts with the sultan's air force at a starting annual salary the equivalent of about US$20,000 plus side benefits and allowances.

According to Qaboos the highest rank held by Omani officers in late 1976 apparently was that of lieutenant colonel, of whom he said there were several. One of these officers, in a television interview in May 1976 stressing the need for Omani pilots, stated that military policy would be to emphasize quality over quantity and, in what may have been an indication of personnel priorities, added: "We urge Omani young men to join the military in any of its branches, be it the air force, the navy, the artillery, or others." A gradual process of Omanization was well under way and probably irreversible but would take time at all levels. Baluchis and Baluchi Omanis have long been the largest single manpower element, if not a majority, in the army. By 1975, however, the all-Baluchi battalions in the north had been desegregated, although some reportedly remained among the forces in the Dhofar combat zone.

Base pay levels, in-grade step increases, and special allowances, though not the highest in the Arab world, were major contributors to morale and loyalty. Reportedly a 12.5-percent increase in military salaries was granted in October 1975. The complexity of the armed forces personnel categories and the multiplicity of special allowances make it difficult to generalize about pay rates. The approximate monthly base pay in United States dollar values for a lieutenant colonel was US$600; a first lieutenant, US$350; a sergeant, US$200; and a private, US$125. As Omanization becomes more general, new codifications of pay and allowances for Omani armed forces personnel may be expected.

The role of the armed forces since 1970 has not been limited to training and combat but has also been one of participation in the development of the country through civic action. The armed forces are the country's principal employer and constitute one of the main means of education. In the interior they are the chief agency of medical aid and disease control. Other contributions have been made in roadbuilding and communications that have benefited the public. All these activities and the establishment of security have tended to develop a sense of national identity among the people and among the forces.

Organization and Administration

Command Relationships

The missions of the armed forces are the defense of the realm by land, sea, and air; the protection of the monarchy; and the maintenance of internal security. Sultan Qaboos acts as his own prime minister and defense minister and, as absolute ruler, is supreme commander of the armed forces. He was trained at Sandhurst and served as a junior officer in the British army; this background and the subsequent intensive years of force development and war against the Dhofar insurgency have made him closely identified with the armed forces. He is active in the role of supreme commander and is the main channel of communication between the civil and military arms of his government.

The operational command and administration of the armed forces are vested in a commander in chief, who has the additional post of defense secretary to the sultan in the latter's capacity as minister of defense. In late 1976 the commander in chief was Major General Kenneth Perkins. He was assisted by a defense ministry staff and operational forces staff of senior British expatriates and Omanis. The Ministry of Defense is at Muscat, and the principal armed forces field headquarters is at Raysut, west of Salalah in Dhofar. The armed forces general staff is concurrently the army general staff; the air force and navy have limited autonomy and are responsible to and dependent on the armed forces general staff. The principal army formations and the air force and navy are commanded by British officers; in the intermediate and junior ranks of all services are other British and Pakistani officers and a smaller number of Arabs and Baluchis (both Omani and non-Omani). British noncommissioned officers and technicians are also found in all arms and services, particularly the air force. The number of Omani officers in 1976 was gradually increasing, in some cases by appointment of selected noncommissioned officers; they were beginning to be found throughout the structure, particularly the army.

The Army (Sultan of Oman's Land Forces—SOLF)

Total strength in 1976 was about 13,200, organized principally into six infantry battalions, one Royal Guards Regiment of three battalions (infantry), three field artillery battalions, one armored car squadron, one signal battalion, and one engineer battalion. Main items of equipment (British unless otherwise indicated) included sixty-eight Saladin heavy armored cars, about twenty United States V–100 light armored cars procured in 1966, twenty Ferret light armored cars, about forty 25-pounder guns, about twenty 5.5-inch guns, a few 120-mm mortars, and ten United States TOW antitank missile launchers. Oman had no tanks, reliance having been placed on the heavy Saladin armored car with its good cross-country mobility and high-velocity 76-mm guns. The thirty-six 105-mm light guns on order, with their rapid rate of fire, mobility, and improved range over the 25-pounder, will make a notable addition to field artillery holdings and suggest either that the 25-pounder and 5.5-inch guns will be retired or, more likely, that two new battalions will be formed.

Army training was advanced in May 1976 by a three-day combined-arms, live-fire maneuver in the Jabal Akhdar area in the north of Oman, between Nizwa and Ibri. The two infantry battalions stationed in the north formed the maneuver force, with air and artillery support. According to the military liaison officer to Sultan Qaboos, these exercises were the first in a series of large-scale maneuvers directed by the sultan to enhance the training and coordination of all forces.

The Navy (Sultan of Oman's Navy—SON)

Oman's small naval element had a strength of about 400 in 1976 and the standard missions of a coast guard, such as the prevention of smuggling and illegal entry. This role was important during the Dhofar insurgency in intercepting infiltration of guerrillas and supplies by sea. The commander in 1976 was a captain. The main operational bases and ports were at Salalah and Raysut and along the northern coast on the Gulf of Oman at Muscat, Matrah, Mina al Fahal, and Sur.

The flagship and largest vessel of Oman's coastal fleet was a 900-ton, 203-foot yacht purchased from Great Britain in 1971, with 40-mm guns, classed as a corvette. Two 370-ton, 150-foot ex-Netherlands minesweepers secured in 1974 were also classed as corvettes. Rated at sixteen knots, these vessels were modified by removing the minesweeping gear and adding three 40-mm guns. In the fast patrol boat category Oman obtained in 1973 three 37.5-meter boats of the Neal class made by Brooke Marine, displacing 135 tons, mounting two 40-mm guns, and rated at twenty-nine knots. Four more of these fast patrol boats were on order in 1976. In addition the fleet included one thirty-foot Fairey Marine Spear fast patrol boat, rated at twenty-six knots, one 1,500-ton training ship, two 1,450-ton logistic support ships, and two sixty-ton Hoodmaster harbor utility boats.

The Air Force (Sultan of Oman's Air Force—SOAF)

Commanded by an air commodore (brigadier), Oman's air force was proportionately the most expensive of the services and the most dependent on expatriates. It grew from about 400 officers and men in 1973 to about 1,000 in 1976 (including about 600 expatriates). Organizational formations included one ground attack fighter and reconnaissance squadron with twenty-nine Hawker Hunter jets secured from Jordan; one counterinsurgency squadron with fifteen BAC–167 jets (at least five more were believed to be in storage); one tactical transport squadron with sixteen Skyvans; two transport squadrons, one having two BAC–111s and two Viscounts and the other eight BN Defenders; and one helicopter squadron with twenty AB–205 and two AB–206 helicopters, variously procured—mostly from Italy. A number of older American and Canadian aircraft were believed to be in storage. The twenty-eight Rapier surface-to-air missile launchers ordered for the air defense mission will reportedly be assigned to the air force and coordinated with the twelve Jaguar fighters by the Blindfire radar system, also on order. The principal air bases are at Muscat and Salalah, and there are five fields along the gulf coast, seven in the central oil fields region, four in Dhofar, three in the south-central area, and a number of emergency strips throughout the country.

The Police and Gendarmerie

The Royal Oman Police, under the Ministry of Interior, are headed by an inspector general who has British advisory assistance. The force was estimated in 1976 to number about 6,000 men, divided between the uniformed metropolitan police of the Muscat-Matrah area and the district police in the smaller towns and villages. The main internal security problem in Oman is not ordinary crime but political terrorism and subversion, and the public security branch has developed a high capability of dealing with these threats. In the mid-1970s a few Omani police officers received training in the United States in such fields as airport security and defusing terrorist bombs.

The Oman gendarmerie of about 1,000 men was a lightly armed

paramilitary force used especially in the northern and central parts of the country as frontier and oil field guards and for rural police and security duties. It was under army control, constituted in effect a national guard or militia, and could be used to reinforce the police in civil disturbances that exceeded their capacity.

Uniforms, Ranks, and Insignia

Uniforms in all services closely resemble the summer uniforms of the corresponding British services, a consequence of the long period of British influence and guidance. Nevertheless, distinctively Arab items of dress and accoutrement are seen in some units, particularly among bodyguards and ceremonial detachments.

The system of officer, noncommissioned officer, and other ranks is the British system with only slight modification. The Arabic words for the various ranks are the same for all services, including the police. Specific distinction for the particular service may be made by adding the appropriate service modifier; for example, *aqid al bahariya* is a captain *(aqid)* of the navy *(bahariya).* In translation into English Oman follows the British usage for air force ranks; for example, the air force rank equivalent to colonel in the army or captain in the navy is

Table 42. Oman, Armed Forces Ranks and Insignia, 1976

Rank	Arabic[1]	Insignia[2]
Major General[3]	Liwa	Crossed sabers and one star
Brigadier[4]	Zaim (Amid)[5]	Crown and three stars
Colonel	Aqid	Crown and two stars
Lieutenant Colonel	Muqaddam	Crown and one star
Major	Raid	Crown
Captain	Rais (Naqib)[5]	Three stars
First Lieutenant	Mulazzim Awwal	Two stars
Second Lieutenant	Mulazzim Thani	One star
Officer Cadet	Dhabit Murasha	One gold bar
Warrant Officer, Grade 1	Wakil Katiba	Crossed sabers in circular wreath
Warrant Officer, Grade 2	Wakil	Crossed sabers in oval wreath
First Sergeant	Naqib (Raqib Awwal)[5]	Star and three chevrons
Sergeant	Shawish (Raqib)[5]	Three chevrons
Corporal	Arif	Two chevrons
Lance Corporal	Naib Arif	One chevron
Private	Jundi	None
Recruit	Jundi Jadid	Do.

[1]Arabic words differ in some cases from usage common elsewhere.
[2]Navy uses British system of wide and narrow sleeve stripes; all stars are five pointed.
[3]Highest rank in use in 1976, except that sultan was field marshal.
[4]Oman usage translates simply as brigadier as in United Kingdom.
[5]Word in parentheses more common usage in Arab world.

group captain. Insignia of rank also follow the basic British system in modified form, employing combinations of a five-pointed star (corresponding to the British pip), a crown, and crossed sabers (see table 42).

* * *

Reliable studies of contemporary social, economic, and political events in Oman are scarce. With few exceptions foreigners were not granted visas to enter the country before 1970. Robert G. Landen's careful study *Oman Since 1856* contains a superb section on the twentieth century to 1966, the year before his publication date. Except for works on the Dhofar Rebellion, most books that have reference to Oman are general works on the Persian Gulf and naturally concentrate on the more newsworthy rich oil states. An instance is David E. Long's *The Persian Gulf: An Introduction to Its People, Politics, and Economics,* which has a small section on Oman, and Fred Halliday's *Arabia Without Sultans,* which offers a controversial approach to gulf affairs. John Duke Anthony has written *Political Dynamics of the Sultanate of Oman.* An anecdotal work that deals only with Oman and that offers superb insights into Omani society in 1975 is Michael Darlow and Richard Fawkes' *The Last Corner of Arabia.*

The government of Oman publishes a picture book called *Oman,* which contains a general history, some data on the society, and a few statistical tables. The government also publishes a *Statistical Year Book;* in late 1976 the most recent was for 1974. The *Middle East Economic Digest* periodically publishes careful special reports on Oman, as does the *Financial Times.*

Reliable and generally available standard references on Oman's military holdings and procurement patterns notably include the annual issues of *The Military Balance* by the International Institute for Strategic Studies and the annual publications of *Jane's Fighting Ships, Jane's Weapon Systems,* and *Jane's All the World's Aircraft.* The publications by the Stockholm International Peace Research Institute (SIPRI) are also highly valuable, especially *The Arms Trade Registers* and the annual *World Armaments and Disarmament: SIPRI Yearbook.*

For the historical background of conflict and earlier military operations and the sultanic succession Wendell Phillips' *Oman: A History* is an important reference, as is Landen's *Oman Since 1856.* Penelope Tremayne's "Guevara Through the Looking Glass: A View of the Dhofar War" and D.L. Price's *Oman: Insurgency and Development* are analyses by informed and experienced researchers covering the progress of that war not only in narrative detail but also in contextual relation to regional trends and insurgency doctrine. Gérard Laliberté's

"La Guérrilla du Dhofar" is a scholarly presentation that, although differing sometimes from Price in details, shows more of the economic factors involved and is a valuable assessment of the origins and progress of the insurgency through mid-1973. (For further information see Bibliography.)

BIBLIOGRAPHY

Abdulla, Saif Abbas. "Politics, Administration, and Urban Planning in a Welfare Society: Kuwait." Unpublished Ph. D. dissertation. Bloomington: Department of Political Science, Indiana University, 1974.

Abir, Mordechai. *Oil, Power and Politics: Conflict in Arabia, the Red Sea, and the Gulf.* London: Frank Cass, 1974.

Abu-Hakima, A.M. "Kuwait and the Eastern Arabian Protectorates." Pages 430–449 in Tareq Y. Ismael, et al., *Governments and Politics of the Contemporary Middle East.* Homewood, Illinois: Dorsey Press, 1970.

Abu-Hakima, Ahmad Mustafa. "The Development of the Gulf States." Pages 31–53 in Derek Hopwood (ed.), *The Arabian Peninsula: Society and Politics.* Totowa, New Jersey: Rowman and Littlefield, 1972.

———*History of Eastern Arabia, 1750–1800.* Beirut: Khayats, 1965.

Adelman, Morris A. "Politics, Economics, and World Oil," *American Economic Review,* 64, No. 2, May 1974, 58–67.

———*The World Petroleum Market.* Baltimore: Johns Hopkins University Press, 1972.

Ajtony, M.A. *The Expanding Role of KNPC in the Oil Business.* Kuwait: Kuwait National Petroleum Company, n.d.

Al Baharna, Husain M. *The Legal Status of the Arabian Gulf.* Manchester, England: Manchester University Press, 1968.

———"Qatar." Pages Q1–Q4 in Viktor Knapp (ed.), *International Encyclopedia of Comparative Law.* The Hague: Mouton, May 1972.

Al-Ebraheem, Hassan A. *Kuwait: A Political Study.* Kuwait: Kuwait University, 1975.

Al-Marayati, Abid A., et al. (eds.) *The Middle East: Its Governments and Politics.* Belmont, California: Duxbury Press, 1972.

Al Rumaihi, Mohammad. "The Reformative Movement of 1938 in Kuwait, Bahrain, and Dubai," *Journal of Gulf and Arabian Peninsula Studies* [Kuwait], 1, No. 4, October 1975, 29–48.

Annual Statistical Abstract, 1975. Kuwait: Central Statistical Office, Planning Board, 1975.

Annual Statistical Bulletin, 1974. Vienna: Organization of Petroleum Exporting Countries, June 1975.

Anthony, John Duke. *Arab States of the Lower Gulf: People, Politics, and Petroleum.* (James Terry Duce Memorial Series, III.) Washington: Middle East Institute, 1975.

______"The Impact of Oil on Political and Socioeconomic Change in the United Arab Emirates." Pages 79–98 in John Duke Anthony (ed.), *The Middle East: Oil, Politics, and Development.* Washington: American Enterprise Institute, 1975.

______. *Political Dynamics of the Sultanate of Oman.* (U.S. Department of State, Office of External Research, Foreign Affairs Research Paper, No. FAR 21070.) Washington: October 1974.

______. "The Union of Arab Amirates," *Middle East Journal,* 26, No. 3, Summer 1972, 271–288.

Anthony, John Duke (ed.). *The Middle East: Oil, Politics, and Development.* Washington: American Enterprise Institute, 1975.

"Arab Military Industries: A Reality," *Middle East* [London], No. 10, July 1975, 105–106.

Aramco Handbook. Dhahran: Arabian American Oil Company, 1968.

Atiyeh, George N. "Middle East Ideologies." Pages 47–68 in William E. Hazen and Mohammad Mughisudden (eds.), *Middle Eastern Subcultures: A Regional Approach.* Lexington, Massachusetts: Lexington Books, D.C. Heath, 1975.

Attia, Abdel Moneim. *Oman.* (Constitutions of the Countries of the World series, Albert P. Blaustein and Gisbert H. Flanz, eds.) Dobbs Ferry, New York: Oceana Publications, December 1974.

Azar, Edward E. *Probe for Peace: Small-State Hostilities.* Minneapolis: Burgess, 1973.

Baaklini, Abdo I. "The Legislature in the Kuwaiti Political System." (Paper presented at Annual Convention of International Studies Association.) Toronto: February 1976 (mimeo.).

Baaklini, Abdo I., and Alia Abdul-Wahab. "The Role of the National Assembly in Kuwait's Economic Development: National Oil Policy." (Paper presented at Carmel Conference on Comparative Legislative Studies.) Carmel, California: August 1975 (mimeo.).

Bacharach, Jere L. *A Near East Studies Handbook (570–1974).* Seattle: University of Washington Press, 1974.

Bahrain. Ministry of Finance and National Economy. *Foreign Trade, 1974.* Bahrain: 1975.

______. *Statistical Abstract, 1973.* Bahrain: September 1974. "Bahrain Enhances Role as a Persian Gulf Center of Communications and Commerce," *IMF Survey,* 4, No. 23, December 15, 1975, 372–374.

Bahrain, Qatar, United Arab Emirates, and the Sultanate of Oman: The Businessman's Guide. London: Standard Chartered Bank Group, 1975.

Bargar, Thomas C. *Arab States of the Persian Gulf.* Newark: University of Delaware, 1975.

Becker, Abraham S. "Oil and the Persian Gulf in Soviet Policy in the

1970s." (Rand Corporation Paper, No. P-4743.) Santa Monica: Rand Corporation, December 1971.
Becker, Abraham S., Bent Hansen, and Malcolm H. Kerr. *The Economics and Politics of the Middle East.* New York: American Elsevier, 1975.
Belgrave, Charles Dalrymple. *The Pirate Coast.* London: G. Bell and Sons, 1966.
Beling, Willard A. (ed.) *The Middle East: Quest for an American Policy.* Albany: State University of New York Press, 1973.
Berger, Morroe. *The Arab World Today.* Garden City: Doubleday, 1964.
Bibby, Geoffrey. *Looking for Dilmun.* New York: Alfred A. Knopf, 1970.
Bidwell, Robin L. *The Arab World, 1900–1972.* London: Frank Cass, 1973.
Bill, James A., and Carl Leiden. *The Middle East: Politics and Power.* Boston: Allyn and Bacon, 1974.
Bill, James A., and Robert W. Stookey. *Politics and Petroleum.* Brunswick, Ohio: King's Court Communications, 1975.
Bishtawi, Kathleen. "Kuwait to Increase Security," *Financial Times* [London], September 4, 1976, 9.
Borthwick, Bruce M. "The Islamic Sermon as a Channel of Political Communication," *Middle East Journal,* 21, No. 3, Summer 1967, 299–313.
Boyd, Douglas A. "The Arab States Broadcasting Union," *Journal of Broadcasting,* 19, No. 3, Summer 1975, 311–320.
Burrell, R.M. *The Persian Gulf.* Beverly Hills: Sage Publications, 1972.
Buxton, James. "Sharjah: Long-Term Prosperity Is the Aim," *Financial Times* [London], December 10, 1976, 17.
——. "United Arab Emirates: Seeking a Formula for Unity," *Financial Times* [London], May 10, 1976.
Chisholm, Archibald H. T. *The First Kuwait Oil Concession Agreement.* London: Frank Cass, 1975.
Churba, Joseph. *Conflict and Tension among the States of the Persian Gulf, Oman and South Arabia.* (Air University Documentary Research Study, No. AU–204–71–IPD.) Maxwell Air Force Base, Alabama: Directorate of Documentary Research, Air University, December 1971.
Clarke, J.I., and W.B. Fisher (eds.). *Populations of the Middle East and North Africa: A Geographical Approach.* New York: Africana Publishing, 1972.
Cooley, John K. "United Arab Emirates Celebrate Growth," *Christian Science Monitor,* December 13, 1976, 6.
Cottrell, Alvin J. "The Political Balance in the Persian Gulf," *Strategic Review,* II, No. 1, Winter 1974, 32–38.

Couhat, Jean Labayle (ed.). *Combat Fleets of the World, 1976–77: Their Ships, Aircraft, and Armament.* (Trans., James J. McDonald.) Annapolis: Naval Institute Press, 1976.

Crowe, Kenneth C. *Kutuayba Alghamim of Kuwait.* (U.S. Department of State, Office of External Research, Foreign Affairs Research Paper, No. FAR 24752–N.) Washington: May 1976.

Daniels, John. *Abu Dhabi: A Portrait.* London: Longmans, 1974.

Darlow, Michael, and Richard Fawkes. *The Last Corner of Arabia.* London: Namara Publications, 1976.

Deakin, Michael. *Ras Al-Khaimah: Flame in the Desert.* London: Namara Publications, 1976.

Demir, Soliman. *The Kuwait Fund and the Political Economy of Arab Regional Development.* New York: Praeger, 1976.

Demographic Yearbook, 1974. (26th issue.) New York: Statistical Office, Department of Economic and Social Affairs, United Nations, 1975.

Dostal, Walter. "The Shihuh of Northern Oman: A Contribution to Cultural Ecology," *Geographical Journal* [London]. 138, No. 1, March 1972, 1–6.

Eilts, Hermann F. "Sayyid Muhammad bin Aqil of Dhufor: Malevolent or Maligned?" *Historical Collections,* 109, No. 3, July 1973, 179–230.

Ellis, Harry B. "Old Rivalries Threaten Unity of Sheikhdoms," *Christian Science Monitor,* May 14, 1976, B2.

The Europa Yearbook, 1975: A World Survey, I. London: Europa Publications, 1975.

Fenelon, K.G. *The United Arab Emirates: An Economic and Social Survey.* London: Longmans, 1973.

Ffrench, G.E., and A.G. Hill. *Kuwait: Urban and Medical Ecology: A Geomedical Study.* (Geomedical Monograph Series, Vol. 4.) Vienna: Springer-Verlag, 1971.

Field, Michael. *A Hundred Million Dollars a Day.* London: Sidgwick and Jackson, 1975.

Financial Times [London], February 10, 1976.

Fisher, W.B. *The Middle East: A Physical, Social and Regional Geography.* London: Methuen, 1971.

"Five Years Ago a New Nation Was Born: The United Arab Emirates," *New York Times,* December 19, 1975, E5.

Freeman, S. David. *Energy: The New Era.* New York: Vintage Books, 1974.

Freeth, Zahra. *A New Look at Kuwait.* London: George Allen and Unwin, 1972.

Friedlaender, Israel. "The Heterodoxies of the Shiites in the Presentation of Ibn Hazm," Pt. 1, *Journal of the American Oriental Society,* 28, 1907, 1–80.

Gibb, H.A.R., and J.H. Kramers (eds.). *Shorter Encyclopedia of Islam.* Ithaca: Cornell University Press, 1953.

Glassman, Jon D. *Arms for the Arabs.* Baltimore: Johns Hopkins University Press, 1975.

Graham, Robert. "Defence Policy," *Financial Times* [London], February 25, 1976, 24.

Green, Timothy. *The Universal Eye: The World of Television.* New York: Stein and Day, 1972.

______. *The World of Gold.* London: Michael Joseph, 1968.

Griffith, William E. "The Great Powers, the Indian Ocean, and the Persian Gulf," *Jerusalem Journal of International Relations* [Jerusalem], 1, No. 2, Winter 1975, 5–19.

"Gulf States Move Closer Together," *Middle East* [London], No. 16, February 1976, 35–36.

Gurfinkel, Mariano. "As Oil Prices Rise," *IMF Survey,* 4, No. 19, October 13, 1975, 297–305.

Habachy, Saba. "A Study in Comparative Constitutional Law: Constitutional Government in Kuwait," *Columbia Journal of Transnational Law,* 3, No. 2, 1965, 116–126.

Haddad, Wadi'd. "The Interaction Between Science and Society in the Arabic Press of the Middle East," *Science Education,* 58, No. 1, 1974, 35–49.

Halliday, Fred. *Arabia Without Sultans: A Political Survey of Instability in the Arab World.* New York: Vintage Books, Random House, 1975.

______. "Imperialism's Last Stand," *New Statesman* [London], 91, January 2, 1976, 6–7.

Harrigan, Anthony. "Security Interests in the Persian Gulf and Western Indian Ocean," *Strategic Review,* I, No. 3, Fall 1973, 13–22.

Hawley, Donald. *The Trucial States.* New York: Twayne Publishers, 1970.

Hazleton, Jared E. *Public Finance Prospects and Policies for Bahrain, 1975–1985.* Beirut: Ford Foundation, 1975.

Hess, Andrew C. "Consensus or Conflict: The Dilemma of Islamic Historians," *American Historical Review,* 81, No. 4, October 1976, 788–799.

Hewins, Ralph. *A Golden Dream: The Miracle of Kuwait.* London: W.H. Allen, 1964.

Hijazi, Ahmad. "Kuwait: Development from a Semi-Tribal, Semi-Colonial Society to Democracy and Sovereignty," *American Journal of Comparative Law,* 13, No. 3, Summer 1964, 428–438.

Hill, A.G. "The Gulf States: Petroleum and Population Growth." Pages 242–274 in J.I. Clarke and W.B. Fisher (eds.), *Populations of the Middle East and North Africa: A Geographical Approach.* New York: Africana Publishing, 1972.

Hitti, Philip K. *History of the Arabs.* (6th ed.) London: Macmillan, 1956.
______. *Islam: A Way of Life.* Minneapolis: University of Minnesota Press, 1970.
Hoagland, Jim. "Oman's Leap into the Present," *Washington Post,* December 29, 1974, B2–4.
Hodgson, Marshall G.S. *The Venture of Islam, I: The Classical Age of Islam.* Chicago: University of Chicago Press, 1974.
Hopwood, Derek (ed.). *The Arabian Peninsula: Society and Politics.* Totowa, New Jersey: Rowman and Littlefield, 1972.
Hourani, George. *Arab Seafaring.* Beirut: Khayats, 1963.
Huneidi, Isa A. "The Transplants That Produced a Democratic Judiciary," *Kuwait Digest* [Kuwait], 4, No. 1, January–March 1976, 23–25.
Hurewitz, J.C. *Diplomacy in the Near and Middle East.* 2 vols. New York: D. Van Nostrand, 1956.
______. *Middle East Politics: The Military Dimension.* New York: Praeger, 1969.
______. "The Persian Gulf: British Withdrawal and Western Security," *Annals of the American Academy of Political and Social Science,* 401, May 1972, 106–115.
______. *The Persian Gulf: Prospects for Stability.* (Headline Series, No. 220.) New York: Foreign Policy Association, April 1974.
International Bank for Reconstruction and Development. *The Economic Development of Kuwait.* Baltimore: Johns Hopkins Press, 1965.
International Institute for Strategic Studies. *The Middle East and the International System,* Pt. II: Security and the Energy Crisis. (Adelphi Papers, No. 115.) London: 1975.
______. *The Military Balance: 1975–1976.* London: 1975.
______. *The Military Balance: 1976–1977.* London: 1976.
International Monetary Fund. *IMF Survey,* August 11, 1975.
______. *International Financial Statistics.* Washington: November 1976, 318.
International Petroleum Encyclopedia, 1970. Tulsa: Petroleum Publishing, 1970.
International Petroleum Encyclopedia, 1971. Tulsa: Petroleum Publishing, 1971.
International Petroleum Encyclopedia, 1972. Tulsa: Petroleum Publishing, 1972.
International Petroleum Encyclopedia, 1973. Tulsa: Petroleum Publishing, 1973.
International Petroleum Encyclopedia, 1974. Tulsa: Petroleum Publishing, 1974.
International Petroleum Encyclopedia, 1975. Tulsa: Petroleum Publishing, 1975.

International Petroleum Encyclopedia, 1976. Tulsa: Petroleum Publishing, 1976.
Ismael, Tareq Y., et al. *Governments and Politics of the Contemporary Middle East.* Homewood, Illinois: Dorsey Press, 1970.
———. *The Middle East in World Politics.* Syracuse: Syracuse University Press, 1974.
Izzard, Ralph. "The Fight for Federation," *Middle East International* [London], No. 1, April 1971, 33–35.
Jane's All the World's Aircraft, 1972–73. (Ed., John W. Taylor.) New York: McGraw-Hill, 1973.
Jane's All the World's Aircraft, 1975–76. (Ed., John W. Taylor.) London: Jane's Yearbooks, 1975.
Jane's Fighting Ships, 1975–76. (Ed., John E. Moore.) London: Jane's Yearbooks, 1975.
Jane's Weapon Systems, 1969–70. (Eds., R.T. Pretty and D.H.R. Archer.) New York: McGraw-Hill, 1970.
Jane's Weapon Systems, 1971–72. (Eds., R.T. Pretty and D.H.R. Archer.) New York: McGraw-Hill, 1972.
Jane's Weapon Systems, 1976. (Ed., R.T. Pretty.) London: Jane's Yearbooks, 1976.
Johns, Richard. "The Emergence of the United Arab Emirates," *Middle East International* [London], No. 21, March 1973, 8–10.
———. "A Hard Road to Unity," *Financial Times* [London], May 22, 1975.
Johns, Richard, and Robert Graham. "Kuwait to Buy 150 Chieftain Tanks in £100m. Deal," *Financial Times* [London], February 16, 1976, 1.
Joyner, Christopher C. "The Petrodollar Phenomenon and Changing International Economic Relations," *World Affairs,* 138, No. 2, Fall 1975, 152–176.
Kabeel, Soraya. *Source Book on the Arabian Gulf States: Arabian Gulf in General, Kuwait, Bahrain, Qatar, and Oman.* Kuwait: Kuwait University Press, 1975.
Kazziha, Walid. *Revolutionary Transformation in the Arab World.* New York: St. Martin's Press, 1975.
Kelly, J.B. "A Prevalance of Furies: Tribes, Politics, and Religion in Oman and Trucial Oman." Pages 107–144 in Derek Hopwood (ed.), *The Arabian Peninsula: Society and Politics.* Totowa, New Jersey: Rowman and Littlefield, 1972.
Kelly, John Barrett. *Britain and the Persian Gulf, 1795–1880.* New York: Oxford University Press, 1968.
———. *Eastern Arabian Frontiers.* New York: Praeger, 1964.
Kennedy, Edward M. "The Persian Gulf: Arms Race or Arms Control?" *Foreign Affairs,* 54, No. 1, October 1975, 14–35.
Kennedy, Gavin. *The Military in the Third World.* New York: Charles Scribner's Sons, 1974.

Kergan, J.L. "Social and Economic Changes in the Gulf Countries," *Asian Affairs* [London], 62 (New Series, VI), Pt. III, October 1975, 282–289.

Key, Kerim K. *The Arabian Gulf States Today.* Washington: Asia Research Center, 1974.

Kilner, Peter and Jonathan Wallace, eds. *The Gulf Handbook 1966–77.* London, MEED, 1976

Knauerhase, Ramon. *The Saudi Arabian Economy.* New York: Praeger, 1975.

Koury, Enver M. *Oil and Geopolitics in the Persian Gulf Area: A Center of Power.* Beirut: Catholic Press, 1973.

"Kuwait," *Financial Times* [London], February 25, 1976 (special supplement).

"Kuwait." Pages 157–168 in Anthony Axon, et al., *Middle East Annual Review: 1975–6.* Great Chesterford, England: Middle East Review, 1975.

"Kuwaiti Crown Tightens Reins," *Middle East Enterprise,* September 9, 1976, 2.

"Kuwait's Efforts to Pace Development Result in Large Rise in Foreign Assets," *IMF Survey,* 4, No. 22, November 24, 1975, 351–353.

Laliberté, Gérard. "La Guérrilla du Dhofar," *Etudes Internationales* [Quebec], IV, Nos. 1 and 2, March-June 1973, 159–181.

Landen, Robert G. "Gulf States." Pages 295–315 in Abid A. Al-Marayati, et al. (eds.), *The Middle East: Its Governments and Politics.* Belmont, California: Duxbury Press, 1972.

———. *Oman Since 1856: Disruptive Modernization in a Traditional Arab Society.* Princeton: Princeton University Press, 1967.

Laost, Henri. *Essai sur les Doctrines Sociales et Politiques de Taki-D-Din Ahmad b Tamiya.* Cairo: Imprimerie de l'Institut Français d'archéologie Orientale, 1939.

Lateef, Abdul. "A Security Pact in the Gulf?" *Middle East International* [London], No. 55, January 1976, 21–23.

Lederer, Ivo J., and Wayne S. Vucinich (eds.). *The Soviet Union and the Middle East: The Post World War II Era.* Stanford: Hoover Institution, Stanford University Press, 1974.

Leemans, W.F. *Foreign Trade in the Old Babylonian Period.* Leiden: E.J. Brill, 1959.

Levy, Reuben. *The Social Structure of Islam.* (2d ed.) Cambridge: Cambridge University Press, 1969.

Lewicki, T. "Ibadiyya." Pages 648–660 in Bernard Lewis, et al. (eds.), *The Encyclopedia of Islam,* III. (New ed.) Leiden: E.J. Brill, 1968.

Liebesny, Herbert J. "Administration and Legal Development in Arabia: The Persian Gulf Principalities," *Middle East Journal,* 10, No. 1, Winter 1956, 33–42.

———. *The Law of the Near and Middle East: Readings, Cases, and Materials.* Albany: State University of New York Press, 1975.

———. *Qatar.* (Constitutions of the Countries of the World series, Albert P. Blaustein and Gisbert H. Flanz, eds.) Dobbs Ferry, New York: Oceana Publications, July 1973.

Lippman, Thomas W. "Jet Airports Are Status Symbols in Wealthy Emirates," *Washington Post,* November 6, 1976, A16.

———. "Kuwait Crackdown Snuffs Last Lively Arab Newspapers," *Washington Post,* September 16, 1976, A17.

———. "Unity of Arab Emirates Is Facing Its Strongest Challenge," *Washington Post,* November 21, 1976, A18.

Long, David E. "Confrontation and Cooperation in the Gulf." (Middle East Problem Paper, No. 10.) Washington: Middle East Institute, 1974.

———. *The Persian Gulf: An Introduction to Its People, Politics, and Economics.* New York: Westview, 1976.

Mackie, Alan. "Kuwait's Housing: A Social and Political Issue," *Middle East Economic Digest* [London], 20, No. 39, September 1976, 5–6.

McLachan, Keith, and Narsi Ghorban. *Oil Production, Revenues, and Economic Development.* (Quarterly Economic Review Special Series, No. 18.) London: Economist Intelligence Unit, 1975.

McLaurin, R.D. *The Middle East in Soviet Policy.* Lexington, Massachusetts: Lexington Books, D.C. Heath, 1975.

Magnus, Ralph H. "Middle East Oil," *Current History,* 68, No. 402, February 1975, 49–53.

Malone, Joseph J. "America and the Arabian Peninsula: The First Two Hundred Years," *Middle East Journal,* 30, No. 3, Summer 1976, 406–424.

———. *The Arab Lands of Western Asia.* Englewood Cliffs: Prentice-Hall, 1973.

Mann, Major Clarence. *Abu Dhabi: Birth of an Oil Sheikhdom.* Beirut: Khayats, 1964.

Maull, Hanns. *Oil and Influence: The Oil Weapon Examined.* (Adelphi Papers, No. 17.) London: International Institute for Strategic Studies, 1975.

Mejido, Manual. "Kuwait: Oil Millions Bring Little Change," *Atlas,* 22, No. 3, March 1975, 36–37.

Melamid, Alexander. "Boundary Disputes in the Persian (Arab) Gulf." (Paper presented at Annual Meeting of Middle East Studies Association.) Boston: 1974 (mimeo.).

The Middle East and North Africa: 1975–76. (22d ed.) London: Europa Publications, 1975.

Middle East Economic Digest [London], January 9, 1976, 9.

Middleton, Drew. "The Sun Sets on Kipling's Domain," *New York Times,* July 21, 1976, 2.

Miles, Samuel Barrett. *The Countries and Tribes of the Persian Gulf.* London: Frank Cass, 1966.

Morris, Joe Alex. "Differences Plague Gulf Federation," *Washington Post,* April 4, 1976, El, E6.
Mosley, Leonard. *Power Play.* New York: Random House, 1973.
Nakhleh, Emile A. *Arab-American Relations in the Persian Gulf.* Washington: American Enterprise Institute, 1975.
______. *Bahrain: Political Development in a Modernizing Society.* Lexington, Massachusetts: D.C. Heath, 1976.
______. "Constitutional Development in the Arab Gulf." (Paper presented at Annual Meeting of Middle East Studies Association.) Boston: 1974 (mimeo.).
"Near and Middle East: Qatar." Pages 179 and 353 in *World Radio TV Handbook, 1976.* Hvidovre, Denmark: World Radio-TV Handbook, 1976.
"Notes: Qatar," *Military Review,* LV, No. 9, September 1974, 97.
"The Oil Crisis in Perspective," *Daedalus,* 104, No. 4, Fall 1975.
Oman. Embassy in Washington. *Oman News.* Washington: January–December 1976.
Oman. Ministry of Information and Tourism. *Agriculture and Fisheries.* N. pl.: N.d.
______. *Economic Development.* N. pl.: N.d.
______. *Oman.* Edinburgh: 1974.
______. *Southern Region—Dhofar.* N. pl.: N.d.
Oman: A MEED Special Report, 1976. London: Middle East Economic Digest, June 1976.
"Oman: A MEED Special Report, 1973," *Middle East Economic Digest* [London], 17, No. 29, July 20, 1973 (special supplement).
"Oman: Problems on All Sides," *Events: Newsmagazine on the Middle East* [London], October 15, 1976, 13–16.
"Oman Seeks Modernization of Economy, Emphasizes Income-Generating Projects," *IMF Survey,* 5, No. 3, February 2, 1976, 46–48.
O'Shea, Raymond. *The Sand Kings of Oman.* London: Methuen, 1947.
"Out of the Fire," *Economist* [London], 255, No. 6873, May 17, 1975 (special supplement).
Owen, Roderic. *The Golden Bubble.* London: Collins, 1957.
Page, Stephen. *The USSR and Arabia: The Development of Soviet Policies and Attitudes Towards the Countries of the Arabian Peninsula.* London: Central Asian Research Centre, 1972.
Peterson, J.E. "Britain and 'The Oman War': An Arabian Entanglement," *Asian Affairs* [London], 63 (New Series, VII), Pt. III, October 1976, 285–298.
Phillips, Wendell. *Oman: A History.* New York: Revnal, 1968.
______. *Unknown Oman.* New York: McKay, 1966.
Polk, William R. *The United States and the Arab World.* (3d ed.) Cambridge: Harvard University Press, 1975.

Population Reference Bureau. *1976 World Population Data Sheet.* Washington: 1976.
Price, D.L. "Building Bridges in the Gulf," *Middle East International* [London], No. 59, May 1976, 24–25.
______. *Oman: Insurgency and Development.* (Conflict Studies, No. 53.) London: Institute for the Study of Conflict, January 1975.
______. *Stability in the Gulf: The Oil Revolution.* (Conflict Studies, No. 71.) London: Institute for the Study of Conflict, May 1976.
"The Provisional Constitution of the United Arab Amirates," *Middle East Journal,* 26, No. 3, Summer 1972, 307–325.
Qatar. Customs Department. Statistics Section. *Yearly Bulletin of Imports and Exports for 1973.* Doha: N.d.
Qatar. Embassy in Washington. *Qatar.* Washington: N.pub., n.d.
______. *Qatar News,* 3, No. 1, January–February 1976.
______. *Qatar News,* 3, No. 2, March–April 1976.
Qatar. Ministry of Information. *Qatar: A Forward Looking Country with Centuries Old Traditions.* Doha: 1974.
"Qatar," *Financial Times* [London], November 18, 1975.
"Qatar." Pages 278–279 in Arthur S. Banks (ed.), *Political Handbook of the World, 1975.* New York: McGraw-Hill, 1975.
"Qatar." Page 319 in Jean Labayle Couhat (ed.), *Combat Fleets of the World, 1976–77: Their Ships, Aircraft, and Armament.* (Trans., James J. McDonald.) Annapolis: Naval Institute Press, 1976.
"Qatar." Pages 1328–1333 in *The Europa Yearbook, 1976: A World Survey,* II. London: Europa Publications, 1976.
"Qatar." Pages 579–586 in *The Middle East and North Africa, 1974–75.* London: Europa Publications, 1974.
"Qatar." Pages 584–591 in *The Middle East and North Africa, 1976–77.* London: Europa Publications, 1976.
"Qatar." Pages 1261–1262 in John Paxton (ed.), *The Statesman's Year-Book, 1976–1977.* New York: St. Martin's Press, 1976.
"Qatar," *Voice of the Arab World* [London], (advanced supplement), No. 53, September 16, 1976, 5–20.
"Qatar: A Special Report," *The Times* [London], June 23, 1975, I–IX.
Qatar Statistical Yearbook, 1974–1975 (in Arabic). Doha: Ministry of Education and Youth Welfare, n.d.
"Qatar, with '74 Oil Income Approaching Past Decade's Total, Stresses Welfare," *IMF Survey,* 3, No. 17, August 19, 1974, 258–260.
Ramazani, Rouhallah K. *The Persian Gulf: Iran's Role.* Charlottesville: University of Virginia Press, 1972.
Rand, Christopher T. *Making Democracy Safe for Oil.* Boston: Atlantic Monthly Press, 1975.
Rentz, George. "A Sultanate Asunder," *Natural History,* 83, No. 3, March 1974, 58–66.
______. "The Wahhabis." Pages 270–284 in A.J. Arberry (ed.), *Religion*

in the Middle East: Three Religions in Concord and Conflict, II: Islam. Cambridge: Cambridge University Press, 1969.

Report of the Directors and Balance Sheet. Doha: Qatar National Bank, 1976.

Report of the Directors and Balance Sheet, 31 December 1974. Doha: Qatar National Bank, 1975.

Rouleau, Eric. "My People Feel the Need of an Absolute Monarch to Protect Their Interests. . . ." (Undated interview with Sultan Qaboos of Oman), *Le Monde* [Paris], November 1976.

"Royal Decree No. 2774 Concerning 1974 Draft Budget," *Al-Jaridah Al-Rasmiyah* [Oman], April 15, 1974, 4–9.

Rubin, Trudy. "What Do the Palestinians Want, III: The Palestinians in Kuwait." (U.S. Department of State, Office of External Research, Foreign Affairs Research Paper, No. FAR 22716.) Washington: June 1975 (mimeo.).

Rubinacci, Roberto. "The Ibadis." Pages 302–317 in A.J. Arberry (ed.), *Religion in the Middle East: Three Religions in Concord and Conflict, II: Islam.* Cambridge: Cambridge University Press, 1969.

Rustow, Dankwart A. "Who Won the Yom Kippur and Oil Wars," *Foreign Policy,* 17, Winter 1974, 166–175.

Sadik, Muhammad T., and William P. Snavely. *Bahrain, Qatar, and the United Arab Emirates: Colonial Past, Present Problems, and Future Prospects.* Lexington, Massachusetts: D.C. Heath, 1972.

Salisbury, Matthew. "End of a Rebellion," *Middle East International* [London], No. 57, March 1976, 18–20.

Sampson, Anthony. *The Seven Sisters: The Great Oil Companies and the World They Made.* New York: Viking Press, 1975.

Sayigh, Yusif A. "Problems and Prospects of Development in the Arabian Peninsula," *International Journal of Middle East Studies* [London], 2, No. 1, January 1971, 40–58.

Scuka, Dario. "OPEC: Background, Review and Analysis." (Library of Congress, Congressional Research Service, HD 9560 Middle East, 74–189E.) Washington: U.S. Library of Congress, October 24, 1975.

Sheehan, Edward R.F. "Unradical Sheiks Who Shake the World," *New York Times Magazine,* March 24, 1974, 13–16, 50–54, 58.

Shehab, Fakhri. "Kuwait: A Super Affluent Society." *Foreign Affairs,* 42, No. 3, April 1964, 461–474.

Shwadran, Benjamin. "The Kuwait Incident," Pt. I, *Middle Eastern Affairs,* XIII, No. 1, January 1962, 2–14.

———. "The Kuwait Incident," Pt. II, *Middle Eastern Affairs,* XIII, No. 2, February 1962, 45–53.

Skeet, Ian. *Muscat and Oman: The End of An Era.* London: Faber and Faber, 1974.

Soysal, Ismail. "Bridging Information Gap Between Muslim States," *Middle East* [London], No. 23, September 1976, 72–75.

The Statesman's Year-Book, 1976–1977. (Ed., John Paxton.) New York: St. Martin's Press, 1976.

Statistical Year Book. Muscat: Development Council, National Statistical Department, Sultanate of Oman, 1974.

Stephens, R. *The Arabs' New Frontier.* London: Temple Smith, 1973.

Stoakes, Frank. "Social and Political Change in the Third World: Some Peculiarities of the Oil Producing Principalities of the Middle East." Pages 189–215 in Derek Hopwood (ed.), *The Arabian Peninsula: Society and Politics.* Totowa, New Jersey: Rowman and Littlefield, 1972.

Stockholm International Peace Research Institute. *Arms Trade Registers: The Arms Trade with the Third World.* Cambridge: MIT Press, 1975.

Stookey, Robert W. *America and the Arab States: An Uneasy Encounter.* New York: John Wiley, 1975.

Stork, Joe. *Middle East Oil and the Energy Crisis.* New York: Monthly Review Press, June 1975.

Sweet, Louise E. "The Arabian Peninsula." Pages 199–226 in Louise E. Sweet (ed.), *The Central Middle East.* New Haven: HRAF Press, 1971.

Tachau, Frank (ed.). *Political Elites and Political Development in the Middle East.* Cambridge: Schenkman Publishing, 1975.

Tahtinen, Dale R. *Arms in the Persian Gulf.* Washington: American Enterprise Institute, 1974.

Tibawi, A. L. *Islamic Education: Its Traditions and Modernization into the Arab National Systems.* London: Luzac, 1972.

Tomkinson, Michael. *The United Arab Emirates.* London: Michael Tomkinson Publishing, 1875.

Tremayne, Penelope. "Guevara Through the Looking Glass: A View of the Dhofar War," *Journal of the Royal United Services Institute for Defence Studies* [London], 119, No. 3, September 1974, 39–43.

United Arab Emirates. Ministry of Information and Culture. *United Arab Emirates.* N. pl.: N. pub., n.d.

"United Arab Emirates." Pages 489–539 in Peter Kilner and Jonathan Wallace (eds.), *The Gulf Handbook, 1976–77.* London: Trade and Travel Publications, 1976.

"United Arab Emirates, Aided by Oil Income, Continues Effort Toward Central Institutions," *IMF Survey,* 4, August 11, 1975, 238–240.

"United Arab Emirates: A MEED Special Report, 1973," *Middle East Economic Digest* [London], 17, No. 26, June 29, 1973.

U.S. Arms Control and Disarmament Agency. *Arms Control Report.* Washington: July 1976.

———. *World Military Expenditures and Arms Transfers, 1965–1974.* (ACDA Publication, No. 84.) Washington: GPO, 1976.

U.S. Central Intelligence Agency. *Communist Aid to Less Developed*

Countries of the Free World, 1975. (No. ER 76–10372U.) McLean, Virginia: July 1976.

______. *Research Aid: Handbook of Economic Statistics, 1975.* (No. A [ER] 75–65.) Washington: August 1975.

U.S. Congress. 90th, 1st Session. House of Representatives. Committee on Foreign Affairs. Subcommittee on Near Eastern and South Asian Affairs. *Arms Sales to Near East and South Asian Countries.* (Hearings, March 14, April 13, 20, 25, and June 22, 1967.) (Document No. 78–872.) Washington: GPO, 1967.

U.S. Congress. 92d, 2d Session. House of Representatives. Committee on Foreign Affairs. Subcommittee on Near East and South Asia. *U.S. Interests in and Policy Toward the Persian Gulf.* (Hearings, February 2, June 7, August 8, and 15, 1972.) (Document No. 45–672.) Washington: GPO, 1972.

U.S. Congress. 93d, 1st Session. House of Representatives. Committee on Foreign Affairs. Subcommittee on Near East and South Asia. *New Perspectives on the Persian Gulf.* (Hearings, June 6, July 17, 23, 24, and November 28, 1973.) Washington: GPO, 1973.

U.S. Congress. 93d, 2d Session. House of Representatives. Committee on Foreign Affairs. Subcommittee on Near East and South Asia. *The Persian Gulf, 1974: Money, Politics, Arms and Power.* (Hearings, July 30, August 5, 7, and 12, 1974.) (Document No. 45–672.) Washington: GPO, 1974.

U.S. Congress. 93d, 2d Session. Senate. Foreign Relations Committee. Subcommittee on Multinational Corporations. *Multinational Corporations and United States Foreign Policy,* Pt. 7. Washington: GPO, 1974.

U.S. Congress. 94th, 1st Session. House of Representatives. Committee on International Relations. Special Subcommittee on Investigations. *The Persian Gulf, 1975: The Continuing Debate on Arms Sales.* (Hearings, June 10, 18, 24, and July 29, 1975.) (Document No. 78–872.) Washington: GPO, 1975.

______. *United States Arms Sales to the Persian Gulf: Report of A Study Mission to Iran, Kuwait, and Saudi Arabia.* (Hearings, May 22–31, 1975.) Washington: GPO, January 19, 1976.

U.S. Department of Agriculture. Foreign Agricultural Service. *Saudi Arabia, Bahrain, and Kuwait: Food Supply.* Washington: August 1974.

U.S. Department of Commerce. *Foreign Economic Trends and Their Implications for the United States.* (FET 76–007.) Muscat: American Embassy, 1976.

U.S. Department of Commerce. Domestic and International Business Administration. *Market Profiles for the Near East and North Africa.* (Overseas Business Reports, OBR 75–40.) Washington: October 1975.

U.S. Department of Commerce. Office of Technical Services. Joint Publications Research Service—JPRS (Washington). The following items are from the JPRS Series:

Translations on Near East.

"Al-Ittihad Follows the National Assembly—Discussions in Reply to the Speech by the Head of State," *Al-Ittihad,* Abu Dhabi, January 17, 1974. (JPRS: 61985, No. 1161, May 14, 1974.)

"Annual Report of Development Ministry," *'Uman,* Muscat, August 10, 1974. (JPRS: 62326, No. 1254, October 31, 1974.)

"Doubling Patrols in City Eliminates Thieves and Reduces Crime," *Al-Ra'i al-'Amm,* Kuwait, February 17, 1974. (JPRS: 61822, April 23, 1974.)

"Federal Bill Defines Police, Security Forces Jurisdiction." (JPRS: 62321, June 25, 1974.)

"Flag of Omani Revolution Will Never Fly at Half-Mast," *Al-Thawri,* Aden, June 12, 1976. (JPRS: 67892, No. 1559, September 10, 1976.)

"Interview with Shaykh Sa'd al Abdullah, Minister of Defense and Interior," *Al Siyasah,* Kuwait, April 17, 1974. (JPRS: 61971, May 13, 1974.)

"Law Concerning Military Pay and Allowances, Kuwait," *Al Kuwait al Yawm,* Kuwait, March 5, 1972. (JPRS: 55731, No. 746, April 17, 1972.)

"1976 Budget Breakdown," *Al-Jaridah Al-Rasmiyah,* Manamah, Bahrain, February 12, 1976. (JPRS: 67309, May 14, 1976.)

"Omani-Saudi Relations Are Strong and Are Becoming Better Established with Time," *'Uman,* Muscat, June 29, 1976. (JPRS: 67892, No. 1559, September 10, 1976.)

"On the Interior Ministry's Five-Year Plan," *Al Adaf,* Kuwait, September 11, 1975. (JPRS: 66031, October 29, 1975.)

"Political Concerns of the Arab Gulf States," *Al Taliah,* Kuwait, November 4, 1975. (JPRS: 66606, January 16, 1976.)

"Salim Al-Ghazali Talks About Armed Forces," *'Uman,* Muscat, May 29, 1976. (JPRS: 67661, No. 1542, July 27, 1976.)

"The Six Draft Induction Law Articles Which the National Assembly Has Approved," *Al-Ra'i al-'Amm,* Kuwait, November 23, 1975. (JPRS: 66606, January 16, 1976.)

U.S. Department of State. *Host Country Financed Technical Assistance: Oman.* (A report prepared for Agency for International Development.) East Orange: Louis Berger International, 1976.

———. *Purchase of Defense Articles and Services: Agreement Between the United States of America and the United Arab Emirates, June 15 and 21, 1975.* (Treaties and Other International Acts, Series 8139.) Washington: GPO, 1975.

U.S. Department of State. Bureau of Public Affairs. Office of Media Services. *Background Notes: Oman.* (Department of State Publication, No. 8070.) Washington: GPO, May 1975.

______. *Background Notes: Qatar.* (Department of State Publication, No. 7906.) Washington: GPO, October 1974.
______. *Background Notes: United Arab Emirates.* (Department of State Publication, No. 7901.) Washington: GPO, June 1975.
U.S. Department of State. Embassy in Muscat. *Major Projects List for Oman.* (A-51.) Muscat: 1976.
"Unveiling Oman's Economic Realities," *Middle East Economic Digest,* London, 20, No. 8, February 20, 1976, 7.
Vaglieri, Laura Veccia. "The Patriarchal and Umayyad Caliphates." Pages 57–103 in P.M. Holt, Ann K.S. Lambton, and Bernard Lewis (eds.), *The Cambridge History of Islam, I: The Central Islamic Lands.* Cambridge: Cambridge University Press, 1970.
Vreede-De Stuers, Cora. "Girl Students in Kuwait," *Bijdragen Tot De Taal-Land-En Volkenkunde* [Amsterdam], 130, 1974, 110–131.
Wall, David. *The Charity of Nations: The Political Economy of Foreign Aid.* New York: Basic Books, 1973.
Wall, Michael. "A Jump of Centuries: A Survey of the Arabian Peninsula," *Economist* [London], June 6, 1970 (special supplement).
"War of the Advisers," *Events: Newsmagazine on the Middle East* [London], October 15, 1976, 17–19.
Wilkinson, J.C. "The Origins of the Omani State." Pages 67–88 in Derek Hopwood (ed.), *The Arabian Peninsula: Society and Politics.* Totowa, New Jersey: Rowman and Littlefield, 1972.
Wilson, Arnold Talbot. *The Persian Gulf: An Historical Sketch from the Earliest Times to the Beginning of the Twentieth Century.* Oxford: Clarendon Press, 1928.
Winder, R. Bayly. *Saudi Arabia in the Nineteenth Century.* New York: St. Martin's Press, 1965.
Winstone, Harry, and Zahra Freeth. *Kuwait: Prospect and Reality.* New York: Crane, Russak, 1972.
World Armaments and Disarmament: SIPRI Yearbook, 1975. (Stockholm International Peace Research Institute.) Cambridge: MIT Press, 1975.
World Communications: A 200-Country Survey of Press, Radio, Television, and Film. New York: UNESCO Press, 1975.
World Health Organization. *World Health Organization Official Records,* Vol. 225. Geneva: 1975.
Wright, Denis. "The Changed Balance of Power in the Persian Gulf," *Asian Affairs* [London], 60, (New Series, IV), Pt. III, October 1973, 255–262.
Yager, Joseph A., and Eleanor B. Steinberg. *Energy and U.S. Foreign Policy.* Cambridge, Massachusetts: Ballinger Publishing, 1974.
Zabih, Sephr. "Iran's Policy Toward the Persian Gulf," *International Journal of Middle East Studies* [London], 7, No. 3, July 1976, 345–358.
Zand, S. "OPEC: Birth, Growth, Achievements." (Paper presented to

American University Seminar on International Oil, April 14, 1975.) Washington: 1975 (mimeo.).

(Various issues of the following periodicals were also used in the preparation of this book: *Arab Report and Record* [London], January 1971–December 1976; *Economist* [London], January–July 1976; *Financial Times* [London], January 1973–December 1976; *Foreign Broadcast Information Services Daily Report: Middle East and North Africa* [Washington], January 1971–December 1976; *International Financial Statistics* [Washington], January–July 1976; *IMF Survey* [Washington], June 1973–November 1975; *Keesings Contemporary Archives* [London], March 1971–September 1976; *Kuwait Digest* [Kuwait], January 1975–July 1976; *Manchester Guardian Weekly* [London], January–September 1976; *Middle East Economic Digest* [London], January 1971–December 1976; *Middle East International* [London], January 1975–July 1976; *Middle East Journal* [Washington], January 1973–September 1976; *Monthly Bulletin of Statistics* [New York], January–December 1976; *New York Times*, January 1973–November 1976; *Oman News* [Washington], January–December 1976; *Petroleum Intelligence Weekly* [New York], January–July 1976; *Translations on Near East and North Africa* [Washington], November 1975-November 1976; and *Washington Post*, January 1973-September 1976.)

GLOSSARY

Al—Uppercased it connotes family of or belonging to, as in Al Sabah, Al Khalifah, Al Thani, Al Nuhayyan, Al Maktum, and Al Qasimi. Lowercased it represents the definite article *the*, as in Ras al Khaymah.

amir (pl., *umara*)—Literally, commander. In many of the Arab states on the Persian Gulf, amir means ruler or king.

amirate—Political entity under the rule of an amir. Analogous to a shaykhdom and, if an independent state, to a kingdom.

Bahraini dinar—Consists of 1,000 fils. Dinar notes were first issued in October 1965 with a gold content of 1.866 grams of fine gold, which remained unchanged in late 1976. The gold definition of the dinar kept it essentially constant at BD1 equal to 2.1 special drawing rights (SDR) *(q.v.)* through 1975. The dinar's value in terms of United States dollars was stable at BD1 equal to US$2.10 through 1971. Since then the yearly average exchange rate has fluctuated from BD1 equal to US$2.28 in 1972, to US$2.50 in 1973, to US$2.53 in 1974 and 1975. Until 1972 the dinar was pegged to the British pound sterling, but in March 1972 the dinar was unofficially pegged to the United States dollar. In 1976 the dinar had full foreign exchange backing and was a strong, stable currency. Bahrain maintained no exchange restrictions and no restrictions on transfers of profits or capital.

barrels per day—Production of crude oil and petroleum products is frequently measured in barrels per day, often abbreviated bpd or bd. A barrel is a volume measure of forty-two United States gallons. Conversion of barrels to metric tons depends on the density of the specific product. About 7.3 barrels of average crude oil weigh one metric ton. Heavy crude would be about seven per metric ton. Light products, such as gasoline and kerosine, would average close to eight barrels per metric ton.

BD—Bahraini dinar *(q.v.)*.

bin—Literally, son of; same as *ibn (q.v.)*.

dirham—Currency of United Arab Emirates (UAE), first issued in May 1973, consisting of 100 fils. Although many writers use the symbol Dh, the symbol UD is more common and is used in this study. The dirham was defined as equal to 0.1866 grams of fine gold, corresponding to 0.21 special drawing rights (SDR) *(q.v.)* and to US$0.25. Its value remained the same in late 1976. The Currency Board continued to peg the dirham to the United States dollar in 1976 even though several neighboring gulf states pegged their currencies to the SDR. Before the appearance of the dirham other currencies—Persian Gulf Indian rupees, Bahraini dinars, and Qatar-Dubai riyals—circulated in the amirates. The various currencies were related. The gold content of the Indian rupee (before 1966), the Qatar-Dubai riyal, and the dirham was the same and was one-tenth the gold value of the Bahraini dinar. In 1976 agreements were reached among the UAE, Qatar, and Bahrain permitting some use of each other's currencies in domestic transactions. Journalists also reported in 1976 that new notes were being designed for a unified lower gulf currency to be introduced, perhaps in 1978. Presumably the UAE, Qatar, and Bahrain would be the initial members if a unified currency system was introduced.

downstream—The oil industry views the production, processing, transportation, and sale of petroleum products as a flow process starting at the wellhead. Downstream includes any stage between the point of reference and the sale of products to the consumer. Upstream is the converse and includes exploration and drilling of wells.

hadith—Tradition based on the precedent of Muhammad's words and deeds that serves as one of the sources of Islamic law (sharia).

hijra (pl., *hujar*)—Literally, to migrate, to sever relations, to leave one's tribe. Throughout the Muslim world hijra refers to the migration of Muhammad and his followers to Medina. In this sense the word has come into European languages as Hegira and is usually and somewhat misleadingly translated as flight. In Saudi Arabia the term also refers to agricultural settlements that combined features of religious missions, farming communities, and army camps. In the area of present-day Yemen (Sana) the term denoted a traditional process by which a holy man accepted an invitation to join a tribe or a group of tribes as judge or mediator.

ibn—Literally, son of; used before proper name to indicate descent from. *Bint* means daughter of; *banu* (or *bani*) is literally sons of and is used to mean tribe or family of.

imam—A word used in several senses. In general use it means the leader of congregational prayers; as such it implies no ordination or special spiritual powers beyond sufficient education to carry out this function. It is also used figuratively by many Sunni *(q.v.)* Muslims to mean the leader of the Islamic community. Among Shiites *(q.v.)* the word takes on many complex and controversial meanings; in general, however, it indicates that descendant of the House of Ali who is believed to be God's designated repository of the spiritual authority believed to be inherent in that line. The identity of this individual and the means of ascertaining his identity have been the major issues causing divisions among Shiites. Among the Ibadites of Oman the imam was elected to office and was regarded by all as the spiritual leader of the community and by some as the temporal ruler as well. Claims of various Omani imams to secular power led to open rebellions as late as the 1950s.

jihad—The struggle to establish the law of God on earth, often interpreted to mean holy war.

KD—Kuwaiti dinar *(q.v.)*.

Kuwaiti dinar—Consists of 1,000 fils. Dinar notes were first put into circulation on April 1, 1961, with a value of 2.488 grams of fine gold. The gold value remained unchanged in late 1976, but the relationship to the United States dollar and other currencies changed. One dinar was equal to US$2.80 through 1970. After the dollar was devalued and currency values began to vary widely, average exchange rates were KD1 equal to US$2.82 in 1971, US$3.05 in 1972, US$3.39 in 1973, US$3.41 in 1974, US$3.45 in 1975, and US$3.40 from January to June 1976.

majlis—Tribal council; in some countries the legislative assembly. Also refers to an audience with an amir or shaykh, open to all citizens for purposes of adjudication.

qadi (pl., qadis)—Judge in sharia *(q.v.)* courts.

Omani rial—Monetary unit of Oman, abbreviated OR or RO, although OR is more common and is used in this study. Divided into 1,000 baizas. Its gold content has remained unchanged since it replaced the Persian Gulf Indian rupee in May 1970. In late 1976 the Omani rial continued to be a freely convertible currency. Average conversion rates to the United States dollar have been: before August 15, 1971, OR1 equal to US$2.40; in 1972 OR1 equal to US$2.60; and since 1973 OR1 equal to US$2.89524.

OR—Omani rial *(q.v.)*.

Qatari riyal—Had a gold value equal to 0.1866 gram of fine gold, the same as the Qatar-Dubai riyal and the Persian Gulf Indian rupee. This made QR1 equal to 0.21 SDR (special drawing rights) *(q.v.)* through 1975. QR1 was equal to US$0.21 through 1971. In 1972 QR1 was equal to US$0.228; in 1973 QR1 was equal to US$0.25; in 1974 QR1 was equal to US$0.253; and in 1975 QR1 was equal to US$0.254. In March 1975 the riyal was pegged to the SDR, which resulted in an appreciation of the riyal of about 4 percent in terms of major currencies; previously the United States dollar had been the peg.

QR—Qatari riyal *(q.v.)*.

SDR—Special drawing rights *(q.v.)*.

sharia—Islamic law.

shaykh—Leader or chief. Word is used to mean either a political leader of a tribe or town or a learned religious leader. Also used as an honorific.

Shiite—A member of the smaller of the two great divisions of Islam. The Shiites supported the claims of Ali and his line to presumptive right to the caliphate and leadership of the world Muslim community, and on this issue they divided from the Sunni *(q.v.)* in the first great schism of Islam. Later schisms have produced further divisions among the Shiites.

special drawing rights (SDR)—An International Monetary Fund unit of account made up of a basket of major international currencies.

Sunni—A member of the larger of the two great divisions of Islam. The Sunni, who rejected the claims of Ali's line, believe they are the true followers of the Sunna, the guide to proper behavior composed of the Quran and the hadith *(q.v.)*.

UD—The dirham *(q.v.)*; currency of the United Arab Emirates.

ulama (sing., alim)—Collective term for Muslim religious scholars.

Wahhabi—Name used outside Saudi Arabia to designate an adherent of Wahhabism *(q.v.)*.

Wahhabism—Name used outside Saudi Arabia to designate official interpretation of Islam in Saudi Arabia. The faith is a puritanical concept of Unitarianism (the oneness of God—Al Dawah al Tauhid) that was preached by Muhammad ibn Abd al Wahhab, whence his Muslim opponents derived the name. The royal family of Qatar and many Qataris are Wahhabis.

INDEX

Abbasid caliphate: 18, 342
Abd al Aziz (king, Saudi Arabia): 27
Abd al Aziz bin Khalifah Al Thani, Shaykh: 257, 258, 259
Abd al Aziz ibn Saud. *See* Ibn Saud
Abd al Hafiz Salim Rajab: 349
Abd al Malik Al Qasimi, Shaykh: 291, 295
Abd al Rahim trading house: 355
Abd al Rahman trading house: 355
Abd Allah al Sabah: 26
Abd Allah al Salim Al Sabah, Shaykh (first amir of Kuwait): 122, 166–167, 169
Abd Allah bin Humayd Al Qasimi, Shaykh: 291, 295
Abd Allah ibn al Saffar: 49
Abd Allah ibn Humayd al Salmi: 53
Abd Allah ibn Nasir Al Suwaydi: 257
Abd Allah ibn Wahb al Rasib: 48
Abu Bakr: 16, 44, 49
Abu Bilal Midras ibn Udaiya al Tamimi: 48, 49
Abu Dhabi (United Arab Emirates): 11, 31, 35, 273, 300, 303, 306, 318, 320, 383; agriculture, 312, 314; banking, 321, 322; budget of UAE, 281–282, 283, 308; capital city, 288, 314; defense forces, 36, 325, 326, 327, 331; development and planning, 37, 307; government, 276–281, 287, 306, 309–312; oil industry, 83, 101, 109–113; oil production, 76, 108; oil reserves, 4, 80, 109, 309; oil revenues, 303; Oman and, 385, 395, 401; Qatar and, 264, 326; social system, 62, 63
Abu Dhabi Defense Force (ADDF): 325, 330, 334
Abu Dhabi Fund for Arab Economic Development (ADFAED): 298, 310–311
Abu Dhabi Marine Areas Company (ADMA): 109, 110, 111, 112
Abu Dhabi National Oil Company (ADNOC): 109, 110, 112
Abu Dhabi Oil Company (ADOC): 109, 112
Abu Dhabi Petroleum Company (ADPC): 109, 110, 111, 113
Abu Musa island: 13, 73, 114, 273, 326
Abu Safah oil field: 102
Abu Zaby (Abu Dhabi city): 109
ADDF. *See* Abu Dhabi Defense Force
Adelman, Morris A.: 82
ADMA. *See* Abu Dhabi Marine Areas Company
Adnani tribe (*see also* Nizari tribe): 15, 16, 28, 30, 32, 65
ADNOC. *See* Abu Dhabi National Oil Company
ADOC. *See* Abu Dhabi Oil Company
ADPC. *See* Abu Dhabi Petroleum Company
adult education: 367
AFESD. *See* Arab Fund for Economic and Social Development
African countries: 180, 265; oil reserves, 80
agriculture (*see also* irrigation): Bahrain, 206, 214, 215, 216–217; Kuwait, 143, 152–154; Oman, 14, 359, 361, 370–374, 377; Qatar, 244–246; United Arab Emirates, 312–316
Ahmad Abd Allah Al Ghazali: 349, 356
Ahmad al Khatib: 169
Ahmad bin Ali Al Thani, Shaykh (former ruler, Qatar): 256, 257, 262, 266
Ahmad bin Muhammad, Shaykh: 352
Ahmad bin Rashid Al Mualla, Shaykh (ruler, Umm al Qaywayn): 278, 294
Ahmad ibn Majid: 21
airports: 206, 208, 236, 238, 284, 306, 368
Ajman (United Arab Emirates): 273, 294, 316; government, 276, 278, 279, 287, 294; history, 31; oil, 4
al (word): vi–vii, 50, 57, 60. *See* glossary
Al Adami: 120
Al Ahmadi: 120, 126, 175, 198
Al Ahmadia school: 131
Al Asfoor trading house: 355
Al Awali: 206
Al Ayn: 63, 273, 288, 312, 313, 317
Al Azd tribe: 15
Al Bu Falah clan: 31
Al Bu Falasah family: 31, 64, 290
Al Bu Koosh Oil Company: 109, 112
Al Bu Said family: 5, 53, 66, 342, 343, 353
Al Bunduq oil field: 105, 264
Al Dafrah tribe: 288
Al Ghanim family: 60, 166
Al Ghubra: 378, 379

Al Hasa: 25
Al Hirth tribe: 66, 352, 353, 354
Al Husayniyah: 236
Al Jabal al Ali: 318
Al Jabir royal family: 122, 166, 169
Al Jahrah oasis: 120, 123
Al Jamaliyah: 236
Al Jufayr naval base: 231
Al Julanda tribe: 15
Al Karama: 281
Al Kathir tribe: 66, 354, 389
Al Khalid family: 166
Al Khalifah ruling family: 25, 26, 27, 28, 29, 52, 54, 61, 207, 230, 266
Al Khawr: 236, 258, 262
Al Kindi family: 290
Al Kiranah: 236
Al Madfa family: 291
Al Maktum family: 64, 290; chart, 292
Al Mualla tribe: 294
Al Muharraq island: 196, 206, 208
Al Nuhayyan family: 63, 288; chart, 289
Al Qasimi family: 291; charts, 293, 295
Al Sabah family: 3, 25, 60, 122, 165, 230; chart, 167
Al Salih family: 166
Al Salim family: 122, 166
Al Saqr family: 166
Al Saud family: 50, 230
Al Sharqi tribe: 294
Al Tanab mosque: 47
Al Thani ruling family: 5, 28, 54, 62, 230, 237, 256, 261; chart, 257
Al Ubaid culture: 11, 12
Al Wakrah: 236, 258
alfalfa: 245, 313, 371, 372
Algeria: 111, 150; oil and, 84, 86, 87, 90
Ali: 44, 45, 48, 49
Ali bin Abd Allah Al Thani, Shaykh (former ruler, Qatar): 256, 257
Ali bin Khalifah Al Khalifah, Shaykh: 207
Ali ibn Ahmad Al Ansari: 257
Ali Sultan trading house: 355
aluminum smelters: Bahrain, 103, 219, 223; Dubai, 114, 318
American Independent Oil Company (Aminoil): 93, 94, 96
amir (office): 57, 258; Bahrain, 225, 227; Dubai, 306; Kuwait, 144, 165, 172–173, 188, 198; Qatar, 252
An Nabi Salih island: 206, 208, 216
ANM. *See* Arab National Movement
Anthony, John Duke: 59, 196, 288, 294, 337, 353
Ar Rayyan: 236, 258
Ar Rifa Al Gharbi: 206
Ar Rumaytha: 206
Arab Fund for Economic and Social Development (AFESD): 148, 265
Arab Gulf Organization for Consultancy Industrial: 249
Arab-Israeli wars: 179, 186; (June 1967), 86, 170; (October 1973), 87, 170
Arab League: 6, 7, 39, 84, 170, 180, 184, 230, 265, 268, 297, 326, 344
Arab Military Industries Organization: 192, 265
Arab National Movement (ANM): 169, 176, 182, 233, 388, 389
Arab States Broadcasting Union: 267
Arabia: 11, 12, 17
Arabian American Oil Company (Aramco): 86, 93, 102
Arabian Oil Company (OAC): 93, 94, 96, 97
Arabic language: vi–vii, 228, 229, 242, 267, 268, 280, 296, 297, 358
archaeology: 11, 122
area: *See* Country profile, ix; Oman, 358; Qatar, 238; United Arab Emirates, 299, 301
armed forces and defense (*see also* Great Britain; military equipment and assistance received; military pay and allowances; military rank and insignia; Sultan's Armed Forces): ix, 20, 36, 57; Bahrain, 213, 231–232; foreign members, 35, 58, 59; Kuwait, 145, 173, 183–188, 192–203; Oman, 356, 393, 399, 405–408; Qatar, 268–271; United Arab Emirates, 4, 282, 308, 323–335
artisans: 349, 376
Ash Shamiyah: 120
Ash Shuaybah: 120, 124, 125; industrial park, 155, 157; refinery, 73, 96, 98
Asim al Jamali: 349
Australia: 230
Awali oil field: 101, 103
Awamir tribal group: 63, 288
awqaf: 43
Az Zannah: 109, 317
Azdi tribal faction: 65
Azraqis: 48

Baath Party of Iraq: 233, 263, 386
Bahr as Salwa: 236, 238
Bahrain: ix, 2, 3, 5, 37, 61, 206–236, 322; Great Britain and, 24, 35; history, 13, 14, 22, 25–27, 29, 36; Iran and, 38; Kuwait and, 180; legal system, 176; oil industry, 83, 90, 91, 92, 100–104, 217–218; oil pro-

duction, 76; oil reserves, 2, 75; Qatar and, 238, 266, 267; religion, 46, 52, 54
Bahrain island: 206, 208, 216
Bahrain National Liberation Front: 226, 234
Bahrain Petroleum Company (Bapco): 101, 102, 103; newspapers, 228–229
Bahwan trading house: 355
balance of payments: Bahrain, 221, 222; Kuwait, 160, 161; Oman, 384; Qatar, 251; United Arab Emirates, 321
Baluchis: 3, 59, 65, 67, 263, 269, 388, 403
Baluchistan: 33
Bandar Abbas: 23, 24, 29
Bangladesh: 251
bani (word): 57
Bani Abd al Qais: 15, 16, 27
Bani Ali family: 256
Bani Amr tribe: 66, 353
Bani Ghafir tribe: 32, 66, 353
Bani Hamad family: 256
Bani Hina tribe: 32, 66, 344
Bani Khalid family: 25, 256
Bani Riyam tribe: 344
Bani Tamin clan: 28
Bani Yas tribal group: 28, 31, 63, 64, 288
banking system: Bahrain, 223–225; Kuwait, 162–163; Oman, 381, 382; Qatar, 254; United Arab Emirates, 321
Banu Umayyah tribe: 44
base-point pricing system for oil: 81–82
Bashur: 350
Basrah: 23, 26
Batinah coast: 340, 350, 359, 361, 362, 371, 372
Beirut College: 301
Belgrave, Charles: 37, 227
birth control. *See* family planning programs
birthrate: 129
boundaries: xiv, 73, 120, 122, 236, 264, 273
British Petroleum (BP): 91, 92, 97, 98, 99, 104, 109, 112
British relationship with Gulf states. *See* Great Britain
Bubiyan island: 120, 122, 185
budget, national: Bahrain, 222, 223; Kuwait, 144, 145, 146, 165, 189, 192; Oman, 379, 380, 399–402; Qatar, 252–254; United Arab Emirates, 280, 284, 304, 307–309, 332
Burayk bin Hamud al Ghaffari: 349
Buraymi Oasis: 15, 54, 63, 273, 288, 298, 303, 313, 316, 328, 340, 344, 350
Burgan oil field: 93, 97, 195
Bushire (Residency): 24, 29
calendar, Muslim: vii, viii, 41
California-Texas Oil Company (Caltex): 101
camels: 18, 371
Canada: oil, 80, 101; Oman and, 375, 383, 400
cattle. *See* livestock
cement production: 125, 154, 156, 247, 248, 255, 284, 316, 317, 376, 383; concrete buildings, 362
censuses: 218, 300, 361
CFP. *See* Petroleum Company of France
children: 70
China, People's Republic of: Dhofar and, 390; trade, 159, 160, 264, 298, 384
China, Republic of: 98
Christianity: 15, 16, 47, 54; missionaries, 39
citizenship: 58; Bahrain, 210; Kuwait, 121, 126, 172, 176; Oman, 354; Qatar, 261, 262; United Arab Emirates, 335
civil service: Bahrain, 222; Kuwait, 129, 145, 171, 174; Oman, 353, 356; Qatar, 262; United Arab Emirates, 308, 310
climate: Bahrain, 209; Kuwait, 123; Oman, 360, 377; Qatar, 239; United Arab Emirates, 299–300
coal: 77, 78, 79, 80
Cochrane, Ronald (Muhammad Mahdi): 269
communications, mass: Bahrain, 228–229; Kuwait, 180–182; Oman, 357–358, 368; Qatar, 267–268; United Arab Emirates, 294–297
communist countries (*see also* names of specific countries): oil industry, 80; Oman and, 390, 398; Qatar and, 264; United Arab Emirates and, 333
Concorde plane: 215
Constitution: Bahrain, 207, 225–226; Kuwait, 122, 169, 172, 175, 176, 188, 201; Oman, 341
Constitution, Provisional: Qatar, 238, 258, 259, 260, 261, 269; United Arab Emirates, 276, 277, 279, 280–296, 329, 331, 336
consumer goods: Kuwait, 140, 156, 160; Oman, 355, 382; Qatar, 250, 265–266; United Arab Emirates, 319, 320
Council of Ministers (*see also* ministries): Kuwait, 172, 173, 181; Oman, 349; Qatar, 259; United Arab Emirates, 277, 278, 279, 284, 288
courts: 173, 177, 227, 260, 279–280, 290, 347
Crescent Petroleum Company: 114
crime and punishment: 197, 201, 202, 208, 336, 351
crude oil production: 76, 79, 80, 89; Abu Dhabi, 109, 110, 111, 113; Bahrain, 101–

102; Dubai, 113; Kuwait, 94–97, 100; Oman, 114, 116, 117
Cuba: 397, 398
currency: ix, 91; Bahrain, 223, 224, 225; Kuwait, 162; Oman, 370, 381; Qatar, 254; United Arab Emirates, 321–322

dagger *(khanjar):* 349, 363
Darwish trading house: 355
Das island: 73, 109, 112, 273
date palms: 10, 18, 153, 217, 244, 315, 371, 372
Dayinah island: 273
desalination plants: Bahrain, 218; Kuwait, 124, 125, 155; Oman, 378; Qatar, 239, 247, 248, 252; United Arab Emirates, 317
development and planning: 249; Bahrain, 218, 223; Kuwait, 147–148; Oman, 364, 366; Qatar, 248–249, 258; United Arab Emirates, 300, 306, 307, 308, 323
development projects: 36; Oman, 368, 373; United Arab Emirates, 4, 306, 317
Dhahirah region: 66, 340, 350, 353
Dhawahir tribal group: 63, 288
Dhofar Charitable Association (DCA): 388, 389
Dhofar insurgency: 6, 385–399
Dhofar Liberation Front (DLF): 388, 389, 390
Dhofar Province: 13, 115, 340, 350, 358, 359, 360, 361, 363, 371, 373, 378; tribes, 66, 354
Dhut Taj Lakit ibn Malik Al Azd: 16
diet and nutrition: Bahrain, 211; Kuwait, 136, 137; Oman, 361–363, 373, 374; Qatar, 241; United Arab Emirates, 304
Dilmun: 11, 12, 13
dinar: 162, 321
diseases: Bahrain, 213; Kuwait, 136, 137; Oman, 363–364; Qatar, 241; United Arab Emirates, 302
divorce: 70
Djabir ibn Zaid al Azdi: 49
Doha: 28, 62, 236, 238, 258, 267, 268
DPC. *See* Dubai Petroleum Company
Drury, William: 30
Dubai (United Arab Emirates): 10, 37, 273, 283, 300, 321, 383; Americans in, 299; banking, 321, 322; defense forces, 36, 325, 327; government, 276–281, 287, 290, 306, 307; history, 31, 34; oil industry, 113–114; oil production, 4, 76, 108, 303; port, 299, 307; Qatar and, 253, 266; social system, 62, 64; trade, 319
Dubai Petroleum Company (DPC): 113, 114
Dukhan: 105, 236
Dura tribe: 66, 353
Dutch East India Company: 23

economy: 57; Bahrain, 214–215, 222; Kuwait, 138–144, 155; Oman, 368–370; Qatar, 237, 243–244; United Arab Emirates, 303–305
Ecuador: 84
education: 2–3, 57, 72; Bahrain, 213–214; Kuwait, 130–134, 135, 145, 174; Oman, 6, 39, 67, 356–357, 358, 364, 365–368; Qatar, 242–243, 262; United Arab Emirates, 2, 301–302, 304
Egypt: 17, 176; loans and grants received, 150, 151, 191, 196, 264, 310; oil, 90, 110; Oman and, 396–397
Egyptians in Gulf states: 58, 62, 127, 132, 240, 262, 263, 367
electricity: Bahrain, 219; Kuwait, 155; Oman, 378; Qatar, 247, 248, 252; United Arab Emirates, 304, 316
Elphinstone inlet: 359
employment (*see also* labor force): non-citizens, 58, 59, 61, 64; oil industry, 101, 139, 164, 370; women, 129, 199, 211, 213, 355–356; skilled labor needs, 107, 300
English East India Company: 22, 23, 24, 29, 30, 34
English language: Bahrain, 229; Kuwait, 131, 182, 183; Oman, 358, 367; Qatar, 242, 267, 268; United Arab Emirates, 296
entrepôt trade: 208, 215, 319
equalization point in oil pricing: 82
exports: ix; Bahrain, 211, 217, 219, 220, 221; Kuwait, 154, 157, 159; Oman, 371, 374; Qatar, 246, 251; United Arab Emirates, 319, 320
exports of oil and gas: Abu Dhabi, 110; Bahrain, 101, 102; Kuwait, 95, 96, 158; Oman, 116, 117, 383; Qatar, 106

Fahad Mahmud Al Bu Said: 349
Fahar bin Taimur Al Bu Said: 349
Fahud oil fields: 73, 116, 340
Faisal (former king, Saudi Arabia): 263
Faisal bin Ali Al Bu Said: 349
Faisal bin Turki Al Bu Said, Sultan: 344
Falna: 31
family planning programs: 130, 209, 239
Farsi language: 59, 358
Fateh oil field: 113
Faylakah island: 122
Federation of the Arab Amirates: 230, 238, 263, 276
fertilizers: 157, 248, 251

Ffrench, G. E.: 135
firqat forces: 394
fishing industry: 57; Bahrain, 61, 214, 215, 216, 217; Kuwait, 2, 143, 153, 158; Oman, 65, 374; Qatar, 244, 246; United Arab Emirates, 63, 315–316
flag of the United Arab Emirates: 291, 326
flared gases: 76, 77, 96, 97, 118
flour mills: 219, 248, 376
food (*see also* diet and nutrition): imports, 152, 211, 216, 244
food prices: Bahrain, 211, 212, 215, 222; Oman, 382; Qatar, 241, 261
foreign aid: loans and grants received; (Bahrain) 222, 223, 230; (Egypt) 150, 151, 191, 196, 264, 310; (Oman) 385, 395; (United Arab Emirates) 305
foreign aid given: 2, 140, 147; by Abu Dhabi, 310, 311, 321; by Kuwait, 146, 148, 149, 150, 151, 152, 161, 179, 184, 191, 307; by Qatar, 251, 264, 265, 266; by the United Arab Emirates, 298–299
foreign nationals: 2, 3, 4, 5, 35, 58–59, 146, 171, 174; in Bahrain, 209, 210, 216, 232; in Kuwait, 121, 126, 127, 134, 138, 158, 164, 181; in Oman, 6, 39, 370; in Qatar, 240, 246; in the United Arab Emirates, 3–4, 285, 287, 300, 301, 304–305, 313
foreign policy: 32, 34–35; Bahrain, 229–231; Kuwait, 171, 178–180; Oman, 388; Qatar, 263–267; United Arab Emirates, 298
foreign trade (*see also* exports; exports of oil and gas; imports): Bahrain, 215, 220–221; Kuwait, 142, 143, 158–161, 178, 189–190; Oman, 370, 382–385; Qatar, 250, 251; United Arab Emirates, 221, 298, 318–320
France: military sales, 189–190, 191, 194, 195, 196, 333; oil, 96, 106, 109, 110, 116; Oman and, 383; Qatar and, 108, 249, 265, 267
frankincense: 13, 373
free services by the government: Bahrain, 208, 211, 213; Kuwait, 130, 138, 140; Oman, 363; Qatar, 237, 240, 241, 242; United Arab Emirates, 297, 302, 309, 315
freedom of the press: Bahrain, 228; Kuwait, 172, 181, 186, 203; Qatar, 261; United Arab Emirates, 283, 296, 297
Friedlaender, Israel: 46
Fujayrah (United Arab Emirates): 273, 294, 299, 315, 316, 318, 354; government, 276, 278, 279, 287, 294; history, 31; oil, 4

Gabon: 84
gas, natural (*see also* liquified petroleum gas): 76, 78; Bahrain, 103, 217, 218; Dubai, 114; Kuwait, 95, 99–100, 157; Oman, 118; Qatar, 106, 107, 244, 246, 249
gas gathering systems: 157, 249
Getty Oil Company: 80, 93, 96
Ghaba oil field: 116
Ghafiri tribe: 30, 32, 65, 343, 344
Ghalib bin Ali: 344, 345, 389
Ghassani, Muhammad Ahmad al: 390, 397
Glubb Pasha: 331
gold smuggling: 10, 162, 224, 254, 319
government (*see also* budget, national; free services by the government; Great Britain): 20; adviser system, 35; Bahrain, 225–227; Kuwait, 122, 142, 144, 165, 170, 172–176; Oman, 5–6, 341, 347–352; Qatar, 238, 240, 258–261; United Arab Emirates, 277, 278
Great Britain: 7, 26, 36, 148, 224, 225; history of Gulf area and, 1, 22, 23, 27, 29, 30, 31, 32, 34, 36, 39, 275; military presence in Kuwait, 170, 179, 183, 184, 185, 189, 190, 191, 195, 197; military presence in Oman, 387, 397, 401; military presence in United Arab Emirates, 323–324, 332; oil imports, 96, 97, 106, 110, 116; relationship with Oman, 35, 342–343; relationship with Qatar, 237–238, 255, 264, 269; trade with Oman, 383; trade with Qatar, 220, 250, 267, 269, 270, 271; trade with United Arab Emirates, 319, 333; workers from, 4, 5, 58, 67, 210, 269
gross domestic product: 215, 216
gross national product: ix; Kuwait, 139, 140, 143–144; Qatar, 240, 243; United Arab Emirates, 303–304
Gulbenkian Foundation (Partex): 105, 115
Gulf Oil Company: 91, 92, 97, 98, 99, 101
Gurdanas trading house: 355

Habrushi family: 288, 290
Habu tribe: 66, 354
haj: 42, 43
Hajar area: 299, 312, 340, 350, 359, 360
Halul island: 73, 105, 264
Hamad, Abdlatif Y. al: 149
Hamad bin Hamad Al Bu Said: 349
Hamad bin Isa Al Khalifah, Shaykh: 207, 228
Hamad bin Khalifah Al Thani: 269
Hamad bin Muhammad Al Sharqi, Shaykh (ruler, Fujayrah): 278, 294
Hamad bin Qasim Al Thani: 257, 269
Hamdan bin Rashid Al Maktum, Shaykh: 290, 292

Hanafi law: 47, 177
Hanbali law: 47, 50, 51, 54, 260
handicrafts. *See* artisans
Hasan Kamil: 255
Hatt: 31
Hawalli: 120, 126, 175, 198
Hawar islands: 208, 236, 238, 266, 267
Hawasin tribe: 66, 353
Hawk air defense system: 187, 190, 192, 194, 196
health services. *See* medical care
Hegira (hijra): 41
Hejaz: 13, 16, 17, 27
higher education: 64, 67; Bahrain, 214; Kuwait, 130, 131, 174; Oman, 364, 367; Qatar, 242, 262; Sharjah, 290; United Arab Emirates, 285, 301, 302
Hilal bin Hamad Al Sammar: 349
Hill, A. G.: 127, 129
Hinawi tribe: 32, 65, 343, 344
Hindus: 53
Holland: 22, 23
Holmes, Frank: 91, 101, 104
Hormuz: 9–10, 18, 22, 23, 29
Hornbeam line: 393, 394
hospitals. *See* medical care
housing: Bahrain, 212, 216, 231; Kuwait, 137–138, 147, 148, 164; Oman, 362, 368; Qatar, 241, 255; United Arab Emirates, 302–303, 304
Hufuf Oasis: 11
Humayd bin Ahmad Al Mualla, Shaykh (Umm al Qaywayn): 294
Husayn: 45, 46, 51
Hussa bint al Murr, Shaykha: 72
Huwaisah oil fields: 116

Ibadi movement: 17, 32, 34, 48–50, 53, 66, 351
Ibn Ibad al Murri al Tamimi, Abd Allah: 48, 49, 342
Ibn Saud (Abd al Aziz ibn Saud): 52
imam (office): 17, 46, 49, 342, 344 *See* glossary
immigration (*see also* foreign nationals): 58, 64, 181, 210, 300, 336
Imperial Iranian Task Force: 6, 395, 396
imports: ix; Bahrain, 102, 211, 216, 220, 221; Kuwait, 152, 159, 160; Oman, 384; Qatar, 243, 244, 250–251; United Arab Emirates, 319, 320
income, per capita: Kuwait, 91, 121, 139; Oman, 368; Qatar, 237
independence: Bahrain, 207, 238; Kuwait, 122, 181; Oman, 342; Qatar, 238; United Arab Emirates, 276
India: 34, 159, 160, 251; aid received, 149, 265
Indian rupee: 162, 254, 321, 381
Indians: 2, 3, 4, 59, 62, 65, 67, 210, 240, 263
Indonesia: 84, 102
industrial complexes: 218, 247, 317
industry (*see also* oil industry): Bahrain, 217–220; Kuwait, 154–158, 162; Oman, 376; Qatar, 108, 246–250, 252; United Arab Emirates, 316–318
inflation: Bahrain, 211, 215, 220; Oman, 382, 385; Qatar, 241, 254–255; United Arab Emirates, 305
interest rates: Abu Dhabi fund, 310; Kuwait, 163; Qatar, 254
internal security (*see also* Dhofar insurgency): Bahrain, 228, 232, 233–234; Kuwait, 186, 188, 201–203; Oman, 349, 406; Qatar, 268; United Arab Emirates, 282–283, 328–329, 335–337
international memberships and organizations: Kuwait, 180; Oman, 385; Qatar, 266; United Arab Emirates, 297–298
International Monetary Fund (IMF): 380, 385
investment authorities: 90, 150, 312
investments: 140, 141, 144–145, 148, 150, 157, 162, 247, 253, 307
IPC. *See* Iraq Petroleum Company
Iran: 6, 46; Bahrain and, 38, 210, 221, 230; Dhofar rebellion and, 395, 398; Iranians in Gulf states, 59, 62, 126, 240, 263; Kuwait and, 159, 180; oil industry, 84, 88, 89, 94, 102, 114; oil reserves, 80; United Arab Emirates and, 294, 298, 319, 326, 327
Iraq: 6, 17, 46, 124, 176, 178; Dhofar rebellion and, 391, 396, 397; Iraqis in Gulf states, 58, 126, 127; Kuwait and, 123, 148, 150, 184–186; oil industry, 80, 84, 87, 90, 91, 92, 101; Qatar and, 265
Iraq Petroleum Company (IPC): 76, 104, 105, 109, 115
irrigation: 124, 152, 216, 245, 313, 372, 378
Isa bin Ali Al Khalifah, Shaykh: 207
Isa bin Salih: 34
Isa bin Salman Al Khalifah, Shaykh (ruler, Bahrain): 5, 27, 207, 225, 228, 231
Isa Gharrim Al Kuwari: 257
Isa Town: 213
Islam: 15–17, 41–44, 52, 69, 258; divisions, 226, 231, 232
Israel (*see also* Arab-Israeli wars): 264, 298
Italy: 21; military sales, 333, 400; oil imports, 98, 106

Jabal Akhdar: 345, 359, 360, 372
Jabal al Qara: 360, 386
Jabal Dukhan: 208
Jabir al Ahmad al Jabir Al Sabah, Shaykh (prime minister and heir apparent, Kuwait): 3, 122, 165, 169, 170, 172, 173, 178
Jabir al Ali Al Sabah, Shaykh: 190–191
Janabah tribe: 66
Japan: 109, 112, 230, 374–375; oil consumption, 76, 79, 80, 93, 96, 103, 110, 116; trade with Bahrain, 220; trade with Kuwait, 159, 160, 180; trade with Oman, 383; trade with Qatar, 249, 251, 265, 268
Jazirat al Hamra: 31
Jesus: 42
Jibrin Oasis: 28
Jiddah islet: 206, 208
jihad: 43, 51
Joasmees: 28
Jordan: loans and grants received, 150, 151, 186, 191, 264; military equipment to Oman, 400, 401
Jordanians: 132; in military forces of Gulf states, 58, 232, 269, 324, 332
Judasim: 15, 47
judiciary. *See* legal system

Kamilla (wife of Sultan of Oman): 346, 352
Karim Ahmad al Harami: 349
Kathiri tribe: 53
Kazimi, Abd al Muttalib ibn Abd al Husayn al: 174, 178
KFAED. *See* Kuwait Fund for Arab Economic Development
Khalfan bin Nasir al Wahaybi: 349
Khalid bin Saqr Al Qasimi, Shaykh (Ras al Kaymah): 291, 295
Khalid bin Hamad Al Thani: 257, 259, 269
Khalid ibn Abd Allah Al Atiya: 257
Khalid Muhammad Al Mani: 257
Khalidi, Walid Muhammad al, Major General: 326, 330
Khalifa ibn Mubarak Al Hinawi: 32, 66
Khalifah bin Hamad Al Thani, Shaykh (ruler, Qatar): 5, 256, 257, 258–259, 261, 262, 263
Khalifah bin Salman Al Khalifah, Shaykh: 207, 226
Khalifah bin Zayid Al Nuhayyan, Shaykh (Abu Dhabi): 288, 289 326
Khalifah family. *See* Al Khalifah ruling family
Khama island: 273
Kharadjite movement: 17, 45, 49, 341
Kharg island: 23
Khawr al Udayd: 236, 238, 328
Khimji Ramdas trading house: 355
Khor Fakkan: 22, 23
kinship patterns. *See* tribalism
KOC. *See* Kuwait Oil Company
Kunj: 28
Kuria Muria islands: 33, 375
Kuwait: ix, 3, 120, 121–202; Bahrain and, 221, 230; foreign aid given, 37, 307; government, 3, 99, 100, 198; history, 25, 26, 35, 36; Iraq and, 6–7; oil industry, 76, 80, 83, 87, 90, 91–100; oil reserves, 2, 75, 80; OPEC and, 84, 89; Qatar and, 270; religion, 53; social system, 59–61, 68
Kuwait City: 120, 125, 175, 195; planned suburbs, 59–60, 121, 138, 146; press, 182; trees, 123, 125
Kuwait Fund for Arab Economic Development (KFAED): 75, 148, 149, 150, 151, 174, 179, 310, 311
Kuwait National Petroleum Company (KNPC): 94, 96, 97, 98, 99, 157
Kuwait Oil Company (KOC): 92, 94, 97, 98, 99, 124; government take-over, 140–141, 171; headquarters, 126, 174; refinery, 96, 156
Kuwait Oil, Gas, and Energy Company: 98
Kuwait Shell Petroleum Development Company: 94

labor courts: 260
labor force (*see also* employment): 58–59, 62, 65, 67, 165; Bahrain, 101, 210, 216, 217, 219, 220; Kuwait, 128, 138, 141, 142, 154, 158, 174; Oman, 370; Qatar, 107, 243, 246; United Arab Emirates, 304, 305
labor organizations: 262, 265
land purchase scheme: Kuwait, 59–60, 137, 146
Landen, Robert Geran: 35
languages (*see also* Arabic language; English language): 59, 66; Kuwait, 182; Oman, 354, 358, 367, 387; Qatar, 242, 267, 268
Laost, Henri: 51
Lebanese in Gulf states: 4, 58, 285
Lebanon: 150, 180
legal system: Bahrain, 228; Kuwait, 122, 166, 173, 176–178, 197; Oman, 347, 351–352; Qatar, 260; Sharjah, 290; United Arab Emirates, 280–281, 330
libraries: 134
Libya: 84, 86, 87, 90; Dhofar rebellion and, 398
life expectancy: 240

Lingeh: 28
liquified petroleum gas: 100, 108, 112, 113, 317
literacy: ix, 3; Bahrain, 214; Kuwait, 130, 199; Oman, 351, 387; Qatar, 243, 268
livestock: 153, 245, 313, 315, 371, 373
Liwa: 273; Oasis, 63, 303
local and regional government: Kuwait, 175; Oman, 350, 351; Qatar, 258
Lock, Ronald: 269
Luft: 28

Madinat Khalifah: 268
Magan (Maazun): 13, 14
Mahrah tribe: 53, 66, 354
Majalla legal system: 176, 177
Maktum bin Rashid Al Maktum, Shaykh: 290, 292
Maliki rite: 47, 52, 54, 61
Maluhha: 13
Manama: 206, 208, 211, 232
Manasir tribe: 288
manufacturing: Bahrain, 215, 217, 218; Kuwait, 142–143, 154, 157; Qatar, 244; United Arab Emirates, 317
marriage: 29, 49, 69, 71–72, 130, 352
Marshall Company (U.S.): 375
Masirah island: 66, 116, 358, 359, 360, 378, 387, 397, 399
Matrah: 65, 67, 340, 350, 360, 368
Mecca: 14, 18, 42, 43
medical care: 2, 4; Bahrain, 213; Kuwait, 135–137; Oman, 363–365; Qatar, 241; United Arab Emirates, 302
Medina: 14, 17, 18, 41
Mellon, Andrew: 92
metric system: 370
Mexico: 80, 84
midwives: 365
military equipment and assistance received: by Kuwait, 187, 189–190; by Oman, 400, 405; by Qatar, 269, 270; by the United Arab Emirates, 332, 333–334
military pay and allowances: Bahrain, 231; Kuwait, 197–198, 199, 200, 202, 203; Oman, 404; United Arab Emirates, 334–335
military rank and insignia: Bahrain, 232, 233; Kuwait, 197, 199; Oman, 407; United Arab Emirates, 334, 335
Mina Abd Allah refinery: 73, 96
Mina al Ahmadi refinery: 73, 96, 99
Mina al Fahal: 73, 116, 118, 340, 378
Mina Qaboos: 368
Mina Salman: 206, 208, 218
Mina Suud refinery: 73, 96
mineral resources (*see also* oil-producing areas): Kuwait, 125; Oman, 375
ministries (Bahrain): 225, 228: Ministry of Defense, 222; Ministry of Education, 222; Ministry of Foreign Affairs, 230; Ministry of Health, 209, 213; Ministry of Housing, 212; Ministry of Interior, 222, 232; Ministry of Labor and Social Affairs, 210; Ministry of Public Works, Electricity and Water, 222
ministries (Kuwait): 173; Ministry of Defense, 192, 196; Ministry of Electricity and Water, 155; Ministry of Health, 135, 173, 174; Ministry of Information, 181, 183, 185; Ministry of Interior, 175, 192, 193, 199, 200, 201; Ministry of Justice, Awqaf, and Islamic Affairs, 177, 201; Ministry of Legal and Legislative Affairs, 172, 174
ministries (Oman): 347, 348, 349; Ministry of Agriculture, Fisheries and Minerals, 358; Ministry of Defense, 358, 404; Ministry of Education, 356, 358, 367; Ministry of Foreign Affairs, 356; Ministry of Health, 365; Ministry of Information and Culture, 357; Ministry of Interior, 406; Ministry of Labor and Social Affairs, 67, 357; Ministry of National Heritage, 348, 349; Ministry of Youth Affairs, 348, 357
ministries (Qatar): 257, 259; Ministry of Defense, 269; Ministry of Education and Youth Welfare, 242; Ministry of Information, 268; Ministry of Interior, 269; Ministry of Municipal Affairs, 258; Ministry of Public Health, 241
ministries (United Arab Emirates): 108, 284–285, 287, 308; Ministry of Education, 301; Ministry of Finance, 312; Ministry of Health and Education, 285; Ministry of Information and Culture, 283, 296, 297; Ministry of Interior, 335, 336
Mobil Oil Company: 105
Morocco: 150
mortality rates: Bahrain, 213; Kuwait, 135, 136; Qatar, 241
motor vehicles: 125, 160, 208–209, 217
Muawiyah: 45
Mubarak al Abd Allah al Jabir Al Sabah, Shaykh: 189
Mubarak al Sabah Al Sabah, Shaykh: 26, 122, 165, 166
Mubarak Khaduri: 349
Mubarak oil field: 73, 114
Muhammad (The Prophet): 16, 41, 44, 51

Muhammad Abd Allah al Khalili, Imam: 344
Muhammad al Zubayr: 349
Muhammad bin Ahmad Al Bu Said: 349
Muhammad bin Khalifah bin Hamad Al Khalifah, Shaykh: 232
Muhammad bin Mubarak Al Khalifah, Shaykh: 229–230
Muhammad bin Rashid Al Maktum, Shaykh: 290, 292, 326
Muhammad ibn Jabr Al Thani: 257
Muhammad ibn Nasir Al Ghafiri: 32, 66
Muhanna ibn Sultan al Yaruba: 32
Musandam peninsula: 66, 116, 340, 350, 354, 361, 371, 373
Musayid: 73, 105, 107, 108, 236, 238–239, 268; industrial park, 247, 248, 249
Muscat: 18, 22, 23, 33, 34, 53, 340, 354, 362, 368, 405, 406; capital, 342, 350; population, 65, 67, 361; press, 358
Mutallah: 206

Nahrawan canal: 48
Najd: 50, 60
names: vi–vii
Nasir bin Khalid Al Thani: 256, 257
Nasser, Gamal Abdul: 386
Natih oil fields: 116, 340
National Democratic Front for the Liberation of Oman and the Arab Gulf: 391
National Liberation Front of Qatar: 263
nationalization of oil companies: 84, 87
Netherlands, 106, 116, 383, 400
Neutral Zone: 93, 94, 100, 123, 125, 184, 273; industry, 96, 154
newspapers: Bahrain, 228–229; Kuwait, 180–182; Oman, 358; United Arab Emirates, 296, 297
Nigeria: 84
Nizari tribe: 65, 66
Nizwa: 49, 53, 340, 342
Non-Kuwaiti Professionals: 193, 198, 199, 203
Norway: 267
Nuaimi family: 294

OAPEC. *See* Organization of Arab Petroleum Exporting Countries
Occidental Oil Company: 86
offshore oil fields: 102, 108, 109, 112, 114, 116
oil industry (*see also* exports of oil and gas; refineries; and names of individual producing countries): 73–118
oil pricing: 77, 78, 81–83, 85, 86, 88, 89–90, 97, 106, 117, 140, 299
oil-producing areas: 73, 79; Kuwait, 125, 141; Oman, 353; Qatar, 238
oil reserves: 2, 79, 80, 85, 92, 99, 101, 106, 108, 115, 246
oil revenues: 4, 37, 76, 85; Abu Dhabi, 111; Bahrain, 103, 104, 216, 222, 223, 224; Dubai, 113; Kuwait, 95, 97, 151, 170; Oman, 115, 117, 347, 379; Qatar, 106, 237, 243, 250, 251; United Arab Emirates, 305, 319, 320
oil revenues and social change: 61, 62–63, 139, 140, 144, 165, 216
Oman: ix, 3, 6, 10, 14, 35, 340–409; history, 12–13, 15, 18, 23, 29, 32–34, 36, 39–40; oil industry, 76, 80, 83, 101, 115–118; oil reserves, 80; religion, 16, 17, 30, 48, 53, 66; social system, 65–67
Omanis abroad: 126, 263, 325
OPEC. *See* Organization of Petroleum Exporting Countries
Organization for the National Struggle of Qatar: 263
Organization of Arab Petroleum Exporting Countries (OAPEC): 90, 106, 298; Bahrain, 103, 219, 220; Kuwait, 171, 179; Qatar, 265
Organization of Petroleum Exporting Countries (OPEC): 1–2, 75, 77, 83–90, 103, 106, 110, 116, 298; foreign aid given, 151; Kuwait and, 174, 178
ownership of land and businesses. *See* property ownership

Pakistan: 159; aid received, 149, 265
Pakistanis: 2, 3, 4, 59, 62, 67; in Bahrain, 232; in Qatar, 240, 263, 269; in United Arab Emirates, 325, 332
Palestinian Liberation Organization (PLO): 3, 179, 203, 264
Palestinians: 3, 58, 62; in Kuwait, 126, 127, 171, 179, 202, 336; in Qatar, 263
Partex. *See* Gulbenkian Foundation
participation agreements with oil companies: 87, 93, 111, 112, 115, 117
PDO. *See* Petroleum Development
pearls: 10, 13, 18, 20, 26, 31, 57, 61, 63, 65, 303
Perkins, Kenneth: 404
Persian Gulf: 19, 73
Persians: 19, 59, 67
petrochemical industries: 78; Kuwait, 99, 100, 147, 157; Qatar, 108, 249; United Arab Emirates, 317
Petroleum Company of France (CFP): 105, 109, 115

Patroleum Development (PDO): 115, 116, 117, 345, 378
Philippines: 98
pipelines: 73, 86, 102, 105, 109, 116, 151
piracy (*see also* Trucial Coast states): 13, 20, 24, 27, 28, 30, 31, 275
police forces: Bahrain, 232; Kuwait, 145, 188, 189, 193, 196, 198, 199, 201, 202; Oman, 356, 388, 406–407; United Arab Emirates, 334, 335–336
political parties: Bahrain, 226; Qatar, 262, 263; United Arab Emirates, 286
Popular Front for the Liberation of Oman (PFLO): 336, 394, 395, 396, 397, 398
Popular Front for the Liberation of the Occupied Arab Gulf States (PFLOAG): 233, 234, 336, 390, 391, 392
population (*see also* foreign nationals): ix, 2, 3, 5; Abu Dhabi, 310; Bahrain, 61, 209–211; Kuwait, 126–130; Oman, 65, 361, 368; Qatar, 239; United Arab Emirates, 300, 304, 331
ports and harbors: ix; Bahrain, 218, 220; deep water terminals, 104, 105; Dubai, 64, 299, 318; Kuwait, 122, 157, 160, 164; Oman, 359, 368; Qatar, 236, 238, 239, 252; 255,; United Arab Emirates, 305, 318
Portuguese: 21–22, 23, 24, 342
postage stamps: 305
posted prices for crude oil: 82–83, 84, 86, 88, 106, 117, 158, 319
Price, D. L.: 388, 390, 392, 403
private schools: Bahrain, 214; Kuwait, 132, 134; United Arab Emirates, 301
property ownership: Abu Dhabi: 318; Bahrain, 218–219; Kuwait, 138, 157, 164; Oman, 369, 370; Qatar, 247
public health and sanitation: Kuwait, 136, 145; Oman, 364; Qatar, 242
publishing (*see also* newspapers): 134, 268

Qaboos bin Said bin Taimur Al Bu Said, Sultan of Oman: 1, 6, 348, 349, 368; 404; biography, 39, 40, 341, 346, 352, 360, 399; Dhofar rebellion, 392, 393, 395, 396, 397, 398
Qahtani Arabs: 15, 16, 32
Qais Abd al Munim Zawawi: 349
Qais island: 18, 19
Qara tribe: 53, 66, 346, 354
Qasim: ?8, 257
Qasim bin Ali Al Thani: 257, 269
Qasimi tribe: 30, 31
Qatar: ix, 3, 5, 62, 192, 221, 236, 237–271, 321, 368; history, 27–28, 35, 36, 38; oil industry, 76, 83, 84, 90, 104–108; oil reserves, 2, 76, 80; oil revenues, 88, 251; religion, 48, 53, 54; territorial dispute, 208
Qatar General Petroleum Company: 107
Qatar Petroleum Company (QPC): 104, 106
Qatitani tribe: 65
Qawasim tribe: 27, 28, 29, 30, 31, 64
Qeshm island: 23, 28, 29
Quran: 42, 44
Quraysh tribe: 17, 41

radio and television: Bahrain, 229; Kuwait, 182–183; Oman, 355, 358; Qatar, 267; United Arab Emirates, 296, 297
railroads: 120, 209
rainfall: Bahrain, 209; Kuwait, 123, 152; Oman, 361, 372, 377; Qatar, 239, 244; United Arab Emirates, 300, 312
Ras al Jabal: 359
Ras al Khafji refinery: 73, 96
Ras al Khaymah (United Arab Emirates): 10, 27, 28, 30, 31, 36, 63, 64, 114, 273, 275, 299, 303, 307, 325, 328, 354; agriculture, 312, 314, 315; defense force, 327; government, 278, 279, 291; history, 38, 276, 291; industry, 316, 317, 318; oil, 303
Rashid bin Humayd Al Nuaimi, Shaykh (ruler, Ajman): 278, 294
Rashid bin Said Al Maktum, Shaykh (ruler, Dubai): 266, 278, 283, 290, 292, 326
Rashid ibn Abd al Aziz al Rashid: 178
Rashid ibn Mattar ibn Qasim: 29
Raudhatain: 123, 125
Raysut: 340, 404
refineries: 73; Abu Dhabi, 109, 112, 317; Bahrain, 102, 217; Kuwait, 96, 97, 126, 142, 156, 157; Qatar, 105, 248, 249
religion (*see also* Islam): 15, 41–54
rial: Oman, 370, 381; Qatar-Dubai, 254
rights and liberties (*see also* freedom of the press): 170, 281
roads: Bahrain, 206, 208; Kuwait, 120; Oman, 340; Qatar, 236, 238, 267; United Arab Emirates, 304, 309, 315, 319
Roberts, Edmund: 33
Rockefeller Foundation: 314
Royal Dutch Shell Oil Company: 104, 105, 115
royalty payments for oil: 83, 85, 86, 88, 97, 103, 104, 110, 117
Rub al Khali: 13, 116, 340, 358
Rustaq: 342
Ruwi: 350; Valley, 368, 377

Sabah al Ahmad al Jabir Al Sabah, Shaykh: 178, 191

Sabah al Salim Al Sabah, Shaykh (amir, Kuwait): 3, 122, 165, 166, 169–172, 173, 178, 186
Sadat, Anwar al (president, Egypt): 386, 396
Saffaniyah oil field: 94
Sahib bin Issa: 344
Said Ahmad al Shanfari: 349
Said Ahmad Ghubash: 291
Said al Abd Allah al Salim Al Sabah: 178, 189, 196
Said bin Abd Allah bin Salman: 291
Said bin Taimur Al Bu Said: 6, 39, 345, 368; Dhofar rebellion and, 385, 388, 389, 390, 403
Salalah: 39, 340, 360, 364, 378, 379, 387, 406
Salim al Muhammad al Jabir Al Sabah, Shaykh: 189
Salman bin Hamad Al Khalifah, Shaykh: 207
Salman ibn Duaij Al Sabah, Shaykh: 172
Saqr bin Muhammad Al Qasimi, Shaykh (ruler, Ras al Khaymah): 278, 291, 295
Saqr bin Muhammad Al Qasimi, Shaykh (deputy ruler, Sharjah): 291, 293
satallite earth stations: 252, 268, 302
Saudi Arabia: 27, 50, 397; foreign aid given, 151, 307, 385, 395; oil industry, 84, 86, 93, 94, 96, 98, 101, 102, 103; oil prices, 88, 89, 90; oil reserves, 2, 80; relations with Gulf states, 40, 230, 263, 265; Saudis in Gulf states, 126, 127, 269; territorial disputes, 328, 344; trade with Gulf states, 159, 220, 221, 250
Sayf bin Ghubash: 291
Sayyid Said: 30, 31, 33
Sayyid Thuwayyni bin Shihab: 347, 348
scouting and girl guides: 357
Sea-Bed Treaty: 268
securities, stocks, and bond markets: 163, 323
Shafii law: 47, 52, 54
Shakhbut bin Sultan Al Nuhayyan, Shaykh: 288, 289
sharia law: 20, 47, 52, 176–177, 228, 260, 350
Sharjah (United Arab Emirates): 10, 28, 64, 273, 299, 300, 301; border dispute, 283; defense forces, 325, 326, 327; government, 276, 278, 279, 287, 290; history, 31, 35; military base, 323, 331; oil, 4, 76, 108, 114, 303; trade, 318, 322
Sharquiyya area: 66, 340, 344, 350, 353
Shatt al Arab: 124
shaykh (office): 57
Shell Oil Company: 346, 364, 367
Shell Oil Company of Qatar (SCQ): 105, 107, 108
Shia Islam: 17, 44, 46, 52, 54, 59, 61, 67
Shihuh tribe: 53, 66, 354
shipbuilding and dry docks: 90, 155, 219, 220, 306, 318
shrimp exports: 154, 217, 246
Sib: 340, 350, 368
Singapore: 96, 116, 383
Sir: 28
Sirri island: 273
Sisco, Joseph J.: 186, 187
Sitrah island: 206, 208, 216
slave trade, 27, 31, 33, 34, 65, 67, 343
social mobility: 58, 61, 63
social systems: 59–61, 68–72
social welfare (*see also* free services by the government; medical care): 10–11; Bahrain, 211; Kuwait, 137, 145, 173; Qatar, 237, 243, 261, 262; United Arab Emirates, 283
Socotra island: 22
Sohar: 18, 22, 23
Soviet Union: Dhofar rebellion and, 386, 388, 392, 396, 397, 398, 399; oil exports, 81; oil production, 78; oil reserves, 80; trade with Kuwait, 190, 191
sports: 357
standard of living: Bahrain, 211, 212; Qatar, 241
Standard Oil Company of California (Socal): 75, 101, 104
Standard Oil Company of New Jersey: 105
Sudan: 150
Sufis: 47, 48
Sufris: 49
Suhayl family: 290
Suhaym bin Hamad Al Thani, Shaykh: 257, 259, 266
sulfur: 103, 105, 110, 112, 117
sultan (title): 33, 39, 53
Sultan bin Muhammad Al Qasimi, Shaykh (ruler, Sharjah): 278, 290, 291, 293
Sultan bin Zayid Al Nuhayyan, Shaykh 288, 289
Sultan's Armed Forces (SAF): 389, 391, 392, 393, 394, 396, 399, 402, 403
Sulyman bin Himyar: 344, 345
Sun Oil Company: 116
Sunni Islam: 17, 52, 53, 59, 61, 67, 226, 231, 232, 342
Sur: 340, 350, 356, 359
Suwaydi family: 288

Syria: 90; loans and grants received, 150, 151, 186, 191, 202, 310
Syrians: 58

Taimur bin Faisal Al Bu Said: 34, 39: 344
Talib bin Ali: 344, 345
tankers: Abu Dhabi; 110, 112; Bahrain, 219, 220; Kuwait, 99, 147; Qatar, 108; terminals, 73, 105, 116
Tariq bin Taimur Al Bu Said: 346, 347, 348, 352
Taryam family: 291
taxes (*see also* zakat): Bahrain, 222, 224, 225; Kuwait, 145, 157; Oman, 380; Qatar, 252, 261; taxes on oil producers, 83, 85, 86, 88, 97, 103, 110, 117; United Arab Emirates; 308, 309, 318
teachers: 58, 132, 214, 301, 367; training, 242, 301
telephones: 368
Thailand: 106
Thakkira: 258
Thamarit: 340
Thani ibn Muhammad ibn Thamir ibn Ali: 28
Theophilus Indus: 15
tourism: 265, 302, 318, 349, 377
trade (*see also* foreign trade): 21, 23, 24, 57; ancient trade routes, 12, 13, 18; merchant families of Oman, 354–355; traditional commerce, 59, 61, 63, 64
transportation (*see also* airports; ports and harbors; roads): 120, 125, 209, 306
treaties: 1, 10, 22, 33, 345; arms control, 268; Bahrain, 230; Kuwait, 26; Oman, 33, 343, 344; Qatar, 28, 265;; Trucial Coast, 31, 275, 276; United Arab Emirates, 323
Treaty of Sib: 344, 345
tribal groups: 63, 64; Oman, 65–66, 352–354, 355
tribalism: 14–15, 20; social organization and, 56, 57, 60, 63, 64, 66–67, 68, 70
Trinidad and Tobago: 116, 383
Trucial Coast states (*see also* United Arab Emirates): 1, 27, 28, 31, 36, 230, 238, 263, 275, 276, 286, 291
Trucial Oman Scouts: 290, 324, 325, 328, 329
Trucial States Council: 37, 276, 324
Tunb islands: 38, 273, 294, 326, 327
Tunisia: 150
Turks: 18, 22, 26, 27, 28

UDF. *See* Union Defense Force
Umm al Qaywayn (United Arab Emirates): 114, 273, 294; government, 276, 278, 279, 287, 294; history, 31; oil, 4
Umm Bab: 236
Umm Nasan island: 206, 208
Umm Salal: 258
unemployment: 210
Union Defense Force: 290, 325, 327, 329, 330, 334
United Arab Emirates (*see also* names of individual amirates): ix, 3, 4, 62, 139, 192, 226, 230, 273, 275–337; foreign relations, 35, 265; independence, 36, 38, 276; oil industry, 2, 76, 80, 88, 108–114; OPEC and, 84, 90; religion, 53, 54; territorial disputes, 238, 294
United Kingdom. *See* Great Britain
United Nations: 179, 180, 230, 266, 297, 366; Food and Agriculture Organization, 245, 246, 266; Oman and, 345–346, 367, 369
United States: Bahrain and, 210, 220, 230, 231; citizens employed in Gulf States, 4, 5, 58, 210; Kuwait and, 96, 148, 159, 160, 162; military sales and aid, 185, 189, 190, 191, 192, 194, 195, 196, 333; oil imports, 106, 110, 116, 383; oil pricing, 81; oil reserves, 80; Oman and, 33, 343, 400; Qatar and, 251, 264–265, 269; United Arab Emirates and, 299, 302, 319
University of Arizona: 314
University of Kuwait: 121, 132, 134
Urdu language: 59, 358
Utaybah family: 288
Uthman: 44
Utub tribe: 25, 26, 29

Venezuela: oil industry, 82, 84, 85, 86; oil reserves, 80
Vietnam: 251

Wadi al Batin: 123, 124
Wadi Samail: 360
wages and salaries (*see also* military pay and allowances): 155, 222, 237, 304
Wahhab, Muhammad ibn Abd al: 50
Wahhabi Islam: 5, 26, 27, 29–30, 48, 50–52, 54, 183, 263, 328
Walid bin Zahir al Hinai: 349
Warbah island: 120, 185
water supply (*see also* desalination plants): Bahrain, 208, 209, 212, 216; Kuwait, 121, 123, 124, 125, 152, 155, 156; Oman, 372, 377–379; Qatar, 239, 245, 247; United Arab Emirates, 303, 304, 312, 314–315, 316
water tunnels (*falaj* system): 312, 313, 372
welfare system. *See* social welfare
West Germany: trade, 251, 319, 333

wildlife: 209, 239, 259, 360
Willoughby, John: 324
women and girls: vii, 127, 128, 129, 210, 211, 355–357, 363, 366; education, (Bahrain) 213, 214; (Kuwait) 130, 131, 133, 134; (Qatar) 242, 243; (United Arab Emirates) 301; religion and, 43; status, 39, 61, 68, 70–71, 198
World Bank (IBRD): Kuwait and, 143, 149, 150, 158; Oman and, 367, 369, 373, 385; Qatar and, 266; United Arab Emirates and, 299, 311
world consumption of oil: 78, 80

Yamani, Ahmad Zaki, Shaykh: 87
Yamaniyah tribe: 65
Yathrib: 14
Yemen (Aden): 325, 354; Dhofar rebellion, 388, 390, 396, 397, 398, 399
Yemen (Sana): 46, 47
the Yemens: 13, 14, 150; Yemenis in Gulf states, 58, 65, 66, 240, 263, 269
Yibal oil fields: 116

zakat: 42, 43, 51, 52, 294
Zanzibar: 33, 343
Zawawi trading house: 355
Zayid bin Sultan Al Nuhayyan, Shaykh (ruler, Abu Dhabi and president, United Arab Emirates): 4, 38, 278, 281, 282, 283, 286, 288, 289, 313–314, 325, 326, 329, 332
Zubarah: 236, 266
Zubayr trading house: 355

PUBLISHED AREA HANDBOOKS

550–65 Afghanistan
550–98 Albania
550–44 Algeria
550–59 Angola
550–73 Argentina

550–169 Australia
550–176 Austria
550–175 Bangladesh
550–170 Belgium
550–66 Bolivia

550–20 Brazil
550–168 Bulgaria
550–61 Burma
550–83 Burundi
550–166 Cameroon

550–96 Ceylon
550–159 Chad
550–77 Chile
550–60 China, People's Republic of
550–63 China, Republic of

550–26 Colombia
550–67 Congo, Democratic Republic of (Zaire)
550–91 Congo, People's Rep of
550–90 Costa Rica
550–152 Cuba

550–22 Cyprus
550–158 Czechoslovakia
550–54 Dominican Republic
550–52 Ecuador
550–43 Egypt

550–150 El Salvador
550–28 Ethiopia
550–167 Finland

550–29 Germany
550–155 Germany, East
550–173 Germany, Federal Republic of
550–153 Ghana
550–87 Greece
550–78 Guatemala
550–174 Guinea

550–82 Guyana
550–164 Haiti
550–151 Honduras
550–165 Hungary
550–21 India

550–154 Indian Ocean Territories
550–39 Indonesia
550–68 Iran
550–31 Iraq
550–25 Israel

550–182 Italy
550–69 Ivory Coast
550–177 Jamaica
550–30 Japan
550–34 Jordan

550–56 Kenya
550–50 Khmer Republic (Cambodia)
550–81 Korea, North
550–41 Korea, Republic of
550–58 Laos

550–24 Lebanon
550–38 Liberia
550–85 Libya
550–163 Malagasy Republic
550–172 Malawi

550–45 Malaysia
550–161 Mauritania
550–79 Mexico
550–76 Mongolia
550–49 Morocco

550–64 Mozambique
550–35 Nepal, Bhutan and Sikkim
550–88 Nicaragua
550–157 Nigeria
550–94 Oceania

550–48 Pakistan
550–46 Panama
550–156 Paraguay
550–92 Peripheral States of the Arabian Peninsula
550–85 Persian Gulf States

550-42	Peru
550-72	Philippines
550-162	Poland
550-181	Portugal
550-160	Romania
550-84	Rwanda
550-51	Saudi Arabia
550-70	Senegal
550-180	Sierra Leone
550-184	Singapore
550-86	Somalia
550-93	South Africa, Republic of
550-171	Southern Rhodesia
550-95	Soviet Union
550-179	Spain
550-27	Sudan, Democratic Republic of
550-47	Syria
550-62	Tanzania
550-53	Thailand
550-178	Trinidad and Tobago
550-89	Tunisia, Republic of
550-80	Turkey
550-74	Uganda
550-97	Uruguay
550-71	Venezuela
550-57	Vietnam, North
550-55	Vietnam, South
550-183	Yemens, The
550-99	Yugoslavia
550-75	Zambia

☆ U.S. GOVERNMENT PRINTING OFFICE : 1977 O—261-035

www.ingramcontent.com/pod-product-compliance
Lightning Source LLC
Chambersburg PA
CBHW020600270326
41927CB00005B/113
* 9 7 8 1 4 3 4 4 6 2 1 0 7 *